FLORA OF SOMALIA

VOLUME 4

Angiospermae (Hydrocharitaceae–Pandanaceae)

Edited by

MATS THULIN

ROYAL BOTANIC GARDENS

KEW

Published by
Royal Botanic Gardens, Kew

© Copyright The Trustees of The Royal Botanic Gardens, Kew 1995

First published 1995

General editor J. M. Lock. Prepared by the Flora of Somalia Project funded by SAREC (Swedish Agency for Research Cooperation with Developing Countries).

Cover photograph: stand of *Livistona carinensis*, near-endemic and endangered palm, near its type locality in Bari Region, Galgala, M. Thulin 1986.

Typeset by Ord och Grafik, Valhallavägen 146 F, S-115 24 Stockholm, Sweden.

ISBN 0 947643 88 5

Printed in Great Britain by
Whitstable Litho Printers Ltd., Whitstable, Kent.

CONTENTS

PREFACE

In January 1995 I got the opportunity to visit Somalia again for field work for the first time in almost five years. Although the country is still without a government and the situation in many respects is problematic, many people are determined to work for a new and better Somalia. For me this visit has provided an inspiring impetus to the completion of this volume.

Although this is the second volume to be completed, it constitutes, with its coverage of the monocotyledons, number four in the sequence of families adopted for the project. With this volume we get for the first time a detailed picture of which grasses, sedges, palms, sea-grasses, aloes, lilies, etc. that grow in Somalia. 585 species are treated in total, 50 of which have been described as new within the project, and many of which are here recorded for the first time from Somalia. In addition, 21 species additional to vol. 1 have been listed in an appendix, four of which are new species.

The project, as before, has been run from the Department of Systematic Botany of Uppsala University, Sweden, with support from the Swedish Agency for Research Cooperation with Developing Countries (SAREC). At present it forms part of a more inclusive project called "Biodiversity of NE African arid regions".

Of the people I particularly wish to thank at the completion of this volume I wish to start with my field companions in Somalia earlier this year, Abdi M. Dahir, former Head of the now destroyed National Herbarium in Mogadishu, now in London, Abdisalam S. Hassan, formerly at the British Forestry Project in Mogadishu, now in Reading, and my wife Gunilla. Their whole-hearted participation, as well as the support we got from numerous local Somalis in the Boosaaso area, not least the staff of the "DANDOR" organization headed by Mohamed A. Hassan, made this trip to the mist-covered escarpments of the eastern Cal Madow Range an unforgettable experience.

It is a pleasure to acknowledge, again, the continuous support from the Herbarium of the Royal Botanic Gardens, Kew, through its Keeper, G. Ll. Lucas and

his staff. Particularly I wish to mention R. M. Polhill, Assistant Keeper and editor of the Flora of Tropical East Africa, J. M. Lock, Head of the Editorial Unit, and T. A. Cope, author of the account of the grasses, the most substantial family in this volume. I am also most grateful to M. G. Gilbert, Kew, and his family, for their hospitality during my regular visits to Kew, and to the late J. B. Gillett for his keen interest and support up to his recent death on March 18, 1995.

Very important has also been the continued collaboration with Erbario Tropicale in Florence, through its Director G. Moggi and his staff.

The contributors to the present volume include twelve botanists from eight countries (see list below). I thank them all for their loyalty and patience, and for all the efforts they have put into their accounts. Many other botanists have also contributed with their expertise in various ways, and I would particularly like to mention the following who read and commented on the accounts of various families or genera: Sasha Barrow (*Phoenix*), J. Dransfield (*Arecaceae*), Demel Teketay and P. K. Mbugua (*Sansevieria*), Susan Holmes (*Aloaceae*), Inger Nordal (*Anthericaceae, Asphodelaceae*), Sebsebe Demissew (*Colchicaceae*), S. Snogerup (*Juncaceae*), Brita Stedje (*Hyacinthaceae*), and J.-J. Symoens (*Ottelia*).

Great benefit and increased efficiency has also been gained through cooperation with major botanical projects in neighbouring countries or regions, notably Flora of Tropical East Africa edited at the Royal Botanic Gardens, Kew, Flora of Ethiopia edited in Addis Ababa and Uppsala, and the Flora of Arabia edited in Edinburgh.

Several of the illustrations of this volume have been made by Kerstin Thunberg and Louise Petrusson, but the majority have been borrowed from various sources. For kind permission to reproduce illustrations I am grateful to the editors of Flora of Tropical East Africa; the editors of Nordic Journal of Botany; the Flora Zambesiaca Editorial Board; The Trustees of the Royal Botanic Gardens, Kew (Kew Bulletin, Hooker's Icones Plantarum, Flora of West Tropical Africa); National Botanical Institute, Pretoria (Flora of Southern Africa, Grasses of southern Africa in

i

Mem Bot. Surv. S. Afr. 58); the Oxford University Press (Agnew, Upland Kenya Wild Flowers); the National Herbarium, Islamabad (Flora of Pakistan); Koninklijke Nederlandse Akademie van Wetenschappen (den Hartog, The Sea-grasses of the world); the publishers of Flora of Palaestina; Laboratoire de Phanerogamie, Museum National d'Histoire Naturelle, Paris (Flore de Cameroun, Flore du Gabon); Oklahoma Agricultural Experiment Station (Burger, Families of Flowering Plants in Ethiopia); National Herbarium, Addis Ababa (Flora of Ethiopia); the Editorial Committee of the Wageningen Agricultural University Papers (Meded. Landbouwhogeschool Wageningen); the Botany Department, Cairo University (Täckholm, Student's Flora of Egypt); Commonwealth Information Services, Australian Government Publishing Service (Flora of Australia); Societa Botanica Italiana (Giornale Botanico Italiano); Smithsonian Institution Press, Washington (Smithsonian Contr. Bot.); Academic Press Ltd, London (J. Linn. Soc. Lond. Bot.). Each source and/or illustrator is also acknowledged in the legends to the illustrations.

The colour photographs are my own except for Plates 1 F and 2 A−C by Susan Holmes. In this connection I wish to correct a most unfortunate slip in vol. 1. Susan Holmes as botanical author is Susan Carter and her treatments of *Euphorbia* and *Monadenium* in vol. 1 are to be attributed to her under the latter name.

Finally, let me comment on the introductory note in vol. 1 on the type specimen of *Aristolochia rigida*, collected by A. Pervillé the 21 March 1842 in Boosaaso, and thus being the first known botanical specimen collected in Somalia. This species was still thriving in Boosaaso at our visit in January 1995, despite the recent enormous growth of the town after the civil war. Maybe this can be seen as a symbol of the perseverance of the plants of Somalia as well as of its people.

Mats Thulin
Uppsala, 13 April 1995.

INTRODUCTION

In the first volume of this Flora (1993) a brief background to the project was given and this will not be repeated here. However, to facilitate for the user, most of the information on the format of the work is given here again.

For the geographical subdivision of Somalia eight regions are used, three northern, two central and three southern (see map). The regions are denoted as follows:

N1 Woqooyi Galbeed and Togdheer
N2 Sanaag and part of Nugaal
N3 Bari and part of Nugaal
C1 Mudug and Galguduud
C2 Hiiraan and Bakool
S1 Gedo and Bay
S2 Shabeellaha Dhexe, Shabeellaha Hoose
   and Banaadir
S3 Jubbada Dhexe and Jubbada Hoose

This regional subdivision, reflecting a historical situation, is used for practical purposes only, and has no political significance.

To facilitate the understanding of the work the following points should also be noticed:

Sequence of families. For the dicotyledons the sequence and circumscription will largely follow Hutchinson, Families of flowering plants, vol. 1, ed. 1 (1926) to facilitate comparison with Flora of Ethiopia and several other tropical African Floras. For the monocotyledons Dahlgren & Clifford, The monocotyledons: a comparative study (1982), is followed with some modification.

Cultivated and introduced species. Commonly naturalized species are treated in the same way as native ones. Established cultivated species and ornamentals grown in public places are included, but usually described more briefly.

Erroneous records and misapplied names. Many species have been recorded erroneously or doubtfully from Somalia. Erroneous records, e.g., in Cufodontis's Enumeratio (1953–1972), are mentioned and clarified as far as possible under the respective family or genus.

Bibliography. Only the year of publication is given for genera, species and infraspecific taxa, wherever the place of publication can be ascertained from the Index Kewensis. If a name has been conserved, this is indicated by "nom. cons.".

Useful references to literature are given when appropriate. Some standard works are abbreviated as follows: Adumbr. = Adumbratio Florae Aethiopicae; Chiov., Fl. Somala = Chiovenda, Flora Somala; Cuf. Enum. = Cufodontis, Enumeratio Plantarum Aethiopiae Spermatophyta; Fl. Trop. E. Afr. = Flora of Tropical East Africa.

Synonyms. All published names based on material from Somalia are included. Synonyms (as well as basionyms) are otherwise generally omitted and only more important ones are given, such as names accepted as correct in Cuf. Enum.

Synonyms based on the same type are placed in chronological sequence in the same paragraph with a semicolon in between, while synonyms based on different types are placed in chronologically arranged, separate, paragraphs. Illegitimate names are indicated by "nom. illeg.".

Types. Types collected in Somalia are indicated, but no others. If a type has not been consulted or is destroyed, this is indicated by respectively "not seen" and "destr.".

Vernacular names and Somali place names. Vernacular names in Somali (Som.), English (Eng.) or Arabic are included when available, but complete lists of orthographic variants are not aimed at. The spellings have often been checked by Somali botanists, but are not claimed to be correct. Sometimes both spellings according to standard Somali orthography and old, but more well-known spellings, are included. If, in a widespread species, different Somali names are used

1

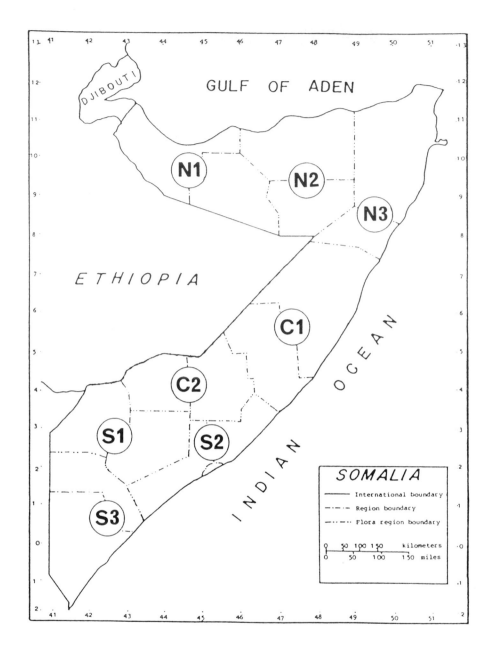

The regions used for the geographical subdivision of Somalia.

in different parts of the country, the names may be divided into the following groups: N = North, regions N1–3; C = Central, regions C1, 2; S = South, regions S1–3.

Modern spellings are given for place names as far as possible, but for type specimens the original spelling is often retained within citation marks.

Habitat. Information on habitat is briefly given for each species and infraspecific taxon, as well as the known altitudinal range within Somalia.

Distribution. Distributions within Somalia are indicated by the designations for the geographical regions used (Fig. 1). If a locality is situated on or close to the border between two regions, the symbols of these regions may be connected by a hyphen.

Distributions outside Somalia are briefly given starting with adjacent countries.

Specimen citation. Reference specimens (normally three when available) are cited for all species and infraspecific taxa.

Notes. Where available and appropriate, notes about properties and uses, observations on variation, circumscription, etc, as well as on lack of information or material are added after the taxa.

LIST OF CONTRIBUTORS TO VOLUME 4

T. A. Cope, The Herbarium, Royal Botanic Gardens, Kew, Richmond, Surrey TW9 3AE, UK (*Poaceae*).

R. B. Faden, Department of Botany, Smithsonian Institution, Washington DC 20560, USA (*Commelinaceae*).

P. Goldblatt, Missouri Botanical Garden, P.O. Box 299, St. Louis, Missouri 63166−0299, USA (*Iridaceae*).

S. Ittenbach & W. Lobin, Botanisches Institut und Botanischer Garten, Rheinische Friedrich-Wilhelms-Universität Bonn, Meckenheimer Allee 171, D-53115 Bonn, Germany (*Araceae: Amorphophallus*).

J. Lavranos, Apartado 243, P-8100 Loulé, Portugal (*Aloaceae*).

K. Lye, Department of Biology and Nature Conservation, Agricultural University of Norway, P.O. Box 5014, N-1432 Ås-NLH, Norway (*Cyperaceae; Juncaceae; Potamogetonaceae*).

Inger Nordal, Department of Biology, University of Oslo, P.O. Box 1045, Blindern, N-0316 Oslo, Norway (*Amaryllidaceae*).

B. Pettersson, Department of Systematic Botany, Uppsala University, Villavägen 6, S-752 36 Uppsala, Sweden (*Orchidaceae*).

G. Sartoni, Dipartimento di Biologia Vegetale, Università di Firenze, Via G. La Pira, 4, I-50121 Firenze, Italy (*Cymodoceaceae; Hydrocharitaceae: Enhalus, Halophila, Thalassia*).

Sebsebe Demissew, The National Herbarium, Addis Abeba University, P.O. Box 3434, Addis Ababa, Ethiopia (*Asparagaceae*).

M. Thulin, Department of Systematic Botany, Uppsala University, Villavägen 6, S-752 36 Uppsala, Sweden (*Agavaceae; Alismataceae; Alliaceae; Anthericaceae; Aponogetonaceae; Araceae: Arisaema, Pistia, Stylochaeton; Arecaceae; Asphodelaceae; Cannaceae; Colchicaceae; Dracaenaceae; Flagellariaceae; Hyacinthaceae; Hydrocharitaceae: Lagarosiphon, Ottelia; Hypoxidaceae; Lemnaceae; Musaceae; Pandanaceae; Typhaceae; Velloziaceae; Zingiberaceae*).

LIST OF FAMILIES INCLUDED IN VOLUME 4

Angiospermae (monocotyledons)

138. Hydrocharitaceae
139. Aponogetonaceae
140. Alismataceae
141. Potamogetonaceae
142. Cymodoceaceae
143. Araceae
144. Lemnaceae
145. Asparagaceae
146. Dracaenaceae
147. Agavaceae
148. Hypoxidaceae
149. Asphodelaceae
150. Aloaceae
151. Anthericaceae
152. Hyacinthaceae
153. Alliaceae
154. Amaryllidaceae
155. Iridaceae
156. Colchicaceae
157. Orchidaceae
158. Velloziaceae
159. Musaceae
160. Zingiberaceae
161. Cannaceae
162. Commelinaceae
163. Typhaceae
164. Juncaceae
165. Cyperaceae
166. Flagellariaceae
167. Poaceae
168. Arecaceae
169. Pandanaceae

KEY TO THE FAMILIES OF MONOCOTYLEDONS IN SOMALIA

The recent split, particularly of *Liliaceae* in a wide sence, into smaller and presumably more natural units, has resulted in many presently recognized families of monocotyledons that are still little known even to many experienced botanists. To facilitate the use of this flora volume a key to the families of monocotyledons occurring in Somalia is therefore given below.

1. Plants marine 2
  − Plants terrestrial, epiphytic or aquatic in fresh water 3
2. Leaf sheath, if present, without ligule, not auriculate; gynoecium of 3 or more united carpels..........
  .... 138. *Hydrocharitaceae*
  − Leaf sheath with ligule, auriculate; gynoecium of 2 free carpels .... 142. *Cymodoceaceae*
3. Minute aquatic herbs, the plant body reduced to a thallus-like frond .... 144. *Lemnaceae*
  − Plants terrestrial, epiphytic or aquatic, with well differentiated leaves 4
4. Leaves pinnately or palmately divided; palms .... 168. *Arecaceae*
  − Leaves simple 5
5. Leaves 1.2 m long or more, with margins and underside of midrib spiny; tree with stilt-roots from lower part of trunk (screw-pines) .... 169. *Pandanaceae*
  − Leaves and habit not as above 6
6. Gynoecium composed of 2 or more free carpels 7
  − Gynoecium composed of 1 carpel or 2 or more united carpels 9
7. Flowers in whorls; tepals 6, the inner petal-like, very different from the outer .... 140. *Alismataceae*
  − Flowers in spikes; tepals 1−4, all similar 8
8. Leaves all basal; spikes 2-branched ....139. *Aponogetonaceae*
  − Leaves inserted on the stem; spikes simple ....141. *Potamogetonaceae*
9. Ovary inferior or semi-inferior; perianth present 10
  − Ovary superior; perianth sometimes absent 21
10. Tepals all petal-like 11
  − Outer tepals sepal-like 17
11. Aquatic herbs .... 138. *Hydrocharitaceae*
  − Terrestrial or epiphytic herbs or shrubs 12
12. Stamen 1; pollen aggregated into pollinia; ovary twisted ....157. *Orchidaceae*
  − Stamens 3 or 6; pollen not aggregated into pollinia; ovary not twisted 13
13. Stamens 3 .... 155. *Iridaceae*
  − Stamens 6 14
14. Flowers in large panicles ....147. *Agavaceae*
  − Flowers solitary or in raceme- or umbel-like inflorescences 15
15. Plant with a bulb; flowers in umbel-like inflorescences, or solitary ....154. *Amaryllidaceae*
  − Plant with a corm or rhizome; flowers solitary or in racemes 16
16. Leaves entire, hairy; flowers yellow ....148. *Hypoxidaceae*
  − Leaves setose-serrate near apex at the margin and on the underside of the midnerve; flowers white, flushed with pink, mauve or purple .... 158. *Velloziaceae*
17. Aquatic herbs; flowers regular .... 138. *Hydrocharitaceae*
  − Terrestrial or epiphytic herbs; flowers ± zygomorphic 18
18. Fertile stamens 5; fruit a banana; tree-like herbs, c. 3−6 m tall ....159. *Musaceae*
  − Fertile stamen 1; fruits various; plants smaller 19
19. Stamen not accompanied by petal-like staminodes; pollen aggregated into pollinia; ovary 1-celled
  .... 157. *Orchidaceae*
  − Stamen accompanied by 1 or more petal-like staminodes; pollen not aggregated into pollinia; ovary 3-celled
 20
20. Outer tepals united into a tube; anther 2-thecous ....160. *Zingiberaceae*
  − Outer tepals free; anther 1-thecous ....161. *Cannaceae*
21. Perianth absent or cup-like or reduced to bristles or scales 22
  − Perianth present, consisting of distinct, free or ± united tepals 25
22. Flowers arranged in small spikes (spikelets) with scale-like bracts, bisexual or unisexual (sedges and grasses) 23
  − Flowers not arranged in small spikes with scale-like bracts 24

23. Flowers each enclosed by a single bract (glume) on the outside, the female flowers sometimes each surrounded by a closed flask-like structure (utricle); stem usually solid, often 3-angled; leaf-sheaths usually closed; sedges .... 165. *Cyperaceae*
— Flowers each enclosed by an outer bract (lemma) and an inner bract (palea), the spikelet usually also with 2 empty bracts (glumes) at the base; stem usually with hollow internodes, terete or compressed; leaf-sheaths usually open; grasses ....167. *Poaceae*
24. Leaves broad, entire or divided; inflorescence a spadix enclosed in a spathe .... 143. *Araceae*
— Leaves narrow, long, linear; inflorescence spike-like, not enclosed in a spathe .... 163. *Typhaceae*
25. Inner tepals petal-like, very different from the outer .... 162. *Commelinaceae*
— All tepals ± similar 26
26. Plant a woody climber to 5 m long; leaves inserted on the stem, ending in a tendril; tepals 2−3 mm long, white; fruit a drupe ....166. *Flagellariaceae*
— Plant herbaceous or, if woody, then not climbing; leaves not ending in a tendril or, if so, then tepals large and brightly coloured; fruit a berry, capsule or schizocarp 27
27. Tepals scale-like, dry (rushes) .... 164. *Juncaceae*
— Tepals petal-like 28
28. Leaves reduced to scales or spines, their function taken over by modified green branches (cladodes) ....145. *Asparagaceae*
— Leaves well developed, sometimes produces after the flowers 29
29. Fruit a berry .... 146. *Dracaenaceae*
— Fruit a capsule or schizocarp 30
30. Style ± deeply 3-branched or styles 3 ....156. *Colchicaceae*
— Style simple, sometimes with a 3-lobed stigma 31
31. Plant with a bulb 32
— Plant with a rhizome 33
32. Inflorescence a raceme or spike ....152. *Hyacinthaceae*
— Inflorescence umbel-like ....153. *Alliaceae*
33. Leaves thick and fleshy, usually with toothed margin; flowers tubular, usually red to yellow .150. *Aloaceae*
— Leaves ± thin, without teeth; flowers not tubular, usually white 34
34. Rhizome yellow when cut (plant producing anthraquinones); anthers dorsifixed ....149. *Asphodelaceae*
— Rhizome not yellow when cut (plant not producing anthraquinones); anthers basifixed ..151. *Anthericaceae*

5

ANGIOSPERMAE (cont.)
Monocotyledons

138. HYDROCHARITACEAE

by M. Thulin (*Lagarosiphon, Ottelia*) and G. Sartoni (*Enhalus, Halophila, Thalassia*).

Cuf. Enum.: 1204−1206 (1968); Fl. Trop. E. Afr. (1989).

Annual or perennial herbaceous freshwater or marine aquatics. Leaves in a basal rosette, whorled along the stem, distichous or rarely opposite, sometimes sheathing at the base. Flowers unisexual or bisexual, regular, solitary, paired or in umbel-like groups, enclosed by 2 separate or fused spathal bracts. Tepals 3 or 6, free, when 6 the outer may be sepaloid and the inner petaloid. Stamens 2−many, the innermost or outermost often sterile (staminodes); anthers 2-thecous, basifixed, longitudinally dehiscent. Ovary inferior, composed of (2−)3−6(−20) carpels, generally 1-celled; styles as many as the carpels; ovules few to many. Fruit a fleshy and berry-like capsule, dehiscent or opening by decay of the pericarp. Seeds small with straight embryo, without endosperm.

Family of some 16 genera and almost 100 species, cosmopolitan.

1. Freshwater plants...2
− Marine plants..3
2. Leaves spirally arranged or in whorls along the stem, sessile...................................1. *Lagarosiphon*
− Leaves all basal, petiolate...2. *Ottelia*
3. Leaves petiolate; blade with midvein and a pair of marginal longitudinal veins connected by cross-veins, the leaves appearing as pinnately veined...5. *Halophila*
− Leaves sessile; blade linear with many longitudinal veins...4
4. Rhizome 10 mm or more in diam., covered by fibrous remains of old leaf-sheaths; roots numerous, crowded ....3. *Enhalus*
− Rhizome 2−5 mm in diam., without fibres; roots 1 per node...4. *Thalassia*

1. LAGAROSIPHON Harvey (1841)
Symoens & Triest in Bull. Jard. Bot. Nat. Belg. 53: 441−488 (1983).

Perennial freshwater herbs, dioecious; roots simple, arising from nodes, without root hairs; stems simple or 1-branched from axils. Leaves sessile, spirally arranged or in whorls, linear to lanceolate, denticulate, 1-veined. Male inflorescence axillary, sessile, many-flowered. Male flowers becoming detached and floating before anthesis; tepals 6. Stamens 3; anthers fixed at right-angles to the horizontally spreading filaments; staminodes often present, joined at the apex and acting as a sail. Female flowers solitary; perianth tube filiform, exserted laterally from the ovary near apex of the spathe, lengthening so that the flower bud reaches the water surface; tepals 6; staminodes 3, minute. Ovary with 5−30 ovules; styles 3; stigmas 6, linear. Capsule ovate, beaked at the apex, irregularly dehiscent. Seeds ellipsoid.

Nine species in Africa and Madagascar. One species, *L. major* (Ridley) Moss, also introduced in Europe and New Zealand.

L. cordofanus Caspary (1855). Fig. 1.
L. crispus Rendle (1895).

Stems filiform. Leaves spirally arranged or in whorls, linear to narrowly linear-lanceolate, 4−30 × 0.25−1 (−1.8) mm, acute, spreading or recurved, with green margins but with a broad band of sclerenchyma fibres along either side of midrib. Male spathe 1.3−2.7 mm long, containing 7−14 flowers. Male flowers white; tepals c. 1 mm long; perianth tube up to 50 mm long; stigmas c. 1.5 mm long, purplish. Capsule ovoid, 2.5−4 mm long. Seeds 1−1.5 mm long.

Pools in shallow water; 30−230 m. C2; S1, 3; widespread in tropical and southern Africa. Puccioni & Stefanini 208; Thulin, Hedrén & Dahir 7508; Gillett & al. 25025.

2. OTTELIA Pers. (1805)

Annual or perennial freshwater herbs, monoecious, dioecious or bisexual; roots simple; stems usually corm-like. Leaves all basal, juvenile and mature ones markedly different; blades wholly or partly submerged, or floating, with usually several longitudinal veins. Spathes solitary in leaf-axils, with 1−40

7

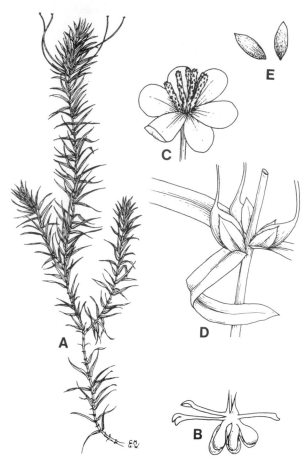

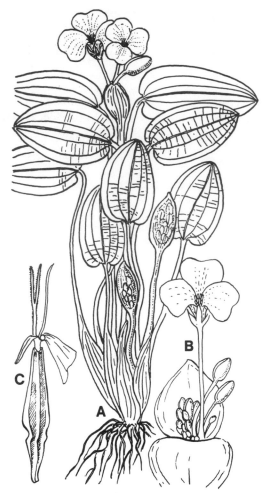

Fig. 1. *Lagarosiphon cordofanus*. A: habit, × 0.45. B: male flower, × 8.5. C: female flower, × 8.5. D: fruits, × 4. E: seeds, × 14. − Modified from Fl. Trop. E. Afr. (1989). Drawn by C. Grey-Wilson.

Fig. 2. *Ottelia exserta*. A: male plant, × 0.4. B: male inflorescence, × 0.8. C: female flower showing spathe, 1 outer tepal, 1 inner tepal (cut off), 1 staminode and 1 style, × 0.8. − Modified from Fl. S. Afr. 1 (1966).

flowers. Male flowers pedicellate; female and bisexual flowers sessile or subsessile. Tepals 6, the 3 inner petaloid; perianth-tube ± narrowly cylindric. Stamens 3−15 or more, in whorls of 3; staminodes (0−)3 or more, often present in female flowers. Ovary of 3 to many carpels; ovules numerous; stigmas 2 per style, linear. Fruit fleshy, dehiscent or opening by decay of the pericarp. Seeds numerous, densely covered by unicellular hairs.

Some 20 species throughout the warmer regions of the world.

O. exserta (Ridley) Dandy (1934); *Boottia exserta* Ridley (1886). Fig. 2.

O. somalensis Chiov. (1932); type: S3, "Uamo Ido", Senni 295 (FT holo.).

Plant dioecious or sometimes bisexual. Leaves petiolate, the juvenile with linear to narrowly elliptic blade; mature leaves floating with blade lanceolate to ovate, 3.9−15 × 1.3−7 cm, rounded to somewhat cordate at the base, with 4−7(−10) longitudinal veins, entire, with petiole 10−30 cm long. Spathe 2.4−9 × 0.5−3.9 cm, 2−6-lobed at the apex, often with longi-

tudinal ribs; peduncle up to 30 cm long. Male flowers up to 20 or more, female or bisexual flowers solitary. Outer tepals green, 9−26 mm long; inner tepals 25−30 × 10−15 mm, white with yellow base; perianth tube 20−25 mm long. Stamens up to 15. Ovary of up to 15 carpels; styles up to 15, deeply bilobed. Fruit ellipsoid, 4−10 cm long, opening by decay of the pericarp.

Ponds; near sea level. S3; widespread in tropical Africa.

The type of *O. somalensis* is the only specimen of *Ottelia* known from Somalia. This appears to be bisexual, and *O. somalensis* is here regarded as a bisexual form of the otherwise normally dioecious *O. exserta*. Apparently bisexual forms of *O. exserta* are also known from Tanzania (treated as *O. somalensis* by Simpson in Fl. Trop. E. Afr.).

O. ulvifolia (Planch.) Walp., with cuneate leafbases, is widespread in tropical Africa and may well occur also in Somalia.

3. ENHALUS L. C. Rich. (1811)

Dioecious perennials with creeping, coarse, simple or sparsely branched rhizomes; rhizome dorsiventral with distichously arranged leaves; internodes very short, obscured by the persistent fibrous remains of leaf sheaths; roots unbranched, numerous. Leaves strap-shaped, with an open sheathing base and with many parallel nerves. Male inflorescences shortly pedunculate, initially enclosed by two fused bracts; flowers numerous, small, pedicellate, breaking off just before anthesis and floating on the surface with the tepals reflexed; tepals 6; stamens 3; anthers subsessile, latrorsely dehiscent; pollen grains globose, very large. Female inflorescence 1-flowered on a long peduncle, enclosed by two overlapping, scarcely fused, strongly keeled bracts, distinctly hairy on the keel; outer tepals 3, green, recurved; inner tepals 3, white, linear-oblong, erect, papillose; ovary of 6 carpels, with several ovules; styles 6, each with 2 long stigmatic branches; flowers floating on the surface at maturity and drawn beneath the water surface after pollination. Fruit fleshy, softly spiny, dehiscing by decay of the pericarp; seeds few, obconical, angular.

Genus with a single species.

E. acoroides (L. f.) Rich. ex Steud. (1840). Fig. 3.

Rhizomes 1−3 cm in diam.; roots whitish, up to 20 cm long and 2−5 mm thick. Leaves 3−6 in a shoot, sheathing each other at the base; sheath colourless, c. 15 cm long; blade bright to dark green, 1−2 cm wide and up to more than 1 m long, with rounded or obtuse apex. Peduncle of male inflorescences terete, 5−10 cm long; bracts broadly ovate-lanceolate, slightly keeled, c. 5 × 3 cm, with somewhat inrolled margins. Male flowers with pedicels 2−12 mm long; tepals c. 2 × 1 mm; anthers 1−2 mm long, white. Peduncle of female flower up to 50 cm long, after anthesis coiled and contracted; bracts 3−6 × 1−2 cm, the margins not inrolled; outer tepals c. 1 × 0.5 cm, reddish; inner tepals 4−5 cm long, 3−4 mm wide, when old with reddish apex; ovary up to 5 × 0.5 cm; stigmas 10−12 mm long. Fruit ovate, 3−5(−7) cm long, green or dark brown to almost black. Seeds 10−15 mm long.

Along shallow sheltered coasts in sandy or muddy areas, with the leaves lying on the water surface at low tide. S3, no doubt occurring also further to the north; in the Red Sea and around coasts bordering the Indian Ocean and the tropical part of the western Pacific. Bavazzano s.n.

4. THALASSIA Banks ex König (1805)

Dioecious perennials; rhizomes creeping, scale-bearing and with extended internodes, giving rise at regular intervals to erect, short, leafy shoots; older erect shoots partially enclosed by persistent remains

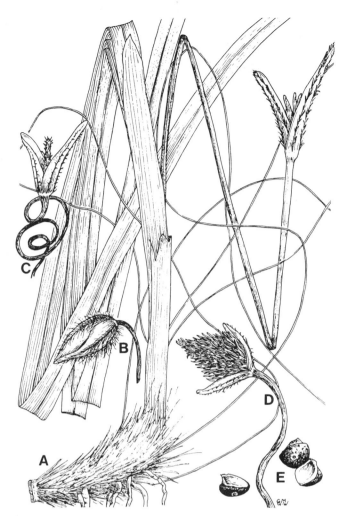

Fig. 3. *Enhalus acoroides*. A: habit of female plant, × 0.45. B: male spathe, × 0.45. C: young fruit, × 0.45. D: mature fruit, × 0.45. E: seeds, × 1.4. − Modified from Fl. Trop. E. Afr. (1989). Drawn by C. Grey-Wilson.

of leaf bases; roots one per node, unbranched, conspicuously septate, covered with sand-binding, fine hairs. Leaves usually 2−6, distichously arranged, linear or falcate, distinctly sheathing at the base; veins 9−17, connected by cross-veins. Inflorescences pedunculate, 1-flowered, in the axils of the leaves, 1−2 in male plants, 1 in female plants. Spathal leaves united on one side only in male plants and on both sides in female ones, margins entire or serrulate. Male flower shortly pedicellate; tepals 3, strongly recurved at anthesis; stamens 3−12; anthers subsessile, oblong, erect, 2−4-celled, latrorsely dehiscent; pollen grains spherical, stuck together into moniliform chains. Female flower subsessile; tepals 3; ovary of 6−8 carpels; styles 6−8, each with 2 filiform stigmas 2−6 times as long as the style. Fruit globose, beaked, echinate, split open at the top by stellate dehiscence of the fleshy pericarp into a

Fig. 4. *Thalassia hemprichii*. A: habit of female plant in young fruit, × 0.45. B: male flower, × 1.4. C: female flower, × 0.45. D: opened fruit, × 1.4. E: seed, × 2.8. — Modified from Fl. Trop. E. Afr. (1989). Drawn by C. Grey-Wilson.

number of irregular valves. Seeds with a thickened basal portion.

Genus of two similar species, *T. testudinum* Banks ex König in the Caribbean and *T. hemprichii* in the Indian Ocean and the western Pacific.

T. hemprichii (Ehrenb. ex Solms) Aschers. (1871). Fig. 4.

Rhizomes terete, greenish to light-brown, 2−6 mm in diam. Leaf sheath persistent, whitish or translucent, up to 9 cm long; blade 4−25(−40) cm long, 2.5−7(−10) mm wide; margin entire, sometimes slightly serrulate near the rounded apex. Peduncle of the male inflorescence c. 3 cm long, that of the female inflorescence 1−1.5 cm long, elongating up to 4 cm after anthesis. Spathal leaves 1.7−2.5 × 0.5−1 cm, entire or rarely serrulate near the apex. Male

flower on a pedicel 2−3 cm long; female flower subsessile. Tepals colourless or light-brown, dotted or striated with reddish brown, 7−10 × c. 3 mm, revolute at anthesis. Stamens (3−)6−9(−12); anthers 7−11 mm long. Ovary of 6 carpels, up to 1 cm long; hypanthium 2−3 cm long; stigmatic branches 10−15 mm long, becoming recurved. Fruit c. 2−2.5 cm in diam., few-seeded.

A common species forming extensive beds on sandy or muddy flats of the upper sublittoral or just above low tide level. S2, 3, no doubt occurring also further to the north; in the Red Sea and throughout the whole tropical region of the Indian Ocean and the western Pacific. Corradi 10502 bis; Sartoni s.n.

5. HALOPHILA Thouars (1806)

Monoecious or dioecious annuals or perennials with creeping rhizomes; rhizome bearing at each node a subopposite pair of scales with a lateral shoot and 1−few unbranched root(s) covered with sand-binding, fine hairs. Leaves in pairs, in pseudo-whorls or distichously arranged, sessile or long-petiolate, usually with a distinct blade, linear to ovate; veins conspicuous, including a thickened midrib and a pair of longitudinal veins usually connected by cross-veins. Inflorescences sessile, usually of a solitary unisexual flower or with 2 flowers of different sex; spathe of 2 sessile imbricate keeled bracts. Male flower pedicellate; tepals 3, imbricate; stamens 3; anthers sessile, linear-oblong, 2−4-celled, latrorsely or extrorsely dehiscent; pollen grains ellipsoid, united in moniliform chains. Female flower usually sessile, with 3 minute tepals at top of hypanthium; ovary of 3−5 carpels; styles 2−6, undivided. Fruit fleshy, thin-walled, beaked, opening by decay of pericarp; seeds several, globose.

Genus of about 10 species, widely distributed in all tropical seas and extending also into warm temperate waters. All species reported from the East African coast belong to section *Halophila*, with very short erect lateral shoots bearing one pair of petiolate leaves at each node, and with leaf blade with ascending lateral veins. *H. decipiens* Ostenfeld and *H. minor* (Zoll.) den Hartog, found respectively in Tanzania and Kenya, have not yet been found in Somalia.

1. Leaf-blades obovate or elliptic, entire; petiole slender, not sheathing; scales up to 6 mm long......
   ....1. *H. ovalis*
 − Leaf-blades linear to oblong, serrulate; petiole short, sheathing; scales 6−15 mm long.............
   ....2. *H. stipulacea*

1. **H. ovalis** (R. Br.) Hook. f. (1858). Fig. 5.
   Plants dioecious; rhizome terete, 0.5−1.5 mm in diam.; internodes 1−5 cm long; scales suborbicular,

often distinctly keeled, emarginate at the apex and auriculate at the base, 2.5−6 mm long. Leaves in pairs, 1 pair on each lateral shoot; leaf blade oblong-elliptic, 1−3 × 0.5−1.5 cm, rounded at apex, cuneate at base, often gradually narrowing into the petiole, entire; midrib and submarginal veins connected by 10−17 pairs of cross-veins ascending at angles of 45−60°; petiole (0.7−)1−4(−6) cm long. Spathal leaves lanceolate, keeled, 3−5 mm long. Male flower with pedicel up to 2.5 cm long at anthesis; tepals elliptic, 3−4 × 2−3 mm, with a prominent central vein, spreading or reflexed, yellowish or whitish; anthers 2−4 mm long. Female flower sessile; ovary of 3 carpels, ovoid, 1−3 mm long; hypanthium 3−10 mm long, with 3 rudimentary petals up to 1 mm long; styles 3, extending into filiform stigmas 10−25 mm long. Fruit ovoid to globose, 3−5 mm long, with a beak up to 6 mm long, 20−30-seeded.

A rather common species in sandy or silty areas of the upper sublittoral or just exposed at low tide in sheltered localities. S2, 3, no doubt occurring also further to the north; in the Red Sea, along the coast of East Africa as far south as Madagascar and temperate South Africa, and around coasts bordering the Indian and western Pacific Oceans. Corradi 10504; Sartoni s.n.

H. ovalis shows a wide range of variation in size, shape and venation of leaves, and small-leaved plants may be confused with the closely related *H. minor*.

2. H. stipulacea (Forssk.) Aschers. (1867).

Plants dioecious; rhizome terete, 0.5−2 mm in diam.; internodes 0.5−4 cm long; scales conspicuous, obovate, markedly emarginate, keeled, 6−15 × 4−10 mm. Leaves in pairs, 1 pair on each lateral shoot; blade oblong or linear-elliptic, often slightly curved, membranous or cartilaginous and becoming bullate with age, 1.5−6 × 0.2−1 cm, serrulate at least near the rounded or obtuse apex, cuneate or gradually decurrent into the petiole at the base; midrib and submarginal veins connected by numerous pairs of inconspicuous lateral veins ascending at angles of 40−60°; petiole 0.5−1.5 mm long, sheathing at the base. Spathal leaves keeled, hairy. Male flower with pedicel 7−15 mm long; tepals ovate, U-shaped in cross-section, 4 × 2.5 mm, with a prominent central vein; anthers 2−3.5 mm long. Female flower sessile; ovary ellipsoid of 3 carpels, 1.5−2.5 mm long; hypanthium 3−5 mm long, with 3 rudimentary petals; styles 3, extending into filiform stigmas 15−20 mm long. Fruit ellipsoid, c. 5 mm long, with a beak up to 6 mm long, 30−40-seeded.

In shallow sandy or silty areas of the sublittoral

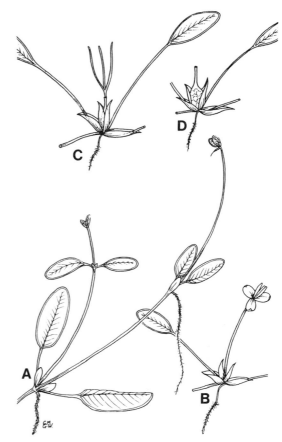

Fig. 5. *Halophila ovalis*. A: habit, × 1.4. B: male spathe and flower, × 1.4. C: female spathe and flower, × 0.7. D: fruit, × 1.4. − Modified from Fl. Trop. E. Afr. (1989). Drawn by C. Grey-Wilson.

subject to slight to moderate wave action. C1; S2, 3, no doubt occurring also further to the north; widespread along the coast of East Africa from the Red Sea to Madagascar, in the Persian Gulf and in the southern part of the Indian Peninsula. Robecchi-Bricchetti s.n. (bullate leaves); Sartoni s.n. (smooth leaves).

Plants with membranous, smooth leaves and more or less deciduous scales have often been treated as a separate species, *H. balfourii* Soler., as opposed to plants with more or less cartilaginous, bullate leaves and stiff, persistent scales. Both forms have been collected along the coast of Somalia, bullate-leaved ones in C1 (Robecchi-Bricchetti s.n.), and smooth-leaved ones in S2 and S3 (Sartoni s.n.). The differences are probably due to environmental conditions, but the relationship between *H. stipulacea* and *H. balfourii* should be studied further.

139. APONOGETONACEAE

by M. Thulin

Cuf. Enum.: 1202–1203 (1968); van Bruggen in Bull. Jard. Bot. Nat. Belg. 43: 193–233 (1973); Fl. Trop. E. Afr. (1989).

Perennial glabrous freshwater herbs from starch-rich rhizome or tuber. Leaves basal, simple, usually long-petiolate. Inflorescence usually a simple or 2-branched spike, at first enclosed in a thin spathe. Flowers bisexual or more rarely unisexual, usually irregular. Tepals 1–6 or absent, petal-like, free. Stamens usually 6, free; filaments narrow; anthers 2-thecous, longitudinally dehiscent. Carpels superior, usually 3–6, free or slightly united at the base, 1-celled; ovules 1–14; style short. Fruit a 1–several-seeded follicle. Seeds with a straight embryo, without endosperm.

Family of a single genus only.

APONOGETON L. f. (1782), nom. cons.

Description as for the family.

Genus of about 40 species in tropical and subtropical parts of the Old World.

Records from Somalia of *A. desertorum* Zeyh. ex Spreng. f. and *A. subconjugatus* Schumach. & Thonn. in Cuf. Enum.: 1202 (1968) are based on material of *A. abyssinicus* var. *cordatus*.

1. Plant dioecious; tepals white or cream, absent in female plants..........................1. *A. nudiflorus*
2. Plant bisexual (stamens may be absent in flowers of apomictic plants); tepals pink, mauve or purplish, always present but sometimes soon falling..........
....2. *A. abyssinicus*

1. **A. nudiflorus** Peter (1928).

Leaf-blades ± long-petiolate, floating, narrowly ovate to narrowly oblong, 2–13 × 0.3–4.5 cm, acute to obtuse at the apex, cuneate to truncate at the base. Dioecious. Spikes 2-branched, pedunculate; male spikes relatively lax, c. 10–80 × 2–5 mm; female spikes dense, 10–50 × 4–8 mm. Male flowers with 2 white or cream tepals and usually 6 stamens. Female flowers without tepals, with 3, 3–6-ovulate carpels. Follicles 3–4 mm long including 1–1.5 mm long beak. Seeds 1.5–2 × 0.4–0.7 mm.

In temporary pools; 200–1000 m. N1, 2; S1; E Ethiopia, Kenya, N Tanzania. Glover & Gilliland 164; Hemming 165; Paoli 642.

2. **A. abyssinicus** Hochst. ex A. Rich. (1850).

Muco (Som.).

Leaf-blades ± long-petiolate, floating, ovate to lanceolate, 3–12 × 0.5–5 cm, acute to obtuse at the apex, cuneate to cordate at the base. Mostly bisexual. Spikes 2-branched with branches 10–50 × 2–8 mm, dense, pedunculate. Flowers mostly with 2 pinkish to purplish tepals, 6 stamens and 3, 4–14-ovulate carpels (rarely stamens absent and carpels up to 7 in apomictic plants). Follicles 3–7 mm long including 1–2 mm long beak. Seeds 0.7–1.5 × 0.2–0.4 mm.

var. **cordatus** Lye in Lidia 1: 75 (1986). Plate 1 A.
Leaf-blades with cordate base.

In temporary pools; 40–200 m. S1–3; E Kenya. Hemming 425; Thulin 6326; Paoli 1170.

Var. *abyssinicus* is distributed in eastern Africa from Ethiopia southwards to south-eastern Zaire. Three further varieties were recognized by Lye in Lidia 1: 73–79 (1986), all in Kenya and Tanzania.

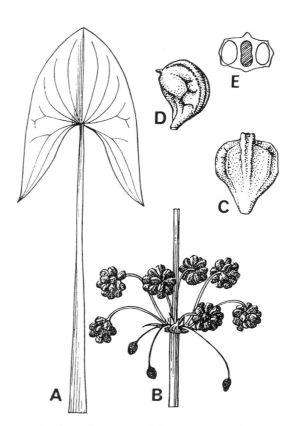

Fig. 6. *Limnophyton obtusifolium*. A: leaf, × 1/3. B: lower whorl of inflorescence with bisexual flowers in fruit, × 1. C: nutlet, dorsal view, × 4. D: nutlet, lateral view, × 4. E: transverse section of nutlet showing air chambers. — Modified from Fl. Trop. E. Afr. (1960). Drawn by E. M. Stones.

140. ALISMATACEAE

by M. Thulin

Fl. Trop. E. Afr. (1960); Cuf. Enum.: 1203 (1968).

Perennial or rarely annual, aquatic or marsh herbs. Leaves in a basal rosette; petiole with a widened base; blade entire. Inflorescence paniculate with whorls of branches, rarely umbel-like or with solitary flowers. Flowers bisexual or unisexual; 3 outer tepals green, persistent; 3 inner tepals petal-like, usually soon falling. Stamens (3−)6(−many); filaments narrow; anthers 2-celled, longitudinally dehiscent. Carpels superior, 3−many, free or united at the base, in a whorl or spiral, 1-celled; ovules 1−many; style apical, lateral or basal. Fruit usually an indehiscent nutlet. Seeds oblong with a strongly curved embryo, without endosperm.

Some 11 genera and 95 species, cosmopolitan, but best represented in the northern hemisphere.

1. Blade of mature leaves sagittate..1. *Limnophyton*
2. Blade of mature leaves with cuneate to rounded base...2. *Burnatia*

1. LIMNOPHYTON Miq. (1856)

Perennial herbs. Leaves erect, sagittate in mature leaves, cuneate at the base in seedlings. Inflorescence of 4−7 whorls of flowers, the lowest sometimes with 2−3 branches; upper whorls with male flowers, lower with bisexual and male flowers. Stamens 6, much smaller in bisexual flowers. Carpels 10−30, free, ovoid; ovules solitary; style short, lateral. Nutlets swollen, with lateral air chambers.

Genus of three species in tropical Africa, and in Madagascar and southern Asia.

L. obtusifolium (L.) Miq. (1856). Fig. 6.

Leaves glabrous; petiole 25−50 cm long, triangular in section; blade up to 25 × 20 cm in outline, sagittate. Peduncle about as long as petioles; lowest whorl of inflorescence with 6−10 bisexual flowers and 0−2 male flowers; pedicels 2−4 cm long, slender in male flowers, thickened and recurved in fruit. Outer tepals c. 6 × 3.5 mm in bisexual flowers, c. 4 × 2 mm in male flowers. Inner tepals larger than the outer, white. Nutlets 4 × 3 mm, ridged, pale brown.

Marshy ground and pools; 10−50 m. S1−3; widespread in tropical Africa, also Madagascar and southern Asia. Gillett & al. 25026, 25293; Maunder 53.

2. BURNATIA Micheli (1881)

Perennial glabrous dioecious herbs. Leaves erect; blade linear-lanceolate to ovate. Inflorescence of whorls of 3 branches or 3 flowers. Male flowers with 3 + 3 tepals, 9 stamens and abortive carpels. Female flowers with outer tepals only. Carpels numerous, free; ovules solitary; style very short, lateral. Nutlets flattened with subcircular lateral flanges, glandular.

Genus of a single species in tropical and southern Africa only.

B. enneandra Micheli (1881). Fig. 7.

Baar, baar-biyood (Som.).

Petioles up to 40 cm long; leaf-blade c. 13−16 × 1−7 cm, very variable in shape, acute at the apex, cuneate to rounded at the base. Male inflorescence c. 20 cm long, of 1−5 whorls of branches; female inflore-

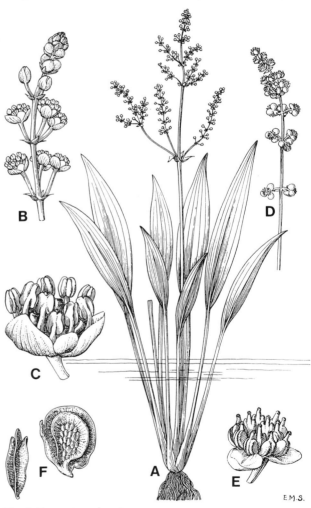

Fig. 7. *Burnatia enneandra*. A: male plant, × 0.2. B: part of male inflorescence, × 1.4. C: male flower, × 6. D: part of female inflorescence, × 1.4. E: female flower, × 6. F: nutlet, dorsal and lateral view, × 6. − Modified from Fl. Trop. E. Afr. (1960). Drawn by E. M. Stones.

scence shorter. Male flowers with outer tepals 2−3 mm long and inner tepals c. 1 mm long. Female flowers with outer tepals c. 1.5 mm long and inner tepals minute or lacking. Nutlets 8−20, 1.5(−2.5) mm long.

Pool; c. 200 m. S1; widespread in tropical Africa south to South Africa. Paoli 1171, 1172.

141. POTAMOGETONACEAE

by K. A. Lye

Cuf. Enum.: 1197−1199 (1968).

Glabrous chiefly perennial herbs of fresh or brackish water rooted in the substrate with creeping sympodial rhizomes; stems elongated, submerged or floating, rarely prostrate on wet mud. Leaves alternate or rarely opposite or in whorls of 3, sheathing at the base; blades either submerged or floating or both; stipules usually present. Flowers in axillary or terminal bractless spikes, inconspicuous, bisexual, regular. Tepals 4, free, valvate, sometimes regarded as appendages of the connectives of the anthers. Stamens 4, sessile, opposite the tepals and basally adnate; anthers 2-thecous, opening by longitudinal slits. Carpels superior, (1−)4, free, sessile, each with a solitary ovule and a short almost sessile terminal style or stigma. Fruit a nutlet or drupe; seed without endosperm.

Family of two genera, widely distributed in both temperate and tropical regions. The second genus, *Groenlandia* Gay, is known from North Africa.

POTAMOGETON L. (1753)

Slender to robust aquatic plants surviving cold or dry periods by rhizomes or specialized buds borne either on the rhizome or on the leafy stem. Submerged leaf-blades usually thin and translucent, linear and grass-like to oblong, 1−many-nerved; floating leaf-blades usually more coriaceous and opaque; stipules either free from the leaf-bases or adnate to them in their lower part (stipular sheath) and free above (ligule), in either case the basal part may be open or tubular. Spikes cylindrical to ovoid, dense or lax, sometimes interrupted, either raised a little above the water and wind-pollinated or submerged and water-pollinated. Tepals rounded, with a stalk (clawed), green or brownish. Fruit brown or greenish, often asymmetric.

About 90 species, but only few in dry and saline regions.

1. Leaves all submerged, narrowly linear to filiform, up to 3 mm wide.........................1. *P. pusillus*
− Leaves submerged or floating, more than 10 mm wide..............................2. *P. schweinfurthii*

1. **P. pusillus** L. (1753). Fig. 8 A, B.

P. panormitanus Biv. (1838).

Submerged perennial with poorly developed slender rhizome, or arising from seeds or detached buds; stems 20−100 cm long, strongly branched from the base; internodes of relatively equal length. Leaves translucent, narrowly linear, 2−8 cm long and 1−2.5 mm wide, with 1 prominent midrib and 1(−2) weaker nerves on each side of the midrib, apiculate at the tip; sheath 5−18 mm long, tubular for more than half the length when young, later splitting almost to the base, light brown, semipersistent. Spikes axillary, 4−12 mm long, 2−15-flowered, rarely interrupted; pe-

duncles 1−3 cm long. Fruit 2−2.5 x 1−1.5 mm, green to olive, obovoid and compressed with convex margins; keel obscure, rounded, smooth; beak 0.3−0.4 mm long, straight or somewhat oblique, almost centrally placed.

In stagnant or slow-flowing water; c. 1800 m. N2 ("Medishe Springs"); scattered in Africa, more common in Europe and temperate parts of Asia and North America. Glover & Gilliland 917.

2. **P. schweinfurthii** A. Benn. (1901). Fig. 8 C, D.

Perennial with creeping woody rhizome and 50−250 cm long and 1−3 mm thick terete strongly branched stems carrying thin submerged leaves and sometimes also well developed lanceolate floating coriaceous leaves in addition to the submerged ones. Submerged leaves linear-lanceolate, usually 10−20 x 1−2 cm when mature, distinctly net-veined, strongly attenuate at base and apex, apiculate at the tip, entire and prominently but finely undulate at the margin; petiole 1−6 cm long, merging gradually into the leaf-blade; stipular sheath 2−5 cm long, folded, persistent. Spikes 2−4 cm long, 15−30-flowered; peduncle up to c. 10 cm long and 2−3 mm thick. Fruit 3−4 mm long with 0.5−0.6 mm long beak, smooth; ventral side fairly straight, dorsal rounded with rounded keel and 2 weaker lateral keels.

In stagnant or slow-flowing water; c. 100−1800 m. N2, 3; C1; also in Sudan, Ethiopia and eastern Africa south to the Cape region. Robecchi-Bricchetti 3; Glover & Gilliland 593; Merla, Azzaroli & Fois s.n.

P. schweinfurthii is a very variable species and some collections are probably hybrids. Plants with predominantly coriaceous floating leaves could be hybrids with *P. thunbergii* Cham. & Schlecht., and at least some plants with submerged leaves only are hybrids with *P. lucens* L.

Fig. 8. A, B: *Potamogeton pusillus,* from de Wilde 6053 and Schou 1990. A: habit, × 0.6. B: stipular sheath, × 5. − C, D: *P. schweinfurthii*, from de Wilde 10882. C: habit, × 0.6. D: fruit, × 9. − E: flower of *Potamogeton*. − Drawn by G. M. Lye.

142. CYMODOCEACEAE

by G. Sartoni

Cuf. Enum.: 1200−1201 (1968); den Hartog, The sea-grasses of the world, Verhand. Kon. Ned. Akad. Wetensch. Nat. 59(1): 144−212 (1970).

Dioecious, marine perennials with creeping, monopodially or sympodially branched rhizomes; rhizome leafy or with scale leaves, herbaceous and rooting at the nodes or woody and rooting from the internodes. Leaves distichous, with distinct blade and sheathing base; sheath ligulate, auriculate; blade 3- to several-nerved, narrowly or broadly ribbon-like or, rarely, terete and subulate. Flowers usually solitary and terminal on short erect shoots or on branches of erect shoots, but forming cymose inflorescences in *Syringodium*. Perianth absent. Male flowers subsessile or stalked, consisting of 2 extrorsely dehiscent, 2-thecous anthers ± dorsally united and attached either at the same height or one slightly above the other (*Halodule*). Pollen grains filamentous and tightly coiled within the anther. Female flowers sessile or shortly stalked, consisting of two free carpels, each carpel having either a simple, long style (*Halodule*) or a style divided into 2 or 3 slender stigmas, and containing one pendulous ovule. Fruit a 1-seeded nut, indehiscent, either with a stony endocarp or viviparous. Seed without endosperm.

Family of five genera and about 20 species; for the most part occurring in tropical and subtropical oceans.

1. Rhizome woody, sympodial; leaf blade and sheath shed together; male flowers sessile.....................
    ....3. *Thalassodendron*
−   Rhizome herbaceous, monopodial; leaf blade shed before the persistent sheath; male flowers stalked........2
2. Rhizome with scale leaves; leaf blade terete; flowers in conspicuous cymose inflorescences............
    ....4. *Syringodium*
−   Rhizome with foliage leaves; leaf blade flat; flowers solitary...3
3. Leaf blade usually less than 3 mm wide, 3-veined; roots unbranched; styles simple; anthers attached at unequal heights..1. *Halodule*
−   Leaf blade usually more than 3 mm wide, 7−17-veined; roots branched; styles 2-branched; anthers attached at the same height..2. *Cymodocea*

1. HALODULE Endl. (1841)

Rhizome with 1−6 unbranched roots and distichously arranged foliage leaves at each node, producing lateral branches but without consistent differentiation between long and short shoots; rhizome scales 2, ovate or elliptic. Leaf sheath 1−6 cm long, scarious, persisting longer than the blade, when shed leaving an annular scar on stem; mouth of the sheath with two lateral rounded auricles; ligule a narrow apical ridge; blade dark green, flat, often narrowed at the base, 3-veined, with prominent midrib, widened or forked at the apex, and inconspicuous lateral nerves; apex very variable in outline, tridentate or bicuspidate, rounded and serrulate or emarginate. Flowers solitary, terminal on a short erect shoot. Male flower stalked; anthers facing opposite directions, joined below, one slightly above the other. Female flower subsessile, with obovoid or globose carpels, each with a long simple style. Fruit with a stony pericarp, ovoid, somewhat compressed, with a short subapical beak.

Genus of six species, pantropical.

Also *H. wrightii* Aschers. has been reported from the coasts of East Africa. However, the presence of this species in the Indian Ocean is not supported by chemotaxonomic data, and previous reports in the literature probably refer to narrow-leaved forms of *H. uninervis*.

H. uninervis (Forssk.) Aschers. (1882). Fig. 9.

Rhizome rather fleshy, 0.5−2 mm in diam., with 1−6 simple roots densely covered with root-hairs and an erect shoot at each node; internodes 0.5−4 cm long; scales elliptic, 6−7 mm long, transparent when young, becoming dark brown with age. Leaf sheath biauriculate, 1.5−3.5 cm long, with inconspicuous ligule; blade linear, straight or somewhat curved, 5−10(−15) cm long, up to 3 mm wide, but not more than 1 mm in the narrow-leaved forms; apex with two short, acute lateral teeth and an obtusely rounded median tooth ending the conspicuous, widening but rarely forked midrib. Male flower on 6−20 mm long stalk; anthers 2−3 mm long, the upper attached c. 0.5 mm above the lower. Female flower with ovate ovary, 1−2 mm long; style 1.2−3 mm long. Fruit globose, 1.5−2 mm in diam., with apical beak 1 mm long.

H. uninervis is a typical pioneer species occurring in shallow waters near and below low tide level on firm sand and soft mud. S2, 3, but no doubt occurring also further to the north; Red Sea and Indo-Pacific from the eastern coast of Africa to Japan, northern Philippines and Queensland. Sartoni s.n.

2. CYMODOCEA König (1805)

Rhizome herbaceous, monopodial, bearing at each node 1−2(−3)-branched roots and distichously arranged foliage leaves, some leaves subtending a

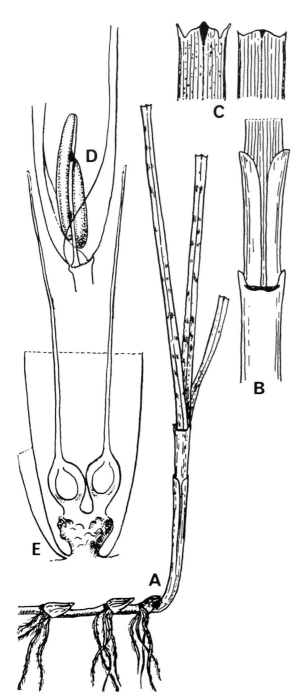

Fig. 9. *Halodule uninervis*. A: habit, × 1. B: leaf sheaths, × 2. C: leaf tips, × 4. D: young male flower, × 9. E: female flower, × 10. – Modified from Fl. S. Afr. 1 (1966).

short leafy shoot; internodes 1−6 cm long. Leaf sheath compressed, persisting longer than the leaf blade, when shed leaving an open or closed circular scar; blade flat, often narrowed at the base, up to 30 cm long, with 7−17 veins joined at the apex by a marginal commissure; apex rounded, obtuse or sometimes emarginate, with either microscopic marginal

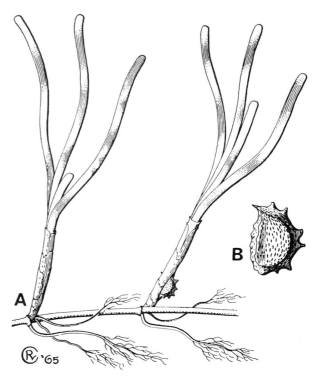

Fig. 10. *Cymodocea rotundata*. A: habit, × 1. B: fruit, × 2. – Modified from den Hartog, The Sea-grasses of the world (1970).

outgrowths or conspicuous marginal teeth. Flowers solitary, terminal on short erect shoots. Male flower stalked; anthers dorsally united and attached at the same level, each with a short apical process. Female flower terminal, sessile or shortly stalked; carpels each with a style divided into 2 slender stigmas up to 3 cm long. Fruit with a stony pericarp, laterally compressed, semicircular or semiovate in outline with dorsal ridges.

Genus of four species with a wide but disjunct distribution. Two species are widely distributed in the tropical and subtropical Indian and western Pacific Oceans.

1. Leaf scars closed; leaf blade 2−4 mm wide, 9−15-nerved; leaf apex rounded to emarginate, entire or serrulate; fruit with dentate dorsal ridges............
....1. *C. rotundata*
− Leaf scars open; leaf blade 4−10 mm wide, 13−17-nerved; leaf apex obtuse with conspicuous marginal teeth; fruit with dorsal blunt ridges........
....2. *C. serrulata*

1. **C. rotundata** Ehrenb. & Hempr. ex Aschers. (1870). Fig. 10.
Rhizome fleshy and rather delicate, with 1(−2)-branched roots and a short leafy shoot at each node; internodes 1−5 cm long. Leaf sheath pale purplish, 1.5−5 cm long; blade linear, often somewhat curved, 6−10(−15) cm long, 2−4 mm wide. Male flower with

17

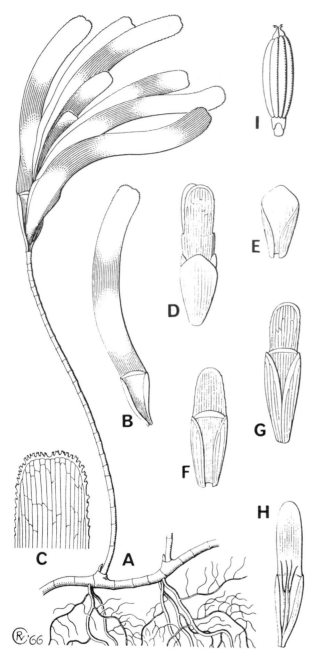

Fig. 11. *Thalassodendron ciliatum*. A: habit, × 2/3. B: leaf showing ligule and open sheath, × 2/3. C: leaf tip, × 2. D: cluster of leafy bracts surrounding the female flowers, × 1. E: bract 1, × 1. F: bract 2, × 1. G: bract 3, × 1. H: bract 4, enclosing the 2 ovaries, × 1. I: male flower, × 4. — Modified from den Hartog, The Sea-grasses of the world (1970).

Common on sheltered sandy flats at or below low tide level. S3, no doubt occurring also further to the north; Red Sea, coasts of the Indian Ocean and western Pacific from the Ryukyu Is. to Queensland. Sartoni s.n.

Flowers and fruits in this species have only been reported in material from the southern hemisphere.

2. C. serrulata (R. Br.) Aschers. & Magnus (1870).

Rhizome fleshy but becoming somewhat tough with age, usually mottled purple, with 2–3 sparsely branched roots and a short leafy shoot at each node; internodes 1.5–5 cm long. Leaf sheath obconical, bright purple, 1.5–3.5 cm long, with 1 mm high ligule; blade linear, often curved, 5–20 cm long, 4–10 mm wide, with smooth margins becoming slightly spinulose towards the apex. Male flower with c. 2 cm long stalk; anthers yellowish, 6–9 mm long, not crowned by a subulate process. Female flower subtended by a single leaf-like bract, on a bifurcate peduncle c. 2 mm long; ovary ovoid, with a curved, flattened style divided in 2 slender stigmas up to 27 mm long. Fruit sessile, elliptic in outline, 7–9 × 4–4.5 mm, with 3 dorsal, parallel, blunt ridges.

Common in sheltered localities below low tide level, on coral sand or on mud-covered coral debris. S2, 3, no doubt occurring also further to the north; Red Sea, coasts of East Africa as far south as Madagascar, eastward in the Indian Ocean to the western Pacific. Sartoni s.n.

Flowering has rarely been observed and flowers have only occasionally been found in the Indo-Pacific.

3. THALASSODENDRON den Hartog (1970)

Rhizome robust, woody, bearing scale leaves, roots, and 1 or 2 erect unbranched or little branched stems at every fourth internode; roots 1–5, ± branched, arising together and associated exclusively with branch-bearing regions of rhizome. Leaf sheath compressed, markedly ligulate, narrowed below; blade flat, linear, 13–27-veined, with spinulose margin and coarsely denticulate, rounded or obtuse apex. Flowers solitary and terminal on short lateral shoots, subsessile, each flower enclosed by 4 leafy bracts, red-pigmented at maturity. Male flower with dorsally united anthers, each with a terminal appendage. Female flower with a short style and two stigmas. Fruit consisting of 1(–2) fertilized carpels and the fleshy innermost bract; germination partly or wholly viviparous to produce a tree-floating seedling.

Genus of two species, one widely distributed in the tropical part of the Indo-Pacific, the other endemic to south-western Australia.

T. ciliatum (Forssk.) den Hartog (1970); *Cymodocea ciliata* (Forssk.) Ehrenb. ex Aschers. (1867). Fig. 11.

Rhizome stout, up to 5 mm in diam., becoming

2–3 cm long stalk; anthers 11–15 mm long, crowned by a subulate process and two acute protuberances. Female flower with ovary and style only 5 mm long; stigmas spirally coiled, 3 cm long. Fruit semicircular in outline, c. 9–10 × 5–6 mm, with dorsal, dentate ridges, varying in colour from light to dark brown.

woody with age; internodes 5−10(−25) mm long; scales obtuse and dark brown, at the nodes of young rhizomes, early deciduous; roots 1−5, on the internode preceding a stem-bearing internode, 0.5−2 mm thick, up to 15 cm long. Erect stems 1(−2), 10−30 cm or more high, with prominent annular leaf scars; lowest internodes up to 10 mm long, the upper shorter and laterally compressed. Leaves in distichous apical tufts, early deciduous; blade and sheath shed together; sheath compressed, cuneate at the base, 15−30 mm long with rim-like ligule 2−2.5 mm high; blade linear, curved, (5−)10−15 cm long, (6−)8−10(−13) mm wide; midrib distinct with about 10 parallel veins on either side; margin irregularly serrate at and near the rounded or obtuse apex; apical teeth acute or trapezoid, 0.5−1 mm high. Male flower 13−15 mm long, with dorsally united anthers on a short common stalk, elongating at maturity. Female flower on c. 0.3 mm long stalk; ovary ellipsoid, 0.5−0.7 mm in diam.; style 4 mm long, divided into 2 subulate stigmatic arms 20−30 mm long. Fruit oblong, 3.5−5 cm long.

Widespread along the coast in deeper pools of the rocky platform and forming extensive beds in the upper part of the sublittoral zone inside the reef. S 2, 3, no doubt occurring also further to the north; Red Sea and western Indian Ocean as far south as Zululand, tropical Asiatic coasts of the Pacific and tropical east coast of Australia. Corradi 10502; Bavazzano s.n.; Sartoni s.n.

4. SYRINGODIUM Kütz. (1860)

Rhizomes creeping, herbaceous, bearing at each node 1−3 branched roots and an erect leafy shoot arising from the axil of the scale; internodes with prominent annular open leaf scars. Erect shoots usually unbranched with short internodes. Leaf sheath ligulate, persisting longer than the blade; blade subulate, succulent, often narrowed at the base and minutely toothed at the apex. Flowers in distinct cymose inflorescences terminating an erect shoot, each flower enclosed in a sheathing bract resembling the foliage leaves but with a shorter sheath and gradually decreasing in size towards the top of the inflorescence. Male flower shortly stalked, with dorsally united anthers without apical processes. Female flower sessile, each carpel with 2 slender styles. Fruit ellipsoid or obovoid, with a fleshy exocarp, a persistent stony endocarp and a short beak.

Genus consisting of two distinct but closely related species, one widespread in the Indo-Pacific, the other in the western tropical Atlantic.

S. isoetifolium (Aschers.) Dandy (1939). Fig. 12.

Rhizome slender, 1.5−3 mm in diam.; internodes 1−3.5 cm long; scales early deciduous. Erect shoots bearing 2−3 distichous leaves and with shorter internodes near the base from which roots may arise. Leaf

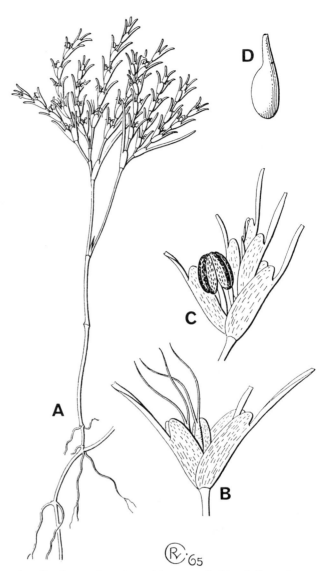

Fig. 12. *Syringodium isoetifolium*. A: habit of flowering plant, × 1. B: female flower, × 4. C: male flower, × 4. D: fruit, × 4. − Modified from den Hartog, The Sea-grasses of the world (1970).

sheath cylindric to obconic, up to 4.5 cm long, c. 4 mm wide, ligulate; blade terete, narrowed at the base, 6−30 cm long, 1−3 mm in diam., with midvein surrounded by 7−12 pericentral veins. Male flower with ovate anthers, 4 mm long. Female flower with elongate carpels about 1 mm in diam., tapering at the apex and forming 2 recurved styles 6−8 mm long. Fruit 3−4 mm long, with rounded base and bifid beak.

Mainly on sandy mud in shallow protected waters, just below low tide level. S2, 3, no doubt occurring also further to the north; widely distributed in the tropical part of the Indian Ocean from the Red Sea south to Madagascar, Mauritius and the Seychelles, and in the Persian Gulf eastward into the western Pacific. Corradi 10500; Moggi & Bavazzano s.n.; Sartoni s.n.

143. ARACEAE

by S. Ittenbach & W. Lobin (*Amorphophallus*) and M. Thulin (*Arisaema, Pistia, Stylochaeton*)

Cuf. Enum.: 1500−1504 (1971); Fl. Trop. E. Afr. (1985).

Perennial herbs, terrestrial, epiphytic or aquatic. Leaves alternate, 1−many, petiolate; petiole often with distinct basal sheath; blade simple or variously lobed, main venation pinnate, palmate, pedate or rarely parallel. Inflorescence pedunculate, consisting of a fleshy spadix (spike) subtended by a bract-like spathe; spadix either uniform with bisexual flowers or monoecious with pistillate flowers at base and staminate flowers above, sterile flowers often present, apical portion of spadix sometimes forming a sterile appendix. Flowers numerous, minute, sessile, bisexual or unisexual, without bracts; tepals 4−9, free or ± united, or forming cup-like perianth, or absent. Stamens free or united into synandria; anthers sessile or with filaments, opening by lateral or apical slits or pores. Ovary normally superior, 1−many-celled, each cell with 1−many ovules; stigma sessile or on usually short conical style. Fruit a 1−many-seeded berry. Seeds minute to large, with or without endosperm.

Some 108 genera and more than 2500 species, the majority in humid tropical regions, some also in dry or temperate areas.

1. Plant aquatic, floating...4. *Pistia*
− Plant terrestrial..2
2. Leaf-blade simple; spadix without sterile appendage...2. *Stylochaeton*
− Leaf-blade compound, pinnately or pedately to radiately divided; spadix with sterile terminal appendix.....3
3. Leaf-blade basically 3-parted, each part much divided.......................................1. *Amorphophallus*
− Leaf-blade pedately to radiately divided to the base...3. *Arisaema*

1. AMORPHOPHALLUS Bl. ex Decne. (1834), nom. cons.

Herbs with subglobose to discoid tubers. Leaf normally solitary; blade 3-parted, main segments widely spreading, highly divided, terminal segments ovate to lanceolate. Inflorescence normally produced before the leaves. Spathe forming a cylindric to bowl-shaped tube with convolute basal portion, upper portion ± expanded and often forming a broad limb. Spadix with basal pistillate part followed by staminate part and an apical sterile appendage. Flowers unisexual, without perianth. Stamens usually densely congested; anthers sessile or with short filaments, dehiscing by apical pores. Pistils usually densely congested, 1−4-celled, usually with 1 basal ovule per cell; stigma globose to variously lobed, sessile or stalked. Berries often brightly coloured. Seeds smooth, without endosperm.

Some 120 species in the Old World tropics and subtropics.

1. Stigma sessile; spathe-tube depressed-globose, clearly constricted at the apex, warty inside towards the base........................1. *A. maximus*
− Stigma on a distinct style; spathe-tube cylindric-ovoid, only slightly constricted at the apex, with longitudinal warty ridges inside towards the base..................................2. *A. laxiflorus*

1. **A. maximus** (Engl.) N.E. Br. (1879). Plate 1 B.
Abeesoole (Som.).
Tuber ± strongly depressed-globose. Leaf: petiole up to 1 m long, mottled; blade up to 1.2 m in diam., with terminal segments lanceolate to obovate, 5−19 x 1.5−2.5 cm. Peduncle 30−70 cm long, usually spotted. Spathe 15−30 cm long; tube depressed-globose, constricted at the apex, warty inside towards the base, greyish-purple to greyish-brown and ± spotted outside, purple inside; limb ovate-triangular with undulate margins, inner surface purple. Spadix 27−50 cm long or more, much longer than the spathe; appendix narrowly elongate-conic, purple to brown; staminate part 2−4.5 cm long; pistillate part 1−3.5 cm long. Stamens 1.4−2 mm wide. Pistils c. 2.5−3 mm long, 2(−3)-celled; stigma sessile, 1−3 mm in diam., shallowly 2(−3)-lobed.

Bushland or woodland, rocky outcrops; 30−300 m. S1−3; Kenya, Tanzania, Zimbabwe. Kilian (2161) & Lobin 7013; SMP 151; Thulin & Warfa 4465.

The fresh tuber is crushed and applied to the body as a remedy against snake bite.

2. **A. laxiflorus** N.E. Br. (1901).
A. gallaensis (Engl.) N.E. Br. var. *major* Chiov., Fl. Somala 2: 431 (1932); type: S3, "Banta", Tozzi 270 (FT holotype not found, photo seen only).
Tuber discoid to depressed-globose. Leaf: petiole 30−40 cm long or more, mottled; terminal segments elliptic to ovate or obovate, 8−12 x 1.5−3 cm. Peduncle 20−50 cm long, spotted. Spathe 20−33 cm long; tube cylindric-ovoid, only slightly constricted at the apex, with longitudinal warty ridges inside towards the base, greyish, purplish or olive green and spotted outside, brownish-purple inside; limb narrowly triangular, erect, with strongly undulate

margins, inner surface dark purple. Spadix 30−90 cm long, much longer than the spathe; appendix narrowly elongate-conic, reddish-brown or purplish-green; staminate part 3−4 cm long; pistillate part 2−4 cm long with pistils ± distant from one another. Stamens c. 2.2 mm wide. Pistils 4−5 mm long, 1-celled; stigma on a 2−3 mm long style, 2-lobed.

Acacia-Commiphora bushland; 30−100 m. S?2, 3; Kenya. Bally 9351A.

The locality for Bally 9351A is between "Jelib" and "Brava" that can be in either S3 or S2.

2. STYLOCHAETON Lepr. (1834)

Herbs with underground rhizome and fleshy roots. Leaves basal, solitary to several; blade lanceolate, ovate, cordate-sagittate, sagittate or hastate-sagittate. Inflorescences borne at or below ground-level on short peduncles, produced before or with the leaves. Spathe forming a tube in lower part, upper part a lanceolate-elliptic ± expanded limb, or unexpanded with a lateral slit. Spadix monoecious with pistillate part near the base and long staminate part above, often with a zone of sterile flowers between. Flowers unisexual, surrounded by glandular cup-shaped perianth. Male flowers with 2−7 stamens; filaments subulate; anthers dehiscing by lateral slits. Female flowers with 1−2-celled ovary; stigma capitate to broadly disc-shaped. Berries borne at or below ground-level in globose to cylindric infructescence, 1−few-seeded. Seeds with endosperm.

Some 15 species in tropical and southern subtropical Africa.

The inflorescences are very inconspicuous, often developing partly below ground and appearing before the leaves. The present account is provisional and more detailed field work is needed.

1. Leaf-blade with sagittate or hastate base; inflorescence often appearing with the leaves..............
....1. *S. borumensis*
− Leaf-blade truncate to broadly cuneate at the base; inflorescence appearing before the leaves...........
....2. *S. grandis*

1. S. borumensis N.E. Br. (1901). Plate 1 C.
Rhizome 0.5−1.5 cm thick. Leaves 2 to several, glabrous; petiole 5−40 cm long, lower part mottled with purple; basal sheath up to half total petiole-length; blade very variable in shape, up to 30 x 12 cm, sagittate to hastate at the base, often lobed with midlobe lanceolate to linear, 10−30 x 0.5−6 cm, acute, side-lobes usually shorter, 0.3−3 cm wide. Inflorescence usually present as leaves emerge, at or near ground level; peduncle 1−6 cm long. Spathe greenish, purplish or brownish, 3.5−15 x 0.5−1.5 cm, cylindric; apex erect, acuminate, with oblique opening. Spadix hidden or slightly protruding from

the mouth of the spathe; male part separated from 0.4−1 cm long female part by a 0.2−1.2 cm wide sterile zone. Male flowers with 2−4 stamens, filaments ± thickened towards the apex; perianth with unlobed to dentate margin. Female flowers 5−10 in 1−2 whorls; ovules c. 20. Seeds several, c. 2.5−3.5 mm long, sulcate.

Deciduous bushland or woodland, usually in seasonally wet places; 40−450 m. C1; S1, 2; East Africa and southwards to Mozambique. Thulin, Hedrén & Dahir 7544, 7680; Thulin 6328 A.

As treated here this is a highly variable taxon in Somalia and more than one species may be involved.

2. S. grandis N.E. Br. (1901).
Rhizome 1 cm or more thick. Leaves 1−3, glabrous; petiole 8−10 cm long; basal sheath c. half to 3/4 total petiole length, forming a pseudostem; blade ± ovate in outline, 10−20 x 6−11.5 cm, truncate to broadly cuneate or subcordate at the base, acute at the apex. Inflorescence appearing before the leaves, at or near ground level, subsessile. Spathe with tube c. 7.5 x 1.3 cm, cylindric; limb ovate, c. 2.5 cm long, apiculate, with oblique opening. Spadix shortly protruding from the mouth of the spathe, c. 9 cm long; male part separated from female part by a 2 cm wide sterile zone. Male flowers with 1−3 stamens, filaments filiform; perianth divided almost to the base into 3 obtuse lobes. Female flowers c. 9, in one whorl.

Acacia-Commiphora bushland on sand; 100−180 m. C1, 2; S2; E Ethiopia. Thulin, Hedrén & Dahir 7314; Kuchar 15761; Gillett & Hemming 24610.

The type of *S. grandis*, James & Thrupp s.n. from eastern Ethiopia, consists of a single inflorescence only, while all other material seen is from Somalia and consists of leaves only. However, these plants are believed to be conspecific as they all come from the same habitat and their rhizomes and roots look very similar.

3. ARISAEMA Mart. (1831)
Mayo & Gilbert in Kew Bull. 41: 261−278 (1986).

Erect herbs with subglobose tubers. Leaves basal, 1−3, with long sheaths often imbricate to form a pseudostem; blade compound, pedately to radiately divided to the base into acuminate lobes. Inflorescence terminal, solitary, appearing with the leaves. Spathe forming a tube in lower part, upper part expanded into a limb. Spadix unisexual or monoecious with basal pistillate part, central staminate part and sterile apical part. Flowers unisexual, without perianth. Male flowers with 2−5 stamens; filaments united; anthers free to united, dehiscing by pores or slits. Female flowers with 1-celled ovary; stigma sessile or on short conical style. Berries orange to red, few-seeded, in cylindrical to subglobose infructescence. Seeds

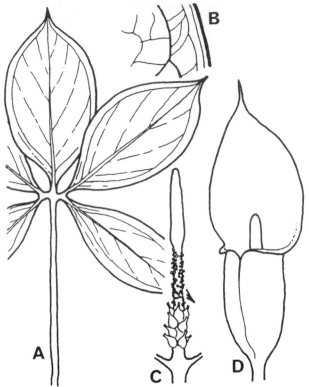

Fig. 13. *Arisaema somalense*. A: leaf. B: leaflet margin. C: spadix. D: inflorescence. − Modified from Kew Bull. 41: 263 (1986). Drawn by S. Mayo.

subglobose, with endosperm.

Some 150 species, the majority in subtropical and temperate East Asia, but also in tropical Asia, North and Central America and in the mountains of Africa and Arabia.

1. Spathe 10.5−14.5 cm long, with a white limb; sterile appendage of spadix 1.4−2.7 cm long........ ....1. *A. somalense*
− Spathe 3.5−7.2 cm long, with a yellow limb with purplish-brown patch at base within; sterile appendage of spadix 0.5−0.7 cm long............... ....2. *A. flavum*

1. **A. somalense** M. Gilbert & Mayo (1986); type: N2, "Sugli", Collenette 250 (K holo., FT iso.). Fig. 13.

Herb up to c. 60 cm tall. Leaf solitary; petiole with basal sheaths forming short pseudostem, free part 12−22 cm long; blade radiately to pedately divided; leaflets (3−)5−7, broadly elliptic, 10−17.5 × 4−10.5 cm, central one largest, acuminate. Inflorescence overtopping leaves, free part of peduncle 15−24 cm long. Spathe 10.5−14.5 cm long; tube cylindric, 3.5−5 × 1.6−2.5 cm, green; limb ovate, 7−9.5 × 3.2−7.2 cm, white. Spadix 5−8 cm long, bisexual or male; sterile appendix subcylindrical to slightly conical, 1.4−2.7 cm long. Anthers dehiscing by

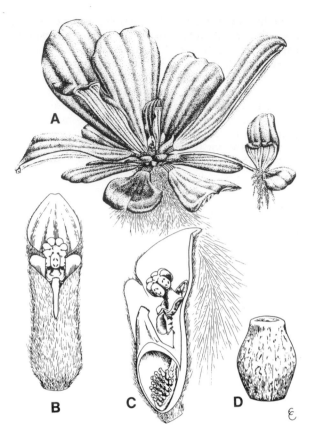

Fig. 14. *Pistia stratiotes*. A: flowering plant with stolon, × 0.5. B: inflorescence, front view, × 3.5. C: inflorescence, side view, in longitudinal section, × 3.5. D: seed, × 1. − Modified from Fl. Trop. E. Afr. (1985). Drawn by E. Catherine.

apical slits. Pistils flask-shaped, 2.5−3 mm long; stigma capitate. Berries unknown.

Juniperus forest on N-facing slope, in shady damp sites among large boulders; 1450−1575 m. N2; not known elsewhere. Bally 11083; Thulin, Dahir & Hassan 8930.

2. **A. flavum** (Forssk.) Schott (1860); *Arum flavum* Forssk. (1775). Plate 1 D.

Herb up to c. 50 cm tall. Leaves 1 or 2; petiole with basal sheaths forming pseudostem up to 30 cm long, free part 2−17 cm long; blade pedately divided; leaflets 5−11, lanceolate to elliptic, 3−10.5 × 0.6−3.8 cm, central one largest, acuminate. Inflorescence normally overtopping leaves, free part of peduncle 6−18.5 cm long. Spathe 3.5−7.2 cm long; tube broadly cylindric, 0.9−2 × 1−1.5 cm, green; limb oblong-ovate, acuminate, yellow with purplish-brown patch at base within, 2.6−5.5 × 1.1−2 cm, folded sharply forward over spathe mouth. Spadix 1.3−2.7 cm long, bisexual; sterile appendage usually clavate, 0.5−0.7 cm long. Anthers dehiscing by apical pores. Pistils subglobose; stigma minute. Berries red, c. 5 mm in diam., in subglobose cluster.

On limestone rocks or in rocky limestone slopes; 800–1250 m. N3; S Ethiopia, S Arabia and eastwards to China. Thulin & Warfa 5613, 5834, 6248.

First record for Somalia.

4. PISTIA L. (1753)

Plants aquatic, floating, stemless, perennial, with fibrous roots, reproducing vegetatively by stolons. Leaves rosetted, subsessile, densely pubescent, with parallel primary veins. Inflorescence small, hidden among leaf-bases. Spathe tubular, apical part erect, expanded. Spadix reduced, adnate to spathe. Flowers unisexual, without perianth. Male flowers 2–8, in a single whorl borne on a short stipe subtended by shallow basal cup, each flower a synandrium of 2 united stamens. Female flower solitary, basal, consisting of an oblique, 1-celled ovary; ovules numerous; style curved. Fruit thin-walled, dehiscing by irregular splits, several-seeded.

Genus of one species only.

P. stratiotes L. (1753). Fig. 14.

Leaf-blade up to 14 × 8 cm, but often much smaller, rounded to truncate at the apex, paler and more densely pubescent beneath; main veins 5–7, winged beneath. Spathe to 1.3 cm long, whitish green, pilose. Fruit to 5 × 3 mm. Seeds to 2 × 1 mm, reddish-brown.

Freshwater, in pools or slow-flowing streams; 50–230 m. S1–3; pantropical. Thulin & Bashir Mohamed 7086; Bavazzano 312; Tardelli 492.

144. LEMNACEAE

by M. Thulin

Daubs, Monogr. Lemnac. (1965); den Hartog & van der Plas in Blumea 18: 355–368 (1970); Cuf. Enum.: 1504–1505 (1971); Fl. Trop. E. Afr. (1973); Landolt in Veröff. Geobot. Inst. ETH, Stift. Rübel, Zürich 71 (1986).

Aquatic small to minute herbs, floating or submerged in fresh water, reduced to a flat to globose glabrous frond without differentiated stems and leaves; vegetative budding common; roots simple, solitary to several, or absent. Monoecious, but flowering erratic. Flowers in reproductive pouches, naked or enclosed by a spathe, each inflorescence with 1–2 male flowers with 1 2-thecous anther and 1 female flower with 1 sessile 1-celled pistil with 1–6 ovules and short terminal style. Fruit 1–6-seeded with a straight embryo, with no or scanty endosperm.

Family of six genera and about 30 species, cosmopolitan, the genus *Wolffia* including the smallest of all angiosperms.

The interpretation of the floral organs is somewhat controversial in *Lemnaceae*. According to the most widespread view, which is followed here, the floral organs represent a reduced inflorescence consisting of 1–2 male flowers and 1 female flower. According to the alternative view the corresponding structure is interpreted as a flower with 1–2 stamens and 1 pistil.

1. Roots present, solitary on each frond..1. *Lemna*
− Roots absent... 2. *Wolffiella*

1. LEMNA L. (1753)

Fronds floating or submerged, separate or connected, flat or swollen. Roots solitary or absent. Floral poach lateral with spathe enclosing 2 male and 1 female flower. Anther thecae 2-celled. Fruit ± globose, 1–6-seeded; seeds usually ribbed.

Cosmopolitan genus of some 17 species.

L. aequinoctialis Welw. (1859). Fig. 15.

L. paucicostata Engelm. (1867).

Fronds floating, pale green, ovate-elliptic, asymmetrical, 1.5–6 mm long, 3-nerved; daughter fronds remaining connected. Root with a winged sheath; root-cap acute. Seed solitary, with 8–26 longitudinal ribs, falling out of the fruit after ripening.

Pools; c. 200 m. S1; throughout tropical and warmer parts of the world. Thulin & Bashir Mohamed 7085.

This has previously often been treated as *L. perpusilla* Torrey, but according to Kandeler & Hügel in Pl. Syst. Evol. 123: 83–96 (1974) and Landolt (1986) *L. perpusilla* is a North American species distinct from the pantropical *L. aequinoctialis*.

First record for Somalia.

2. WOLFFIELLA (Hegelm.) Hegelm. (1895)
Pseudowolffia den Hartog & van der Plas (1970).

Fronds floating or submerged, flat. Roots absent. Floral pit dorsal, without a spathe, with 1 male and 1 female flower. Anther thecae 1-celled. Fruit slightly compressed, 1-seeded; seed smooth.

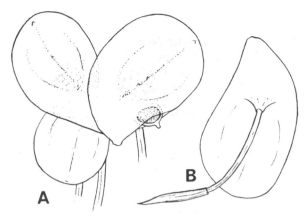

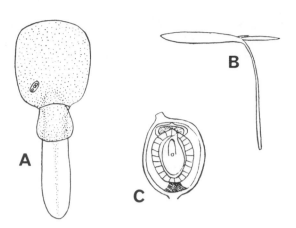

Fig. 15. *Lemna aequinoctialis*. A: fronds, one with fruit, × 8. B: frond, ventral surface, × 8. — Modified from Fl. Trop. E. Afr. (1973). Drawn by F. N. Hepper.

Fig. 16. *Wolffiella hyalina*. A: dorsal view of fertile frond, × 12. B: frond in floating position. C: median section of fruit and seed, × 25. — Modified from Fl. Trop. E. Afr. (1973). Drawn by F. N. Hepper.

Eight species in tropical or subtropical Africa and America.

W. hyalina (Del.) Monod (1949); *Lemna hyalina* Del. (1813); *Pseudowolffia hyalina* (Del.) den Hartog & van der Plas (1970). Fig. 16.

Fronds 1−3 mm long, asymmetrically 4-sided, floating, with a hyaline oblong appendage, 0.5−5 × 0.6−1.8 mm, at right angles to the frond and suspended in the water; daugther fronds remaining connected. Flower-pit to one side of mid-line of frond.

Pools; c. 200 m. S1; northern and central Africa, introduced in India. Paoli & Stefanini 677 bis.

145. ASPARAGACEAE

by Sebsebe Demissew

Cuf. Enum.: 1562−1566 (1971), *Liliaceae*, in part.

Scandent or erect shrubs or subshrubs, often spiny; roots often swollen and fusiform. Leaves normally reduced and scale-like, but leaf-like modified green branches (cladodes) often present. Flowers solitary or clustered, or arranged in racemes or umbel-like inflorescences, unisexual or bisexual (in Somalia), regular, small. Tepals 6, ± equal, free or fused at the base, white, yellow or green. Stamens 6, inserted at the base of the tepals, non-functional in female flowers; filaments free from each other; anthers 2-thecous, introrse, dorsifixed, longitudinally dehiscent. Ovary superior, (2−)3-celled with 2−12 ovules in each cell; placentation axile; style short with capitate or lobed stigma. Fruit a ± globose berry, 1−several-seeded. Seeds black, with endosperm; embryo straight or slightly curved.

Family with a single genus only.

ASPARAGUS L. (1753)
Jessop in Bothalia 9(1): 31−96 (1966).

Description as for the family

More than 300 species in Africa, parts of Europe, Asia and Australia, mainly in areas with arid tropical or Mediterranean climates.

Some recent authors have subdivided *Asparagus* into three genera, *Asparagus*, *Protasparagus* Oberm. and *Myrsiphyllum* Willd., but Malcomber & Sebsebe Demissew in Kew Bull. 48(1): 63−78 (1993) argue for a single genus with two subgenera, *Asparagus* and *Myrsiphyllum* (Willd.) Bak. All Somali representatives fall within subgen. *Asparagus* that is characterized by having free, spreading filaments.

The species of *Asparagus* are generally called "argeeg" or "ergeg" in Somali.

1. Flowers solitary or clustered.........................2
− Flowers in racemes or condensed umbel-like inflorescences.......................................5
2. Flowers solitary or paired; tepals usually tinged pinkish outside; ovary with 1−2 ovules in each cell....................................3. *A. flagellaris*

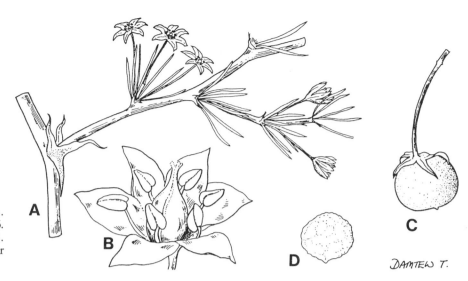

Fig. 17. *Asparagus africanus*.
A: Flowering branch, × 1.6.
B: flower, × 8. C: fruit, × 2.4.
D: seed, × 4.8. — Drawn for
Fl. Ethiopia by Damtew T.

— Flowers in clusters of 3−8; tepals whitish outside;
   ovary with 5−8 ovules in each cell..................3
3. Flowering branches without cladodes; inflorescences always axillary, congested; flowers soon
   falling............................6. *A. leptocladodius*
— Flowering branches commonly with cladodes;
   inflorescences axillary and/or terminal; flowers
   persistent..4
4. Terminal branches often puberulous, without
   spines; cladodes 3−12 mm long, rounded or
   angled, stiff............................1. *A. africanus*
— Terminal branches glabrous, always with spines;
   cladodes 15−26 mm long, flattened or with
   grooves above, flexible..............2. *A. scaberulus*
5. Cladodes linear, flattened, 15−85 × 1.25−3 mm....
   .... 4. *A. falcatus*
— Cladodes subulate or only slightly flattened, 8−35
   × less than 1 mm......................................6
6. Racemes simple; pedicels articulated above the
   middle or at the apex................................7
— Racemes branched or condensed and umbel-like;
   pedicels articulated at the middle or below.........8
7. Young branches grey, scabrid to puberulous; style
   0.75−1 mm long; anthers black....7. *A. aspergillus*
— Young branches pale brown, smooth, glabrous;
   style c. 0.5 mm long; anthers cream to yellowish...
   ....8. *A. buchananii*
8. Flowering branches commonly with cladodes;
   racemes lax, 1.5−10 cm long; flowers persistent;
   bracts 2−4 mm long.................5. *A. racemosus*
— Flowering branches without cladodes; racemes
   condensed, umbel-like, 1−2.5 cm long; flowers
   soon falling; bracts 1−1.5 mm long..................
   ....6. *A. leptocladodius*

1. **A. africanus** Lam. (1783). Fig. 17.
  A. sennii Chiov. (1932); types: S3, "Cu Daio", Senni
158 (FT syn.), and "Gobuin" and "Chisimaio",
Gorini 316 (FT syn.).

A. asiaticus sensu Cuf. Enum.: 1563 (1971), non L.
(1753).

Low erect shrublet or stems climbing or scrambling
up to 5 m; branches glabrous to puberulous, terete to
angled, with spines 3−5 mm long, the terminal
branches without spines. Cladodes clustered, 5−25
together, subulate, stiff, 3−12 mm long. Flowers
clustered, 2−10 together, axillary and terminal;
pedicels 3−8 mm long, articulated below the middle.
Bracts lanceolate, c. 1.5 mm long, falling off quickly.
Tepals white to pale yellowish, 3.5−5 mm long,
entire. Stamens shorter than tepals; anthers yellow.
Ovary with 6−8 ovules in each cell; style c. 1 mm
long, 3-branched. Berry red, 5−6 mm in diam.,
1-seeded. Seeds rugose.

Hillsides, dunes, *Juniperus* forest, limestone
plateaus, open *Acacia* woodland on alluvial soil;
15−2050 m. N1, 2; S2, 3; Ethiopia, Sudan, Uganda,
Kenya, Tanzania and southwards to South Africa,
and in Arabia and eastwards to India. Thulin, Hedrén
& Dahir 7198; Glover & Gilliland 658; Wood
S/72/79.

2. **A. scaberulus** A. Rich. (1850); *A. asiaticus* L. var.
scaberulus (A. Rich.) Engl., Hochgebirgsfl. Trop.
Afr.: 169 (1892).

Erect to scandent shrub up to 2 m high; branches
glabrous, terete, smooth or lined with down-curved
spines 1−3 mm long, the terminal branches also with
spines. Cladodes clustered, 2−25(−35) together,
flexible, straight or bent, 15−26 mm long, flattened,
angled, sometimes grooved on the upper side.
Flowers clustered, (2−)3−6 together, axillary or
terminal; pedicels 5−8 mm long, articulated at the
middle or below. Bracts ovate, c. 1 mm long,
membranous. Tepals white, c. 3 mm long. Stamens
shorter than tepals; anthers yellow. Ovary with 5
ovules in each cell; style c. 1 mm long, 3-branched.
Berry red, 4−5 mm in diam., 1-seeded.

Acacia or *Acacia-Commiphora* bushland; 230—770 m. N1; S1; Ethiopia, Sudan, Kenya and Tanzania. Becket & White 1595; Gillett & Watson 23742; O'Brien 145.

3. A. flagellaris (Kunth) Bak. (1875).

A. somalensis Chiov. (1916); type: S1, "Aden Caboba", Paoli 924 (FT holo.).

Erect or scandent shrub up to 2 m high; branches glabrous, terete or grooved, smooth to lined with straight or curved spines 2—4 mm long. Cladodes clustered, 1—8 together, subulate, stiff, 5—20 mm long, sometimes absent in flowering and fruiting specimens. Flowers axillary, solitary or paired; pedicels 4—5 mm long, articulated towards the base. Tepals white to purple, 2.5—3 mm long. Stamens shorter than tepals. Ovary with 1—2 ovules in each cell; style c. 1 mm long, slender, 3-branched. Berry orange-red, 5—7 mm in diam., 1(—3)-seeded. Seeds rounded, rugose.

Acacia-Commiphora bushland in shallow soil over limestone, or on limestone outcrops; 230—430 m. S1; Ethiopia, Sudan, Uganda, Kenya, Tanzania and westwards to West Africa. Beckett & White 1567, 1762.

4. A. falcatus L. (1753).

Climbing or scrambling shrub up to 3 m high; branches glabrous to shortly puberulous, with well developed recurved spines 3—7 mm long, the terminal branches also with spines. Cladodes clustered, 3—6(—12) together, flattened, straight or curved, with a distinct vein, 15—85 × 1.25—3 mm. Racemes 2—9 cm long, glabrous to puberulous, branched or unbranched; pedicels 3—4 mm long, articulated at or above the middle. Tepals white to cream, obovate, 3—3.5 mm long. Stamens shorter than tepals; anthers yellow. Ovary with 4—5 ovules in each cell; style 0.7—1 mm long. Berry red or white flushed purple, c. 7—13 mm in diam., 1—3-seeded.

1. Cladodes 35—85 mm long, thin and flexible.........
....var. *falcatus*
— Cladodes less than 35 mm long, thick and ± stiff....................................var. *ternifolius*

var. falcatus.

Semi-evergreen bushland with grassy glades on sand; c. 40 m. S3; Ethiopia and south to South Africa, and in Asia. Gillett & al. 25185; Senni 37.

var. ternifolius (Bak.) Jessop in Bothalia 9: 70 (1966).

Acacia-Euphorbia open woodland, *Juniperus-Buxus* forest or dunes; 15—1730 m. N1; ?2; S2; widespread in tropical and southern Africa, and in Arabia and east to India. Gillett 4846; Hemming 2117; Thulin, Hedrén & Dahir 7313.

Both glabrous and more or less puberulous forms are found, but they do not seem to differ in any other characters or in ecological preferences.

This variety is superficially similar to *A. aethiopicus* L. var. *angusticladus* Jessop, a taxon widespread in Africa, but not yet recorded from Somalia. In *A. aethiopicus* var. *angusticladus* the cladodes are lighter and translucent when dry, and the pedicels are articulated at the middle or below.

5. A. racemosus Willd. (1799).

Climbing shrub up to 7 m high; branches terete, lined or angled, glabrous, with 2—8 mm long spines. Cladodes clustered, 2—6 together, subulate to flattened, 8—35(—40) × 0.5—0.7 mm. Racemes 1.5—10 cm long, branched, glabrous; pedicels 4—6 mm long, elongating up to 10 mm in fruit, articulated at or below the middle. Bracts ovate, concave, 2—4 mm long, glabrous, membranous, sometimes falling quickly. Tepals white to greenish white, (3—)4—5 mm long. Stamens shorter than tepals; anthers orange to red. Ovary obovate, with 6—7 ovules in each cell; style 1—1.25 mm long, 3-branched. Berry green turning red at maturity, 8—10(—13) mm in diam., 1(—3)-seeded.

Deciduous bushland; c. 50 m. S2, 3; Ethiopia, Sudan, Kenya, Tanzania, Mozambique, Angola, and in Asia. Puccioni 38; Socco s.n.

6. A. leptocladodius Chiov. (1940).

A. racemosus var. *ruspolii* Engl. in Ann. R. Ist. Bot. Roma 9: 245 (1902).

A. gillettii Chiov. (1941); type: N1, "Elmis", Gillett 4501 (K holo., FT iso.).

Erect or scandent shrub up to 2 m high; branches glabrous to puberulous, terete, white, peeling off, with straight spines 4—17 mm long. Cladodes clustered, 2—15 together, flattened, curved, 10—60 mm long, triangular. Racemes 1—2.5 cm long, often condensed and umbel-like; pedicels 3—6 mm long, articulated at or below the middle. Bracts ovate, 1—1.5 × 0.5 mm, white, falling quickly. Tepals white, 2.5—4 mm long. Stamens shorter than tepals; anthers black. Ovary with 6—8 ovules in each cell; style 0.3—0.7 mm long, 2—3-branched. Berry red, 6—9 mm in diam., 1-seeded. Seeds 4—5 mm in diam.

Deciduous bushland and woodland, rocky slopes, dunes; 15—1800 m. N1—3; S1, 2; Djibouti, Ethiopia, Kenya. Carter 933; Thulin, Hedrén & Dahir 7215; Gillett & Hemming 24793.

7. A. aspergillus Jessop (1966).

Climbing or erect herb or shrub up to 2 m high; branches scabrid to puberulous, pale grey, with spines 3—10 mm long. Cladodes clustered, subulate, 10—20 mm long and less than 0.5 mm thick, absent during the flowering period. Racemes simple, 1.2—4.5 cm long, scabrid; pedicels solitary 1.5—3.5 mm long, articulated at the apex. Bracts ovate, c. 1 mm long. Tepals white, oblong to obovate, 2—3 mm long.

Stamens slightly shorter than tepals; anthers black. Ovary with 4−6 ovules in each cell; style 0.75−1 mm long, 3-branched. Berry red, globose, c. 6 mm in diam., 1−2-seeded.

Deciduous bushland; c. 200 m. S1; Kenya and south to South Africa and Namibia. Paoli 623.

This is similar to *A. racemosus*, but is easily distinguished by the apical articulation of the pedicel, the black anthers, and the shorter tepals. The single collection seen from Somalia was made in "El Magu".

8. A. buchananii Bak. (1893).

Climber up to 5 m high; branches glabrous, pale brown to greyish, smooth, shiny, with spines on main branches up to 4 cm long and dorsally flattened towards the base. Cladodes clustered, 3−5 together, subulate, 10−18(−27) × c. 0.5 mm. Racemes simple, 1.5−4 cm long, glabrous; pedicels solitary or paired, 2−5 mm long, articulated above the middle. Bracts ovate, 1−2 mm long. Tepals white to cream, elliptic to obovate, 2−3 mm long. Stamens shorter than tepals; anthers yellow. Ovary obovate, with 6−8 ovules in each cell; style c. 0.5 mm long, 3-branched. Berry red, c. 5 mm in diam., 1−2-seeded.

Acacia scrub; c. 40 m. S3; Uganda, Sudan, Kenya and south to South Africa. Maunder 134.

Maunder 134 from the bank of Juba R. south of Buaale has a single young fruit only, but agrees with *A. buchananii* in having an unbranched raceme and bracts c. 1 mm long. The sterile specimen Paoli 712 from "Bur Meldac" in S1 was cited as *A. buchananii* by Chiovenda in Result. Scient. Miss. Stefanini-Paoli: 170 (1916) and may well belong here too. Better material is needed to confirm the presence of *A. buchananii* in Somalia.

146. DRACAENACEAE

by M. Thulin

Cuf. Enum.: 1567−1572 (1971), *Agavaceae*, in part.

Plants with a ± woody erect stem, or with creeping rhizomes and with or without aerial stems. Leaves spirally arranged, often in rosettes, sometimes with the leaves arranged in 2 ranks, linear to ovate, sessile or petiolate, sometimes half-cylindric or cylindric. Inflorescences axillary, pedunculate, raceme-, panicle- or sometimes head- or umbel-like; pedicels articulated. Flowers bisexual, regular; tepals 6, similar, united at the base. Stamens 6, inserted at the top of the tube formed by the tepals; anthers 2-thecous, dorsifixed, longitudinally dehiscent. Ovary superior, 3-celled with septal nectaries; style simple; ovules 1 per cell. Fruit generally a red or orange berry, sometimes hard and woody. Seeds up to 3, globose or elongate, endospermous.

Two genera and some 110 species, in tropical to subtropical regions of the Old World.

1. Plant a tree with leaves in dense rosettes...1. *Dracaena*
− Plant not a tree, if having aerial stems, leaves not in dense rosettes.................................2. *Sansevieria*

1. DRACAENA L. (1767)

Plants with a ± woody trunk, sometimes trees; roots usually orange. Inflorescence a panicle with short or long branches; flowers 2 or more to each floral bract. Flowers usually small, fragrant and opening at night; free parts of tepals spreading or recurved, white or greenish, sometimes with a reddish midrib. Filaments slender or thickened. Fruit a globose, sometimes lobed, red, orange or yellow berry containing 1−3 seeds.

Some 60 species, the majority in Africa. Several species grown as ornamentals, particularly for their often patterned leaves.

D. ombet Kotschy & Peyr. (1867). Plate 1 E.

D. schizantha Bak. (1877); type: "Ahl" Mts, near "Meid", Hildebrandt 1472 (K holo., WAG iso.).

Mooli (Som., tree); xanjo-mooli (Som., resin).

Tree, 2−8 m tall, with a forked trunk, producing a red resin. Leaves in dense rosettes at ends of branches, linear from a wide base, 40−60 × up to 3 cm, narrowed gradually to the acute tip, thick and rigid, with smooth margins, flat to concave on the upper side, rounded on the back in the lower half and keeled in the upper part. Panicle to c. 0.5 m long, much branched, glabrous or pubescent; bracts minute, ovate-lanceolate; pedicels clustered, 2−4 mm long, articulated at the middle, glabrous or pubescent. Tepals whitish, 4−6 mm long, linear, almost free.Stamens somewhat shorter than tepals; filaments flattened. Ovary oblong, shortly stipitate. Berries c. 10−12 mm in diam.

Bushland or woodland, usually on limestone; 1000−1800 m. N1−3; Djibouti, Eritrea, Ethiopia, Sudan. Gillett & Watson 23462; Collenette 338; Barbier 962.

The Somali plant has previously been called *D. schizantha*, but it does not seem possible to separate this from *D. ombet* with type from Sudan. *D. ombet*, along with the closely related *D. cinnabari* Balf. f. on Socotra and *D. draco* (L.) L. on the Canary Is., produces a red resin used in traditional medicine.

2. SANSEVIERIA Thunb. (1794), nom. cons.
Brown in Kew Bull. 1915: 185−261 (1915).

Plants with creeping rhizomes, with or without aerial stems. Inflorescence a spike- or head-like raceme or panicle; flowers solitary or 2 or more to each floral bract. Flowers fragrant, opening at night; free parts of tepals spreading or recurved, usually white, cream or dull pink. Filaments slender. Fruit a berry containing 1−3 bony seeds.

Some 50 species in the Old World, the majority in Africa. Some species, in Somalia especially *S. ehrenbergii*, yield hemp ("bowstring hemp") and some are often grown as tolerant foliar ornamentals. *Sansevieria* is very close to *Dracaena* and perhaps not generically distinct (see Bos, Belmontia 17, 1985).

The vernacular names "xaskul" ("haskul") or "xig" ("hig") are generally used for all species of *Sansevieria* in Somalia.

S. canaliculata Carrière (1861), with a type of unknown origin, has been associated with Somalia by various authors (see Brown 1915). However, no material from Somalia agreeing with the description of *S. canaliculata* has been seen.

The present account is preliminary and more field work on the genus in Somalia and elsewhere is needed before a satisfactory treatment can be made. Most species flower only rarely and for a short time, which adds to the difficulties. Leaf characters in the descriptions refer to mature leaves if not otherwise indicated, young leaves often being markedly different.

1. Plant with stem at least c. 1 m tall..... 1. *S. powellii*
− Plant stemless or with aerial stem to c. 25 cm tall..2
2. Leaves flat..........................7. *S. forskaoliana*
− Leaves cylindrical or almost so.......................3
3. Leaves with a channel only in the basal part above.................................. 3. *S. phillipsiae*
− Leaves, at least some, with a channel running all along above...4
4. Leaves up to 18 cm long, 2−3 together...............
....6. *S. eilensis*
− Leaves much longer, solitary or several together.. 5
5. Leaves thicker than wide, with a triangular channel above; inflorescence a much branched panicle .....
.... 2. *S. ehrenbergii*
− Leaves wider than thick, with a concave channel above; inflorescence spike-like.....................6
6. Leaves solitary, with c. 4−6 furrows on the sides and back.................................4. *S. fischeri*
− Leaves several together, with c. 30−40 furrows on the sides and back.....................5. *S. volkensii*

1. S. powellii N.E. Br. (1915); *Acyntha powellii* (N.E. Br.) Chiov. (1932). Plate 1 F.
Aerial stem to c. 2 m high, leafy throughout or with ring-like leaf scars at the base. Leaves many, in 2 ranks, spreading, recurved, 24−70 × up to 3 cm, slightly rough, rounded on the back, with a concave channel all along above, gradually tapering to the hard pale brown very acute spine-like tip, without markings; margins of channel with a narrow red-brown line edged with white. Panicle to c. 0.5 m high, with ascending-spreading branches; flowers 4−6 in a cluster, greenish-white with dull purple lines; pedicels 2−3 mm long, articulated at or slightly above the middle. Free parts of tepals c. 10 mm long, linear, obtuse; tube c. 6 mm long.

Bushland and woodland; 0−200 m. S1−3; Kenya. Rose Innes & Trump 1034; Alstrup & Michelsen 160; Moggi & Bavazzano 1431.

2. S. ehrenbergii Schweinf. ex Bak. (1875). Fig. 18.
Sanseverinia rorida Lanza (1910); *Sansevieria rorida* (Lanza) N.E. Br. (1915); *Acyntha rorida* (Lanza) Chiov. (1916); type: S2, near Mogadishu, Macaluso s.n. (PAL holo., not found).

Aerial stem to 25 cm high. Leaves 5−9, crowded, in 2 ranks, erect or spreading, 75−180 × 2.5−4 cm, laterally compressed with flattened sides, rounded on the back, with a triangular channel all along above and with 5−12 shallow grooves or lines down the sides and back, tapering upwards to the spine-like tip, dark green with blackish-green longitudinal lines; margins of channel acute, reddish-brown with white membranous edges. Inflorescence a panicle to 2 m high, much branched; flowers 4−7 in a cluster; pedicels 2.5−4 mm long, articulated above the middle. Free parts of tepals c. 7 mm long, linear, obtuse; tube 5−6 mm long.

Open bushland, semidesert grassland, dunes; 0−1200 m. N1−3; S1−3; Djibouti, Sudan, Eritrea, Ethiopia, Kenya, Tanzania. Aronson & al. 26; Gillett 4885; Paoli 500.

S. ehrenbergii and *S. powellii* differ mainly in habit (*S. powellii* with a more or less tall stem and *S. ehrenbergii* more or less stemless), and are often virtually indistinguishable on herbarium material.

3. S. phillipsiae N.E. Br. (1913); type: plant collected in N1 by Lort Phillips, cultivated in Kew and illustrated in Hook. Ic. Pl. 30: 3000 (1912).
S. hargeisana Chahinian (1994); type: cultivated plant originating from N1, WSW of Hargeisa, Lavranos 7382 (MO holo., UPS iso.).

Plant usually with short erect aerial stems branching at or above ground level. Leaves usually 5−10, ± irregularly directed, when young ascending or suberect, later spreading and slightly recurved, rigid, 10−45 × c. 1−1.8 cm, smooth to rough, cylindric, with 5−10 longitudinal lines or slight furrows extending from base to apex and a deeply concave sheathing portion to 9 cm long at the base with acute white edges, rather suddenly narrowed at the apex into a hard brown spine-like tip, young leaves faintly marked with transverse bands of paler green, old leaves dark and slightly bluish-green. Inflorescence a spike-like raceme, up to 45 cm tall including peduncle; flowers 3−6 in a cluster, white or greenish; pedicels 2.5−3 mm long, articulated at or slightly above the middle. Free parts of tepals c. 12 mm long, linear, obtuse; tube c. 10 mm long.

Bushland and woodland; 400−1500 m. N1; S1; Ethiopia. Gillett 3997; Carter 867; Bally 8037.

The material from S1, Moggi & Bavazzano 1256 and Beckett & White 1450A, has more slender leaves and smaller paired flowers compared to typical *S. phillipsiae* and seem to intergrade with *S. gracilis* N.E. Br. (1911) in East Africa that is an earlier name. On the other hand flowers with free part of tepals 15−17 mm long and tube 20−28 mm long has been reported in material from Ethiopia (Dyer & Bruce in Fl. Pl. Afr. 28: pl. 1090 (1950−1951). The relationship between *S. phillipsiae* and *S. gracilis* obviously needs to be further studied.

4. S. fischeri (Bak.) Marais (1986); *Boophane fischeri* Bak. (1898).

S. singularis N.E. Br. (1911).

Plant without aerial stem. Leaves solitary, erect, rigid, 45−240 × 2−4 cm, slightly rough, cylindric, with 4−6 furrows on the sides and back extending from base to apex and with a narrow concave channel all along above, suddenly narrowing to a stout whitish spine-like tip, marked with transverse pale green bands when young. Inflorescence a spike-like to head-like raceme. Free parts of tepals 5−13 mm long, linear, obtuse; tube 20−50 mm long.

Habitat unknown but probably deciduous bushland at c. 60 m. S3; S Ethiopia, Kenya, N Tanzania. Senni 131; Calvino s.n.; Guidotti s.n. (all FT).

Calvino s.n. was identified as *S. stuckyi* Godefr.-Leb. (1903) by N.E. Brown according to Chiovenda, Fl. Somala 2: 422 (1932), a species otherwise only known from Mozambique. However, the leaves of the sterile Somali material shows well-marked furrows along the sides and back and therefore agree better with *S. fischeri* in Kenya and Tanzania, and I provisionally refer it to this species. Flowering material from Somalia is needed for conclusiveness.

The description of the flowers above is based on East African material, and is taken from Rauh in Sukkulentenkunde 7−8: 108−127 (1963) and Pfennig in Bot. Jahrb. Syst. 102: 175 (1981). According to Pfennig (op. cit.: 176, 1981) *S. stuckyi* has a large head-like inflorescence with free parts of tepals c. 40 mm long and tube 90−100 mm long.

The record of *S. robusta* N.E. Br. in Cuf. Enum.: 1571 (1971) is based on Guidotti s.n. (FT). This consists of sterile material from southern Somalia that belongs here.

5. S. volkensii Gürke (1895).

S. intermedia N.E. Br. (1914).

Acyntha polyrhitis Chiov. (1932); *S. polyrhitis* (Chiov.) Cuf. (1971); type: S3, "Gelib", Tozzi 410 (FT holo., K iso.).

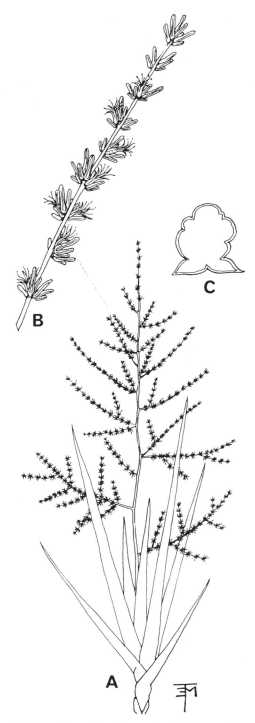

Fig. 18. *Sansevieria ehrenbergii*. A: habit. B: flowering branch. C: cross-section of leaf. − Modified from Agnew, Upland Kenya wild flowers: 723 (1974).

Plant without aerial stem. Leaves 5−8(−12), spirally arranged, erect, rigid, 50−120 × c. 1.5 cm, smooth to rough, subcylindric, with 30−40 fine but distinct furrows extending from base to apex on the sides and back and with a concave furrowed channel all along

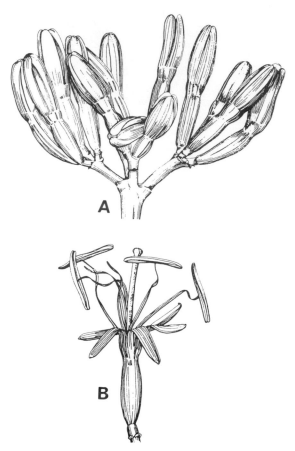

Fig. 19. *Agave sisalana*. A: portion of inflorescence, × 0.7. B: flower, × 0.7. — Modified from Fl. Austr. 46: Fig. 15 (1986).

above, ± curved and obtusely cuspidate at the tip, without markings; margins of channel white or with a red-brown line edged with white. Inflorescence a spike-like raceme, up to 60 cm high including peduncle; flowers 3−6 in a cluster, white or greenish-white; pedicels c. 2 mm long, articulated near the tip. Free parts of tepals 12−20 mm long, linear, obtuse; tube 14−32 mm long.

Habitat not known but probably deciduous bushland at c. 30 m. S3; Kenya, Tanzania.

This is in Somalia only known from the sterile type collection of *S. polyrhitis*. Fertile material is needed to prove the identity of the Somali plant.

6. **S. eilensis** Chahinian (1995); type: N3, Eyl Pass. 4 km NNW of Eyl, Lavranos & Horwood 10178 (MO holo., UPS iso.).

Plant without aerial stem. Leaves from underground rhizome, 2−3, ascending, 7−12 x 1.9−2.5 cm, slightly rough, cylindric, with up to 12 longitudinal lines or grooves and often with a channel all along above, ± curved and abruptly tapering to a cuspidate hard tip, light green with grey-green cross-banding; margins of channel red-brown with a white edge. Spike-like raceme c. 34 cm high including peduncle; flowers 2−4 in a cluster, greenish-white; pedicels 4−5 mm long, articulated above the middle. Free parts of tepals c. 14 mm long, linear, reflexed; tube c. 8 mm long.

In shattered limestone; c. 120 m. N3; not known elsewhere. Lavranos & al. 23374.

7. **S. forskaoliana** (Schult. f.) Hepper & Wood (1983); *Smilacina forskaoliana* Schult. f. (1829, as *"forskaliana"*.

S. abyssinica N.E. Br. (1913).

Acyntha elliptica Chiov. (1932); *Sansevieria elliptica* (Chiov.) Cuf. (1971); type: plant collected by Guidotti in S2 between "Merca" and "Genale", and cultivated in Modena, illustrated in Chiovenda, Fl. Somala 2: Fig. 239 (1932).

A. abyssinica (N.E. Br.) Chiov. var. *sublaevigata* Chiov., Fl. Somala 2: 419 (1932); *S. abyssinica* var. *sublaevigata* (Chiov.) Cuf. Enum.: 1569 (1971); type: S3, "Uamo Ido", Senni 260 (FT holo.).

Plant without aerial stem. Leaves 2−3(−4), ± erect, rigid, up to at least 60 × 7.5 cm, sometimes rough with transverse ridges on both sides or below only, flat, lanceolate, with a hard brown spine-like tip, tapering below into a petiole-like base, often variegated; margins with hardened reddish-brown edges. Inflorescence a spike-like raceme, up to 75 cm or more high including peduncle; flowers 4−5 in a cluster, whitish; pedicels 3−8 mm long, articulated at the middle or near the top. Free parts of tepals c. 10 mm long, linear, obtuse; tube c. 15 mm long.

Bushland and woodland; 40−1550 m. N1, 2; C1; S1−3; Djibouti, Ethiopia, Sudan, East Africa, also in Arabia. Gillett 4041; Bally 11116, 11751.

A record of *S. nilotica* Bak. in Kuchar, Plants of Somalia (1988) is probably based on material of *S. forskaoliana*.

147. AGAVACEAE

by M. Thulin

Large rosette herbs or trees. Leaves fleshy or tough and fibrous, spirally set, linear to lanceolate, often with a sharp point at the tip, and with spiny margins. Inflorescences usually large and many-flowered. Flowers mostly bisexual, regular or almost so; tepals free or ± united into a tube. Stamens 6, inserted at the base of the tepals; anthers dorsifixed, longitudinally dehiscent. Ovary superior or inferior, 3-celled; style simple; ovules several to many. Fruit a capsule or berry. Seeds with endosperm.

Some eight genera and 400 species, all native in America.

AGAVE L. (1753)

Short-stemmed rosette plants. Leaves fleshy or tough and fibrous, spine-tipped, often with toothed margins. Inflorescence a terminal raceme or panicle with flowers in umbellate clusters. Flowers with tubular to shallowly funnel-shaped straight or curved perianth; tepals ± similar or unequal, united in lower part. Stamens exserted. Ovary inferior with many ovules in two rows within each cell; style filiform; stigma 3-lobed. Fruit a loculicidal capsule. Seeds flattened, black.

Genus of some 150 species, native in America. Several species cultivated for fibre or as ornamentals.

A. sisalana Perrine (1838). Fig. 19.

Sisal (Eng.).

Rosettes to 2 m tall, with a short trunk. Leaves 9−130 × 9−12 cm; margins minutely toothed when young, toothless in mature leaves; terminal spine 2−2.5 cm long, dark brown. Inflorescence a 5−7 m tall panicle, producing bulbils after flowering. Flowers 5−6.5 cm long, greenish-yellow; perianth-tube 15−18 mm long; tepals 17−18 mm long, equal, hooded at apex. Filaments 50−60 mm long.

Occasionally cultivated, at least in S2; native in Mexico, widely cultivated for fibres elsewhere. Sacco, Sappa & Ariello 148.

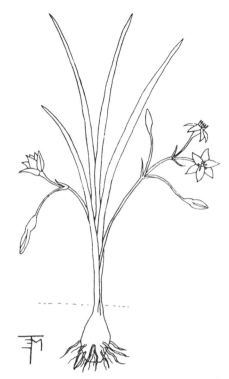

Fig. 20. *Hypoxis angustifolia*, habit. − Modified from Agnew, Upland Kenya wild flowers: 725 (1974).

148. HYPOXIDACEAE

by M. Thulin

Cuf. Enum.: 1577−1578 (1971).

Herbs with a corm covered by a coat of fibers. Leaves in a basal rosette, mostly in 3 rows, linear to lanceolate, prominently parallel-nerved, often hairy. Peduncle leafless, usually hairy. Flowers in spikes, racemes or umbel-like clusters, or solitary, bisexual, regular; tepals 4−6, free or ± united, usually hairy outside. Stamens (3−)6, inserted at the base of the tepals; filaments linear; anthers 2-thecous, longitudinally dehiscent. Ovary inferior, 2−3-celled; style simple, short; ovules numerous. Fruit generally a capsule. Seeds small, globose, black, with endosperm.

Some seven genera, mainly in the Southern Hemisphere.

A record of *Curculigo pilosa* (Schumach.) Engl. from north-western Somalia in Cuf. Enum.: 1577 (1971) is based on Lort Phillips s.n. (K), a sterile plant originally identified as *C. gallabatensis* Schweinf. ex Bak. However, the Somali plant is glabrous without a basal fibrous cover and does not belong in *Hypoxidaceae*.

HYPOXIS L. (1753)

Nordal & al. in Nord. J. Bot. 5: 15−30 (1985).

Leaves in 3 rows, usually hairy. Flowers yellow; tepals persistent, the outer often somewhat narrower, longer and more acute than the inner ones. Stamens 4−6; anthers dorsifixed. Fruit a capsule or dehiscing by a lid.

Perhaps 100 species, particularly in the Southern Hemisphere.

H. angustifolia Lam. (1789). Fig. 20.

Corm subglobose to cylindrical. Leaves linear, grass-like, erect or reflexed, 3−45(−70) × c. 0.2−0.6 cm, hairy. Inflorescence 1−5(−6)-flowered, usually shorter than the leaves; bracts 0.4−1.3 cm long. Tepals 6, 3.5−9(−11) mm long. Stamens 6. Capsule loculicidal, 3-celled. Seeds strongly papillate.

In forest clearings or clay flushes; c. 1450−1500 m. N1−3; widespread in tropical and southern Africa. Bally 11076; Newbould 896; Thulin & al. 8938.

149. ASPHODELACEAE

by M. Thulin

Cuf. Enum.: 1529−1530 (1971), *Liliaceae* in part.

Rhizomatous herbs, generally producing anthraquinones; rhizome yellow when cut. Leaves all basal, ± thin or occasionally subterete, usually spirally set, sheathing at the base, entire. Inflorescences simple or branched; bracts 1 per flower. Flowers bisexual, usually white but sometimes pink, yellow or red, regular; tepals 6, usually free or almost so. Stamens 6, free; filaments linear, glabrous or hairy; anthers 2-thecous, dorsifixed, longitudinally dehiscent. Ovary superior, 3-celled; style simple, slender; ovules 2 to several per cell. Fruit generally a loculicidal capsule. Seeds sometimes arillate, with a straight embryo, with endosperm.

  Some eight genera in the Old World.

1. Filaments glabrous; pedicels articulated near middle ....1. *Asphodelus*
 − Filaments hairy; pedicels not articulated or articulated at the very tip only......................................2
2. Filaments scabrid with short retrorse hairs... 2. *Trachyandra*
 − Filaments densely hairy with hairs at least 1 mm long...3
3. Tepals yellow, all 1-nerved...3. *Bulbine*
 − Tepals white, the outer 3−5-nerved... 4. *Jodrellia*

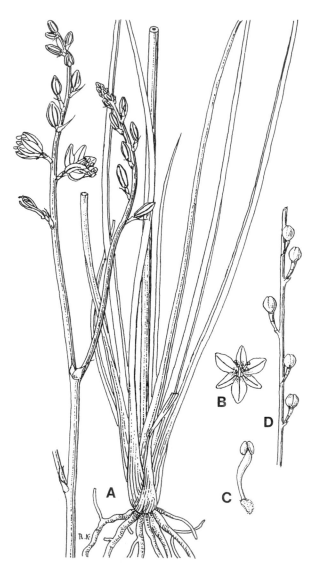

A

B

D

C

R.K.

1. ASPHODELUS L. (1753)

Leaves linear. Inflorescence a many-flowered raceme or panicle; bracts persistent, scarious; pedicels articulated, often thickened above in fruit. Tepals subequal, white or pale pink, free or united only at the base, spreading, 1-nerved, soon falling and leaving a basal cup below ovary. Stamens equal or almost so, shorter than tepals; filaments expanded at the base, glabrous. Capsule 3−6-seeded. Seeds angled, transversely sulcate at back.

  Some 18 species from the Canary Is. and the Mediterranean area eastwards to Himalaya.

A. fistulosus L. (1753). Fig. 21.
  A. tenuifolius Cav. (1801); *A. fistulosus* var. *tenuifolius* (Cav.) Bak., J. Linn. Soc. Lond. Bot. 15: 272 (1876).
  Annual or short-lived perennial with numerous slender roots. Leaves up to 40 × 0.4 cm, hollow, subterete. Inflorescence 15−70 cm tall, simple or branched, with a hollow peduncle; bracts 4−7 mm long; pedicels articulated near middle. Tepals 5−12 mm long, white or pink with brownish midnerve. Capsule 4−7 × 4−5 mm, transversely wrinkled.
  Grassy patches in evergreen bushland, rocky places; c. 600−1150 m. N2, 3; Canary Is., the Mediterranean area, Arabia, Eritrea and Socotra, naturalized in Australia. Azzaroli s.n; Thulin, Dahir & Hassan 8922.
  First record for Somalia. The two collections seen represent the annual small-flowered form that is sometimes recognized as *A. tenuifolius* (see Ruiz Rejon & al. in Pl. Syst. Evol. 169: 1−12, 1990).

Fig. 21. *Asphodelus fistulosus*. A: habit, × 0.7. B: flower, × 0.7. C: stamen, × 2. D: fruits, × 0.7. − Modified from Fl. Palaest. 4: pl. 28 (1986).

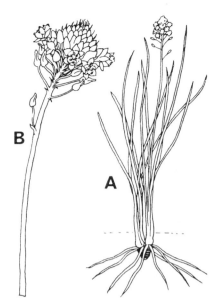

Fig. 23. *Bulbine abyssinica*. A: habit. B: inflorescence. – Modified from Agnew, Upland Kenya wild flowers: 675 (1974).

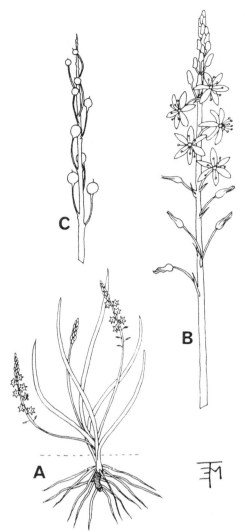

Fig. 22. *Trachyandra saltii*. A: habit. B: inflorescence. C: fruits. – Modified from Agnew, Upland Kenya wild flowers: 680 (1974).

2. TRACHYANDRA Kunth (1843)

Obermeyer in Bothalia 7: 711–759 (1962).

Leaves linear. Inflorescence a many-flowered raceme or panicle; bracts scarious; pedicels not articulated or articulated at the very tip, often thickened in fruit. Tepals subequal, usually white, free or united only at the base, spreading or recurved, 1-nerved, soon falling and leaving a small rim or cup below the ovary. Stamens usually equal, slightly shorter than tepals; filaments scabrid with short retrorse hairs. Capsule few- to many-seeded. Seeds angled.

Some 50 species in Africa, the large majority in South Africa.

1. Leaves flat or inrolled; ovary with c. 8 ovules per cell; seeds smooth.........................1. *T. saltii*
– Leaves triangular in cross-section; ovary with 2 ovules per cell; seeds warty..........2. *T. triquetra*

1. **T. saltii** (Bak.) Oberm. (1962); *Anthericum saltii* Bak. (1876). Fig. 22.

Anthericum lanzae Cuf. (1939).

A. harrarense Poelln. (1941).

Roots many, thin but fairly stout. Leaves filiform to linear, 5–50 × 0.1–2.5 cm, flat or inrolled, glabrous or pubescent. Inflorescence a simple raceme, 6–40 cm long, the peduncle curved near the base and protruding outside the leaf-rosette; bracts narrow; pedicels mostly 8–15 mm long. Tepals white, c. 10 mm long, with brownish midnerve. Capsule subglobose, c. 5 mm in diam., constricted at the base, several-seeded. Seeds angled, smooth.

Rocky ground in woodland or along wadis; 1350–1500 m. N1; Ethiopia and southwards to the Cape. Gillett 4996; Wood 5/72.

2. **T. triquetra** Thulin (1995); type: N2, escarpment S of Xidid along path to Dool, 11°00' N, 48°37' E, Thulin 9229 (UPS holo.).

Roots several, somewhat fleshy. Leaves 30–45 × 0.2–0.3 cm, soft and hanging, triangular in cross-section with minutely denticulate edges, glabrous. Inflorescence a simple very lax raceme, 10–30 cm long; bracts ovate-acuminate to ovate-aristate; pedicels 10–17 mm long, spreading and ± recurved. Tepals whitish, c. 6.5 mm long, with brownish midnerve. Ovary with 2 ovules per cell. Capsule depressed-globose, c. 3.5–4 × 5 mm, few-seeded. Seeds angled, warty.

In deep shade in crevices of limestone rocks; 800–1300 m. N2; not known elsewhere. Thulin, Dahir & Hassan 9125.

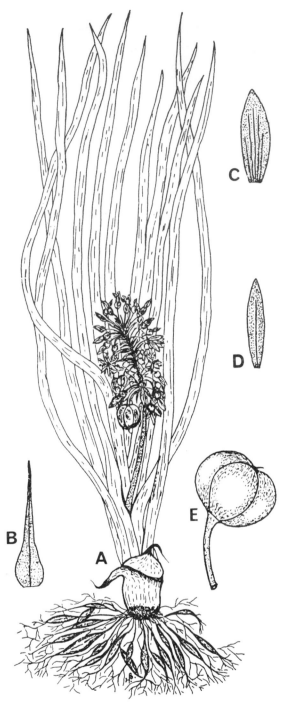

Fig. 24. *Jodrellia macrocarpa*. A: habit, × 2/3. B: bract, × 4. C: outer tepal, × 5. D: inner tepal, × 5. E: capsule, × 1. — Modified from Kew Bull. 32: 575 (1978). Drawn by H. Baijnath.

3. BULBINE Wolf (1776), nom. cons.

Baijnath, Taxonomic studies in the genus *Bulbine* Wolf. PhD thesis, University of Reading (unpublished).

Leaves fleshy, mostly linear and subterete. Inflorescence a many-flowered raceme; bracts scarious; pedicels not articulated. Tepals subequal, yellow, free or almost so, spreading or recurved, 1-nerved, cohering apically when withered, soon falling and leaving a small rim below the ovary. Stamens shorter than the tepals; filaments densely hairy in upper half. Capsule few- to many-seeded. Seeds angled.

Some 40 species, the great majority in South Africa, a few also in Australia.

B. abyssinica A. Rich. (1850). Fig. 23.

B. asphodeloides sensu auctt., non (L.) Spreng. (1825).

Roots many, fleshy. Leaves up to c. 40 × 0.5 cm. Inflorescence up to c. 50 cm tall, dense-flowered in upper part; bracts c. 1 cm long; pedicels up to 2.5 cm long. Tepals bright yellow, c. 8 mm long; filaments with yellow hairs. Capsule subglobose, c. 4–5 mm long, constricted at the base.

Rocky ground, forest glades, usually on limestone; 1350–2000 m. N1, 2; Ethiopia and southwards to South Africa, and in southern Arabia. Bally 10293, 11757; Wood S/73/73.

4. JODRELLIA Baijnath (1978)

Leaves linear, terete or subterete. Inflorescence a many-flowered raceme; bracts scarious; pedicels not articulated. Tepals subequal, white, free, spreading, the inner 1-nerved, the outer 3–5-nerved, soon falling and leaving a small rim below the ovary. Stamens shorter than the tepals; filaments densely hairy. Capsule 3–9-seeded, sometimes inflated and indehiscent with papery wall.

Poorly known genus close to *Bulbine* with three species in tropical Africa.

1. Capsules 5–5.5 × 4–5 mm, not inflated; ovary with 3–4 ovules per cell............ 1. *J. migiurtina*
 — Capsules 8–15 × 12–22 mm, inflated; ovary with 2 ovules per cell......................2. *J. macrocarpa*

1. J. migiurtina (Chiov.) Baijnath (1978); *Bulbine migiurtina* Chiov. (1928); type: N3, "Balli Scillin" to "Bur Inaoshin", Puccioni & Stefanini 777 (FT holo.).

Roots many, fleshy. Leaves up to 15 × 0.3 cm. Inflorescence shorter than the leaves, erect, dense; bracts 5–6 mm long; pedicels c. 10 mm long, becoming slightly recurved. Tepals c. 5 mm long. Capsule 5–5.5 × 4–5 mm. Seeds apparently smooth.

Rocky or sandy ground; c. 400 m. N3: not known elsewhere. Bavazzano & Lavranos s.n. (FT).

2. J. macrocarpa Baijnath (1978). Fig. 24.

Roots many, fleshy. Leaves 15–38 × 0.2–0.6 cm. Inflorescence shorter than the leaves, erect, dense; bracts 5–9 mm long; pedicels 12–20 mm long, becoming slightly recurved. Tepals c. 5 mm long. Capsules becoming inflated and 8–15 × 12–22 mm. Seeds warty.

Rocky or sandy ground; c. 150 m. C1; S1; E Ethiopia, N Kenya. Thulin & Dahir 6699; Paoli 1059.

150. ALOACEAE

by J. Lavranos

Cuf. Enum.: 1542−1552 (1971), *Liliaceae* in part.

Rosette herbs, shrubs or trees, containing anthraquinones. Leaves fleshy, spirally set or distichous, sheathing at the base, usually with toothed or spiny margins. Inflorescences simple or branched. Flowers bisexual, ± tubular, often red or white, regular or 2-lipped; tepals 6, often ± united into a tube. Stamens 6, free; filaments linear, glabrous; anthers 2-thecous, dorsifixed, longitudinally dehiscent. Ovary superior, 3-celled; style simple, slender; ovules several to many. Fruit a loculicidal capsule. Seeds arillate, with a straight embryo, endospermous.

Six or seven genera and some 500 species in S Arabia, Africa and Madagascar, with a marked concentration in Southern Africa. *Aloaceae* are close to and often included within *Asphodelaceae*.

ALOE L. (1753)

Reynolds, Aloes Trop. Afr. Madag. (1966).

Perennial herbs, shrubs or trees. Leaves glabrous, sinuate-dentate to spiny at the margins, rarely entire. Inflorescences terminal or lateral, simple or branched, with racemes capitate to long cylindrical. Flowers with tubular straight or ± curved perianth; tepals about equal in length, usually united in their lower part. Stamens often ± exserted. Ovary oblong to subglobose, with numerous ovules in two rows within each cell; style filiform, usually longer than the stamens; stigma capitate, small. Capsule papery to almost woody. Seeds irregularly triangular to flattened, usually narrowly winged.

About 300 species in Africa south of the Sahara, Madagascar, and in southern Arabia.

1. Plants stemless or stem usually shorter than diameter of rosette....................................2
− Plants with stem always longer than diameter of rosette..22
2. Inflorescence simple..................................3
− Inflorescence branched..............................4
3. Leaves to 6 cm long....................1. *A. jucunda*
− Leaves 10−15 cm long..............2. *A. hemmingii*
4. Perianth with pronounced basal swelling...........5
− Perianth without pronounced basal swelling.......6
5. Leaves spotted; flowers orange..........6. *A. grisea*
− Leaves striate; flowers dull pink....7. *A. albovestita*
6. Leaves shiny, white spotted; perianth glabrous.....7
− Leaves not shiny, usually unspotted; perianth sometimes pubescent......................................9
7. Perianth yellow; bracts 10−12 mm long..............
....3. *A. peckii*
− Perianth red; bracts 5−8 mm long..................8
8. Leaves 20−40 cm long, rigid, bright green; inflorescence less than 90 cm tall...4. *A. somaliensis*
− Leaves 40−50 cm long, leathery, grey or brownish-green; inflorescence more than 100 cm tall... ..
.... 5. *A. parvidens*
9. Leaves rugose......................8. *A. scobinifolia*
− Leaves smooth.....................................10

10. Inflorescence a panicle with many spreading branches..11
− Inflorescence with up to 10 ascending to erect branches..15
11. Leaves ascending to erect; marginal teeth many, white, c. 0.5 mm long...............9. *A. ruspoliana*
− Leaves spreading to recurved; marginal teeth various..12
12. Leaves 50−70 cm long; margins with distict hard brown-tipped teeth................................... 13
− Leaves less than 50 cm long; margins entire or sinuous or with minute teeth of the same colour as the margin.. 14
13. Leaf margins with teeth 1−2 mm long; racemes oblique, much elongated, rather lax; perianth c. 23 mm long..........................10. *A. microdonta*
− Leaf margins with teeth 5−6 mm long; racemes erect, subdense, cylindrical-conical; perianth 28−30 mm long.................11. *A. megalacantha*
14. Leaves grey to brownish; flowers dull pink; bracts to 4 mm long.............................12. *A. luntii*
− Leaves yellowish with thin, brown longitudinal lines; flowers yellow; bracts 5−6 mm long.........
.... 13. *A. brunneostriata*
15. Perianth pubescent to tomentose...................16
− Perianth glabrous....................................19
16. Perianth densely tomentose......15. *A. molederana*
− Perianth pubescent..................................17
17. Leaves white spotted, erect; marginal teeth c. 1.5 mm long.................................20. *A. citrina*
− Leaves unspotted; marginal teeth c. 1−6 mm long
....18
18. Leaves green, erect, 60−80 cm long; marginal teeth 4−6 mm long......................16. *A. rigens*
− Leaves glaucous-green with reddish tinge, ascending to spreading, 40−50 cm long; marginal teeth c. 1−3 mm long......................17. *A. glabrescens*
19. Leaves with dark green lines.....19. *A. bargalensis*
− Leaves uniformly coloured or with a few white spots..20
20. Leaves bright green with a few white spots, to 70 cm tall...............................14. *A. officinalis*

- Leaves uniformly coloured, grey-green or pinkish brown, less than 50 cm long........................21
21. Leaves pinkish brown, c. 50 cm long; raceme capitate, dense........................... 21. *A. bella*
- Leaves glaucous green with reddish tinge, 30−35 cm long; racemes cylindrical, lax...................
....18. *A. breviscapa*
22. Inflorescence simple..............................23
- Inflorescence branched...........................24
23. Leaves lanceolate, unspotted; marginal teeth robust; inflorescence rather dense.................
.... 22. *A. cremnophila*
- Leaves linear-lanceolate, copiously spotted; marginal teeth small; inflorescence lax....23. *A. gillettii*
24. Small shrubby plants to 50 cm tall; leaves to 18 × 2.5 cm.............................. 24. *A. ambigens*
- Plants taller than 75 cm; leaves more than 20 × 4 cm..25
25. Plants densely shrubby, many-stemmed; racemes capitate, subglobose............... 31. *A. rabaiensis*
- Plants few-stemmed or monopodial; racemes cylindrical or conical, rather lax.................. 26
26. Trees, branching from a central stem and forming a crown...............................30. *A. eminens*
- Plants of lax, shrubby growth.....................27
27. Stems less than 100 cm long, decumbent or erect; leaves with sinuate-dentate margins; teeth 2−3 mm long, pungent.................................. 28
- Stems always erect, more than 100 cm long; leaves with straight white margins; teeth 1 mm long or less, white, rather soft............................. 29
28. Stems decumbent to ascending; leaves green; teeth reddish-brown; perianth 26−30 mm long ...
....................................25. *A. hildebrandtii*
- Stems stout, erect; leaves glaucous with reddish-violet tinge; teeth white; perianth 20−22 mm long
....................................26. *A. heliderana*
29. Stems less than 150 cm tall; bracts c. 5 mm long...
....................................27. *A. retrospiciens*
- Stems to 200 cm tall or more; bracts 2−3 mm long
....30
30. Stems 200−400 cm tall; leaves 50−60 × 8 cm, grooved on upper surface; perianth yellow, outer tepals free for 12 mm............28. *A. gracilicaulis*
- Stems up to 200 cm tall; leaves 30 × 5−6 cm, flat or convex on upper surface; perianth red, outer tepals free for 5−6 mm.........29. *A. medishiana*

1. A. jucunda Reynolds (1953); type: N1, "Gaan Libah", E of "Hargeisa", Bally 7157 (K holo.).

Plants small, stemless, suckering from base and forming small groups. Leaves about 12, triangular, spreading, 4−6 × 2−5 cm, shiny dark green, copiously spotted; margins armed with hard triangular teeth. Inflorescences simple, to 35 cm high; racemes cylindric, fairly lax; bracts 5 × 3 mm; pedicels 6−7 mm long. Flowers rose-pink, pendulous; perianth 20 mm long, 7 mm across the ovary, outer tepals free for c. 7 mm. Anthers exserted

2 mm. Stigma exserted 3 mm. Ovary 5 × 2.5 mm.

Juniperus-Buxus forest on limestone; 1100−1700 m. N1; not known elsewhere. Carter 862.

2. A. hemmingii Reynolds & Bally (1964); type: N1, "Sheikh Pass, Golis Range", Bally 7146 (EA holo.).

Plants stemless, solitary or in small groups. Leaves about 10, lanceolate, spreading, 10−12 × 3−4 cm, shiny dark green with numerous elongated white spots; margins with hard dark brown triangular teeth, 1.5 mm long, 5 mm apart. Inflorescence simple, to 40 cm tall; racemes cylindric, to 15 cm long, lax; bracts 8 × 3 mm; pedicels 6−8 mm long. Flowers reddish pink, pendulous; perianth 24 mm long, 8 mm across the ovary, outer tepals free for 8 mm. Anthers exserted 2−3 mm. Stigma exserted 3−4 mm. Ovary 5 × 2 mm.

Semi-deciduous dwarf scrub on limestone; 700−1000 m. N1, 2; not known elsewhere. Reynolds 7207; Bally 9961.

3. A. peckii Bally & Verdoorn (1956); type: N2, "Erigavo", Peck in Bally 4283 (EA holo., PRE iso.).

Plants stemless, solitary or less frequently in groups. Leaves 14−16, ascending or spreading, c. 20 × 6 cm, very fleshy, shiny, olive green with very numerous, mostly confluent, whitish-green longitudinal spots and stripes; margins with triangular brown teeth, 3−4 mm long, 6−10 mm apart. Inflorescences erect with 8−10 branches, to 80 cm tall; racemes cylindric, lax; bracts 10−12 × 4 mm; pedicels 10 mm long. Flowers greenish yellow, pendulous; perianth 25−30 mm long, 7 mm across the ovary, outer tepals free for c. 14 mm. Anthers exserted 2−3 mm. Stigma exserted 4 mm. Ovary green, 6 × 3 mm.

Buxus-Acokanthera bushland with *Acacia etbaica*, on limestone; 1500−1750 m. N2; not known elsewhere. Bally 10359; Hemming 1948.

Plants collected in the eastern "Al Madow" (Bally 11005, 11119; Lavranos & Horwood 10318) may represent a distinct taxon, as yet undescribed.

4. A. somaliensis Watson (1899); type: N1, without precise locality, Cole 261/1895 (K holo.).

Dacar biyo (Som.).

Plants stemless, single or forming small groups. Leaves 12−20, spreading or ascending, 20−40 × 6−8 cm, shiny, brownish to dark green with numerous whitish spots and, in var. *marmorata*, darker green markings; margins with brown teeth, 4 mm long, 4−10 mm apart. Inflorescences with 5−15 branches, 60−85 cm tall; racemes cylindric, lax, to 20 cm long; bracts 6−8 × 3−4 mm; pedicels 8−10 mm long. Flowers pinkish-scarlet to bright red, pendulous but subsecund on lateral racemes; perianth 26−30 mm long, 9 mm across the ovary, outer tepals free for 10−13 mm. Anthers exserted 1−2 mm. Stigma exserted 2−3 mm. Ovary olive green, 6 × 3 mm.

1. Leaves spreading, 15−25 cm long, brownish-green, with only whitish spots; perianth pinkish, 28−30 mm long....................var. *somaliensis*
− Leaves ascending, to 40 cm long, dark green with both whitish and darker green longitudinal spots; perianth bright red, 26 mm long....var. *marmorata*

var. **somaliensis**.
Bushland and rocky slopes on limestone; 700−1000 m. N1; Djibouti. Bally 9669; Reynolds 8363.

var. **marmorata** Reynolds & Bally in J. S. Afr. Bot. 30: 222 (1964); type: N1, near "Upper Sheikh", Bally 11793 (K holo., EA PRE iso.).
Acacia-Commiphora bushland on limestone; c. 800 m. N1; not known elsewhere.

5. **A. parvidens** M. Gilbert & Sebsebe (1992).
   A. pirottae sensu auctt., non Berger.
   Daar, dacar (Som.).
   Plants with very short stem, usually in small groups. Leaves 12−16, sword-shaped, spreading or recurved, up to 40−50 × 8−9 cm, rather shiny, brown to greenish-brown, with numerous whitish spots; margins with whitish triangular teeth, 2−3 mm long, 10−15 mm apart. Inflorescences 100−150 cm long, with 5−10 branches; racemes cylindric, lax, to 15 cm long; bracts 5 × 4 mm; pedicels 5−7 mm long. Flowers pinkish, pendulous; perianth slightly curved, 30 mm long, 8 mm across the ovary, outer tepals free for 8 mm. Anthers exserted 2−3 mm. Stigma exserted 3−4 mm. Ovary green, 6 × 2.5 mm.
   Acacia-Commiphora bushland or semi-deciduous forest, on deep sand; 25−1300 m. N1−3; C1, 2; S1−3; Ethiopia, N and E Kenya. Bally 10194, 10198; Reynolds 7103, 8392.

6. **A. grisea** S. Carter & P. Brandham (1983); type: N1, "Sheikh Pass", Bally 9662 (K holo.).
   Plants stemless, solitary or in small groups. Leaves spreading, triangular, 25 × 15 cm, glaucous, heavily white spotted, not shiny; margins horny, brown with sharp teeth, 1−3 mm long, 2−8 mm apart. Inflorescences erect, to 60 cm tall with 2−3 branches; racemes erect, conical to subcapitate, rather dense; bracts to 20 × 5 mm; pedicels to 30 mm long. Flowers obliquely spreading, orange-scarlet with yellow mouth; perianth cylindric with pronounced basal swelling, 23 mm long, 6 mm across the ovary, abruptly constricted to 3.5 mm, outer tepals free for 5 mm. Anthers and stigma scarcely exserted.
   Juniperus-Buxus forest and scrub, *Acacia etbaica* scrub, frequently much degraded, on limestone or basement rocks; 1200−1700 m. N1, 2; Djibouti. Bally & Melville 16247; Carter 793.

7. **A. albovestita** S. Carter & P. Brandham (1983); type: N2, top of "Mait Pass", Bailes 214 (K holo.).
   Plants stemless, forming small groups. Leaves lanceolate, spreading, 20−30 × up to 12 cm, glaucous green, longitudinally striated with a few indistinct pale spots on upper surface; margins narrowly cartilaginous with red-brown teeth, 1.5 mm long, 5−10 mm apart. Inflorescences to 75 cm tall with 2−3 branches; racemes erect, cylindrical to subcapitate, rather lax; bracts 15 × 3 mm; pedicels to 18 mm long. Flowers obliquely spreading, dull pink with very pronounced bloom; perianth cylindric with pronounced basal swelling, 25−33 mm long, 7 mm across the ovary, abruptly constricted to 4.5 mm, outer tepals free for 7 mm. Anthers and stigma hardly exserted.
   Juniperus forest and shady rock faces, on limestone; 1400−2000 m. N2; Djibouti. Bally 10351, 11110.

8. **A. scobinifolia** Reynolds & Bally (1958); type: N2, near "Erigavo", Reynolds 8403 (PRE holo., K EA iso.). Plate 2 A.
   Plants stemless, solitary or forming small groups. Leaves 16−20, curved-ascending, lanceolate, 30 × 7 cm, rather thick, rugose, olive-green; margins cartilaginous without teeth. Inflorescences erect, 60−70 cm tall; racemes subcapitate, dense; bracts 8 × 2 mm; pedicels 15−18 mm long. Flowers red or yellow, nutant; perianth cylindric-clavate, 22−25 mm long, 4−5 mm across the ovary, outer tepals free for 9−10 mm. Anthers exserted 3−4 mm. Stigma exserted 5 mm. Ovary pale green, 6 × 2.5 mm.
   Stony flats with dwarf shrubs and grasses on gypsum or rocky arid slopes on limestone; 1100−1650 m. N2, 3; not known elsewhere. Reynolds 8402; Lavranos, Carter & al. 24759.

9. **A. ruspoliana** Bak. (1898).
   A. stefaninii Chiov. (1916); type: S1, "Hemin-Gurei", Paoli 687 (FT holo.).
   Dacar (Som.).
   Plants stemless or with stems to 50 cm long, forming small to large dense groups. Leaves c. 16, ascending-erect, 50−60 × 12 cm, uniformly olive-green, very occasionally with a few white spots; margins very narrowly cartilaginous, with minute white teeth, 0.5−1 mm long, 2−8 mm apart, becoming obsolete towards the apex. Inflorescences with many spreading branches, to 180 cm tall; racemes flatly capitate, broader than long; bracts 3 × 1.5 mm; pedicels 5 mm long. Flowers yellow, ± horizontal; perianth 16−20 mm long, 5 mm across the ovary, outer tepals free for 6−8 mm. Anthers exserted 2−3 mm. Stigma exserted 3−4 mm. Ovary 4 × 2 mm.
   Acacia-Commiphora dry woodland and bushland, mainly on sand; up to at least c. 300 m. C1, 2; S1−3; Ethiopia (Ogaden), Kenya. Robecchi-Bricchetti 18; Reynolds 7083, 7085.
   According to Kuchar 17157 this is poisonous to sheep and camel, and used for poisoning hyena.

10. **A. microdonta** Chiov. (1928); type: S2, "Genale" to "Audegle", Puccioni & Stefanini 49 (FT holo.).

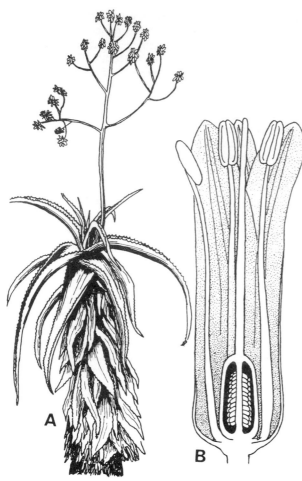

Fig. 25. *Aloe megalacantha.* A: habit. B: flower. — Modified from Burger, Fam. Flow. Pl. Ethiopia (1967).

Dacar qaraar (Som.).

Plants with slender, procumbent stems up to 100 cm long, proliferating from the base and lower part of stems. Leaves c. 16 at the apex of stems, 50–70 × 9–11 cm, spreading, recurved, deeply furrowed above, dull olive-green; margins armed with 1–2 mm long brown-tipped teeth, 5–14 mm apart. Inflorescences spreadingly branched, to 130 cm high; racemes oblique, elongate, to 15 cm long, fairly lax; bracts 2–4 × 2 mm; pedicels 5–6 mm long. Flowers scarlet, secund; perianth cylindric, 23 mm long, 7 mm across the ovary, outer tepals free for 10 mm. Anthers exserted 3 mm. Stigma exserted 5 mm.

Acacia-Commiphora bushland, usually on sand or limestone; 250–500 m. C2; S1–3; not known elsewhere. Bally 9500; Reynolds 7090, 7116.

The sap is used against jaundice.

A. defalcata Chiov. (1932), based on Guidotti 64 (not seen) from near "Uar Scek" (S3), appears to be a mixture of species comprising *A. microdonta* and *A. ruspoliana*, the only two species that occur in that area.

11. **A. megalacantha** Bak. (1898). Fig. 25.

Plants in groups with stems 50–100 cm tall. Leaves 24 or more on upper 50 cm of the stems, 60–80 × 13–15 cm, spreading, much recurved with tips pointing downwards, green, shiny, deeply furrowed above; margins with pink horny edge and red-brown teeth, 5–6 mm long, 20 mm apart. Inflorescences with many spreading branches, c. 100 cm tall; racemes conical, erect at the end of oblique branches, to 8 cm long, fairly lax; bracts 5 × 2–3 mm; pedicels 10–15 mm long. Flowers red or yellow, nutant; perianth 28–30 mm long, 8 mm across the ovary, outer tepals free for 15 mm. Anthers exserted 3 mm. Stigma exserted 4 mm. Ovary pale green, 6 × 3 mm.

Open, often degraded *Acacia-Commiphora* bushland, on limestone or sand; N1, 2; Ethiopia. Bally 9915; Reynolds 6244, 8332.

12. **A. luntii** Bak. (1894).

A. inermis sensu auctt., non Forssk.

Plants suckering and forming small groups; stems to 30 cm long, obliquely ascending. Leaves 12–16, lanceolate, 20–25 × 5–7 cm, horizontally spreading or recurved, grey or brownish-green with a waxy coating, smooth; margins entire, with rounded, cartilaginous edge. Inflorescences with up to 15 spreading branches, to 40 cm high; racemes elongate, oblique, to 15 cm long, lax; bracts 3–4 × 2–3 mm; pedicels 4–6 mm long. Flowers dull red or pink with conspicuous waxy bloom, subsecund; perianth 24–28 mm long, 7 mm across the ovary, outer tepals free for 7 mm. Anthers exserted 2–3 mm. Style exserted 3–4 mm. Ovary brownish, 5 × 2 mm.

Semi-desert flats on gypsum; c. 900 m. N2; Yemen (Hadhramaut), Oman (Dhofar). Bally 10879; Lavranos, Carter & al. 24602.

13. **A. brunneostriata** Lavranos & S. Carter (1992); type: N3, 128 km E of "Gardo" towards "Iskushuban", Lavranos & Horwood 10187 (K holo.).

Plants suckering and forming small groups; stems to 40 cm tall, obliquely ascending. Leaves c. 10, to 30 × 7 cm, lanceolate, horizontally spreading to recurved, creamy yellow, with numerous longitudinal red-brown lines; margins cartilaginous, sharply edged, with or without blunt, cartilaginous teeth less than 0.5 mm long. Inflorescences with up to 12 branches, 50–60 cm high; racemes elongate, lax; bracts 5–6 mm long; pedicels 5–7 mm long. Flowers yellow, subsecund; perianth 16–20 mm long. Anthers exserted 5 mm. Stigma exserted 5 mm.

Semi-desert flats, on limestone; c. 640 m. N3; not known elsewhere. Bally & Melville 15592; Lavranos, Carter & al. 24685.

14. **A. officinalis** Forssk. (1775).

Leaves erect, green, spotted white, 70 × 12 cm, smooth with prominently toothed margins. Inflorescences with up to 3 branches, to 175 cm tall; racemes conical, fairly dense; bracts 10 mm long; pedicels 6−8 mm long. Flowers red or yellow, nutant; perianth cylindrical, somewhat curved.

Cultivated in N1 and S2; native of SW Arabia.

15. A. molederana Lavranos & Glen (1989); type: N2, "Moledera", 58 km SSW of "Erigavo", Lavranos & Horwood 10379 (PRE holo.).

A. tomentosa sensu auctt., non Deflers.

Plants stemless or with short decumbent stems, forming groups. Leaves 16−20, 35 × 9 cm, lanceolate-triangular, spreading to falcately upcurved, bluish-grey to reddish; margins with very small teeth, 0.5 mm long, 20−80 mm apart, sometimes entire. Inflorescences 60−70 cm tall with up to 4 branches; racemes elongate-conical, to 18 cm long, subdense; bracts 7 × 4 mm; pedicels 6−9 mm long. Flowers rose-pink or reddish, tomentose, oblique-nutant; perianth 24−28 mm long, 7−8 mm across the ovary, outer tepals free for 9 mm. Anthers exserted 3−4 mm. Stigma exserted 4−5 mm. Ovary 6 × 3 mm.

Low desert scrub, on gypsum; c. 1450 m. N2; not known elsewhere. Glover & Gilliland 643; Hemming 2037; Reynolds 8445.

16. A. rigens Reynolds & Bally (1958); type: N1, "Bawn", N of "Borama", Reynolds 8369 (PRE holo., EA K iso.).

Daar merodi, dacar maroodi (Som.).

Plants stemless, solitary or forming fairly large groups. Leaves c. 24, 60−80 × 12−15 cm, very rigid, lanceolate, arcuate-ascending to erect, pale to dark green, smooth; margins with pungent reddish-brown teeth, 4−6 mm long, 20−35 mm apart. Inflorescences to 175 cm tall, usually with 2 branches; racemes elongate-conical to 40 cm long, rather lax; bracts 15 mm long; pedicels 5−6 mm long. Flowers rose-pink, shortly pubescent, nutant to pendulous; perianth 30−35 mm long, 7 mm across the ovary, outer tepals free for 10−12 mm. Anthers exserted 2−3 mm. Stigma exserted 4 mm. Ovary 6 × 3 mm.

Grassy or shrubby flats, or *Acacia-Commiphora* bushland, mainly on basement rocks; 1200−1400 m. N1, 2; not known elsewhere. Bally 9661, 11769; Carter 888.

17. A. glabrescens (Reynolds & Bally) Carter & Brandham (1983); *A. rigens* var. *glabrescens* Reynolds & Bally in J. S. Afr. Bot. 24: 179 (1958); type: "Barror" plains, 22 km N of "Qaradag" on road to "Erigavo", Reynolds 8390 (PRE holo., EA K iso.). Plate 2 B.

Plants in groups of various sizes, stemless or with decumbent stems to 50 cm. Leaves c. 20, 40−50 × 10−12 cm, lanceolate-triangular, ascending to spreading, grey green with reddish tinge; margins with brown pungent teeth, 3 mm long, 10−20 mm apart. Inflorescences to 100 cm tall with 2 to 4 branches; racemes cylindric-acuminate, subdense, to

25 cm long; bracts 9 × 4 mm; pedicels 5 mm long. Flowers bright reddish-pink, minutely pubescent, pendent; perianth 32 mm long, 6 mm across the ovary, outer tepals free for 9 mm. Anthers exserted 3−4 mm. Stigma exserted 5 mm. Ovary 7 × 3 mm.

Semi-desert flats, mainly on gypsum or limestone; 800−1000 m. N2; not known elsewhere. Bally 11877; Bally & Melville 16251; Brandham 2775.

Plants recorded from near "Las Anod" in N2 (Lavranos & Horwood 10140; Brandham 2769) and from near "Eil" in N3 (Lavranos & Horwood 10177 A; Hemming 1682; Bally & Melville 15545) and attributed by Brandham & al. (1983) to *A. glabrescens* may well represent an undescribed taxon with glaucous, white-spotted, incurved leaves c. 30 cm long and consistently simple inflorescences.

18. A. breviscapa Reynolds & Bally (1958); type: N2, 77 km E of "Erigavo", on road to "Hadaftimo", Reynolds 8542 (PRE holo., EA K iso.).

Plants forming small or large groups. Leaves c. 24, in dense rosettes, 30−35 × 8−10 cm, lanceolate-attenuate, ascending-spreading, rigid, bluish-grey with red tinge; margins entire or with a few very short teeth near the base. Inflorescences to 50 cm tall, with 3−6 spreading branches; racemes cylindric, lax, 20−25 cm long; bracts 6 × 3 mm; pedicels 10−14 mm long. Flowers scarlet, glabrous, nutant; perianth 26−30 mm long, 8 mm across the ovary, outer tepals free for 10 mm. Anthers exserted 2−3 mm. Stigma exserted 3−4 mm. Ovary 6 × 3 mm.

Semi-desert, on gypsum; c. 1400 m. N2; not known elsewhere. Bally 11188; Hemming 1982.

19. A. bargalensis Lavranos (1973); type: N3, "Defho Ghezani" plateau, W of "Bargal", c. 300 m. Lavranos & Bavazzano 8459 (FT holo.).

Plants solitary or suckering sparingly. Leaves up to 20, 30−60 × 6−20 cm, narrowly lanceolate, somewhat curved, dark glaucous green, prominently white striate on both sides; margins almost entire or with rigid 1−2 mm long teeth, sometimes as close as 12−15 mm. Inflorescences to 150 cm tall, simple or with 1−2 branches; racemes very narrowly cylindric, 30−40 cm long, fairly lax; bracts to 15 × 5 mm; pedicels 5−7 mm long. Flowers reddish with yellow mouth, pendulous; perianth narrow, 30 mm long, 5 mm across the ovary, outer tepals free for 16 mm. Anthers exserted 3 mm. Stigma not exserted. Ovary 5 × 1.5 mm.

Hills and mountains, on limestone; 0−1100 m. N2, 3; not known elsewhere. Lavranos & Bavazzano 8495, 8530.

20. A. citrina S. Carter & P. Brandham (1983); type: C2, 3 km S of "Bulo Burti", Bally & Melville 15287 (K holo.).

Dacar gabarey (Som.).

Plants solitary or with a few offshoots. Leaves c. 20,

to 60 × 15 cm, lanceolate, ascending-erect, rigid, glaucous green, heavily white spotted; margins with brown teeth, 1.5 mm long, 15−35 mm apart. Inflorescences to 200 cm tall, with 2−6 branches; racemes cylindric, to 50 cm long, fairly dense; bracts 12 × 4 mm, pubescent; pedicels 10 mm long, pubescent. Flowers greenish to lemon yellow, tomentose, pendent; perianth 28−30 mm long, 6 mm across the ovary, outer tepals free for 18 mm. Anthers hardly exserted. Stigma exserted 4 mm. Ovary 6 mm long.

Acacia-Commiphora bushland or dry forest, usually on deep sand; 150−700 m. C2; S1−3; SE Ethiopia, NE Kenya. Reynolds 7136; Brandham 2739.

The sap is used for treatment of eye diseases.

21. **A. bella** G. Rowley (1974); *A. pulchra* Lavranos (1973), nom. illeg.; type: N3, 10 km S of "Yibirti" gorge, between "Hordio" and "Bargal", Lavranos & Bavazzano 8458 (FT holo.).

Plants solitary or forming fairly large groups. Leaves c. 20, 50 × 11 cm, lanceolate-triangular, ascending-erect, somewhat curved, rigid, pinkish-brown; margins with red teeth, 2 mm long, 20 mm apart. Inflorescences 100 cm tall, with up to 5 branches; racemes shortly conical to subcapitate, 6−11 cm long, dense; bracts 8−10 × 4 mm; pedicels 6−7 mm long. Flowers brilliant red, shiny, pendent; perianth 27 mm long, 8−9 mm across the ovary, outer tepals free for 7−9 mm. Anthers exserted 3−5 mm. Stigma exserted 6 mm. Ovary 7 × 3 mm.

Coastal semi-desert with low shrubs and scattered low trees, mainly *Commiphora*, on limestone; 0−100 m. N3; not known elsewhere. Lavranos, Carter & al. 24802.

22. **A. cremnophila** Reynolds & Bally (1961); type: N2, near "Daloh", 21 km N of "Erigavo", Reynolds 8450 B (PRE holo.). Plate 2 C.

Plants pendent, with stems to 40 cm long, branched mainly from the base. Leaves 6−8, 10−15 × 2−3 cm, grey-green, without spots; margins with hard light brown teeth, 2 mm long, 3−5 mm apart. Inflorescences 25−30 cm long, simple; racemes cylindrical-conical, to 12 cm long, fairly dense; bracts 10 × 5 mm; pedicels 10−12 mm long. Flowers scarlet with yellowish-green mouth, pendent; perianth slightly clavate, 25 mm long, 5 mm across the ovary, outer tepals free for 5 mm. Anthers exserted up to 1 mm. Stigma exserted 1−2 mm. Ovary 5 × 2.5 mm.

Cliff faces in *Juniperus-Pistacia* forest, generally above 2000 m. N2; not known elsewhere. Hemming 1947; Carter 912.

23. **A. gillettii** S. Carter (1994); type: N3, "Ahl Mescat" Mts, 10°55'N, 49°26'E, Gillett & Watson 23457 (K holo., EA MOG iso.).

Plants decumbent, sparsely branched, the stems to 20 cm or more long. Leaves 6 or 7 in a lax rosette, to 15 × 1.5 cm, dark grey-green with very numerous white spots; margins with cartilaginous teeth to 1 mm long, 3−5 mm apart. Inflorescences to 25 cm tall, simple; racemes 3−5 cm long, lax; bracts 4−5 × 2 mm; pedicels to 10 mm long. Flowers coral red; perianth 20−26 mm long, 7 mm across the ovary, outer tepals free for 6−8 mm. Anthers exserted by 2.5 mm. Stigma exserted by 2.5 mm.

Rocky slopes, on limestone with *Juniperus* in open *Commiphora* bushland at 1350−1650 m. N3; not known elsewhere. Thulin & Warfa 6099 A.

24. **A. ambigens** Chiov. (1928); type: C1, between "Attodi" and "Dolobscio", Puccioni & Stefanini 447 (FT holo.).

Plants shrubby, with stems to 40 cm long. Leaves 5−15 in a lax rosette, to 20 × 2−3 cm, fleshy and rigid, glaucous green, sparsely spotted at times; margins with white cartilaginous teeth, to 1 mm long, 10−20 mm apart. Inflorescences to 100 cm long, with up to 8 spreading branches; racemes 4−12 cm long, lax; bracts 5 × 3 mm; pedicels 4 mm long. Flowers red or yellow, nutant; perianth 20 mm long, 7 mm across the ovary, outer tepals free for 6 mm. Anthers not exserted. Stigma exserted 1−2 mm. Ovary 6 × 2 mm.

Acacia-Commiphora bushland on steep limestone rock faces; c. 250 m. C1 (W of Hobyo). Lavranos, Carter & al. 23390.

25. **A. hildebrandtii** Bak. (1888); type: N2, without precise locality, Hildebrandt s.n. (K holo.).

A. gloveri Reynolds & Bally (1958); type: N1, "Gaan Libah", 19 km NW of "Ghor", Reynolds 8358 (PRE holo., EA K iso.).

Plants shrubby, branched from the base, with stems 50−100 cm long, decumbent to erect. Leaves 10−20, in a lax rosette covering the apical 30−35 cm of the stem, to 30 × 6 cm, spreading, dull green or glaucous with a few spots; margins with triangular sharp teeth, 2−3 mm long, 8−10 mm apart. Inflorescences to 60 cm tall, spreadingly branched; racemes cylindric, to 18 cm long, lax; bracts 3 × 2 mm; pedicels 10−15 mm long. Flowers red or yellow, nutant to subsecund; perianth 26−30 mm long, 8 mm across the ovary, outer tepals free for 12 mm. Anthers exserted 3 mm. Stigma exserted 4 mm. Ovary 6 × 3 mm.

Juniperus-Pistacia-Buxus forest, on limestone; 1150−1700 m. N1, 2; not known elsewhere. Hemming 1967; Bally 5709.

The view has been expressed that plants from the "Gaan Libah-Sheikh" area, described as *A. gloveri*, differ sufficiently from what is considered typical *A. hildebrandtii* to be recognised as a distinct species. The treatment of Reynolds (1966) is followed here, as the considerable range of *A. hildebrandtii* justifies a fair degree of variation, both laterally and altitudinally.

Plate 1

A *Aponogeton abyssinicus var. cordatus*
C *Stylochaeton borumensis*
E *Dracaena ombet*

B *Amorphophallus maximus*
D *Arisaema flavum*
F *Sansevieria powellii*

Plate 2

A *Aloe scobinifolia*
C *Aloe cremnophila*
E *Crinum stuhlmannii*

B *Aloe glabrescens*
D *Aloe eminens*
F *Ammocharis tinneana*

Plate 3

A *Pancratium tenuifolium*
C *Gladiolus somalensis*
E *Merendera schimperiana*

B *Romulea fischeri*
D *Littonia revoilii*
F *Habenaria socotrana*

Plate 4

A *Eulophia petersii*

B *Xerophyta schnizleinia*

C *Aneilema obbiadense*

D *Cyperus chordorrhizus*

E *Hyphaene compressa*

F *Hyphaene reptans*

26. A. heliderana Lavranos (1973); type: N3, c. 50 km SW of "Iskushuban" on road to "Gardo", Lavranos & Bavazzano 8456 (FT holo.).

Plants shrubby, sparingly branched from the base; stems erect, to 100 cm tall. Leaves c. 20, 25 × 6 cm, grey-green with violet tinge, variously white spotted; margins with hard, white, triangular teeth, 2 mm long, 5−10 mm apart. Inflorescences to 60 cm tall, spreadingly branched; racemes subcapitate, 5−8 cm long, lax; bracts 3−5 × 2 mm; pedicels 7 mm long. Flowers usually yellow, rarely red, nutant to subsecund; perianth 20−22 mm long, 6 mm across the ovary, outer tepals free for 5−7 mm. Anthers exserted 5 mm. Ovary 4 × 2 mm.

Semi-desert, on limestone; c. 500 m. N2, 3; not known elsewhere. Bally 10892; Barbier 950.

27. A. retrospiciens Reynolds & Bally (1958); type: N1, "Darburruk", 90 km from "Hargeisa" on road to "Berbera", Reynolds 8482 (PRE holo., EA K iso.).

A. ruspoliana var. *dracaeniformis* Berger in Engl., Pflanzenr. IV.38III.ii.: 266 (1908); type: probably N1, near "Hargeisa", Ruspoli & Riva 227 (?B holo., FT iso.).

Daar burruk, dacar burruq (Som.).

Plants usually branched from ground level, with 2−6 stems to 125 cm high. Leaves c. 12 in a lax rosette covering the upper 10−20 cm of stems, c. 25 × 5−6 cm, lanceolate with subobtuse apex, bluish-grey with reddish tinge, unicoloured or with very few pale spots; margins with cartilaginous white border and firm white teeth, 1 mm long, 5 mm apart. Inflorescences 45 cm high, with 6−12 spreading branches; racemes 3−5 cm long, fairly dense; bracts 5 × 2.5 mm; pedicels 5−6 mm long. Flowers yellow, nutant to secund; perianth 20 mm long, 6 mm across the ovary, outer tepals free for 10 mm. Anthers exserted 3 mm. Stigma exserted 4 mm. Ovary 3 × 2 mm.

Rocky ground with shrubby *Acacia* and *Commiphora*; 300−1100 m. N1−3; not known elsewhere. Gillett 4597; Beckett 607.

28. A. gracilicaulis Reynolds & Bally (1958); type: N2, 30 km N of "Erigavo" on "Mait" road, Reynolds 8428 (PRE holo., EA K iso.).

Daar der, dacar dheer (Som.).

Plants sparsely branched from base with stout stems to 400 cm tall. Leaves c. 20, crowded in a rosette at apex of stems, sword-like, furrowed above, acute, to 60 × 8 cm, grey-green, unspotted; margins with white cartilaginous border and blunt white teeth, 1 mm long, 2−10 mm apart. Inflorescences 60 cm high, with 8−10 spreading branches; racemes cylindric, 5−6 cm long, fairly dense; bracts 3 × 2 mm; pedicels 5−6 mm long. Flowers yellow, nutant; perianth 18 mm long, 5 mm across the ovary, outer tepals free for 12 mm. Anthers exserted to 4 mm. Stigma exserted to 5 mm. Ovary 5 × 2 mm.

Juniperus-Buxus-Dracaena forest; c. 1250 m. N2; not known elsewhere. Barbier 1007; Hemming 2039.

Plants observed by Bally at Agasur, near 49°E, at about 1600 m altitude, seem to belong to *A. gracilicaulis*.

29. A. medishiana Reynolds & Bally (1958); type: N2, "Medishe", 38 km NE of "Erigavo", Reynolds 8441 (PRE holo., EA K iso.).

Plants usually branched from the base, with slender stem to 200 cm tall. Leaves c. 24, crowded in a rosette at apex of stems, to 30 × 6 cm, sword-like, recurved, grey-green, unspotted; margins with white cartilaginous border and teeth less than 1 mm long and 5−10 mm apart. Inflorescences to 50 cm tall, with 3−8 ascending branches; racemes cylindric, 5−6 cm long, fairly dense; bracts 2−3 × 2 mm; pedicels 9 mm long. Flowers dull red, nutant; perianth 19 mm long, 5 mm across the ovary, outer tepals free for 5−6 mm. Anthers exserted 1−2 mm. Stigma exserted 3 mm. Ovary 4 × 2.5 mm.

Hillsides with small shrubs and succulents, on limestone; up to c. 1800 m. N2, 3; not known elsewhere. Bally 11140; Gillett & Watson 23506.

30. A. eminens Reynolds & Bally (1958); type: N2, "Surud", W side of "Tabah Pass", Reynolds 8435 (PRE holo., EA K iso.). Plate 2 D.

Daar der, dacar dheer (Som.).

Irregularly branched shrub or tree to 15 m tall; stem up to 150 cm wide at base. Leaves 16−20, to 45 × 5 cm, lanceolate, recurved, furrowed above, green; margins with cartilaginous white teeth, 1−3 mm long, 3−10 mm apart. Inflorescences 50−60 cm tall, with 2−5 branches; racemes cylindric, 12−20 cm long, fairly dense; bracts 6 × 3 mm; pedicels to 12 mm long. Flowers red, fleshy, nutant; perianth 30−42 mm long, 12 mm across the ovary, outer tepals free for 25−32 mm. Anthers exserted 4 mm. Stigma exserted 5 mm. Ovary 8 × 4 mm.

Juniperus forest or evergreen bushland, on limestone; 1300−1800 m. N2; not known elsewhere. Popov 1158; Bally 11142; Thulin, Dahir & Hassan 9100.

31. A. rabaiensis Rendle (1895).

Plants densely shrubby with stems to 200 cm tall, erect to ascending. Leaves in a fairly dense rosette on upper 30 cm of stems, spreading to recurved, to 45 × 8 cm, lanceolate-attenuate, grey-green, usually unspotted; margins with brownish teeth, 4 mm long, 5−15 mm apart. Inflorescences 60 cm high, with 6−8 branches; racemes capitate, to 8 cm long; bracts 7 × 3 mm; pedicels 18 mm long. Flowers scarlet, pendulous to nutant; perianth 32 mm long, 8 mm across the ovary, outer tepals free for 9 mm. Anthers exserted 2−3 mm. Stigma exserted 4 mm. Ovary 7 × 3 mm.

Woodland; c. 120 m. S3; Kenya, N Tanzania. Bally 9501.

Lavranos & Bauer 22847, a low-growing shrubby

plant allied to the species of the group to which *A.rabaiensis* belongs, but which has short stems not exceeding 40 cm in height and glossy, triangular, often white-spotted leaves, may be an undescribed taxon. It was found on granite, at the top of "Bur Heybe" (S1).

Described species of uncertain position

A. ellenbeckii Berger (1905); type: S3, "Fereschit",

Ellenbeck 2340 (B holo., ? destr.).

Berger refers this tentatively to his sect. *Saponariae*, but the long narrow leaves with thin cartilaginous margins and very small, distant teeth rather suggests an affinity to species such as *A. ambigens* or *A. gillettii*. Further search in the rock gorges along the Juba river around "Bardera" could lead to the rediscovery of this plant.

151. ANTHERICACEAE

by M. Thulin

Cuf. Enum.: 1530−1538 (1971), *Liliaceae* in part; Nordal & Thulin in Nord. J. Bot. 13: 257−280 (1993).

Perennial herbs with a rhizome, not producing anthraquinons and rhizome not yellowish inside. Leaves mostly basal, spirally set or in 2 rows, sheathing at the base. Inflorescences simple or compound racemes, spikes or panicles. Flowers generally bisexual and regular; tepals 6, free or united into a basal tube, usually white. Stamens 6 or rarely 3, mostly free; filaments generally glabrous; anthers 2-thecous, generally basifixed, longitudinally dehiscent. Ovary superior, 3-celled; style 1; ovules 2 to many per cell. Fruit a loculicidal capsule. Seeds black, with straight or slightly curved embryo, with endosperm.

Family of some 30 genera and 600 species, distributed in most parts of the world.

1. Flowers solitary at each node of the inflorescence, supported by 1 bract; roots without tubers; pedicels not articulated; seeds ± swollen...1. *Anthericum*
− Flowers often more than 1 at each node of the inflorescence or, if only 1, then supported by 2 bracts; roots often with tubers; pedicels most often articulated; seeds thin, flat or folded ....2. *Chlorophytum*

1. ANTHERICUM L. (1753)

Roots without tubers. Leaves terete to linear. Inflorescence simple or branched, with flowers solitary at each node, supported by 1 bract; peduncle without leaves or sterile bracts; pedicels not articulated. Stigma terminal, small. Capsule 3-angled. Seeds ± swollen.

Mainly temperate-subtropical genus of medium size, distributed in Europe, the Mediterranean Region, the Middle East and NE and E tropical Africa.

For the distinction between *Anthericum* and *Chlorophytum*, see Nordal & Thulin (1993), where also further references are given.

1. Leaves c. 5 mm wide; inflorescence never branched; capsules slightly ridged, but not warty ....1. *A. corymbosum*
− Leaves c. 10 mm wide; inflorescence often with 1−3 basal branches; capsules warty...2. *A. jamesii*

1. **A. corymbosum** Bak. (1877); type: N2, "Meid, Surrut Mt", Hildebrandt 1471 (K holo.). Fig. 26.
   A. gregorianum Rendle (1895).
   A. corymbosum var. *floribundum* Chiov., Result. Scient. Miss. Stefanini-Paoli: 173 (1916); type: C2,

"Uegit", Paoli 1084 (FT lecto.).
Baharor, baror (Som.).

Plant up to c. 30 cm tall, from a very short fibrous rhizome with fleshy roots. Leaves narrowly linear, up to c. 5 mm wide, ciliate. Inflorescence 3−30-flowered, unbranched; bracts long, white, membranous; pedicels spreading, up to 40 mm long. Tepals white with green 3-nerved median stripe, 8−10 mm long. Capsule transversely veined, otherwise smooth. Seeds angular, c. 1.6 mm across.

Grassland or in shallow soil overlying rocks; 250−1800 m. N2; C2; S1; Ethiopia, Kenya, N Tanzania, and in Yemen. Paoli 936, 1057; Thulin, Hedrén & Abdi Dahir 7521.

2. **A. jamesii** Bak. (1898).
   A. verruciferum Chiov. (1916); type: S1, "El Uré", Paoli 1068 (FT lecto.).

Plant from a very short rhizome, with fleshy roots. Leaves linear, c. 10 mm wide, ciliate to obscurely papillose along margins. Inflorescence several-flowered, often with 1−3 basal branches; bracts small, white, membranous; pedicels spreading, up to c. 18 mm long. Tepals white with green 3-veined median stripe, c. 8 mm long. Capsule transversely veined and warty. Seeds angular, c. 2.4 mm across.

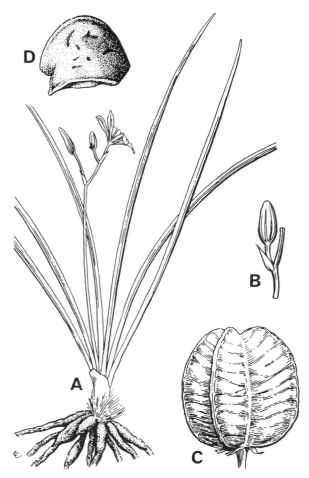

Fig. 26. *Anthericum corymbosum*. A: habit, × 2/3. B: inflorescence node, × 1. C: capsule, × 2. D: seed, × 8. − Modified from Nord. J. Bot. 13: 60 (1993). Drawn by E. Catherine.

Alluvial flats in *Acacia-Commiphora* bushland; c. 400 m. C1; S1; E Ethiopia, NE Kenya. Hemming 1721.

Of the three original syntypes of *A. verruciferum* Paoli 1068 is conspecific with *A. jamesii*, while Paoli 942 and 1254 both are *Chlorophytum bifolium* Dammer. Also the plate published for *A. verruciferum* by Chiovenda in Result. Scient. Miss. Stefanini-Paoli: Pl. 19 B (1916) is *C. bifolium*.

2. CHLOROPHYTUM Ker-Gawl. (1807)

Roots often with tubers. Leaves linear to ovate. Inflorescence simple or branched, with flowers often more than 1 at each node or, if only 1, then supported by 2 bracts; peduncle with or without leaves or sterile bracts; pedicels usually articulated. Stigma terminal, small. Capsule 3-angled or ± deeply 3-lobed. Seeds thin, flat or folded.

Large genus mainly in the Old World tropics.

1. Peduncle with leaves and sterile bracts along all its length ....1. *C. petraeum*
− Peduncle without leaves or bracts, except occasionally for 1 sterile bract just below the inflorescence 2
2. All nodes of the inflorescence with a single flower only 3
− At least some of the lower nodes of the inflorescence with 2 or more flowers 8
3. Pedicels very short, articulated at the apex; perianth papillate, bell-shaped, with erect 1-nerved tepals ....2. *C. silvaticum*
− Pedicels distinct, articulated below the apex; perianth glabrous with spreading 3-nerved tepals 4
4. Flowers zygomorphic with tepals at least 10 mm long; capsule usually at least 10 mm long..........
....3. *C. somaliense*
− Flowers regular with tepals less than 10 mm long; capsule less than 8 mm long 5
5. Perianth forming a persistent beak at the apex of the capsule 6
− Perianth not forming a beak at the apex of the capsule 7
6. Leaves 18−40 mm wide, flat on the ground.........
.... 4. *C. applanatum*
− Leaves 3−8 mm wide, ± erect or spreading........
.. .. 5. *C. hiranense*
7. Pedicels articulated in the lower half to a little above the middle; rhachis of inflorescence usually distinctly puberulous; leaves up to c. 10 mm wide .... 6. *C. bifolium*
− Pedicels articulated near the apex; rhachis of inflorescence glabrous; leaves usually well over 10 mm wide .... 7. *C. littorale*
8. Leaves 1−3 mm wide, folded; tepals up to 5 mm long 9
− Leaves more than 3 mm wide; tepals 5 mm or more long 10
9. Capsules c. 3.5 mm long, clearly wider than long .... 8. *C. filifolium*
− Capsules 5−8 mm long, longer than wide..........
.... 9. *C. inconspicuum*
10. Tepals at least 6 mm wide and with 9 or more vascular bundles ....10. *C. tuberosum*
− Tepals less than 6 mm wide and with 3−7 vascular bundles 11
11. Pedicels articulated in upper part, the lower ones at least 15 mm long 12
− Pedicels articulated near the middle or in the lower half, less than 10 mm long 13
12. Leaves 2.5−4 mm wide; perianth forming an apical beak on the capsule ...11. *C. ramosissimum*
− Leaves 20−80 mm wide; perianth not forming a beak on the capsule .... 12. *C. zavattarii*
13. Leaves with very prominent papillose nerves ....13. *C. parvulum*
− Leaves without very prominent papillose nerves..
....14

14. Root tubers mostly on lateral short branches; flowers often greenish; leaves 17−60 mm wide ....14. *C. gallabatense*
− Root tubers as swellings at the ends of the roots, or roots thick and fleshy without tubers; flowers whitish; leaves narrower 15
15. Roots fleshy, without tubers 16
− Roots wiry with distal tubers 17
16. Tepals 5−6 mm long .... 6. *C. bifolium*
− Tepals c. 12 mm long .... 15. *C. subpetiolatum*
17. Tepals 10 mm or more long, with 5−7 vascular bundles; leaves mostly more than 10 mm wide ....16. *C. zanguebaricum*
− Tepals up to 10 mm long, with 3 vascular bundles; leaves up to 10 mm wide 18
18. Capsule with very prominent raised transversal nerves; persistent perianth tending to form an apical beak on the capsule ....17. *C. nervosum*
− Capsule with weak transversal nerves; perianth spreading in fruit, not forming an apical beak ....18. *C. tordense*

1. **C. petraeum** Nordal & Thulin (1993); type: N2, "Agasur", Bally 11022 (K holo., FT iso.). Fig. 27.

Plant up to at least 32 cm tall, from a short slightly fibrous rhizome; roots fleshy, without tubers. Leaves several in a rosette, lanceolate, up to at least 20 mm wide, with the margin distinctly ciliate usually along all its length. Inflorescence usually unbranched, glabrous; peduncle with leaves and sterile bracts along all its length, glabrous; bracts ovate-lanceolate; flowers 1−4 at each node; pedicels 3−8 mm long, articulated near the middle or in the lower half. Tepals white, medially greenish, 5−6.5(−8) mm long, 3-nerved. Stamens subequal. Style 4−5 mm long. Capsule 4−5 × 5−7 mm, expanding laterally through the persistent perianth that forms an apical beak.

In crevices of rocks, grassy patches in evergreen bushland, or in stony soil, often in shady places; 400−1600 m. N2, 3; not known elsewhere. Collenette 265; Glover & Gilliland 697; Thulin & Warfa 5821.

2. **C. silvaticum** Dammer (1912).

Dasystachys debilis Bak. (1898), non *Chlorophytum debile* Bak. (1878); *C. bakeri* Poelln. (1946).

Plant up to c. 30 cm tall, from an erect shortly fibrous rhizome with fleshy roots, without tubers. Leaves several in a rosette, linear-lanceolate, glabrous, with ± wavy margin. Inflorescence a dense subspicate raceme, glabrous; peduncle leafless; bracts ± projecting between the flowers; flowers 1 at each node; pedicels up to c. 1 mm long, articulated at the apex. Tepals white, c. 5 mm long, 1-nerved, papillate, forming a bell-shaped perianth. Stamens exserted. Style c. 8 mm long. Capsule wider than long, c. 3.5 mm long.

In deciduous bushland; c. 400 m. S1; S Ethiopia and south to Zimbabwe and Mozambique. Tardelli 209.

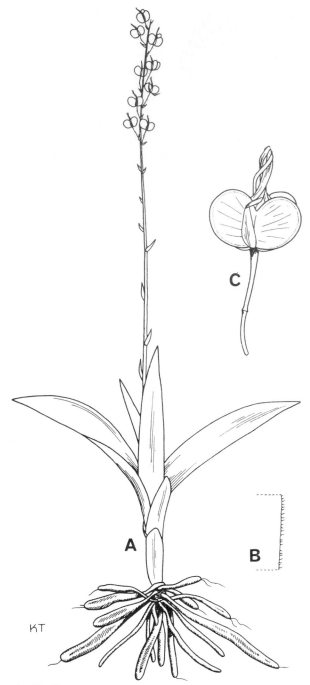

Fig. 27. *Chlorophytum petraeum*. A: habit, × 0.75. B: leaf margin, × 4.5. C: capsule with pedicel, × 4.5. − Modified from Nord. J. Bot. 13: 269 (1993). Drawn by K. Thunberg.

3. **C. somaliense** Bak. (1893); type: N2, "Mt Ahl" near "Mait", Hildebrandt 1468 (B holo.).

C. tenuifolium Bak. (1895); type: N1, "Wadaba", Cole s.n. (K lecto.).

C. baudi-candeanum Chiov. (1911).

C. pauciflorum Dammer (1946); type: S1, "Arbarone

in Gara Libin", Ellenbeck 2214 (B lecto.).

C. boranense Chiov. (1951).

C. tertalense Chiov. (1951).

Plant up to at least 40 cm tall, from a densely fibrous rhizome with wiry roots ending in ellipsoid tubers. Leaves several in a rosette, linear-lanceolate, glabrous, folded, with ± wavy margins. Inflorescence usually a simple raceme, lax, glabrous; peduncle leafless; flowers 1 at each node; pedicels c. 5–15 mm long, articulated near or above the middle. Tepals white, 10–12 mm long, 3-nerved, spreading or reflexed. Stamens 3 long and 3 short. Style c. 8 mm long, downcurved. Capsule rounded to broadly oblong in outline, c. 10–15 × 10 mm.

Deciduous bushland, often in alluvial soil or in rocky places, also on gypsum; 200–1500 m. N1, 2; S1; Ethiopia, Kenya. Gillett & Hemming 24313; Tardelli 95; Wood S/73/36.

4. **C. applanatum** Nordal & Thulin (1993); type: C1, Mudug Region, E of Gawaan, 29–30 km on road between Hobyo and Wisil, 5°19'N, 48°19'E, Thulin & Dahir 6674 (UPS holo., K iso.). Fig. 28.

Plant up to at least 22 cm tall, from a short fibrous rhizome; roots short, ending in ellipsoid tubers. Leaves 1 or 2, basal, flat on the ground, ovate to broadly elliptic, 18–40 mm wide, glabrous; margin flat and with a c. 0.3 mm wide hyaline border. Inflorescence a simple or few-branched raceme, glabrous; peduncle leafless, papillose-puberulous below, glabrous above; flowers 1 at each node; bracts 1.5–4 mm long; pedicels 7–18 mm long, filiform, spreading, articulated 1–3 mm below the flowers, glabrous. Tepals white, medially greenish, 7–8 mm long, 3-nerved. Stamens subequal. Style c. 5.5 mm long. Capsule c. 3–3.5 × 5–5.5 mm, expanding laterally through the persistent perianth that forms an apical slightly twisted beak.

Acacia-Commiphora bushland in shallow loamy sand over limestone; 200–340 m. C1; not known elsewhere. Gillett, Hemming & Watson 22462; Wieland 4264.

5. **C. hiranense** Nordal & Thulin (1993); type: C2, Hiiraan Region, 8 km NNW of Maxaas, Kuchar 16825 (K holo. and iso.).

Plant up to at least 23 cm tall, from a short, fibrous rhizome; roots short, ending in ellipsoid-clavate tubers. Leaves 1 or 2, basal, linear-lanceolate, erect or spreading, 3–8 mm wide, glabrous; margin undulate and with a c. 0.15 mm wide hyaline border. Inflorescence a simple or few-branched raceme, glabrous; peduncle leafless, glabrous; flowers 1 at each node; bracts 0.8–1.5 mm long; pedicels 8–20 mm long, filiform, spreading or becoming deflexed, articulated c. 1 mm below the flowers, glabrous. Tepals c. 5–6 mm long in young fruit. Capsule c. 4 × 8 mm, expanding laterally through the persistent perianth that forms a slightly twisted apical beak.

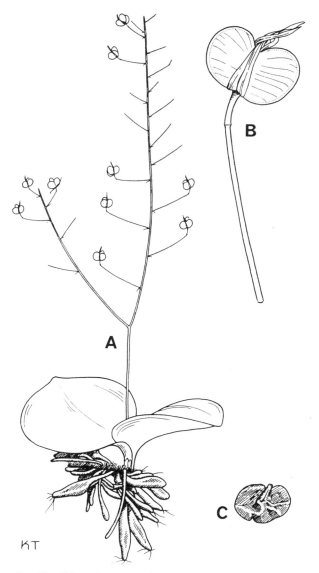

Fig. 28. *Chlorophytum applanatum*. A: habit, × 0.75. B: capsule with pedicel, × 4.5. C: seed, × 7.5. – Modified from Nord. J. Bot. 13: 259 (1993). Drawn by K. Thunberg.

Rocky limestone slope, in shade; c. 200 m. C2; only known from the type.

6. **C. bifolium** Dammer (1905).

Diado (Som.).

Plant up to c. 20 cm tall, from a short fibrous rhizome; roots short, swollen, without tubers. Leaves few in a rosette, linear-lanceolate, up to c. 12 mm wide, glabrous. Inflorescence a simple or branched raceme, fairly dense, rhachis puberulous or occasionally glabrous; peduncle leafless; flowers 1 or sometimes 2 at each node; pedicels 2–4 mm long, articulated near middle or in lower half. Tepals white, c. 5–6 mm long, 3-nerved, spreading.

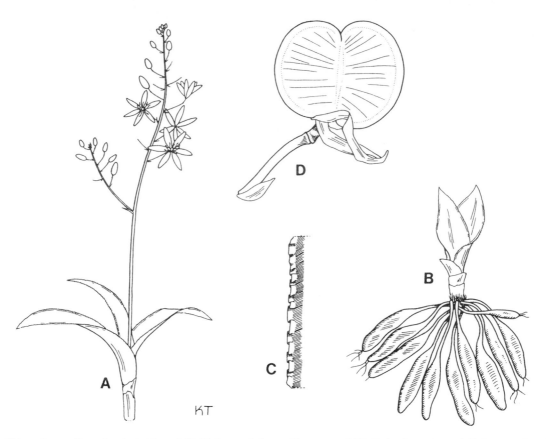

Fig. 29. *Chlorophytum littorale*. A: habit, × 0.75. B: base of plant with roots, × 0.75. C: leaf margin, × 4.5. D: capsule with pedicel and bract, × 4.5. – From Nord. J. Bot. 13: 266 (1993). Drawn by K. Thunberg.

Stamens subequal. Style c. 4–6 mm long. Capsule c. 4 × 5 mm, smooth or warty.

Deciduous bushland, often in alluvial soil, also in fallow fields; 250–1000 m. N1; S1; SE Ethiopia, NE Kenya. Corradi 4637; O'Brien 48; Terry 3452.

The only specimen seen from N1, Glover & Gilliland 1021, has a glabrous inflorescence, but appears to be conspecific. Some collections from S1 have ± warty capsules.

A record of *C. micranthum* Bak. from southern Somalia in Kuchar (1988) is based on material of *C. bifolium*.

7. C. littorale Nordal & Thulin (1993); type: S2, Shabeellaha Hoose Region, 35 km NE of Marka just N of Gandershe, 1°50'N, 44°58'E, Alstrup & Michelsen 113 (K holo., C iso.). Fig. 29.

Plant up to at least 32 cm tall, from a short fibrous rhizome; roots short and slender, ending in ellipsoid tubers. Leaves 2–4, basal, narrowly lanceolate to narrowly ovate, 10–60 mm wide, with a usually finely undulate-crispate, 0.3–0.4 mm wide hyaline margin, glabrous. Inflorescence a simple or few-branched raceme, glabrous; peduncle glabrous, leaf-

less; bracts lanceolate, 2.4–4 mm long; flowers 1 at each node; pedicels 2.5–12(–25) mm long, filiform, spreading, articulated c. 1–2(–3) mm below the flower, glabrous. Tepals white, medially greenish, 6–9.5 mm long, 3-nerved. Stamens subequal. Style 5.5–6.5 mm long. Capsule c. 6–6.5 × 7–8 mm, with the spreading persistent tepals at the base.

Coastal dunes, usually on white coral sand; 5–60 m. S2, 3; not known elsewhere. Thulin & Warfa 4511; Friis, Alstrup & Michelsen 4635; Thulin & Hedrén 7154.

A record of *C. pusillum* Schweinf. ex Bak. by Raimondo & al. in Webbia 35: 221 (1981) was based on a collection of *C. littorale*.

8. C. filifolium Nordal & Thulin (1993); type: S2, Shabeellaha Dhexe Region, 39 km NE of Muqdisho along road to Warshiikh, 2°12'N, 45°38'E, Thulin, Hedrén & Dahir 7186 (UPS holo.). Fig. 30.

Plant slender, from a short rhizome; roots white, 5–14 cm long, somewhat fleshy and to 2 mm thick, without tubers. Leaves several, basal, filiform or very narrowly linear, 50–130 × 1–2 mm, folded, glabrous except for minutely denticulate margins, often strongly curved. Inflorescence a very lax simple raceme, 5−24 cm long, prostrate or subprostrate, glabrous; flowers 1−2 at each node; pedicels 3−4 mm long, articulated near the middle or in the lower half, ± curved. Tepals white, medially greenish, c. 2.8 mm long, 3-nerved, spreading. Stamens subequal. Style c. 0.6−0.8 mm long, fairly long persistent. Capsule c. 3.5 × 5 mm, with the spreading persistent tepals at the base.

Coastal dunes on white or pale orange sand; 10−30 m. C1; S2; not known elsewhere. Thulin & Dahir 6601; Thulin, Hedrén & Dahir 7239; Moggi & Bavazzano 375.

9. **C. inconspicuum** (Bak.) Nordal (1993); *Anthericum inconspicuum* Bak. (1877); type: N2, "Ahl Mts" above "Mait", Hildebrandt 1469 (K holo.).

Plant up to c. 20 cm tall, from a short fibrous rhizome; roots thick, fleshy. Leaves several in a rosette, very narrowly linear, up to c. 2 mm wide, folded, glabrous, except for minutely denticulate margins. Inflorescence a simple slender raceme, very lax, 5−14 cm long, prostrate to ascending, glabrous, the peduncle often much reduced and flowers/fruits then found within the leaf rosette; flowers 1−2 at each node; pedicels 6−8 mm long, articulated near the middle or in the lower half. Tepals white, 3−5 mm long, 3-nerved, spreading. Stamens subequal. Style c. 1.5 mm long, persistent. Capsule about as long as wide to usually longer than wide, c. 5−8 × 5 mm, with spreading persistent tepals at the base.

Deciduous bushland, usually in rocky places; c. 250−1000 m. N1−3; S1; Ethiopia, Kenya. Gillett 4936; Bally & Melville 15799.

Plants from Yemen and Oman identified as *C. laxum* R. Br. are close to *C. inconspicuum*, but differ somewhat in capsule shape and leaf width. The relationship between *C. inconspicuum* and *C. laxum* (with type from Australia) needs further study.

A record of *Anthericum inconspicuum* from Jasiira near Muqdisho by Raimondo & al. in Webbia 35: 221 (1981) was based on material of *C. filifolium*.

10. **C. tuberosum** (Roxb.) Bak. (1876); *Anthericum tuberosum* Roxb. (1800).
C. russii Chiov. (1916); type: S3, "Giumbo", Paoli 275 (FT lecto.).
C. kulsii Cuf. (1969).
Da-ai (Som.).

Plant robust, up to at least 50 cm tall, with a fibrous rhizome; roots wiry, ending in ellipsoid tubers. Leaves several in a rosette, linear-lanceolate, up to at least 30 mm wide, ciliate. Inflorescence usually a simple raceme, dense, glabrous; peduncle leafless; flowers usually 2 at each node; pedicels up to c. 6 mm long, articulated near the middle. Tepals white, at least 15 × 6 mm, with 9 or more nerves, forming a cup-shaped perianth. Stamens shorter than tepals. Style c. 10 mm long. Capsule oblong-obovoid, up to c. 10 mm long.

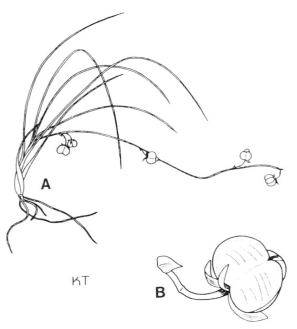

Fig. 30. *Chlorophytum filifolium*. A: habit, × 0.75. B: capsule with pedicel and bract, × 4.5. − From Nord. J. Bot. 13: 264 (1993). Drawn by K. Thunberg.

In grassland and bushland, often in wet depressions on clay; 30−400 m. C2; S1−3; Ethiopia and west to Nigeria and south to N Tanzania, also in India. Thulin & Warfa 4461; Thulin 6327; Kuchar 16971.

A record of *C. longifolium* Schweinf. ex Bak. from southern Somalia in Cuf. Enum.: 1535 (1971) was based on Corradi 4635 that is *C. tuberosum*.

11. **C. ramosissimum** Nordal & Thulin (1993); type: C2, Hiiraan Region, 17 km S of "Mugakori", 3°57'N, 46°11'E, Gillett & Beckett 23292 (K holo.). Fig. 31 A−C.
Da'ai (Som.).

Plant 15−45 cm tall, from an elongated, moniliform, fibrous rhizome; roots slender, ending in ellipsoid tubers. Leaves several in a rosette, linear, 120−250 × 2.5−4 mm, usually folded, minutely scabrous along margins and nerves, or glabrous. Inflorescence a much-branched compound raceme; peduncle minutely scabrous or glabrous, leafless; flowers 1−4(−8) at each node; lower pedicels 15−50 mm long, upper pedicels shorter, stiffly spreading, articulated 1−1.5 mm below the flowers, papillose or glabrous. Buds globose. Tepals white, 5.5−7 mm long, 3-nerved, spreading. Stamens subequal. Style c. 2.5−4.8 mm long. Capsule 2.2−3 × 4.5−5.5 mm, expanding laterally through the persistent perianth that forms a slightly twisted apical beak.

Deciduous bushland on sand; 150−300 m. ?N1; C2; E Ethiopia. Thulin & Warfa 5340; Thulin & Dahir 6471; Kuchar 17550.

Fig. 31. A−C: *Chlorophytum ramosissimum*. A: habit, × 0.75. B: flower, × 4.5. C: capsule, × 4.5. − D: *C. nervosum*, capsule with pedicel and bract, × 4. − Modified from Nord. J. Bot. 13 (1993). Drawn by K. Thunberg.

Thulin & Dahir 6644, from open coastal dunes at 5°20'N, 48°15'E near Hobyo in C1, differs from *C. ramosissimum* in having up to 5 mm wide leaves distinctly ciliate near the base, more robust pedicels up to 15 mm long only, and larger flowers with 7−10 mm long tepals. Mature fruits are not known. More material may show that this is a distinct taxon.

12. C. zavattarii (Cuf.) Nordal (1993); *Anthericum zavattarii* Cuf. (1939).

Plant up to at least 60 cm tall, from a short fibrous rhizome; roots fleshy, without tubers. Leaves in a rosette, lanceolate to broadly elliptic, 20−80 mm wide, glabrous. Inflorescence much branched, lax, glabrous; peduncle leafless; flowers usually 2 at each node; pedicels 12−40 mm long, articulated in upper part. Tepals white, c. 8 mm long, 3-nerved, spreading. Stamens subequal. Style c. 6 mm long. Capsule c. 3−3.5 × 4−4.5 mm, with the spreading persistent tepals at the base.

Acacia-Commiphora bushland; c. 250 m. S1; S Ethiopia, N Kenya. Thulin & Bashir Mohamed 6925.

The only collection seen from Somalia was made 5 km from Garbaharrey along road to Luuq.

13. C. parvulum Chiov. (1928); type: N3, "Hafun", Puccioni & Stefanini 14 (FT holo.).

Baar (Som.).

Plant up to c. 12 cm tall, from a densely fibrous rhizome; roots ending in ellipsoid tubers. Leaves in a rosette, lanceolate, 5−20 mm wide, with very prominent papillose nerves, ciliate. Inflorescence a simple raceme, dense, glabrous or papillose; peduncle leafless; flowers 2−4 at each node; pedicels 5−8 mm long, articulated in lower half. Tepals white, 7−9 mm long, 3-nerved, spreading. Style c. 6 mm long. Capsule rounded to broadly oblong in outline, 7−13 × 7−11 mm.

Dunes, stony places, silty plains; up to 530 m. N3; not known elsewhere. Beckett 511; Kazmi, Mohamed & Hussein 5723.

14. C. gallabatense Schweinf. ex Bak. (1876).

C. ukambense Bak. (1898).
C. rivae Engl. (1902).
C. ginirense Dammer (1905).
C. elachistanthum Cuf. (1969).

Plant up to at least 60 cm tall, from a short rhizome, with slender roots bearing tubers usually on lateral short branches. Leaves in a rosette, several, lanceolate, 17−60 mm wide, glabrous. Inflorescence usually branched, glabrous; peduncle leafless; flowers often 2 at each node; pedicels 1−4 mm long, articulated near the middle or in upper part. Tepals greenish or whitish, 5−6 mm long, 3-nerved, spreading-reflexed. Stamens subequal. Style c. 5.5 mm long. Capsule c. 5 × 8 mm.

Bushland or woodland; up to c. 500 m. S1, 3; Ethiopia, widespread in tropical Africa south to Zimbabwe. O'Brien 164, Pauli 1255; Senni 576.

A record of *C. schweinfurthii* Bak. from southern Somalia in Cuf. Enum.: 1536 (1971) is based on material of *C. gallabatense*.

15. C. subpetiolatum (Bak.) Kativu (1993); *Anthericum subpetiolatum* Bak. (1876).

A. monophyllum Bak. (1878).

Plant up to at least 40 cm tall, from a densely fibrous sometimes moniliform rhizome with fleshy roots without tubers. Leaves ± in 2 rows, linear, up to at least 10 mm wide, the outer ones often much shorter than the others, glabrous or ciliate. Inflorescence a simple raceme, fairly dense, glabrous, papillose or pubescent; peduncle leafless; flowers 1−3 at each node; bracts c. 6−10 mm long; pedicels 4−7 mm long, articulated near the middle. Tepals white, c. 12 mm long, 3-nerved, spreading. Stamens subequal. Style c. 10 mm long. Capsule 5−7 x 5−6 mm.

Degraded bushland around granitic outcrop; 300 m. S1 (Buur Heybo); Ethiopia and west to Nigeria and south to Zimbabwe and Mozambique. Thulin, Hedrén & Dahir 7483.

16. **C. zanguebaricum** (Bak.) Nordal (1993); *Anthericum zanguebaricum* Bak. (1876).
C. calateifolium Chiov. (1932); type: S3, "Dalleri", Senni 218 (FT holo.).
Anthericum subpapillosum Poelln. (1944).

Plant up to at least 100 cm tall, from a densely fibrous short rhizome with wiry roots ending in ellipsoid tubers. Leaves in 2 rows, linear to linear-lanceolate, up to 50 mm wide, glabrous. Inflorescence a simple raceme or with some short branches below, dense, glabrous; peduncle flattened and winged, leafless; flowers 1−3 at each node; pedicels 8−10 mm long, articulated in lower half. Tepals white, 10−15 mm long, 5−7-nerved, spreading. Stamens subequal. Style c. 10 mm long, downcurved. Capsule subglobose, c. 8 mm in diam., with numerous prominent transverse ridges.

In fixed reddish sand or shallow soil overlying rocks; c. 20−50 m. S3; East Africa and west to Central African Republic. Gillett & al. 24925C; Bavazzano & Tardelli 900.

C. zanguebaricum is very close to *C. cameronii* (Bak.) Kativu and hardly differs in anything but the lack of purplish spottings on the outer leaf bases. Further studies of populations in the field may show that the two are conspecific. Also the the relation between *C. zanguebaricum* and *C. pterocaulon* (Welw. ex Bak.) Kativu should be further studied.

17. **C. nervosum** Nordal & Thulin (1993); type: N2, Sanaag Region, "Shimber Beris (Surud)", Newbould 7687 (K holo.). Fig. 31 D.

Plant up to 30 cm tall, from an elongated moniliform fibrous rhizome; roots slender, ending in elongate tubers. Leaves 5−9, ± in 2 rows, linear, 70−240 x 2−6 mm, folded, usually ± curved, minutely scabrous along margins and sometimes the major veins. Inflorescence an unbranched raceme, rhachis glabrous; peduncle terete, not winged, glabrous or very sparsely papillate, sometimes with a few sterile bracts just below the inflorescence; flowers 1−2 at each node; pedicels 3.5−8 mm long, erect, articulated near the base, glabrous. Tepals white or with pink tips, 6−10 mm long, 3-nerved. Stamens subequal. Style c. 5.5 mm long. Capsule 3.5−6 x 5−6.5 mm, with 4−8 very prominent raised transverse ridges, expanding laterally through the persistent perianth that tends to form a slightly twisted apical beak.

In hollows between limestone blocks; c. 2040 m. N2 (NW of Cerigaabo); Yemen.

In Somalia only known from the type collection that was cited as *Anthericum subpapillosum* in Cuf. Enum.: 1532 (1971).

18. **C. tordense** Chiov. (1916); type: S3, "Torda", Paoli 3213 (FT holo.).

Plant up to at least 25 cm tall, from a fibrous ± horizontal moniliform rhizome with wiry roots ending in ellipsoid tubers. Leaves ± in 2 rows, linear, up to c. 8 mm wide, often ciliate. Inflorescence a simple raceme, or with short branches below, fairly lax; peduncle leafless, ± curved below, ± densely papillose-pubescent; flowers 1−3 at each node; pedicels 3−12 mm long, articulated near the base. Tepals white, c. 8−10 mm long, 3-nerved, spreading. Stamens subequal. Capsule erect, subglobose to obovoid, with weak transversal nerves, with remnants of the perianth at the base.

Acacia-Commiphora bushland; up to c. 350 m. S1, 3; S Ethiopia, Kenya, Uganda. Thulin, Hedrén & Dahir 7665.

152. HYACINTHACEAE

by M. Thulin

Cuf. Enum.: 1552−1562 (1971), *Liliaceae* in part.

Herbs with a bulb. Leaves all basal, flat, spirally set, sheathing at the base, entire. Inflorescence simple or branched; bracts 1 or 2 per flower. Flowers bisexual, regular, in Somalia whitish, greenish, yellowish, brownish, pinkish or bluish; tepals 6, free or united. Stamens 6, inserted at the base of the tepals; filaments often broad and flattened, sometimes toothed; anthers 2-thecous, dorsifixed, longitudinally dehiscent. Ovary superior, 3-celled with septal nectaries; style simple; ovules (1−)2−numerous per cell. Fruit a loculicidal capsule. Seeds rounded to angular, with a straight or slightly curved embryo, with endosperm.

Some 40 genera and 900 species, widespread but best represented in Southern Africa and in a region from the Mediterranean to South-West Asia.

1. Bracts, at least the lower ones, with a basal spur; leaves usually produced after the flowers ....3. *Drimia*
  − Bracts not spurred, or absent; leaves produced ± at the same time as the flowers 2
2. Inner tepals erect, the outer ± spreading, 8 mm or more long 3
  − All tepals similar or, if inner erect and outer ± spreading, tepals less than 6 mm long 4
3. Tepals free, persistent, cream to yellow, without appendages .... 1. *Albuca*
  − Tepals united at the base, falling off after flowering, dull green or brown, the outer usually with a tail-like appendage .... 2. *Dipcadi*
4. Style c. 1 mm long, deciduous; pedicels up to 1 mm long; bracts absent .... 5. *Drimiopsis*
  − Style much longer, persistent; pedicels usually much longer; bracts present 5
5. Leaves linear, not spotted .... 4. *Ornithogalum*
  − Leaves ovate to oblanceolate or elliptic, often spotted purplish or dark green .... 6. *Ledebouria*

1. ALBUCA L. (1762)

Knudtzon & Stedje in Nord. J. Bot. 6: 773−786 (1986).

Leaves produced at the same time as the flowers, filiform to lanceolate, glabrous or pubescent. Inflorescence a raceme; peduncle glabrous; bracts lanceolate to ovate-acuminate, not spurred; pedicels erect to ascending. Tepals free, persistent, hooded at the apex, the inner ones erect, the outer ± spreading. Stamens free, at least the inner with an expanding base clasping the ovary. Ovary 3-celled with many ovules per cell; style slender to thick, with terminal stigma. Capsule ovoid. Seeds black, flattened, subcircular.

Some 50 species in tropical and southern Africa, and in SW Arabia.

A record of *A. subspicata* Bak. in Chiov., Result. Sci. Miss. Stefanini-Paoli: 175 (1916) is based on a narrow-leaved form of *A. abyssinica*.

A. abyssinica Jacq. (1783). Fig. 32.
  Ornithogalum melleri Bak. (1873); *Albuca melleri* (Bak.) Bak. (1898).
  A. wakefieldii Bak. (1879).
  A. chaetopoda Chiov. (1928); types: N3, Puccioni & Stefanini 751, 810, 825, 858 (FT syn.).
  A. asclepiadea Chiov. (1932); type: S3, "Afmadu", Gorini 88 (FT holo.).
  Baar, ged adais (Som.).

Plant 10−150 cm tall; bulb-scales with or without fibrous apex. Leaves linear to lanceolate, ± ciliate, pubescent or glabrous. Racemes 4−many-flowered; pedicels 2−20 mm long. Tepals 8−35 mm long, cream to yellow with a green midrib outside. Style slender, 1.5−3 times as long as the ovary. Capsule 10−20 mm long. Seeds 3−7 mm in diam.

Bushland, woodland, grassland in rocky, sandy or silty places; 10−2175 m. N1−3; C1, 2; S1−3; widespread in tropical Africa, and in SW Arabia. Gillett 4724; Thulin, Hedrén & Dahir 7187; Bally 4099.

Very variable, particularly in size, robustness and leaf width.

2. DIPCADI Medik. (1790)

Leaves produced at the same time as the flowers, linear to linear-lanceolate, glabrous or pubescent. Inflorescence a few- to many-flowered, often 1-sided raceme; rhachis drooping in bud, becoming erect during anthesis; bracts soon falling or persistent, not spurred; pedicels short to long. Tepals usually dull green or brown, falling off after flowering, united at the base, the 3 inner erect, the 3 outer usually spreading and often with a tail-like appendage at the tip. Stamens 6, united to tepals at the base, included. Ovary 3-celled with several ovules per cell; style short or long with terminal stigma. Capsule ovoid. Seeds black, flat, rounded in outline.

Some 30 species in Africa, Madagascar, the Mediterranean Region, and eastwards to India.

D. viride (L.) Moench (1802); *Hyacinthus viridis* L. (1762). Fig. 33.
  Baar (Som.).

Glabrous plant of very variable size. Leaves 1−several per shoot, narrowly linear to linear-lanceolate, sometimes spirally twisted or with crinkly margins. Raceme few- to many-flowered; peduncle smooth, terete; bracts lanceolate; pedicels to 18 mm long in fruit. Tepals green or brownish, c. 8−13 mm long, with a short to long tail-like appendage. Style slender, about as long as the ovary. Capsule up to c. 15 × 15 mm. Seeds c. 5−6 mm across.

In a wide variety of habitats in bushland, woodland or grassland, also on dunes; 10−1200 m. N1−3; C1; S1−3; Djibouti, widespread in tropical Africa. Thulin & Bashir Mohamed 6766; Gillett 4894; Alstrup & Michelsen 19.

As conceived here, this is an extremely variable species, sometimes with two or more seemingly distinctive forms occurring within a small area, as for example around Mogadishu. However, to delimit more narrowly circumscribed taxa is scarcely possible at present. Some of the material from southern Somalia has been identified as *D. lanceolatum* Bak. (Cufodontis 1971).

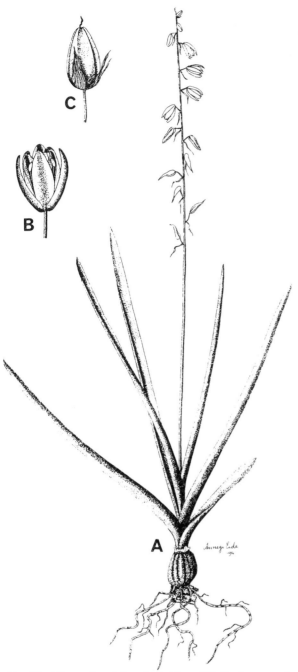

Fig. 32. *Albuca abyssinica*. A: habit, × 0.3. B: flower, × 1.
C: capsule, × 1. – From Nordal 2228. Drawn by A. Eide.

3. DRIMIA Jacq. (1797)
Urginea Steinh. (1834)
Stedje in Nord. J. Bot. 7: 655–666 (1987).

Leaves usually produced after the flowers, filiform to lanceolate, sometimes ciliate. Inflorescence a lax to dense raceme; peduncle glabrous, erect; bracts, at least the lower ones, with a basal spur; pedicels erect

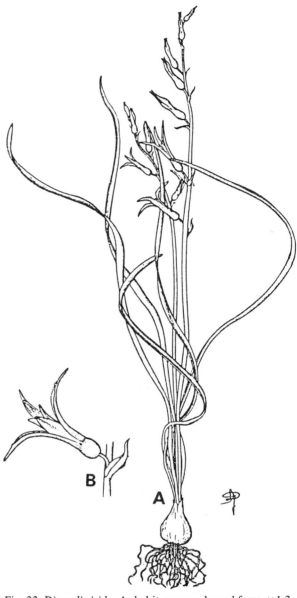

Fig. 33. *Dipcadi viride*. A: habit, narrow-leaved form, × 1.3.
B: flower, × 2.5. – Drawn in Palermo from Somali material.

to spreading. Tepals free or united. Stamens free or united to tepals at the base. Ovary 3-celled with many ovules per cell; style filiform with small terminal stigma. Capsule ovoid to globose. Seeds black, usually flat, subcircular in outline and winged.

More than 100 species in Africa, the Mediterranean Region and eastwards to India.

1. Raceme lax, with up to 25 flowers and less than 20 cm long; tepals brownish .... 1. *D. indica*
– Raceme dense, with numerous flowers and more than 20 cm long; tepals whitish with dark midnerve outside ....2. *D. altissima*

Fig. 34. *Drimia altissima*. A: bulb and inflorescence. B: flower. C: stamen. D: fruit. — Modified from Fl. W. Trop. Afr. 3(1): Fig. 352 (1968). Drawn by W. E. Trevithick.

1. D. indica (Roxb.) Jessop (1977); *Scilla indica* Roxb. (1824); *Urginea indica* (Roxb.) Kunth (1843).

Plant up to 50 cm tall. Leaves linear, up to 30 cm long, 1–18 mm wide. Inflorescence a very lax raceme with 5–25 flowers; bracts up to 2 mm long with spur up to 2 mm long; pedicels 12–30 mm long. Tepals free or united for up to 1.5 mm, 6–11 mm long, greenish-brown or pale brown, becoming reflexed. Filaments linear, free or basally united to tepals. Ovary ovoid, 3–4 mm long; style 4–5 mm long. Capsule ellipsoid, 8–18 mm long. Seeds 7–10 mm long.

Sandy or stony soil in *Acacia-Commiphora* bushland; c. 250 m. S1; Djibouti, widespread in tropical Africa and eastwards to India. Thulin & Bashir Mohamed 6928.

The single collection seen from Somalia consists of very depauperate few-flowered plants.

2. D. altissima (L.f.) Ker-Gawl. (1808); *Ornithogalum altissimum* L. f. (1782); *Urginea altissima* (L. f.) Bak. (1873). Fig. 34.
Drimia paolii Chiov. (1916); type: S1, "El Ualac", Paoli 1096 (FT holo.).
Baar (Som.).

Plant 35–200 cm, usually robust. Leaves lanceolate, 20–50 × 2–7.5 cm. Inflorescence a dense raceme up to 80 cm long with up to 700 flowers; bracts up to 14 mm long with spur up to 3 mm long; pedicels 8–30 mm long, spreading. Tepals free or united for up to 2 mm, 5–12 mm long, whitish with dark midnerve outside, spreading. Filaments linear to very narrowly triangular, free or basally united to tepals. Ovary ovoid, 2–5 mm long; style about as long as ovary. Capsule subglobose, 8–15 mm long. Seeds 5–8 mm long.

Bushland and woodland, usually on rocky ground; 200–1800 m. N2, 3; C2; S1; Djibouti, widespread in tropical and southern Africa. Boaler 51; Kuchar 16962; Sammicheli s.n.

4. ORNITHOGALUM L. (1753)
Stedje & Nordal in Nord. J. Bot. 4: 749–759 (1985).

Leaves produced at the same time as the flowers, linear to ovate, glabrous or pubescent. Inflorescence a raceme; peduncle erect, glabrous; bracts lanceolate to ovate-acuminate, not spurred. Tepals free or almost so, hooded at the apex, ± spreading. Stamens with filaments flattened. Ovary sessile or shortly stipitate, 3-celled with 2–many ovules per cell; style slender, erect or directed to one side, or absent; stigma terminal. Capsule ovoid. Seeds black, flattened and subcircular or angular.

More than 100 species in Africa, Europe and Asia.

1. Tepals 12–16 mm long, midrib of 3 or 5 distinct vascular bundles; ovary with a short wide stipe; pedicels 20–45 mm long ....1. *O. donaldsonii*
− Tepals 5–12 mm long, midrib of many vascular bundles; ovary sessile; pedicels up to 15 mm long .... 2. *O. tenuifolium*

1. O. donaldsonii (Rendle) Greenway (1969); *Albuca donaldsonii* Rendle (1896). Fig. 35.
Urginea somalensis Chiov. (1932); types: S3, "Afmadu", Gorini 40 (FT syn.) & "Gobuin" and "Chisimaio", Gorini 309 (FT syn.).

Plant up to 1.5 m tall, glabrous; bulb up to 7.5 cm across. Leaves linear, up to 60 cm long. Raceme 25–300-flowered; pedicels 20–45 mm long. Tepals 12–16 mm long, whitish; midrib green, of 3 or 5 vascular bundles. Filaments 8–11 mm long, narrowly triangular. Ovary ovoid, with a 0.5–1 mm long stipe. Capsule 12–16 mm long. Seeds 6–8 mm in diam.

Bushland, woodland or grassland, often on sand; 50–350 m. C1; S1, 3; E Ethiopia, Kenya, N Tanzania. Thulin & Dahir 6571; Thulin, Hedrén & Dahir 7358, 7661.

2. O. tenuifolium Delaroche (1811).

O. sordidum Bak. (1895); *O. tenuifolium* subsp.
sordidum (Bak.) Stedje in Nord. J. Bot. 4: 758 (1984);
type: N1, "Golis range, Woob", Cole s.n. (K holo.).

Plant up to 2 m tall, glabrous; bulb up to 15 cm
across. Leaves linear, up to 130 cm long. Raceme
15—300-flowered; pedicels 2—15 mm long. Tepals
5—12 mm long, whitish; midrib of many vascular
bundles. Filaments 3—8 mm long, ovate, often with a
tooth on each side. Ovary ellipsoid, sessile. Capsule
5—13 mm long. Seeds 4—6 mm across.

Bushland or woodland; c. 450—1500 m. N1; S1;
Ethiopia, widespread in tropical and southern Africa.
Thulin & Bashir Mohamed 6842.

This has often been called *O. longibracteatum* Jacq.
in East Africa, but this apparently is a species confined to the Cape. *O. tenuifolium* is very variable and
has been divided into several subspecies, some of
them differing in chromosome number (see Stedje &
Nordal 1984). The type of *O. sordidum* is a small
plant, while Thulin & Bashir Mohamed 6842, a plant
growing in bushes, is almost 2 m tall. They would fall
in subsp. *sordidum* and subsp. *robustum* Stedje, respectively, on morphological grounds, but chromosome numbers are not known from any of them.

5. DRIMIOPSIS Lindl. & Paxt. (1851)
Stedje in Nord. J. Bot. 14: 45—50 (1994).

Leaves produced ± at the same time as the flowers,
linear-lanceolate to ovate, sometimes spotted,
glabrous. Inflorescence a many-flowered spike or
raceme; peduncle erect; bracts obsolete; pedicels
very short or absent. Tepals united at the base,
hooded at the tips, forming a campanulate often
persistent perianth. Stamens united to tepals at the
base, included. Ovary sessile, 3-celled with 1—2 basal
ovules per cell; style terete with small terminal
stigma. Capsule subglobose, thin-walled. Seeds
black, shiny, subglobose.

Some 15 species in Africa.

1. Leaves linear-lanceolate, up to 2 cm wide, both
   outer and inner tepals erect ....1. *D. barteri*
– Leaves ± ovate or broadly elliptic, usually 4 cm or
   more wide; outer tepals spreading, inner erect
   ....2. *D. botryoides*

1. D. barteri Bak. (1870). Fig. 36.

Leaves erect or ascending, linear-lanceolate, up to
20 × 2 cm or more, sometimes spotted. Inflorescence
a dense spike-like raceme; peduncle to c. 15 cm long;
pedicels c. 0.5 mm long. Tepals c. 2.5—3.5 mm long,
greenish-white, all erect. Filaments triangular. Ovary
subglobose; style c. 1 mm long, deciduous. Capsule
subglobose-obovoid, 3-angled, c. 4 mm long. Seeds
c. 3 mm long.

Bushland or woodland, usually on sandy ground;

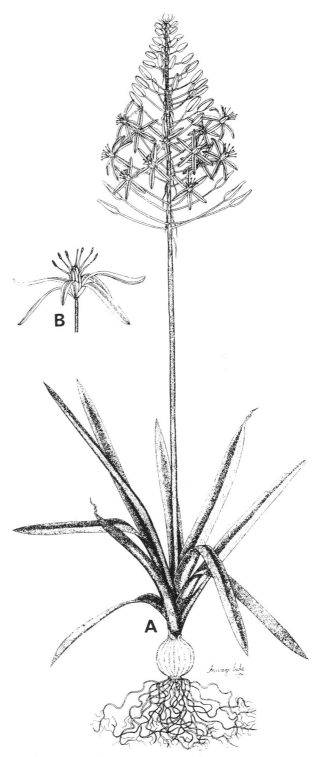

Fig. 35. *Ornithogalum donaldsonii*. A: habit, × 0.4. B:
flower, × 1.5. – From Nordal 2283. Drawn by A. Eide.

15—300 m. S1—3; tropical Africa westwards to Ghana
and Nigeria. Thulin & Warfa 4493; Tardelli 56;
Thulin, Hedrén & Dahir 7207.

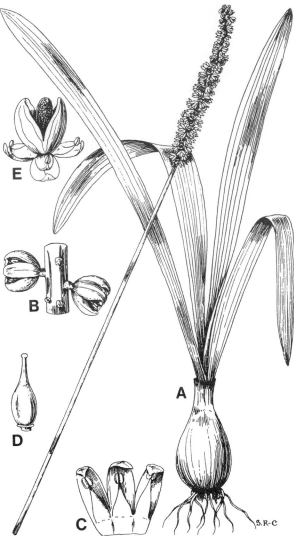

Fig. 36. *Drimiopsis barteri*. A: habit. B: portion of inflorescence. C: tepals and stamens. D. pistil. E: capsule and seed. – Modified from Fl. W. Trop. Afr. 3(1): Fig. 353 (1968). Drawn by S. Ross-Craig.

2. D. botryoides Bak. (1874).

D. erlangeri Dammer (1905).

Leaves erect to ± flat on the ground, ovate to broadly elliptic, narrowing below into a petiole-like base, up to 18 × 8 cm or more, spotted. Inflorescence a dense spike-like raceme; peduncles erect or ± curved, to 15 cm or more long; pedicels c. 1 mm long. Tepals c. 3–6 mm long, greenish-white, the outer spreading, the inner erect. Filaments triangular. Ovary subglobose; style c. 1 mm long, deciduous. Capsule subglobose-obovoid, 3-angled, c. 4–5 mm long. Seeds c. 3–4 mm long.

Bushland or woodland; 60–500 m. S1–3; S Ethiopia, East Africa. Friis, Alstrup & Michelsen 4675; Tardelli 195; Corradi 4639.

6. LEDEBOURIA Roth (1821)

Jessop in J. S. Afr. Bot. 36: 233–266 (1970).

Leaves produced at the same time as the flowers, often with purplish or dark green spots, especially above. Inflorescence a simple raceme, usually flexuose; bracts single or paired, not spurred. Tepals ± united at the base and forming a shallow cup, narrowing upwards and often reflexed during anthesis. Stamens united to tepals near the base, free from one another, the inner ones longer than the outer. Ovary subtruncate above, ± 6-angled, abruptly narrowing below into a short stipe, 3-celled with 2 basal ovules per cell; style slender with small terminal stigma. Capsule depressed-globose to obovoid. Seeds obovoid, black.

Genus of some 15–20 species in Africa and S Asia, the majority of the species in South Africa.

Ledebouria has often been included in *Scilla* L., but differs in the usually spotted leaves, the cup-shaped base of the perianth, the longer inner stamens, and in the truncate, stipitate ovary with 2 basal ovules per cell only.

1. Tepals at least 9 mm long 2
 - Tepals up to 7 mm long 3
2. Pedicels at least 5 mm long ....1. *L. kirkii*
 - Pedicels up to 3 mm long ....2. *L. somaliensis*
3. Leaves lanceolate, narrowing below into a petiole-like base, often spotted but without papillae.........
 .... 3. *L. revoluta*
 - Leaves ± ovate, cordate and clasping at the base, often with purplish papillae above ..4. *L. cordifolia*

1. **L. kirkii** (Bak.) Stedje & Thulin (1995); *Scilla kirkii* Bak. (1873).

Drimia hildebrandtii Bak. (1892).
D. angustitepala Engl. (1892).
Scilla hildebrandtii Bak. (1898).
Urginea corradii Chiov. (1951); type: S1, "Baidoa", Corradi 4632 (FT holo.).

Bulbs large, up to 8 cm or more in diam. Leaves several, ovate to oblanceolate or elliptic, up to 40 × 9 cm or more, narrowing below into a petiole-like base, often spotted purplish or dark green, glabrous. Inflorescence many-flowered, up to 45 cm long including up to 25 cm long peduncle, ± lax; bracts ± subulate from a broader base, c. 0.8–2 mm long, often paired; pedicels 5–15(–20) mm long, ± thick, often pink when young. Tepals united at base, narrowly oblong, reflexed at anthesis for most of the length, greenish and sometimes tinged with purple, 9–14 mm long, obtuse at the tip. Stamens almost as long as tepals; filaments filiform, united to tepals at the base, purplish or white. Ovary deeply 3-lobed, 6-angled and sometimes with 6 horn-like appendages; style about as long as tepals. Capsule depressed-globose.

Deciduous bushland; 0–400 m. S1–3; Ethiopia,

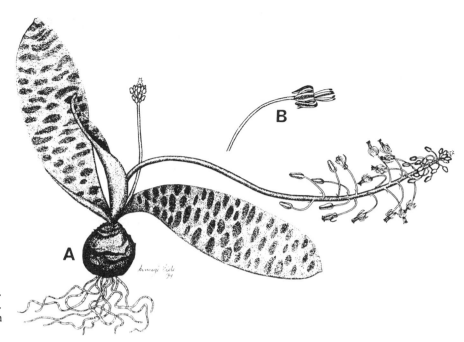

Fig. 37. *Ledebouria revoluta.*
A: habit, × 0.8. B: flower, × 2.
− From Nordal 1123. Drawn
by A. Eide.

Kenya, Tanzania. Thulin & Warfa 4456; Lavranos & Carter 23185; Paoli 932.

2. **L. somaliensis** (Bak.) Stedje & Thulin (1995); *Scilla somaliensis* Bak. (1892); type: N2, near "Meid", Hildebrandt 1470 (K holo.).

Drimia coleae Bak. (1897); type: N1, "Golis range", Cole s.n. (K holo.).

D. confertiflora Dammer (1905); type: S1, "Malkare", Ellenbeck 2150a (B holo.).

Dhego looyo, diado (Som.).

Bulbs large, up to 8 cm or more in diam. Leaves several, ovate to oblanceolate or elliptic, up to 20 × 7.5 cm, narrowing below into a petiole-like base, often spotted purplish or dark green, glabrous. Inflorescence many-flowered, up to 30 cm long including up to 15 cm long peduncle, ± dense; bracts ± subulate from a broader base, c. 2−3.5 mm, often paired; pedicels 2−3 mm long, ± thick. Tepals united at base, narrowly oblong, reflexed at anthesis for most of the length, greenish, 9−13 mm long, obtuse at the tip. Stamens almost as long as tepals; filaments filiform, united to tepals at the base, purplish or white. Ovary deeply 3-lobed, 6-angled and sometimes with 6 horn-like appendages; style about as long as tepals. Capsule depressed-globose.

Deciduous bushland; 50−1000 m. N1−2; C1; S1, 2; Ethiopia, Kenya. SMP 217; O'Brien 74; Gillett 4188.

Used against snake bite.

3. **L. revoluta** (L.f.) Jessop (1970); *Hyacinthus revolutus* L.f. (1782). Fig. 37.

L. hyacinthina Roth (1821); *Scilla indica* Bak. (1870), nom. illegit.; *S. hyacinthina* (Roth) J. F. Macbride (1918).

Drimia brevifolia Bak. (1898); type: near S1/ Ethiopia border, "Dolo", Riva 1251 (B holo., FT iso.)

Bulbs up to 5 cm or more in diam. Leaves 1 to several, lanceolate or narrowly elliptic, up to 15 × 3 cm, narrowing below into a petiole-like base, often spotted purplish or dark green, glabrous. Inflorescence many-flowered, up to 25 cm long including up to 15 cm long peduncle, ± lax; bracts ± subulate from a broader base, c. 1 mm, often paired; pedicels 1−10 mm long, often pink when young. Tepals united at base, narrowly oblong, reflexed at anthesis for most of the length, greenish and sometimes tinged with purple, 3−7 mm long, obtuse at the tip. Stamens almost as long as tepals; filaments filiform, united to tepals at the base, purplish or white. Ovary deeply 3-lobed, 6-angled and sometimes with 6 horn-like appendages; style about as long as tepals. Capsule depressed-globose, c. 5 mm or more long.

Deciduous bushland; 0−1700 m. N1−3; C2; S1; widespread in tropical and southern Africa, also in India and Sri Lanka. Tribe 22; Bally 7240; Gillett 4753.

4. **L. cordifolia** (Bak.) Stedje & Thulin (1995); *Scilla cordifolia* Bak. (1898).

Eriospermum somalense Schinz (1896).

Scilla carunculifera Chiov. (1916); type: S1, between "Usciacca Guràn" and "El Uré", Paoli 1045 (FT holo.).

S. carunculifera var. *glandulosa* Chiov., Fl. Somala 2: 429 (1932); *S. glandulosa* (Chiov.) Chiov. (1951); type: S3, near "Giumbo", Senni 203 (FT holo.).

S. glandulosa forma *major* Chiov. in Webbia 8: 29 (1951); type: S1, Corradi 4628 (FT lecto.).

Balla bioto (Som.).

Bulbs small, usually c. 1—1.5 cm in diam. Leaves 1(—2), often flat on the ground, ovate, up to 9 x 7 cm, cordate and clasping at the base, often with prominent globose to elongate ± purplish papillae above. Inflorescence slender, many-flowered, up to c. 15 cm long including up to 5 cm long peduncle, glabrous; bracts subulate, 0.5—1.5 mm long; pedicels c. 4—12 mm long. Tepals united at the very base, oblong, pale pink, 3—4 mm long. Stamens almost as long as tepals; filaments subulate, united to tepals at the base. Ovary deeply 3-lobed, 6-angled; style about as long as tepals. Capsule depressed-globose, 1.5—4 mm long.

Shallow soil over limestone, sandy plains; 10—400 m. S1, 3; Ethiopia, Kenya, Tanzania, Zambia, Malawi, Angola. Tardelli 19, 178, 187.

153. ALLIACEAE

by M. Thulin

Perennial herbs, usually with a bulb with membranous or fibrous outer scales. Leaves basal, flat, angular, terete or fistulose, usually spirally set, sheathing at the base, entire. Inflorescences usually umbel-like, subtended by an involucre of membranous spathal bracts. Flowers generally bisexual and regular; tepals 6, free or often basally united, persistent. Stamens generally 6, inserted at the base of the tepals or in the perianth tube; filaments ± flat, sometimes lobed; anthers 2-thecous, dorsifixed, longitudinally dehiscent. Ovary superior, 3-celled; style 1; ovules 2 to several per cell. Fruit a loculicidal capsule. Seeds with straight or curved embryo, with endosperm.

Some 20 genera in most parts of the world.

ALLIUM L. (1753)
De Wilde-Duyfjes in Meded. Landbouwhogeschool Wageningen 76—11 (1976).

Plants bulbous (in Africa), with characteristic smell. Leaves flat or terete and hollow, usually linear. Umbel with flowers sometimes partly or entirely replaced by sessile bulbils. Tepals 1—3-nerved. Style gynobasic. Seeds usually 3—6, mostly angular, black.

Genus of some 700 species, mainly in the northern hemisphere, including cultivated edible plants like *A. cepa* L. (onion) and *A. sativum* L. (garlic), as well as ornamentals.

A. subhirsutum L. (1753).

Bulb globose to ovoid-oblong. Leaves linear, to 50 cm long, ciliate. Umbel few- to many-flowered, without or occasionally with a few bulbils; peduncle 10—60 cm long; spathe 1, persistent, to 25 mm long; pedicels 10—40 mm long. Flowers subcampanulate; tepals spreading, 5—8.5 mm long, the outer slightly wider than the inner. Filaments simple. Capsules subglobose, 3—6 mm in diam. Seeds c. 2.5—3.5 mm long.

subsp. **spathaceum** (Steud. ex A. Rich.) Duyfjes in Meded. Landbouwhogeschool Wageningen 76—11: 142 (1976); *A. spathaceum* Steud. ex A. Rich. (1850). Fig. 38.

Protective scales of bulb after decay of outer epidermis with conspicuous sinuate structure. Tepals white with reddish midnerve. Filaments c. 2/3 the length of the tepals.

Probably in rocky places at c. 1500 m. N1; Djibouti, Eritrea, Ethiopia, N Sudan. Donaldson Smith s.n. (BM).

Two further subspecies, subsp. *subhirsutum* and subsp. *subvillosum* (Salzm. ex Schult.) Duyfjes, both in the Mediterranean area, were recognized by De Wilde-Duyfjes (1976).

The single collection known from Somalia was made near "Adadle" in 1899.

154. AMARYLLIDACEAE

by I. Nordal
Cuf. Enum.: 1574—1577 (1971), *Amaryllidaceae* in part; Fl. Trop. E. Afr. (1982).

Bulbous perennial herbs. Leaves in basal rosette or double fan; petioles, when present, forming a false stem; blade simple, entire, linear to lanceolate or strap-shaped. Scape leafless, lateral in relation to the leaves, with 1—many flowers in an umbel-like inflorescence subtended by an involucre of 1—several, usually free bracts, and with ephemeral hyaline bracts between the flowers. Flowers showy, bisexual, regular or slightly irregular; tepals 6, equal to subequal, fused to a prominent tube. Stamens 6, inserted at the top of the tube; filaments free or united into a cup ("false corona") at the base; anthers dorsifixed, versatile, introrse, opening by longitudinal slits. Ovary superior, 3-celled, each cell with 1—several axile ovules; style long and slender, with capitate or slightly 3-lobed stigma. Fruit a capsule, with loculicidal or irregular dehiscence, or a berry. Seeds black or greyish-greenish, globose or flattened, with fleshy endosperm and small embryo.

About 60 genera and some 800 species, in warm temperate and tropical areas around the world.

A family of great horticultural importance.

1. Leaves with sheathing petioles forming a false stem; involucral bracts 4 or more; flower tube less than 3 cm long; fruit a berry ....1. *Scadoxus*
   — Leaves without petioles; involucral bracts 1—2; flower tubes more than 5 cm long; fruit a capsule, dry with thin pericarp or fleshy with thick pericarp 2
2. Leaves up to 1 cm wide; filaments fused to a false corona; capsules dry, loculicidal with glossy black seeds ....4. *Pancratium*
   — Leaves more than 2 cm wide; filaments not fused to a false corona; capsules ± berry-like with greyish/greenish fleshy seeds 3
3. Leaves rosulate; flowers irregular with curved tube and free part of tepals connivent to a funnel or bell; stamens and style curved downwards ....2. *Crinum*
   — Leaves arranged in 2 prostrate fans; flowers regular with straight tube and free part of tepals recurved; stamens and style straight or almost so ....3. *Ammocharis*

1. SCADOXUS Raf. (1838)
Friis & Nordal in Norw. J. Bot. 23: 64 (1976).

Bulbs with distinct rhizomatous part. Petioles sheathing to form a false stem. Many-flowered inflorescence subtended by 4 or more, free or partly fused involucral bracts. Flowers pink to red, on long pedicels, with a distinct narrowly cylindrical tube and spreading linear segments. Filaments red, filiform and semipatent; anthers yellow, small. Fruits red berries with 1—3 rather large seeds with pale testa.

Nine species in tropical Africa south to Natal, the only species in Somalia also reaching the southern part of the Arabian Peninsula.

The species of *Scadoxus* were up to recently referred to *Haemanthus* L., which is, however, a distinct genus restricted to South Africa.

S. multiflorus (Martyn) Raf. (1831); *Haemanthus multiflorus* Martyn (1795). Fig. 39.
   H. bivalvis Beck (1888).
   H. somaliensis Bak. (1895); type: N1, "Golis Mts.", Cole s.n. (K lecto.)
   H. zambesiacus Bak. (1898).
   Geed caddays, geed-jinni (Som.).
   Herb up to c. 50 cm tall. Leaves produced at the same time or after the flowers; false stem 5—20 cm long, often with reddish to brownish spotting at the base; blade lanceolate, acute. Scape 12—40 cm tall; involucral bracts membranous, colourless or tinged with reddish, early drooping or suberect until anthesis; inflorescence semiglobose to globose, 10—50-flowered; pedicels 1—3 cm long. Perianth, filaments and style scarlet, turning more pink when fading, segments often paler than filaments. Perianth

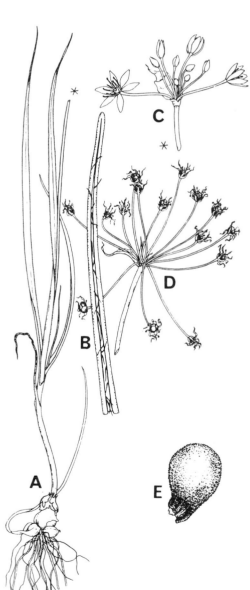

Fig. 38. *Allium subhirsutum* subsp. *spathaceum*. A: habit, inflorescence detached, × 1/3. B: tip of leaf, × 4. C: young inflorescence, × 2/3. D: old inflorescence, × 2/3. E: seed, × 8. — Modified from Meded. Landbouwhogeschool Wageningen 76—11: Fig. 24 (1976).

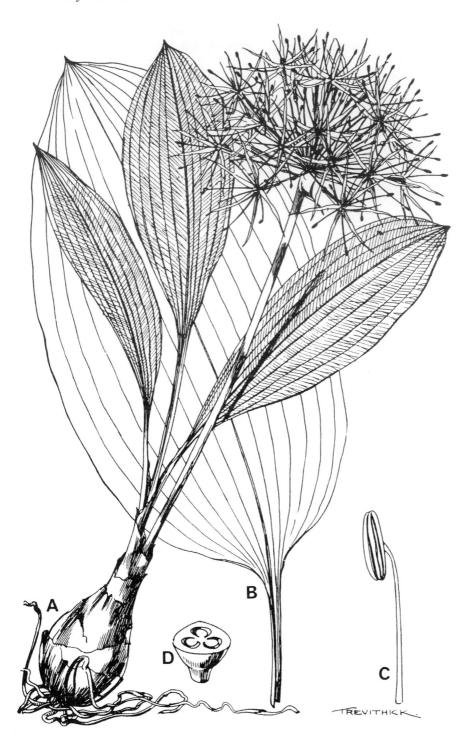

Fig. 39. *Scadoxus multi-florus*. A: habit. B: leaf. C: stamen. D: cross-section of ovary. — Modified from Fl. W. Trop. Afr. 3(1): Fig. 363 (1968). Drawn by W. E. Trevithick.

tube 0.5−1.5 cm long; segments 1.2−2.5 cm long, 0.5−2 mm broad, 1−3-nerved. Filaments 1.5−3 cm long; anthers 1−3 mm long. Berries 0.5−1 cm in diam.

Acacia-Commiphora woodland or bushland, also in *Buxus* scrubland, on blackish to greyish, sometimes loamy soils; 0−1950 m; N1−3; S1−3; Djibouti, widespread in tropical Africa, also in the SE Arabian

Peninsula. Friis, Alstrup & Michelsen 4677; Gillett 4399; Thulin & Warfa 4500.

2. CRINUM L. (1753)

Verdoorn in Bothalia 11: 27 (1973); Nordal in Norw. J. Bot. 24: 179 (1977); Nordal & Wahlstrøm in Nord. J. Bot. 2: 465 (1982).

Plants large. Leaves in rosette, strap-shaped or lanceolate, with or without a thickened midrib. Inflorescence of 1—10 flowers subtended by 2 free involucral bracts. Flowers sessile or on relatively short pedicels; tube up to 12 cm long, narrowly cylindrical, curved; segments white, with or without a reddish to pinkish dorsal streak, or with pinkish flush, narrowly to broadly lanceolate, connivent to a bell or a funnel. Filaments white or tinged pink, filiform and curved downwards; anthers light brown to grey or black, 5—10 mm long. Fruits capsules with fleshy pericarp, bursting irregularly or indehiscent, with several large greyish or greenish subglobose to irregularly compressed seeds, 5—10 mm in diam., sometimes germinating in the fruit.

Some 100 species, pantropical with about 50 species in Africa. Both native and exotic species are grown as ornamentals.

1. Leaves ± erect, with glabrous margin, mostly with intact tips; flowers 2—5, sessile, subtended by ± erect involucral bracts until anthesis 2
- Leaves prostrate or semiprostrate, with scabrid to distinctly ciliate margin, only few young leaves in the middle with intact tips; flowers (5—)6—10, subsessile or with a distinct pedicel, subtended by early withering and drooping bracts 3
2. Leaves erect, glaucous; perianth segments pure white or flushed with pink outside in apical parts .... 1. *C. abyssinicum*
- Leaves semi-erect, with drooping tips; perianth segments white with a sharply bordered broad dark reddish band, visible on both sides ....2. *C. ornatum*
3. Leaves glaucous, spreading, with scabrid undulate margin; perianth segments 2—3 cm wide; anthers grey to black .... 3. *C. macowanii*
- Leaves light green, not glaucous, prostrate, with distinctly ciliate not undulate margin; perianth segments c. 1.5 cm wide; anthers yellowish brown 4
4. Perianth strongly pigmented with red, also with reddish tube; fruits vivid orange to red; seeds smooth, greyish ....4. *C. stuhlmannii*
- Perianth white or tinged with pink only, with greenish tube; fruits green fading to yellow; seeds papillose, green, blackening when exposed to air ....5. *C. papillosum*

1. **C. abyssinicum** Hochst. ex A. Rich. (1850).
    C. schimperi Vatke ex K. Schum. (1889).
Bulbs up to 15 cm in diam., with a considerable neck, often propagating vegetatively to form dense clusters. Leaves glaucous to greyish green, erect, linear to narrowly lanceolate, mostly with intact apices, 40 × 1—3.5(—5) cm, with a distinct midrib. Scape 40—80 cm long, produced with the leaves or somewhat earlier; bracts erect for a while, papery, greyish. Flowers 2—6, sessile or almost so, heavily

scented, red to pink in bud, fading to pure white, sometimes tinged pink during anthesis, only rarely with a pink dorsal streak; tube curved, (3—)6—10 cm long; segments broadly lanceolate, 8—10 × c. 2 cm, forming a bell, with reflexed outer parts during anthesis. Filaments white, 4—6 cm long, of unequal length; anthers 6—10 mm long, curved, black to brownish. Style white, as long as the perianth segments. Fruits greenish, sometimes tinged red, with a thick fleshy pericarp, subglobose without an apical beak. Seeds not seen.

In highlands, most common in waterlogged valley grasslands and swampy depressions or on stream banks, sometimes in fallow fields, on black clayey and loamy soils. N1—3; Ethiopia. Gillett 4441; Glover & Gilliland 949; Glover in Bally 4624.

C. abyssinicum and *C. schimperi* have up to now been regarded as two distinct species, *C. abyssinicum* with 3—4 cm long perianth-tubes, and *C. schimperi* with tubes more than 6 cm long. However, this gap is bridged by some specimens and, also, the tube elongates during the development of the flower. Therefore, the two taxa are here considered conspecific.

2. **C. ornatum** (Ait.) Bury (1834—37); *Amaryllis ornata* Ait. (1789).
    C. zeylanicum sensu Fl. Trop. E. Afr.: 15 (1982), non L.
Bulbs up to 10 cm in diam., often with a considerable neck, often propagating vegetatively to form clusters. Leaves not glaucous, spreading to erect, narrowly lanceolate, mostly with intact apices, 30 × 2.5—5 cm, with distinct midrib. Scape 20—50 cm long, produced with the leaves; bracts persistent, greenish tinged red. Flowers 3—6, dark greenish red in bud; tube greenish red, curved, 8—10 cm long; perianth segments white with a broad sharply bordered dark red band, visible on both sides, broadly lanceolate, 8—10 × c. 2 cm, forming a bell, with reflexed outer parts during anthesis. Filaments white tinged red, shorter than the perianth segments; anthers dark, 8—10 mm long, curved. Style tinged red distally. Fruits greenish tinged red, with thick pericarp, subglobose without or with a very short apical beak. Seeds light green, not particularly smooth, closely stacked and irregularly compressed, 15—45 per fruit.

Along streams; c. 1000 m. N2; widespread in a transition zone between forest and savanna in tropical Africa. Collenette 415.

This was in Fl. Trop. E. Afr. regarded as conspecific with *C. zeylanicum* (L.) L., but recent studies have shown that the African *C. ornatum* and the Asian *C. zeylanicum*, although morphologically very similar, should be kept apart. *C. zeylanicum* should therefore not be lectotypified by West African material, as was done by Nordal & Wahlstrøm (1982).

Fig. 40. *Crinum macowanii*, habit, × 1/3. – From Fl. Trop. E. Afr. (1982). Drawn by A. Aarhus.

3. **C. macowanii** Bak. (1878). Fig. 40.

C. corradii Chiov. (1951); type: S1, "Baidoa", Corradi 4634 (FT holo.).

Bulbs 10–25 cm in diam., often with a considerable neck. Leaves glaucous, broadly lanceolate, 10–60 × 4–10 cm at anthesis, ± prostrate, without a distinct midrib, most leaves without intact apices. Scape 10–40 cm long, produced with the leaves; bracts greyish and papery, early drooping. Flowers 6–10, subsessile to distinctly pedicellate, heavily scented, greenish pink in bud; tube greenish to crimson, curved, 8–12 cm long; perianth segments white with a faint pink not sharply bordered dorsal band, broadly lanceolate, 8–11 × c. 2–3 cm, forming a bell, with reflexed outer parts during anthesis. Filaments white, shorter than the perianth segments; anthers dark grey to black, c. 10 mm long, curved. Style white, tinged pink distally. Fruits green, fading to dull yellow, with thin pericarp closely investing the seeds, giving an irregular undulate surface, often distinctly beaked by the remains of the perianth tube. Seeds greenish, covered with a silvery-grey water repellent membrane making them very smooth, 20–60, variable in shape and size, often flattened, often germinating already within the fruit.

In grassland and open *Acacia* woodland, on sandy soils, gritty silt, or ± heavy black soils; up to c. 1000 m. N2; S1, 3; Ethiopia, widespread in the eastern parts of Africa southwards to South Africa. Burne 93; Gorini 424; Gillett & Beckett 23540.

The type specimen of *C. corradii* is extremely depauperate and difficult to interpret. It is also similar to *C. ornatum*, but geographical considerations support the chosen synonymy, as *C. ornatum*, in Somalia, has been recorded only from the extreme north.

4. **C. stuhlmannii** Bak. (1898). Plate 2 E.

C. somalense Chiov. (1916); type: S1, "Goriei" to "El Magu", Paoli 641 (FT syn.), "Gololonle" to "Uenèio", Paoli 790 (FT syn.).

Dacar (Som.).

Bulbs 12–20 cm in diam., often with a distinct neck. Leaves broadly strap-shaped up to 90 × 8–10 cm at anthesis, spreading on the ground, without midrib, distinctly ciliate at the margin, most leaves without intact apices. Scape 25–65 cm long, produced with the leaves. Bracts greyish and papery, early drooping. Flowers 10–20, with pedicels 2–5 cm long; tube red, curved, 8–12 cm long; perianth segments white with a red to purplish central band fading into white or pink near the margin, linear to narrowly lanceolate, 8–12 × c. 1.5 cm, forming a funnel, with reflexed outer parts during anthesis. Filaments white to pink, shorter than the perianth segments; anthers yellow to light brown, c. 5 mm long, curved. Style deep rose. Fruit vivid orange to deep red, subglobose, without a beak, with thick pericarp. Seeds 3–20, subglobose or somewhat flattened, light green, covered with a silvery grey water repellent membrane making them very smooth.

Sandy coastal plains and up to 200 km inland in open low bushland; up to 160 m. C1, 2; S1, 2; otherwise in a broad belt along the coast south to

Fig. 41. *Ammocharis tinneana*, habit, × 1/4. — From Fl. Trop. E. Afr. (1982). Drawn by A. Aarhus.

Mozambique. Thulin & Warfa 4735; Kuchar 16648.

5. C. papillosum Nordal (1977).

Leaves, broadly strap-shaped, ± prostrate, without midrib, distinctly ciliate at the margin. Scape 10−30 cm long, produced with the leaves. Flowers 2−12, subsessile or with pedicels up to 2 cm long; perianth segments white with very faint pink flush down the center, linear to narrowly lanceolate, forming a funnel. Anthers yellowish. Fruit yellowish green, usually with a beak, with thick pericarp. Seeds papillose, green, turning black when exposed to air.

Occurrence in Somalia uncertain. However, Bally 7670 from Somalia, without locality indicated, might represent this species, which is otherwise distributed in dry, sandy areas from northern Kenya south to Zambia.

3. AMMOCHARIS Herb. (1821)

Bulbous plants with leaves in two opposite fans (biflabellate). Leaves falcate to strap-shaped, without intact apices and midrib. Scape with a many-flowered inflorescence, subtended by two free bracts. Flowers pedicellate, regular, salver-shaped, with a long narrow tube and linear spreading and reflexed segments. Filaments filiform. Ovary with many ovules. Capsule fleshy, opening irregularly. Seeds fleshy, subglobose, pale green.

Three species, all African, one widespread from Botswana north to Sudan and Ethiopia, one from Angola to Kenya and one restricted to South Africa.

A. tinneana (Kotschy & Peyr.) Milne-Redh. & Schweick. (1939); *Crinum tinneanum* Kotschy & Peyr. (1867). Fig. 41; plate 2 F.

Bulb up to 12 cm in diam., with a distinct neck. Leaves spreading on the ground in two opposite fans, 1−3 cm wide, length varying with age, appearing before to after the flowers. Scapes 5−25 cm long. Inflorescence 10−30-flowered; pedicels 1−4 cm long. Flowers pink fading to magenta with age, or white, sweetly scented; tube 6−10 cm long; segments 4−8 × 0.3−0.5 cm, spirally recurved towards the apex at anthesis. Stamens slightly spreading, pink or white, 3−6 cm long; anthers 4−9 mm long, curved. Fruit reddish, subglobose, 2−2.5 cm in diam.

Coastal areas on open plains on coral sand, or on fixed dunes; up to c. 200 m. C1; S1, 3; Ethiopia, westwards to Chad and southwards to Namibia and Botswana. Popov 1014; Raimondo s.n.; Thulin & Warfa 4510.

4. PANCRATIUM L. (1753)
Bjørnstad in Norw. J. Bot. 20: 281 (1973).

Rather small bulbous plants, with linear, often twisted leaves. Inflorescence 1−few-flowered, subtended by 1−2 partly fused bracts. Flowers white or sometimes tinged greenish, regular, sessile, with a long cylindrical perianth-tube, funnel-shaped in the apical part; segments linear, spreading. Filaments united into a conspicuous cup at the base, filiform and free above. Style long, with a small capitate stigma. Fruit a dry loculicidal capsule, with many black angular seeds.

About 20 species in Africa, the Mediterranean region and southern Asia, three species in tropical Africa.

P. tenuifolium A. Rich. (1850). Fig. 42; plate 3 A.
P. trianthum sensu Cuf. Enum.: 1577 (1971), non Herb.

Bulb globose, up to 4 cm in diam., narrowed into a neck. Leaves 35 × 0.3—1 cm, developing after anthesis, finely pubescent near the base. Peduncle 1—10 cm long, finely pubescent, bearing 1 flower enclosed in a pale membranous bifid involucrum. Perianth-tube 9—12 cm long, slender, pale green; segments 5—10 cm long, up to 1 cm wide, white with greenish median stripe. False corona 2—4 cm long, with 2 triangular lobes between each pair of stamens. Filaments 1—2 cm long; anthers 5—8 mm long. Capsule subglobose to cylindrical, up to 2.5 cm long. Seeds up to 30, glossy, black, subglobose to angular, c. 4 mm in diam., furnished with a white appendage.

In open bushland on sandy soils; sea level to c. 1000 m. N1, 3; S2; Djibouti, Ethiopia, widespread in tropical and southern Africa. Gillett & Hemming 24068; Moggi & Tardelli 201; Thulin & Warfa 4432.

Fig. 42. *Pancratium tenuifolium*. A: habit. B: stamen. — Modified from Fl. W. Trop. Afr. 3(1): Fig. 365 (1968). Drawn by W. E. Trevithick.

155. IRIDACEAE

by P. Goldblatt

Cuf. Enum.: 1584—1592 (1972).

Seasonal perennial herbs, with rhizomes, bulbs or corms, rarely annuals or shrubs. Leaves basal and cauline, sometimes the lower 2—3 membranous, without blades, and barely reaching above ground (thus cataphylls); foliage leaves mostly arranged in 2 opposite rows, usually unifacial and sword-shaped, parallel-veined, plane, plicate or rarely terete (bifacial and channeled to flat in a few genera outside Somalia). Flowering stems aerial or subterranean, simple or branched, terete, angled or winged. Inflorescences either spikes of sessile flowers, or solitary flowers enclosed by a pair of opposed bracts (or composed of umbellate clusters (rhipidia) enclosed in opposed leafy to dry bracts (spathes) with flowers usually pedicellate and each subtended by one bract in *Iridoideae*, not occurring in Somalia). Flowers bisexual, with a petaloid perianth of two equal or unequal whorls (rarely one whorl absent), regular or zygomorphic. Tepals usually large and showy, free virtually to the base or united in a tube. Stamens 3, opposite the outer tepals, symmetrically arranged or unilateral; filaments filiform, free or variously united; anthers extrorse, usually dehiscing longitudinally. Ovary inferior (but superior in the Tasmanian *Isophysis*), usually 3-celled with axile placentation, ovules many to few; style filiform, usually 3-branched, the branches either filiform, distally expanded, or sometimes forked, or the branches thickened or flattened and petaloid. Fruit a loculicidal capsule, rarely indehiscent; seeds globose to angular or discoid, sometimes broadly winged, usually dry, with hard endosperm; embryo small.

A family of c. 80 genera and 1750 species, more or less worldwide, but rare in tropical lowlands; best represented in southern Africa.

The family is currently divided into four subfamilies (Goldblatt, Ann. Missouri Bot. Gard. 77: 607, 1990). Subfamily *Isophysidoideae* Takhtajan is monotypic comprising the Tasmanian *Isophysis* T. Moore with a superior ovary. Of the remaining *Nivenioideae* Schulze ex Goldbl., *Iridoideae* and *Ixioideae* Klatt only the last is represented in Somalia.

Iridaceae are of considerable economic importance in horticulture and the cut flower industry, especially *Iris*, *Gladiolus* and *Freesia*. Several other genera (*Dietes*, *Crocus*, *Watsonia*) are cultivated in gardens in both tropical and temperate areas.

1. Flowers solitary on aerial axes; leaves linear (nearly filiform), in section oval to round with 4 longitudinal grooves, and all inserted below the ground; perianth tube shorter than the tepals ....1. *Romulea*
 — Flowers arranged in a spike or panicle of 2—many flowers; leaves usually with an expanded plane blade, some at least inserted above ground level; perianth tube shorter or longer than the tepals 2
2. Corms campanulate with a flat base; stems somewhat compressed to angled or winged; the 3 style branches usually divided for half their length; anthers without acute terminal appendages; perianth tube much longer than the tepals; seeds ± globose .... 2. *Lapeirousia*
 — Corms globose to depressed globose with a rounded base; stems rounded in section; style branches usually expanded terminally, never divided; anthers with or without acute terminal appendages; perianth tube shorter or longer than the tepals; seeds broadly winged .... 3. *Gladiolus*

1. ROMULEA Maratti (1772), nom. cons.

Perennials with small globose corms with woody to cartilaginous or papery tunics. Leaves few to several, the lower 2—3 entirely sheathing (cataphylls), membranous or firm and green, foliage leaves all basal, 1—several, linear to ± filiform, the margins and midribs raised and often winged, thus with 2 narrow to longitudinal grooves on each surface, the blade oval to terete in transverse section with 2 sinuses on each surface between the margin and the midrib, occasionally nearly plane with lightly thickened margins and midrib. Stem short, aerial or subterranean, sometimes above ground in fruit, simple or branched, the branching usually below the ground. Inflorescence composed of solitary flowers terminal on the peduncles, the flowers each subtended by 2 opposed bracts; bracts green, often the margins membranous to scarious and pale or ferrugineous, occasionally the inner bract dry entirely. Flowers regular, usually cup-shaped, variously coloured, often yellow in the centre; tepals united in a short to long tube, subequal, usually ascending below and spreading above. Filaments erect, ± adjoining, sometimes united; anthers diverging or adjoining. Ovary ovoid, style dividing at or above the level of the anthers, the branches short, usually divided for half their length. Capsules oblong. Seeds lightly angled, hard, glossy or matte.

Some 95 species in Africa, southern Europe and the Middle East, centred in the SW Cape, and with a secondary centre in the western Mediterranean Basin.

1. R. fischeri Pax (1892). Plate 3 B.
Plants (1—)7—12 cm high excluding the leaves. Corm ovoid, tapering below to an oblique rounded base, 7—10 mm in diam.; tunics woody or cartilaginous, red-brown. Foliage leaves (2—)3—5,

mostly 8—15 mm long, oblong in transverse section, the midrib raised and with a narrow central hyaline ridge, straight and erect to falcate, occasionally spreading on the ground in depauperate plants, c. 1 mm in diam. Flowering stems 1—4 per plant, ± erect, becoming slightly falcate after anthesis, 0.6—1 mm in diam.; outer bracts green, usually with narrow membranous margins, the inner bracts with broad scarious margins streaked with brown, occasionally entirely scarious, 12—18 mm long, as long as or slightly shorter than the outer. Flowers blue, purple or violet, rarely almost white, cream to yellow in the centre, the tepals with bands of darker pigment over the 3 main veins, particularly marked on the reverse; perianth tube 4—5 mm long; tepals lanceolate, (10—)16—20 x 4—6 mm, erect below, curving outwards above, thus forming a cup. Filaments 4—6 mm long, erect, free, exserted 3—4 mm from the wide part of the tube; anthers 4—6 mm long. Ovary ovoid, c. 3 mm long; style dividing between the upper third and slightly beyond the apex of the anthers, the branches c. 3 mm long. Capsules ovoid-oblong, 7—10 mm long.

Rocky sites or patches of grassland in evergreen bushland; 1300—1400 m. N2; Ethiopia, Eritrea, Kenya, Uganda, and Saudi Arabia. Thulin, Dahir & Hassan 8931, 9074.

2. LAPEIROUSIA Pourret (1788)

Perennials with bell-shaped flat-based corms with densely compacted fibrous or woody tunics. Leaves several, the lower 2—3 membranous and sheathing the base; foliage leaves few, sometimes solitary, the lowermost longest and inserted on the stem near ground level, blade either plane, or shallowly plicate-corrugate, or terete, upper leaves cauline and progressively smaller. Stem somewhat compressed and angular, sometimes entirely subterranean. Inflorescence either panicle-like, or a simple to branched spike, or flowers clustered at ground level, the bracts green to membranous, the outer sometimes ridged, keeled, crisped or toothed. Flowers blue, purple, red, white or pink, regular or zygomorphic, tube short to long, tepals subequal or unequal. Stamens symmetrically disposed around the style or unilateral and arcuate. Style filiform, the branches usually forked for up to half their length, sometimes entire or barely bifid. Capsules membranous to coriaceous, globose. Seeds globose.

Some 40 species widespread across sub-Saharan Africa, from Ethiopia and Nigeria to South Africa.

L. schimperi (Aschers. & Klatt) Milne-Redh. (1934). Fig. 43 A, B.

L. cyanescens Welw. ex Bak. (1878).

L. cyanescens var. *minor* Chiov., Fl. Somala 2: 416 (1932); type: S3, "Chisimaio", Senni 565 (FT holo.).

Plants (20−)30−80 cm high. Corm 18−22 mm in diam., tunics of compacted fibres, light to dark brown, the outer layers ultimately becoming loosely fibrous and reticulate. Foliage leaves linear, 3 or more, the lower 2 longest and usually slightly exceeding the inflorescence, decreasing in size above, narrowly lanceolate, 5−10(−15) mm wide, the midrib lightly raised. Stem rounded below to nearly square and 4-angled to 4-winged above, laxly branched. Inflorescence a lax pseudopanicle, the branches with 1−3 sessile flowers; bracts (10−)20−35(−45) mm long, green, becoming membranous above or completely dry and papery, then light to dark brown, apices dark brown, the inner bract about as long as the outer. Flowers zygomorphic, white to cream, rarely pale violet, when whitish sometimes fading or drying lilac especially on the tube, opening in the evening and then sometimes scented; perianth tube 10−14(−15) cm long, straight, cylindric; tepals 18−22 × 6−7 mm, subequal, lanceolate, extended ± at right angles to the tube. Filaments unilateral, erect, exserted 5−7 mm from the tube; anthers 6−7 mm long, cream. Style dividing near to or up to 3 mm beyond the anther apices; branches c. 2 mm long, forked for ± 1/3 their length. Capsules 8−12 mm long.

Mostly wetter places like seasonal marshes, or among rocks at low altitude. S3; Ethiopia, Sudan, Kenya, Tanzania, Malawi, Zimbabwe, Angola, Namibia, Botswana. Moggi, Tardelli & Bavazzano 322; Paoli & Stefanini 351; Senni 399.

The large white flower with a perianth tube 10−15 cm long and the laxly branched inflorescence are distinctive, but confusion is possible with *Gladiolus gunnisii* and *G. candidus*, which have similar long-tubed white flowers. Both the latter have terete, unbranched stems, and *G. candidus* has, in addition, anthers with long acute terminal appendages.

The corms of *L. schimperi* are edible.

3. GLADIOLUS L. (1753)

Acidanthera Hochst. (1844).
Homoglossum Salisb. ex Bak. (1878).
Oenostachys Bullock (1930).

Perennials with corms with coriaceous to fibrous and reticulate tunics. Leaves few to several, the lower (2−)3 entirely sheathing and mostly below ground; foliage leaves usually contemporary with the flowers (or leaves produced after flowering, then on the same shoot or on separate shoots), few to several, basal or cauline, blades well developed or reduced and largely to entirely sheathing, lanceolate to linear and plane or filiform and terete, the margins, midrib and sometimes other veins thickened and hyaline (margins sometimes winged). Stem terete, simple or branched. Inflorescence a spike, the flowers turned to one side or sometimes in 2 rows; bracts usually green, soft to firm, sometimes dry and brown at anthesis, relatively large, the inner usually smaller than the outer. Flowers zygomorphic or sometimes regular; tepals united in a well developed, sometimes very long tube, subequal to unequal with the uppermost broader and arching to hooded over the stamens, the lower three narrower, shorter or longer than the dorsal. Filaments unilateral and curved, included or exserted from the tube; anthers unilateral (symmetrical in regular flowers). Style exserted, the branches simple, usually expanded above and sometimes apically bilobed. Capsules large and slightly inflated. Seeds usually many, usually with a broad membranous wing.

Some 255 species, centred in southern Africa and extending through tropical Africa and Madagascar to Europe and the Middle East.

1. Perianth tube at least twice as long as the tepals and much exceeding the bracts; flowers white to cream, sometimes with red to purple marks on the lower tepals 2
− Perianth tube shorter to slightly longer than both the upper tepal and the bracts; flowers variously coloured but not white or cream 3
2. Flowering stem with leaves with short blades; floral bracts 25−40(−45) mm long and subequal; anther apices obtuse .... 4. *G. gunnisii*
− Flowering stem with well developed foliage leaves with long blades; floral bracts (25−)40−50(−80) mm long, the outer exceeding the inner; anther apices with acute terminal appendages 1.3−1.8 mm long .... 3. *G. candidus*
3. Dorsal tepal c. 15 mm long and about twice as long as the upper lateral tepals; lower tepals reduced to linear cusps .... 2. *G. schweinfurthii*
− Dorsal tepal as large as or only slightly larger than the upper lateral tepals; lower tepals not much shorter than the three upper tepals................... .... 1. *G. somalensis*

Fig. 43. A, B: *Lapeirousia schimperi*. A: habit, × 0.4. B: floral detail, × 0.8. − C: *Gladiolus candidus*, habit, × 0.4, flower, × 0.8.

1. **G. somalensis** Goldblatt & Thulin (1995); type: N2, escarpment S of Laasqoray, near Ragad, 11°00'N, 48°29'E, Thulin, Dahir & Hassan 9079 (UPS holo.). Plate 3 C.

Plants (7−)12−30 cm high. Corm obconic, c. 12 mm in diam., the tunics of softly textured layers, these decaying with age into fine netted fibres. Leaves 4 or 5, the lower 3 basal and longest, reaching at least to the base of the spike and 1 or more often slightly exceeding the spike, the blade ± linear, (1−)2−4 mm wide, the upper 1 or 2 leaves inserted on the lower half of the stem. Stem erect, simple or with 1 or 2 branches, c. 1.2 mm in diam. below the base of the spike. Spike lightly flexuose, 2−10-flowered; bracts green and soft-textured, the outer (7−)12−17(−23) mm long, the inner c. two-thirds as long as the outer. Flowers zygomorphic, orange, the lower lateral tepals (or lower median tepal) bright yellow in the lower half; perianth tube funnel-shaped, 6−8 mm long; tepals unequal, lanceolate, the upper 3 larger

65

than the lower, the dorsal inclined over the stamens, 16–18 × 8 mm, the upper laterals about as long, the lower tepals ± parallel to the ground, the lower lateral tepals c. 15 x 5.5 mm, the lower median c. 12 x 5 mm, when pigmented the margins raised below and the surface channeled in the lower half. Filaments 8–10 mm long, exserted 4–6 mm from the tube; anthers 3–5 mm long, yellow. Ovary ovoid, 2–3 mm long; style dividing c. 1.5 mm beyond the anther apices, the branches c. 2.5 mm long, filiform, evidently not expanded apically. Capsules and seeds unknown.

Evergreen bushland in rocky places on limestone; c. 1350 m. N2, 3; not known elsewhere. Azzaroli 6; Sacco s.n.

Although capsules and seeds are unknown and critical for placing the species in *Gladiolus*, the softly textured bracts and the secund spike conform to the genus. The style branches are linear and do not appear to be apically expanded as is the case with most species. *G. somalensis* may be related to Ethiopian species such as *G. calcicola* Goldblatt and the Eritrean *G. mensensis* Bak., but also a relationship with species in southern Africa is equally likely.

2. G. schweinfurthii (Bak.) Goldblatt & de Vos (1989); *Homoglossum schweinfurthii* (Bak.) Cuf. (1972).

Plants (30–)50–75 cm high. Corm 8–15 mm in diam., tunics soft-membranous, fragmenting into narrow vertical strips, rarely almost fibrous, light brown. Cataphylls 1–2, herbaceous above. Leaves (3–)4–5, at least the lower 2 basal and largest, the upper 1–2 cauline and reduced, lanceolate to nearly linear, plane, half to two-thirds as long as the stem, not reaching the base of the spike, 4–12(–20) mm wide, usually rather soft-textured, the margins and midribs not thickened. Stem simple or with 1–2 branches, c. 3 mm in diam. at the base of the main spike. Spike 2–7(–12)-flowered; bracts green or flushed red to purple, 18–24(–28) mm long, the inner about half as long as the outer. Flowers bright red (to orange-red) on the upper tepals, greenish fading to yellow on the lower tepals, the throat and tube, in life the tube included in the bracts; perianth tube 11–16 mm long, slender below and erect, fairly abruptly expanded and curved outward into a cylindrical, horizontal upper part, 6–8 mm long; tepals unequal, the dorsal largest, extended horizontally, 12–18(–22) mm long, the upper laterals directed forwards, lanceolate, 8–12(–14) mm long, the lower tepals reduced, the laterals narrowly lanceolate, 6–8 mm long, the lowermost a linear cusp 3–6 mm long. Filaments 16–20 mm long, exserted 5–8 mm from the tube; anthers 4.5–8 mm long, reaching 1–2 mm below the apex of the upper tepal. Ovary 3–4 mm long; style ultimately reaching near to the apices of the anthers, the branches c. 3 mm long, extended beyond the anthers and much expanded above. Capsules globose-obovate, (7–)9–12 mm long; seeds c. 2.5 mm long, somewhat angular, with vestigial wings.

In rocky habitats; 1100–1800 m. N1–3; N and E Ethiopia. Thulin & Warfa 5856; Glover & Gilliland 1092; Collenette 270.

Closely related to *G. abyssinicus* of Ethiopian and West Arabian highlands; more study may show them to be conspecific. The flowers of *G. schweinfurthii* and *G. abyssinicus* are similarly specialized, having the dorsal tepal twice as long as the upper laterals, while the lower tepals are reduced to small cusps. Flowers of *G. schweinfurthii* are 25–36 mm long and the bracts 18–28 mm long, whereas the flowers of *G. abyssinicus* are (40–)50–70 mm long, and the bracts (30–)50–60 mm long.

3. G. candidus (Rendle) Goldblatt (1995); *Acidanthera candida* Rendle (1895). Fig. 43 C.

Acidanthera laxiflora Bak. (1887).

G. ukambanensis (Bak.) Marais (1973).

Geed caddays (Som.).

Plants 20–40 cm high. Corm globose, 12–25 mm in diam., tunics firm–papery, breaking into vertical fibers above and below. Cataphylls 2–3, membranous, the uppermost extending above the ground and green. Leaves 2–3, all ± basal, narrowly lanceolate, about half as long as the stem, 5–10 mm at the widest. Stem unbranched, c. 2.5 mm in diam. below the first flower. Spike 2–4-flowered; bracts (2.5–)4–5(–8) cm long, the inner shorter and narrower than the outer. Flowers white (or palest pink), sometimes with purple median streaks in the lower midline of the lower tepals, sweetly scented; perianth tube (7–)8–10 cm long, ± straight and cylindric; tepals subequal, broadly lanceolate-elliptic, (20–)25–30 mm long, c. 15 mm wide. Filaments c. 2 cm long, included in the tube (or exserted 1–2 mm); anthers 8–10 mm long, with a rigid apiculus 1.3–1.8 mm long. Ovary oblong, c. 5 mm long, the style dividing opposite the anther apices, the branches 5–7 mm long often notably broad and fringed. Capsules narrowly elliptic to obovate, 18–22 mm long; seeds variously winged to wingless.

Rock outcrops, woodland, bushland; 50–800 m. N2; C1; S3; Djibouti, S Ethiopia, Kenya, N Tanzania, and in SW Oman. Hemming 2309; Beckett 643; Thulin & Abdi Dahir 6697.

The corms are edible, and taken as a remedy for snake bite in northern Kenya.

4. G. gunnisii (Rendle) Marais (1973); *Acidanthera gunnisii* Rendle (1898); type: N1, top of "Wagga Mt.", Lort Phillips s.n. (BM lecto., K isolecto.).

Plants 25–35(–45) cm high. Corm globose-conic, 11–14 mm in diam., the tunics of fine(–medium textured) compacted fibers. Cataphylls pale and membranous. Leaves usually 3, the lower 1–2 basal, the uppermost cauline and shortest, about a third as long as the stem, linear, 2–3(–4.5) mm wide, the midribs and sometimes the margins lightly thickened. Stem unbranched, c. 1.5 mm in diam. at the base of the spike. Spike (1−)2−3-flowered; bracts green, 2.5−4(−4.5) cm long, the inner nearly equal or 2−4 mm shorter than the outer. Flowers white(−pale yellow), strongly fragrant; perianth tube slender, 8−12 cm long, expanding in the upper 10 mm; tepals evidently subequal, nearly elliptic, the dorsal probably horizontal, remaining tepals spreading, 25−30 mm long. Filaments c. 9 mm long, included in the tube or barely exserted for c. 1 mm; anthers 8−9 mm long, the bases sometimes within the tube. Ovary 4−5 mm long; style dividing c. 5 mm beyond the anther apices, the branches c. 5 mm long. Capsules ellipsoid, apically emarginate, 15−20 mm long; seeds oblong, 5−6 × c. 2.5 mm, light brown, broadly

winged, more so at the distal and proximal ends. Limestone hills in rocky grassland; up to c. 1800 m. N1, 2; S Ethiopia, N Kenya. Bally & Melville 15706; Bally 11132; Lavranos & Horwood 10243.

The white to cream flower with a 8−11 cm long perianth tube and included filaments, combined with short, narrow, grass-like leaves make the species unmistakable. The superficially similar *G. candidus*, which also has long-tubed white flowers, may be distinguished by its anthers with long acute terminal appendages and unequal floral bracts. Similarly long-tubed and white-flowered *Lapeirousia schimperi* has bell-shaped corms with flat bases, and branched, angular stems.

156. COLCHICACEAE

by M. Thulin

Cuf. Enum.: 1525−1529 (1971), *Liliaceae* in part.

Erect or climbing herbs with a corm. Leaves concentrated to the base of the plant or scattered along the stem, flat, mostly linear or lanceolate, sheathing at the base, sometimes ending in a tendril. Inflorescence a raceme or flowers sometimes solitary. Flowers generally bisexual, regular; tepals 6, free or basally united, usually with nectaries at the base. Stamens 6, free or inserted at the base of the tepals; filaments narrow, or broad at the base, glabrous; anthers 2-thecous, generally dorsifixed, longitudinally dehiscent. Ovary superior, 3-celled; style 3-branched or styles 3; ovules several to numerous per cell, placentation axile. Fruit usually a septicidal capsule. Seeds generally globose, with a straight embryo, endospermous.

Some 17 genera and 170 species in Africa, the Mediterranean area to Asia, and in Australia, best represented in South Africa.

1. Leaves attenuate into a recurved tendril at the tip; flowers usually red, orange or yellow 2
 − Leaves not ending in a tendril; flowers purplish- brown or pinkish-purple 3
2. Tepals reflexed; style bent sharply outwards at base .... 1. *Gloriosa*
 − Tepals spreading; style straight .... 2. *Littonia*
3. Plant with leafy stem; flowers purplish brown with tepals 4−12 mm long .... 3. *Iphigenia*
 − Plant stemless; flowers pinkish-purple with tepals up to c. 50 mm long .... 4. *Merendera*

1. GLORIOSA L. (1753)

Field in Kew Bull. 25: 243 (1972) and The genus Gloriosa, Lilies and other Liliaceae 1973: 93−95 (1972).

Climbing or erect herbs with abruptly bent corm. Leaves spirally arranged or some opposite or whorled, linear to ovate, attenuate into a recurved tendril at the tip. Flowers ± nodding, on long pedicels in upper part of plant. Tepals 6, similar, free, generally yellow or red, reflexed, persistent. Stamens with filiform filaments. Ovary ovoid to oblong; style filiform, bent sharply outwards at the base, 3-branched. Capsule ovoid to oblong, septicidal. Seeds globose, fleshy, red or orange.

One species only, in Africa and Asia.

G. superba L. (1753). Fig. 44.

G. simplex L. (1767).

G. abyssinica A. Rich. var. *graminifolia* Franch. (1882); *G. graminifolia* (Franch.) Chiov. (1916); type:

N3, "Karoma", Révoil s.n. (P holo., not seen).

Littonia baudii Terr. (1892); *G. baudii* (Terr.) Chiov. (1916).

G. minor Rendle (1896).

G. graminifolia var. *heterophylla* Chiov. in Result. Scient. Miss. Stefanini-Paoli: 176 (1916); types: S3, near "Chisimaio", Paoli 146 (FT syn.) & near "Torda", Paoli 302 (FT syn.).

G. aurea Chiov. (1928); types: N3, "Nogal" valley, Puccioni & Stefanini 855, 934 (FT syn.).

Adin tuki, dabalole, tamaior (Som.-N); faraji diil, gheloac (Som.-S).

Plant erect or scandent, up to several meters long, but sometimes less than 0.4 m tall, glabrous or almost so. Leaves spreading, linear to ovate, 60−250 × 2−50 mm. Flowers usually yellow, orange or red with yellow centre, rarely white on pedicels 45−200 mm long. Tepals 35−80 × 5−30 mm, acute, often with undulate margins. Filaments 12−50 mm long; anthers 5−13 mm long. Ovary oblong, glabrous; style 12−50

Fig. 44. *Gloriosa superba*, habit, × 0.5. − Modified from Burger, Fam. Flow. Pl. Ethiopia (1967).

mm long with branches 2.5−10 mm long. Capsule 20−60 × 10−25 mm. Seeds subglobose, c. 4−5 mm in diam.

Bushland, woodland, semi-desert plains, roadsides, often in sandy or rocky places; 20−850 m. N1−3; C2; S1−3; widespread in Africa and S Asia, also cultivated elsewhere as an ornamental. Kazmi, Elmi & Rodol 642; Thulin & Warfa 4626; Hemming 1635. A highly variable species. Three forms can roughly be distiguished in Somalia, a scandent more or less broad-leaved and large-flowered plant in S1−3 agreeing with typical *G. superba* (e.g. Kazmi, Elmi & Rodol 642), a more or less erect, narrow-leaved and small-flowered plant in all parts of the country agreeing with *G. baudii* and *G. minor* (e. g. Thulin & Warfa 5819), and an erect, fairly broad-leaved and large-flowered plant in N3 agreeing with *G. aurea* (e.g. Hemming 1635). However, several specimens are difficult to place and the subdivision breaks down outside Somalia (for example erect, narrow-leaved and large-flowered plants are found in Tanzania). I therefore follow Field (1972) in recognizing a single species only.

Used as a cure against snake-bites.

2. LITTONIA Hook. (1853)

Erect or climbing herbs with a lobed corm. Lower leaves in whorls, the upper alternate, linear to ovate, usually attenuate into a ± recurved filiform tip.

Flowers on long pedicels in upper part of plant. Tepals 6, similar, free, yellow, spreading, persistent. Stamens with filiform filaments. Ovary ovoid to oblong, 3-angled, with many ovules; style straight, 3-branched. Capsule ovoid to oblong, septicidal. Seeds subglobose, fleshy, red.

Some five species in Africa and Arabia.

L. revoilii Franch. (1882); type: N3, "Barroz" valley, Révoil s.n. (P holo., not seen). Plate 3 D.
 L. minor Defl. (1885).
 L. hardeggeri Beck (1888).
 L. obscura Bak. (1895).
 Derodile, mijereri, wela arobis (Som.).
Plant glabrous to densely papillose-pubescent, up to c. 30 cm tall. Leaves spreading, linear, up to 100 × 6 mm, attenuate into an often recurved tip. Flowers yellow or cream, often flushed purplish or brownish; pedicels up to c. 40 mm long. Tepals c. 15−40 mm long, narrow, acute. Filaments c. 10−20 mm long; anthers 4−8 mm long. Ovary oblong, glabrous; style c. 5−30 mm long with lobes 0.5−1.5 mm long. Capsule up to 17 × 11 mm. Seeds angular, c. 3 mm across.

Sandy or stony ground; c. 10−570 m. N1−3; C1; Djibouti, NE Ethiopia, Yemen. Hemming 1621; Thulin & Warfa 5793; Thulin, Hedrén & Abdi Dahir 7294.

The type of *L. revoilii* is papillose-pubescent as are also other specimens from the "Darror" depression in N3. However, there is an intergradation to the glabrous forms found elsewhere.

The variation in style-length is remarkable, long- and short-styled flowers being found both in north-eastern Africa and in Yemen.

3. IPHIGENIA Kunth (1843), nom. cons.

Slender herbs with an ovoid corm covered by thin dark tunics; stems straight or flexuose, with 1−2 tubular basal spathes. Leaves spirally arranged, linear to filiform, the upper subtending flowers. Flowers in a bracteate raceme or solitary; pedicels erect or recurved. Tepals 6, similar, free, linear, without nectaries, soon falling. Stamens with narrow filaments, soon falling. Ovary with many ovules; styles 3, free or united at base. Capsule ovoid to cylindrical, loculicidal. Seeds globose or ovoid, with a distinct raphe.

Some 15 species in Africa, Madagascar, Socotra, India, Australia and New Zealand.

I. oliveri Engl. (1893). Fig. 45.
 I. somaliensis Bak. (1895); type: N1, "Wadaba", Cole s.n. (K lecto.).
 Gurguri (Som.).
Erect herb, up to c. 30 cm tall; stems ± flexuose. Leaves linear, up to c. 4 mm wide, gradually narrowing towards the acute tip. Flowers few to several;

pedicels 5–15(–20) mm long, recurved in fruit. Tepals c. 5–15 × 0.5–1 mm, purplish brown. Stamens c. 1 mm long; anthers c. 0.5 mm long. Ovary ovoid; styles 3, 0.5–1 mm long, recurved. Capsule 3-lobed, c. 8–10 × 6–8 mm, many-seeded. Seeds c. 1.5–2 mm across.

Rocky ground, sometimes in pockets of soil in holes of limestone rocks; 600–1100 m. N1–3; Djibouti, Ethiopia, Kenya, N Tanzania. Thulin & Warfa 5855, 6210; Godding 215.

4. MERENDERA Ramond (1798)
Stefanoff in Sbornik Bulgariskata Akademiya na Naukita 22: 1–100 (1926).

Perennial stemless herbs with an oblong-ovoid corm enclosed by a dark tunic that is usually extended into a short neck. Leaves basal, produced at the same time as or after flowering; lower part of leaves and flowers enclosed within a membranous cylindric sheath. Flowers solitary or in clusters, subsessile, usually pinkish-purple. Tepals 6, similar, free, with a long narrow claw and a shorter and wider blade. Stamens inserted close to base of the blade. Ovary subterranean; styles 3, free to the base. Capsule oblong, septicidal, with 3 apiculate valves, maturing at or just above ground-level by the elongation of the pedicel. Seeds numerous, globose or almost so.

Some 10 species extending from S Europe and N and NE Africa through W Asia to Afghanistan.

Merendera is sometimes included in *Colchicum* L., but differs in the free tepals.

M. schimperiana Hochst. (1842). Plate 3 E.
M. longispatha Hochst. (1842).
M. abyssinica A. Rich. (1850).
M. longifolia Hutch. (1931); type: N2, "Sugli", Collenette 254 (K holo.).

Corm up to c. 3 cm long. Leaves 2–6, produced at the same time as the flowers, 5–20 × 0.3–1 cm, linear, acute, ascending or spreading, glabrous or with scabrid-serrulate margins. Flowers 1 or 2, pinkish-purple. Tepals up to c. 5 cm long. Stamens with free parts of filaments up to c. 6 mm long; anthers c. 2–5 mm long. Ovary oblong; styles up to c. 30 mm long. Capsule c. 1.5 cm long; seeds c. 1.6 mm across.

Semi-evergreen bushland or woodland, in sandy soil or limestone gravel; 1200–1425 m. N2, 3; Ethiopia, Yemen. Barbier 976; Thulin & Warfa 6240; Bavazzano & Lavranos s.n.

M. schimperiana and *M. longispatha* were validly published on the printed labels of Schimper's exsiccatae.

M. longifolia was distinguished on the basis of its large leaves with scabrid-serrulate margins and its large flowers. However, other material now available bridges the morphological gap between the Somali and the Ethiopian populations.

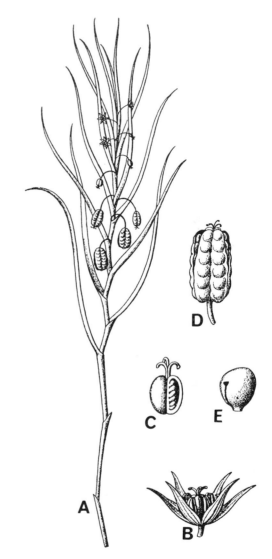

Fig. 45. *Iphigenia oliveri*. A: habit, × 0.5. B: flower, × 4. C: pistil, wall partly removed to show placentation, × 6. D: capsule, × 1.5. E: seed, × 5. — Modified from Nat. Pflanzenfam. 15a: 271 (1930).

157. ORCHIDACEAE

by B. Pettersson

Cuf. Enum.: 1597−1622 (1972); Fl. Trop. E. Afr. (1968, 1984, 1989); Cribb in Kew Bull. 33: 651−678 (1979).

Terrestrial or epiphytic perennial herbs or sometimes shrubs, sometimes saprophytic. Perennating organs in terrestrials: root or stem tubers, rhizomes or pseudobulbs; in epiphytes: pseudobulbs, rhizomes or entire plants. Inflorescence a spike, raceme or panicle. Flowers bisexual, zygomorphic, usually twisted through 180° (resupinate). Tepals 6, arranged in 2 whorls, both whorls similar or outer whorl (sepals) calyx-like and inner (petals) corolla-like; median petal (lip) almost always different from the 2 lateral ones, entire or variously lobed, laciniate or fimbriate, often brightly coloured, spotted or otherwise ornamented, often bearing crests or cushions of hairs, often produced backwards into a spur (rarely 2), often with nectar at apex. Stamen 1 (rarely 2), ± united with the style to form the column; pollen usually aggregated into 2 (sometimes 4, 6 or 8) pollinia per flower; pollinia mealy, waxy or horny, often divided into a number of smaller portions (sectile). Ovary inferior, 1-celled. Fruit a capsule opening by lateral slits. Seeds minute and very numerous with mycotrophic growth.

A worldwide family of about 725 genera and over 20000 species, most species in areas with high rainfall.

Many epiphytes and a few terrestrial species exhibit some succulence (pseudobulbs, thick or terete leaves) or other adaptations (e.g. leathery leaves, velamen radicum) to cope with dry periods.

Four of the terrestrial orchids in Somalia are typical geophytes flowering during the wet season or in arid areas after occasional rains, surviving the dry periods as dormant tubers.

The type of *Aerangis somalensis* (Schltr.) Schltr. collected by Ruspoli & Riva in "Somalia" is almost certainly from Ethiopia (see Cuf. Enum.).

1. Plants terrestrial 2
− Plants epiphytic 6
2. Lip without spur .... 4. *Epipactis*
− Lip with spur 3
3. Leaves, scape and bracts villose .... 1. *Holothrix*
− Whole plant glabrous 4
4. Spur shorter than rest of lip; pollinia 2, subglobose, non-sectile, on joint viscidium; without subterranean root tubers but sometimes with pseudobulbs above ground .... 5. *Eulophia*
− Spur much longer than rest of lip; pollinia 2, sectile, on separate viscidia; with subterranean root tubers ... 5
5. Spur less than 8 cm long .... 2. *Habenaria*
− Spur more than 9 cm long .... 3. *Bonatea*
6. Lip without spur .... 9. *Acampe*
− Lip with spur 7
7. Plant leafless .... 6. *Microcoelia*
− Plant with leaves 8
8. Stem several dm long; leaves evenly spread along (part of) stem .... 7. *Solenangis*
− Stem less than 1 dm long; leaves congested .... 8. *Angraecum*

1. HOLOTHRIX Rich. ex Lindl. (1835), nom. cons.

Terrestrial herbs with small ovoid root tubers and 1 or 2 sessile ovate to orbicular radical leaves. Scape an erect raceme with sessile or shortly stalked and often secund flowers. Sepals entire. Lateral petals entire or divided into 3 to many lobes. Lip similar to lateral petals but broader with more lobes and spurred at base. Column very short with 2 sectile pollinia.

About 20 species in tropical and subtropical Africa with two species extending to Arabia.

H. arachnoidea (A. Rich.) Rchb. f. (1881); *Peristylus arachnoideus* A.Rich. (1840). Fig. 46.

H. vatkeana Rchb. f. (1876); type: N2, "Sérrut" Mts, "Meid", Hildebrandt 1465 (W holo., K iso.).

Perennial herb up to 30 cm tall or slightly more. Leaves 2, basal, broadly lanceolate to orbicular, up to 4 cm long and broad. Leaves, scape, bracts and often also ovaries with long hairs. Inflorescence erect, many-flowered. Flowers small, secund, green ± flushed with violet or purple; tepals c. 2−3 mm long; lip slightly longer, 3-lobed; lobes subequal in length; spur ± 1.5 mm long.

In shade under shrubs on limestone; c. 1800 m. N2; Ethiopia, Kenya, Sudan, Yemen, Saudi Arabia. Bailes 100; Hildebrandt 1465.

2. HABENARIA Willd. (1805)

Terrestrial herbs with globose to elongated root tubers. Leaves variously arranged along the stem, sometimes 1 or 2 radical and appressed to the ground. Inflorescence terminal, erect, few−many-flowered. Flowers usually white and/or green. Sepals entire, the laterals spreading, the dorsal often forming a hood over the column. Lateral petals entire or 2-lobed, often nearly to the base. Lip entire or variously divided or lobed, spurred at base; spur usually long and slender, rarely short and sac-like. Column tall or short with 2 sectile pollinia; stigmatic processes ± club-shaped, often very long.

A genus with about 600 species, widespread in tropical and subtropical areas.

1. Leaves scattered along stem; side lobes of lip dissected .... 2. *H. macrantha*
   − Leaves basal; side lobes of lip entire 2
2. Leaves 3−5, elongate, ± erect in a basal rosette; spur ± 1 cm long ....1. *H. socotrana*
   − Leaves 2, subcircular, appressed to the ground; spur 5−7 cm long ....3. *H. subarmata*

1. **H. socotrana** Balf. f. (1882). Plate 3 F.

H. socotrana Rchb. f. ex Kraenzl. (1893), nom. illeg.

Perennial herb 23−46 cm tall. Leaves 3−5 in a basal rosette, oblanceolate to oblong, 4−14 × 1.5−2.5 cm. Inflorescence slender, rather sparsely many-flowered. Flowers green, sessile; bracts acuminate, shorter than ovary. Sepals and lateral petals entire, subequal, c. 3 × 2 mm. Lip 3-lobed to the base; midlobe c. 4 mm long; side lobes slightly shorter; spur slender, c. 8.5 mm long. Stigmatic processes short and rounded. Ovary c. 5 mm long. Capsule 8.5−12.5 mm long.

In crevices of limestone rocks or in leaf mould in shade of bushes and trees; 800−1600 m. N2, 3; Socotra. Collenette 253; Thulin & Warfa 5830, 6242.

First record for Somalia.

2. **H. macrantha** Hochst. ex A. Rich. (1850).

Perennial herb 20−50 cm tall; stem erect with 2 sheath-like and 3−6 lanceolate or ovate-lanceolate leaves along stem. Inflorescence rather laxly 2−9-flowered. Flowers suberect, greenish; pedicels with ovary 2.3−3.8 cm long. Dorsal sepal erect, ovate to narrowly lanceolate, acute, green, 20−26 × 7−11 mm; laterals spreading or deflexed, obliquely lanceolate, acuminate, subequal in size and colour to the dorsal. Lateral petals adherent to the dorsal sepal, lanceolate, curved, white with green veins, 20−25 mm long. Lip 3-lobed from a 9−15 mm long undivided base, green; midlobe linear, 14−23 mm long; side lobes dissected, subequal in length to the midlobe; spur slender, ± incurved, 2−3.5 cm long. Anther connective 12−15 mm broad; stigmatic processes 10−17 mm long.

Evergreen bushland on limestone cliff; c. 1400 m. N2; Ethiopia, Uganda, Kenya, Yemen. Thulin, Dahir & Hassan 8983.

First record for Somalia. The only Somali collection differs from other material of the species in that the midlobe of the lip is distinctly longer than the side

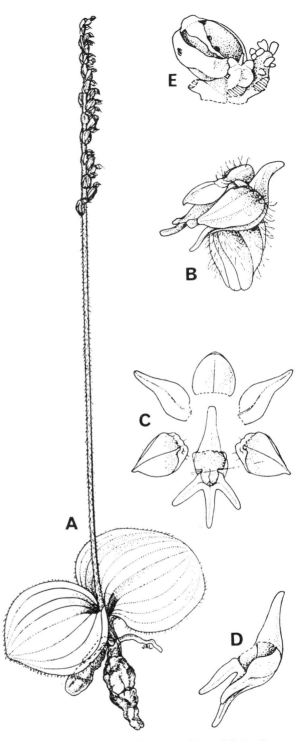

Fig. 46. *Holothrix arachnoidea*. A: habit, × 2/3. B: flower showing spur and ovary, × 8. C: sepals and petals displayed, × 8. D: side view of lip and spur, × 10. E: column with pollen masses on stigma lobes, × 16. — Modified from Kew Bull. 33: 658 (1979). Drawn by J. Lowe.

lobes (not slightly shorter), and the length of the pedicel with ovary is 2.8−3.8 cm versus 2.3−3 cm.

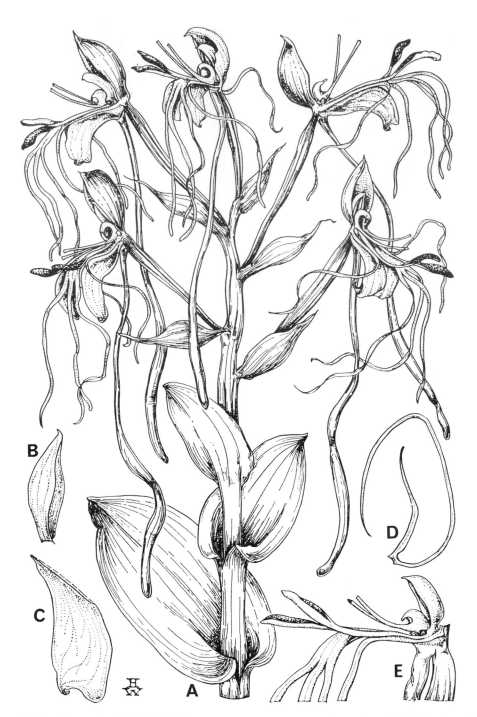

Fig. 47. *Bonatea steudneri*.
A: part of plant, × 2/3. B:
dorsal sepal, × 1. C: lateral
sepal, × 1. D: lateral petal,
× 1. E: column, with part
of lip, anterior petal-lobe,
spur and ovary, × 1. Modified from Fl. Trop. E. Afr.
(1968). Drawn by H.
Wood.

3. H. subarmata Rchb. f. (1881).

Perennial herb 40−65 cm tall; stem erect with 2
large leaves at the base and several much smaller
along the stem. Basal leaves opposite, broadly ovate
to subcircular, appressed to the ground, 7−19 cm long
and broad. Inflorescence rather closely many-
flowered. Flowers spreading outwards, white, faintly
fragrant; pedicel with ovary often slightly curved,
3−4 cm long. Dorsal sepal erect, hooded, 7.5−11 ×
6−7.5 mm; laterals spreading, 9−12 × 4.5−6 mm.

Lateral petals 2-lobed nearly to the base; upper lobe
7.5−11 mm long; lower lobe 25−40 mm long. Lip
3-lobed from a short undivided base; midlobe 11 16.5
mm long; side lobes 25−50 mm long; spur slender,
hanging, 5−7.5 cm long. Anther erect, 4.5−7.5 mm
high; stigmatic processes 5−7 mm long.

Probably semi-evergreen bushland at c. 50 m. S3
("Dallei"); Kenya, Tanzania, Zaire, Mozambique,
Malawi, Zambia, Zimbabwe. Senni 269.

First record for Somalia.

3. BONATEA Willd. (1805)

Terrestrial herbs with elongated fleshy root tubers. Leaves arranged all along the stem. Inflorescence terminal, erect, few−many-flowered. Flowers green or yellow and white. Sepals entire, the laterals partly united to the base of the lip, the dorsal often forming a hood over the column. Lateral petals 2-lobed, the upper lobe usually adherent to the dorsal sepal, the lower partly united to the base of the lip and the stigmatic arm. Lip 3-lobed, spurred at base; spur 2.5−21 cm long, cylindrical. Column with 2 sectile pollinia; stigmatic processes club-shaped, very long, partly united to perianth parts; rostellum standing out in front of the anther, convex and usually hooded.

About 20 species in tropical and subtropical Africa with one species extending to Arabia.

A record of *Bonatea rabaiensis* (Rendle) Rolfe (= *Habenaria rabaiensis* Rendle) from Somalia in Chiov., Fl. Somala 2 (1932) is based on Senni 269 (FT) which is *Habenaria subarmata*.

B. steudneri (Rchb. f.) T. Durand & Schinz (1895); *Habenaria steudneri* Rchb.f. (1881). Fig. 47.

Habenaria phillipsii Rolfe (1895); *Bonatea phillipsii* (Rolfe) Rolfe (1898); type: N1, "Golis range", at "Dara-as", Lort Phillips s.n. (K holo.).

Perennial herb 25−125 cm tall; stem erect, leafy. Leaves 10−20, the largest 7−19 × 3−5.5 cm. Inflorescence rather loosely 3−30-flowered; bracts 2−5.5 cm long, shorter than the ovary with pedicel. Flowers spreading, green to yellowish green and white, strongly fragrant; pedicel with ovary 4−7 cm long. Dorsal sepal erect, hooded, 20−30 × 10−20 mm; laterals deflexed, 20−30 × 10−15 mm. Lateral petals 2-lobed nearly to the base; upper lobe 17−27 mm long; lower lobe 30−70 mm long. Lip 3-lobed from a long narrow undivided base 15−30 mm long; midlobe 20−35 mm long; side lobes 25−85 mm long; spur slender, hanging, 10−21 cm long. Anther erect, 10−18 mm high; stigmatic processes 22−32 mm long; rostellum prominent in front of the anther, midlobe hooded.

Evergreen bushland on limestone rock, north-facing slope of gneiss bedrock, or deep gorge near water; c. 1650−1800 m. N1, 2; Ethiopia, Sudan, Yemen, Saudi Arabia, widespread in E and C Africa. Gillett 4848; Thulin, Dahir & Hassan 8942.

4. EPIPACTIS Zinn (1757), nom. cons.

Terrestrial herbs from rhizomes with fleshy roots, and with erect leafy stem; stem, ovary, outside of sepals and sometimes also the leaves ± densely pubescent. Leaves ovate to lanceolate along most of stem. Bracts similar to foliage leaves and sometimes almost as large as them. Flowers in racemes. Tepals entire, usually reddish and/or greenish. Lip spurless,

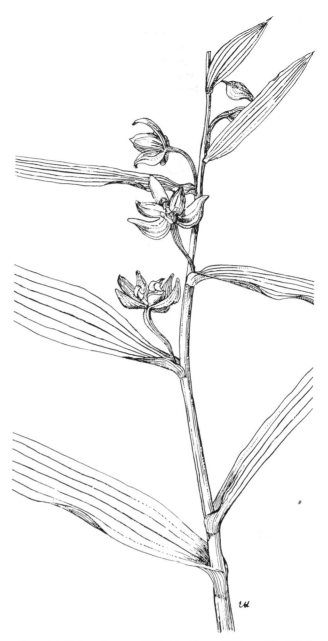

Fig. 48. *Epipactis veratrifolia*, habit, × 0.75. − Modified from Fl. Palaest. 4: 498 (1985).

often articulated in the middle; basal half cup-shaped, nectariferous. Column short; pollinia 2, mealy, consisting of loose pollen masses.

About 25 species, mainly in temperate parts of the northern hemisphere but with three species in tropical Africa.

E. veratrifolia Boiss. & Hohen. ex Boiss. (1854). Fig. 48.

E. somaliensis Rolfe (1897); types: N1, "Golis range, Woob", Lort Phillips s.n. and Cole s.n. (K syn.).

E. abyssinica Pax (1907).

Perennial herb 25–150 cm tall; upper part of stem, bracts, ovaries and outside of sepals pubescent. Leaves 3–10, ovate to lanceolate, arranged all along the stem, the largest 8–25 × 1–5 cm. Inflorescence erect to slightly bent, loosely few–many-flowered; bracts up to 25 cm long, but sometimes equal in length to ovaries or even shorter. Flowers large and fairly open for the genus, spreading or pendulous, green to yellowish green with purplish or reddish radial stripes and white tip to the lip. Sepals ovate-lanceolate, 13–21 mm long; lateral petals distinctly shorter. Lip curved, divided by middle joint into two parts; basal half boat-shaped, nectariferous, 10–12 mm long; apical half triangular, 9–11 mm long.

Close to streams and wadis; c. 400–1800 m. N1–3; Ethiopia, also widespread in SW Asia from Yemen, Oman, Cyprus and Egypt (Sinai peninsula) to the Himalayas. Cole s.n.; Lanza s.n.; Lort Phillips s.n.

5. EULOPHIA R. Br. ex Lindl. (1823), nom. cons.

Terrestrial or lithophytic herbs from perennial rhizome with roots mostly covered by a distinct white velamen; rhizome or above-ground stem often condensed to form several-noded tuber-like or pseudobulbous storing organs. Leaves soft and thin to fleshy and coriaceous from flowering or separate shoot. Inflorescence a lax to rather dense many-flowered raceme or panicle, with small to large, rather showy flowers. Tepals entire; sepals greenish and calyx-like or brightly coloured, corolla-like and similar to the petals. Lip 3-lobed, shortly spurred and with ridges and papillae on upper surface; spur with few exceptions conical, nectarless. Column mostly rather long; pollinia 2, subglobose, waxy.

A genus with about 250 species, widespread in tropical and subtropical areas.

The existence of two Somali collections of *E. clavicornis* Lindl. var. *nutans* (Sond.) A. V. Hall was mentioned by Cribb in Kew Bull. 42: 462–463 (1987), but the occurrence of this species in Somalia has not been substantiated and it is omitted here.

E. petersii (Rchb. f.) Rchb. f. (1865). Plate 4 A.

E. coleae Rolfe (1897); types: N1, "Golis range, Woob", Lort Phillips s.n. and Cole s.n. (K syn.).

E. phillipsiae Rolfe (1897); types: N1, "Golis range, Woob", Lort Phillips s.n. and Cole s.n. (K syn.).

Dhegoweyn (Som.); hangey (Som.-N); schkul (Som.-S).

Perennial herb up to 3 m tall; pseudobulbs large, ovoid to cylindric, 3–24 × 1.2–4 cm, 4–6-noded, green to yellow, 2–4-leaved towards apex; roots thick with white velamen. Leaves stiff, erect or spreading, succulent-leathery, linear-ligulate, acute, 15–55 × 0.7–4.5 cm, with toothed margins. Inflorescence an up to 7-branched panicle, laxly many-flowered, erect to almost horizontal; peduncle up to 90 cm long. Flowers fleshy, fragrant; sepals and petals green with purple-brown stripes; lip whitish with purple markings. Sepals 20–30 and petals 16–22 mm long; lip 3-lobed, 16–30 × 8–15 mm; all segments recurved and rolled at apex; spur 2–8 mm long, incurved; column 10–13 mm long.

Acacia-Commiphora bushland, sandy and rocky places; 10–1600 m. N1, 2; C1, 2; S1, 2; widespread in Africa from Ethiopia, Sudan and Zaire to South Africa, also in Yemen, Oman and Saudi Arabia. Gillett 4613; Paoli 1178; Thulin & Dahir 6440.

6. MICROCOELIA Lindl. (1830)

Leafless perennial epiphytes with short stems; roots greenish or greyish, elongate, unbranched or with a few branches. Inflorescences few to many, spicate, few–many-flowered. Flowers small to minute, white sometimes tinged with green, pink or brown. Sepals and lateral petals entire, subsimilar; lip entire to vagely 3-lobed, spurred; spur rather short. Column fleshy; pollinia 2, subglobose to pyriform, waxy.

About 28 species in tropical and subtropical Africa, including Madagascar.

M. sp.

Semi-evergreen bushland on hard yellowish sand; c. 65 m. S2, 3. Gillett & al. 25163 (?EA); Peveling s.n. (?MOG).

The two known collections appear to be lost. Most probably they both represent *M. exilis* Lindl. (1830), which is known to occur in coastal north Kenya just south of the border with Somalia. It is recognized by its minute, 2 mm long flowers with a globose, 1 mm long spur.

First record for Somalia of the genus.

7. SOLENANGIS Schltr. (1918)

Monopodial perennial epiphytes with long, often scandent stems, with or without leaves; roots produced all along the stem, greenish or greyish, elongate, unbranched or branching. Leaves if present distichous, fleshy or coriaceous. Inflorescences 1–many, few–many-flowered, spreading. Flowers minute to medium-sized, mainly white. Sepals and lateral petals entire, subsimilar; lip entire to 3-lobed, spurred; spur short to very long. Column short; pollinia 2, ovoid, waxy.

Five species in tropical Africa, including Madagascar and adjacent islands.

S. wakefieldii (Rolfe) P. J. Cribb & J. L. Stewart (1985). Fig. 49.

Epiphytic herb with up to 1 m long scandent leafy

stem; roots produced all along the stem, elongate, branching, used for climbing. Leaves widely apart, coriaceous, oblong-lanceolate or lanceolate, unequally acutely bilobed at apex, 1.5−3 × 0.5−1.3 cm, sheathing, articulated. Inflorescences several, 4−6 cm long, laxly 4−6-flowered, spreading. Flowers white, fragrant; pedicel and ovary 1−2 cm long. Dorsal sepal 3 × 1.5−2 mm; lateral petals subequal; lip narrowly 3-lobed in the apical half, 10 × 10 mm. Spur filiform, pendent, 6−7 cm long.

Evergreen forest; less than 50 m. S3; Kenya, Tanzania. Moggi & Bavazzano 1604.

First record for Somalia.

8. ANGRAECUM Bory (1804)

Monopodial perennial epiphytes or lithophytes; stems short to elongate, simple or branched, covered by leaf-bases. Leaves distichous, thin-textured, fleshy or coriaceous, unequally bilobed at apex. Inflorescences axillary, 1−several, 1−many-flowered, racemose or sometimes paniculate. Flowers small to large, fleshy, white, greenish or yellow, resupinate or non-resupinate. Sepals and lateral petals entire, subsimilar; lip entire to 3-lobed, spurred; spur short to very long. Column very short, fleshy; pollinia 2, globose, waxy.

About 200 species in tropical and subtropical Africa, including Madagascar and adjacent islands, and Sri Lanka.

A. dives Rolfe (1897).

Erect epiphyte 20−30 cm tall; stem short, woody with 2−6(−10) coriaceous leaves. Leaves 5−26 cm long, 4−16 mm wide, linear, unequally bilobed at apex and overlapping with their bases. Racemes 2 or more from the lower leaf axils or the stem below, 8−25-flowered, exceeding the leaves in height. Flowers greenish to yellow, sometimes almost white, ± 10 mm in diam., rather closely set on rhachis. Sepals and lateral petals 4−5 mm long; lip concave, ovate, acuminate; spur 2.5−3 mm long, curved with slightly inflated tip.

Coastal evergreen forest and scrub; less than 50 m. S3; Socotra, Kenya, Tanzania. Moggi & Bavazzano 1484, 1771.

First record for Somalia.

9. ACAMPE Lindl. (1853), nom. cons.

Monopodial perennial epiphytes or lithophytes; stem leafy, simple or branched. Leaves distichous, fleshy, coriaceous, mostly unequally bilobed at apex. Inflorescence axillary, many-flowered, racemose, corymbose or paniculate. Flowers rather small, fleshy, non-resupinate. Sepals and lateral petals entire, similar; lip variously lobed, saccate or with a short spur. Column short, fleshy; pollinia 4, globose, unequal, waxy.

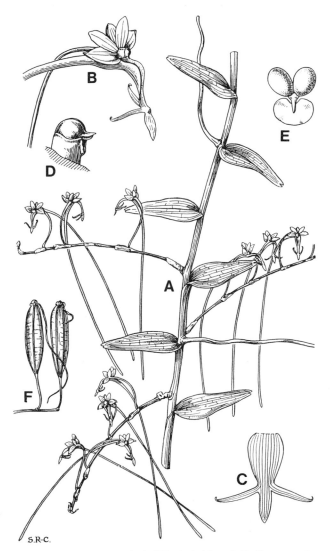

Fig. 49. *Solenangis wakefieldii*. A: habit, × 1. B: flower, × 4. C: lip, spread out, × 4. D: column, side view, × 10. E: pollinarium, × 16. F: fruits, × 1. − Modified from Hook. Ic. Pl. 35: t. 3445 (1943). Drawn by S. Ross-Craig.

About 10 species, mainly in SE Asia with one species in tropical and subtropical Africa.

A. pachyglossa Rchb. f. (1881). Fig. 50.

Epiphytic or lithophytic erect or pendulous herb 18−30 cm tall; stem leafy, simple or rarely branched. Leaves ligulate, thick, coriaceous, retuse to unequally bilobed at apex, 15−20 × 1.5−2.5 cm. Inflorescence erect, racemose, corymbose or shortly paniculate, 3−8 cm long, rather densely few−many-flowered. Flowers fragrant, especially towards the evening, yellowish with reddish markings. Sepals and lateral petals fleshy, strongly curved, the sepals 10−14 × 5−8 and the petals 8−12 × 3−14 mm; lip fleshy, 3-lobed, upper surface variously papillose, saccate at

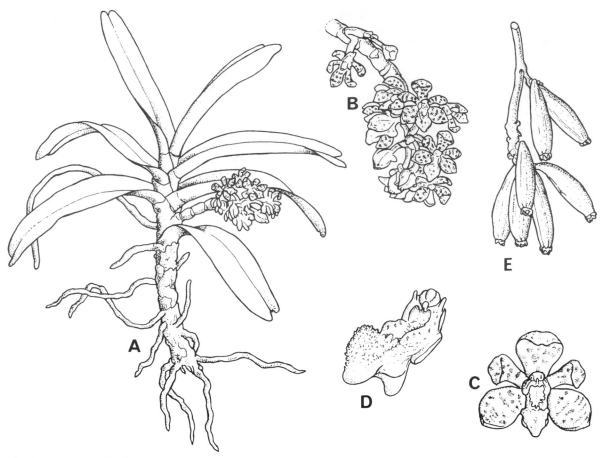

Fig. 50. *Acampe pachyglossa*. A: habit, × 1/3. B: inflorescence, × 2/3. C: flower, × 2. D: lip and column, × 2.7. E: fruiting branch, × 2/3. − Modified from Fl. Trop. E. Afr. (1989). Drawn by C. A. Lavrih.

base, 8−12 × 3−4 mm. Column 1.5−3.5 mm long.

Coastal bushland and woodland; c. 50 m. S3; widespread in tropical Africa from Kenya and Zaire to South Africa, Madagascar and adjacent islands. Moggi & Bavazzano 1512.

First record for Somalia.

158. VELLOZIACEAE

by M. Thulin

Cuf. Enum.: 1580−1581 (1972); Fl. Trop. E. Afr. (1975).

Perennial herbs or shrubs; stems covered with persistent fibrous leaf-sheaths. Leaves clustered, narrow, grass-like. Flowers solitary, conspicuous, usually bisexual, regular; tepals 6, petal-like, equal or almost so, ± united at the base. Stamens 6 or sometimes many; filaments free or adnate to tepals, sometimes flattened and forming a corona-like ring; anthers 2-thecous, longitudinally dehiscent. Ovary inferior, 3-celled; style slender; ovules numerous in many rows. Fruit a tardily and irregularly dehiscent capsule. Seeds with a small embryo, with endosperm.

Family of five to six genera and some 300 species.

XEROPHYTA Juss. (1789)
Smith & Ayensu in Kew Bull. 29: 184−205 (1974).

Leaf-blades usually deciduous along a regular transverse line. Perianth-tube slightly exceeding the ovary; tepals white, blue, mauve or yellow, without corona. Filaments flattened and almost wholly adnate to the tepals. Stylar branches linear, erect, equalling or longer than style-base.

Some 28 species in tropical and southern Africa, Madagascar, and in southern Arabia.

1. Hairs of ovary non-glandular, but enlarged at the base .... 1. *X. schnizleinia*
− Hairs of ovary with a dark globular gland at the apex ....2. *X. acuminata*

1. **X. schnizleinia** (Hochst.) Bak. (1875); *Hypoxis schnizleinia* Hochst. (1844); *Vellozia schnizleinia* (Hochst.) Martelli (1886); *Barbacenia schnizleinia* (Hochst.) Pax (1892). Plate 4 B.

Barbacenia hildebrandtii Pax (1892); *Xerophyta hildebrandtii* (Pax) Th. Dur. & Schinz (1895); *Vellozia hildebrandtii* (Pax) Bak. (1898); type: N2, "Serrut" Mts., above "Meid", Hildebrandt 1466 (B holo., not seen, K photo.).

Vellozia schnizleinia var. *somalensis* Terracc. in Bull. Soc. Bot. Ital. 1892: 425 (1892); *V. somalensis* (Terracc.) Chiov. (1911); *Xerophyta somalensis* (Terracc.) N. Menezes (1971).

Vellozia schnizleinia var. *brevifolia* Chiov., Fl. Somala 1: 314 (1929); *Xerophyta schnizleinia* var. *brevifolia* (Chiov.) Cuf. Enum.: 1581 (1972); type: C2, near "Tigieglo", Stefanini & Puccioni 247 (FT holo.).

Stems usually simple and 1−6(−12) cm high. Leaf-sheaths brownish, after abscission of the blade dividing into course fibres with few or no fine cross-fibres; blades (3−)10−30 × 0.2−0.6 cm, setose-serrate near apex on the margins and on the underside of the midnerve, otherwise entire and glabrous. Peduncles 1−3 at the apex of the stem, setose near apex. Tepals white or flushed with mauve or pink, linear-lanceolate, (5−)18−40 mm long, setose at the base; anthers linear-lanceolate. Ovary subglobose, covered with broad-based subulate stiff hairs. Capsule globose, up to 10 mm wide.

Rocky places, also on gypsum; 200−1200 m. N1−3; C2; S1, 3; Ethiopia, N Kenya, N Uganda, Nigeria. Gillett 4790; Kuchar 16931; Hemming SRS 263/1.

2. **X. acuminata** (Bak.) N. Menezes (1971); *Vellozia acuminata* Bak. (1895); type: N1, "Golis Range", near "Woob", Cole s.n. (K holo.).

Stems short. Leaf-sheaths brownish, after abscission of the blade dividing into course fibres; blades to 30 x 0.6 cm, setose-serrate near apex on the margin and on the underside of the midnerve, otherwise entire and glabrous. Peduncles at apex of the stem, glandular-hairy, particularly above. Tepals white or flushed with purple, linear-lanceolate, 16−22 mm long, glandular outside; anthers linear-lanceolate. Ovary oblong, covered with short dark gland-tipped hairs.

Rocky places; c. 1500 m. N1−3; not known elsewhere. Bally 11046.

This is closely related to *X. arabica* (Bak.) N. Menezes in southern Arabia.

159. MUSACEAE
by M. Thulin
Cuf. Enum.: 1592−1594 (1972).

Robust perennial herbs with false stems made up of overlapping leaf-sheaths. Leaves spirally arranged, petiolate, pinnately veined, entire. Inflorescences terminal, racemose, with flowers arranged in 1 or 2 rows in the axils of large spirally arranged, often coloured bracts; peduncle included in pseudostem for most of its length. Flowers usually unisexual, the basal one usually female (or bisexual), the terminal ones usually male. Tepals 6, 5 of them ± united and forming an upper lip, the 6th tepal free and directed downwards. Stamens 5(−6); anthers 2-thecous, basifixed, longitudinally dehiscent. Ovary inferior, 3-celled; ovules numerous; style filiform. Fruit a berry. Seeds many or absent in cultivated edible forms, testa hard, embryo straight, with endosperm.

Tropical family of two genera.

Ensete ventricosum (Welw.) Cheesman, with base of false stem distinctly swollen, persistent basal bracts and large seeds, was said to occur in southern Somalia in Cuf. Enum.: 1593 (1972). This species grows wild from Ethiopia in the north to Angola in the south and is extensively cultivated in Ethiopia for its starch-rich leaf-bases. It certainly does not grow wild in Somalia, and no evidence of it being cultivated there has been seen.

MUSA L. (1753)
Cheesman in Kew Bull. 2: 97−117 (1947), 3: 11−28 (1948); Stover & Simmonds, Bananas, ed. 3 (1987).

Plants suckering and often forming large clumps; false stem cylindric and only slightly swollen basally. Bracts usually deciduous. Perianth ± tubular, the upper lip 5-toothed. Seeds rarely more than 5 mm in diam.

Genus of some 35 species from south-east Asia, some extensively cultivated for fruit, fibres or ornament.

M. sp. (cultivars).
Banana (Eng.); moos, muus (Som.).
Plant 3−6 m tall. Leaves very large, becoming torn by the wind. Inflorescence pendent; apical bud pear-shaped, acute; bracts usually dull purple outside,

each with 2 rows of flowers. Flowers white, cream or yellow, the free tepal with a marked subapical wrinkle and the point turned inwards. Fruit variable, usually ripening yellow.

Cultivated particularly in southern Somalia and an important export crop; of SE Asian origin, the cultivated bananas being seedless diploids, triploids or tetraploids derived either from *M. acuminata* Colla or from hybrids between this and *M. balbisiana* Colla. The often used names *M. paradisiaca* L. and *M. sapientum* L. refer to different hybrid cultivars.

Detailed descriptions of the banana cultivars grown in Somalia are given by Ciferri in Atti Ist. Bot. Pavia, ser. 4, 10: 73−123 (1937).

160. ZINGIBERACEAE

by M. Thulin

Fl. Trop. E. Afr. (1985).

Rhizomatous herbs with aromatic oil cells, usually unbranched. Leaves entire, in 2 rows. Inflorescence terminal on leafy shoot or borne directly on rhizome. Flowers bisexual, zygomorphic. Perianth of 3 outer sepaloid tepals forming a tube or spathe ("calyx"), and 3 inner ± petaloid tepals, longer than the outer, tubular in lower part ("corolla"); large petaloid lip or "labellum" formed from 2 fused staminodes; fertile stamen 1, often petaloid; anther 2-thecous, longitudinally dehiscent. Ovary inferior, (1−)3-celled; ovules many; style 1, linear, usually held between the thecae of the anther. Fruit a berry or capsule. Seeds arillate.

Pantropical family of more than 40 genera and c. 1000 species.

SIPHONOCHILUS Wood & Franks (1911)
Burtt in Notes Roy. Bot. Gard. Edinb. 40: 369−373 (1982).
Cienkowskia Schweinf. (1867), nom. illeg.
Cienkowskiella Kam (1980).

Perennial herbs from a short rhizome; roots often tuberous. Leafy shoots often produced after flowering; blade narrowly to broadly elliptic, glabrous. Inflorescences racemose, 2−20-flowered; peduncle very short or elongated, with several bracts at the base and pedicelled flowers each subtended by a bract but without bracteoles. Calyx 3-lobed at the apex. Corolla with 3 subequal lobes. Labellum 3-lobed. Fertile stamen with basal anthers and a long terminal lobe. Ovary subglobose; stigma glabrous. Fruit often ± subterranean.

Some 15 species in tropical Africa, previously included in the Asian genus *Kaempferia* L.

S. aethiopicus (Schweinf.) B.L. Burtt (1982); *Cienkowskia aethiopica* Schweinf. (1867); *Kaempferia aethiopica* (Schweinf.) Ridley (1887). Fig. 51.

Rhizome 3−5 cm long, ovoid; roots with elongate tubers. Leaves developing after flowering; blade narrowly elliptic, 17−36 × 2−4.5 cm, acuminate at the apex, narrowly cuneate at the base; sheaths sulcate, forming a well developed false stem to 70 cm tall. Inflorescences 4−12-flowered, with a very short peduncle, mostly below ground; floral bracts 3−4 × 1 cm, obtuse. Calyx 2.7−3.5(−5) cm long. Corolla-tube 2.5−5.5 cm long, 1−2 mm in diam.; corolla-lobes elliptic, 2.8−5.5 cm long, whitish and translucent. Labellum mauve to purple, 3-lobed, 5−11.5 cm long, the central lobe rounded and ± emarginate, with a yellow mark at the base. Stamen obovate to oblong, up to 5 cm long; anthers 8−9 mm long. Fruit ± subterranean, subglobose, ± 1.5 cm in diam. Seeds 6 × 2 mm, pale, shiny.

Bushland and woodland; c. 30 m. S3; widespread in tropical Africa from Ethiopia to Senegal in the west and Zimbabwe in the south. Hemming 84/12.

First record for Somalia (21 km N of Kaambooni at 1°27'S, 41°34'E).

161. CANNACEAE

by M. Thulin

Rhizomatous herbs, usually unbranched. Leaves entire, spirally set. Inflorescence terminal, simple or branched, usually with 2 flowers in the axil of each bract. Flowers bisexual, showy, zygomorphic. Perianth of 3 outer sepaloid free tepals ("calyx"), and 3 inner petaloid tepals longer than the outer and united at the base ("corolla"). Showy part of flower formed by usually 4 basally united petaloid staminodes, the inner one, the lip or "labellum" shorter and reflexed; fertile stamen 1, petaloid; anther 1-thecous, attached at the margin, longitudinally dehiscent. Ovary inferior, warty, 3-celled; ovules many; style 1, flat, petaloid. Fruit a capsule.

Family of a single genus only.

CANNA L. (1753)

Description as for the family.

Some 50 species in tropical and subtropical America. Several species grown as ornamentals and numerous hybrids have been produced.

C. × generalis L. Bailey (1923).

Plant to 2 m tall. Leaves elliptic, acute; blade to 45 cm long. Calyx c. 2 cm long. Corolla tube up to 1.5 cm long; lobes 4—5 cm long. Outer staminodes 3, to 10 × 5.5 cm, white to yellow to red or pink, sometimes with markings in other colours; labellum to 8 × 3 cm, streaked or blotched. Fertile stamen c. 1.5 mm wide. Style c. 4 mm wide. Capsule 2—3 cm long.

The cultivated Cannas in Somalia probably belong to this horticultural hybrid group of unclear origin. Often called *C. indica* L., which is one of the putative parents of *C. × generalis*. *C. indica* has smaller flowers with outer staminodes less than 1.5 cm wide.

Fig. 51. *Siphonochilus aethiopicus*. A: base of plant and leaves, × 2/3. B: flower, × 1. C: base of stamen showing anthers, × 2. — Modified from Fl. Cameroun 4: pl. 4 (1965).

162. COMMELINACEAE

by R. B. Faden

Cuf. Enum: 1507—1522 (1971).

Annual or perennial herbs. Leaves basal or cauline, alternate, sheaths closed, lamina simple, entire, often succulent. Inflorescences terminal, terminal and axillary, or all axillary, sometimes becoming leaf-opposed, cymose, composed of scorpioid cymes, thyrsiform or variously reduced, sometimes enclosed in spathes. Flowers bisexual or bisexual and male, rarely bisexual and female or polygamous, the plants then andromonoecious or polygamomonoecious; sepals 3, free or connate, usually subequal and sepaline, occasionally petaline; petals 3, free or connate, equal or unequal, petaline, deliquescent; stamens 6, all fertile or some staminodial or lacking (rarely all stamens lacking), filaments glabrous or bearded, anthers longitudinally (rarely poricidally) dehiscent; ovary 2- or 3-celled, cells 1—many-ovulate, ovules uniseriate or biseriate, style simple, usually slender, stigma simple or rarely 3-lobed, enlarged or not. Fruits loculicidal capsules, rarely indehiscent or berries. Seeds 1—many per cell, hilum punctiform to linear, embryotega (lidlike thickening in the seed coat that is detached on germination) present.

Cosmopolitan family with about 40 genera and 650 species, mostly tropical.

The flowers are delicate and open for only a few hours after which they liquefy (deliquesce). Some species are recorded as being eaten by domestic stock. Some are weeds in cultivation.

Most if not all *Commelinaceae* have the same name in Somali. This has been variously recorded as "Baar", "Bar", "Barr", "Bhar" and "Bh'ar". For uniformity, all of the variants have been changed to Baar where alternate spellings were used.

1. Flowers actinomorphic (regular); petals fused into a tube; stamens 6, fertile, equal; filaments bearded; seeds with a terminal embryotega .... 1. *Cyanotis*
 — Flowers actinomorphic to strongly zygomorphic; petals free; stamens 2–3; staminodes 2–4; filaments glabrous or bearded; seeds with a lateral to dorsal embryotega 2
2. Leaves all or mainly in a basal rosette; flowers actinomorphic; stamens 3, alternating with 3 staminodes; capsule cells 8–12-seeded .... 3. *Anthericopsis*
 — Leaves usually all or mostly cauline, rarely in a rosette; flowers slightly to strongly zygomorphic; stamens and staminodes not as above; capsule cells 1–2(–4)-seeded 3
3. Inflorescences enclosed in spathes, often becoming leaf-opposed; staminodes with 4–6-lobed antherodes (occasionally reduced) ....5. *Commelina*
 — Inflorescences never enclosed in spathes, not becoming leaf-opposed; staminodes with 2- or 3-lobed antherodes 4
4. Bracteoles soon falling; petals equal, not clawed; stamens 2, staminodes 4, antherodes 3-lobed; capsules 3-celled, 3-valved .... 2. *Murdannia*
 — Bracteoles persistent; 1 petal differentiated from other 2, the 2 clawed; stamens 3, staminodes 2–3, antherodes 2-lobed; capsules 2–3-celled, 2-valved .... 4. *Aneilema*

1. CYANOTIS D.Don (1825)

Perennial or annual succulent herbs, the perennial often with subterranean storage organs. Leaves distichous or spirally arranged, blade sessile. Inflorescences terminal, terminal and axillary, or all axillary, cincinni single or in small groups, sessile with an abbreviated axis, often closely subtended by foliaceous bracts; bracteoles large, herbaceous, persistent. Flowers bisexual, regular, subsessile; sepals free, equal, usually pilose; petals united below into a tube, lobes usually blue to violet or mauve; stamens 6, equal, filaments with a subapical swelling, bearded distally, anthers with a narrow connective, pollen sacs with a longitudinal dehiscence but functionally basally poricidal; ovary 3-celled, usually pubescent, cells 2-ovulate. Capsules 3-celled, 3-valved, cells 1–2-seeded. Seeds uniseriate, hilum punctiform, embryotega terminal.

Genus of c. 50 species in the Old World tropics, especially diverse in Asia. The annual *C. cristata* (L.) D.Don, native to Asia, has been collected in Ethiopia and Socotra.

C. somaliensis C.B. Cl. (1895); types: N1, "Hammar, Golis Range", Cole s.n. (K syn.); "Darra-as", Lort Phillips s.n. (K syn.). Fig. 52.

Baar (Som.).

Perennial; roots thin, fibrous. Shoots with small to large sterile rosettes at the base with rosette leaves lanceolate oblong, to 15 × 1.5 cm, densely pilose; flowering shoots to 30 cm long, with bases thick, covered by overlapping sheaths and clearly perennial; cauline leaves distichous, sheaths to c. 5 mm long, pilose, blade oblong to lanceolate-ovate, 1.5–24 ×

0.6–1 cm, apex acute, margin ciliate, surfaces pilose or the upper glabrous; flowering portion of shoot zigzag, to 17 cm long, the leaves gradually reduced distally and becoming more and more reflexed. Inflorescences sessile, solitary in the axils of the uppermost 3–7 leaves; bracteoles herbaceous, asymmetric, 7–9 mm long, ciliate. Flowers c. 1 cm wide; fruiting sepals oblanceolate, c. 8 mm long, pilose especially apically; corolla pale violet; filaments with a fusiform swelling below the anther, anther oblong; style ± equalling the stamens in length, colour and form, glabrous. Capsules c. 4 mm long, glabrous except for an apical tuft of hairs. Seeds ellipsoid, 1.5–2 mm long, testa brown with small, shallow depressions.

Under shrubs among rocks, or in *Acacia-Euphorbia abyssinica* open woodland on gneiss mountain slopes with scrub; 915–1770 m. N1; Kenya?, Yemen? Gillett 4847; Drake-Brockman 331; Bally & Melville 16212B.

This species is questionably distinct from *C. foecunda* Hochst. ex Hassk. (1870) of tropical Africa. It differs chiefly in having persistent shoot bases and rosettes and in lacking bulbs. *C. nyctitropa* Deflers from Yemen is probably a synonym of *C. somaliensis*. A collection from eastern Kenya is similar to *C. somaliensis* in its perennial base but lacks rosettes. The plant in cultivation under the name *C. somaliensis* has indeterminate growth and no base. It is related to but perhaps not conspecific with this species.

2. MURDANNIA Royle (1840), nom. cons.

Perennials and annuals. Leaves spirally arranged or distichous, blade sessile. Inflorescences thyrses of 2–many cincinni, or reduced to clusters of 1-flowered

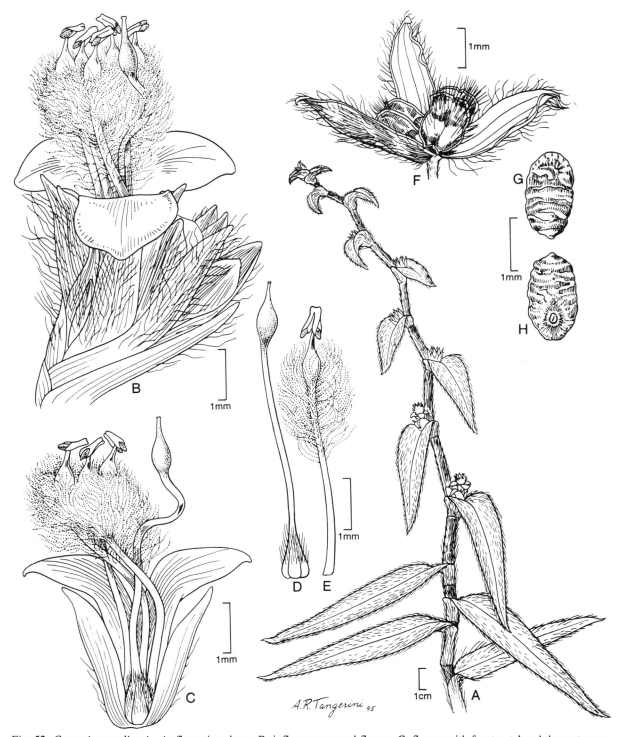

Fig. 52. *Cyanotis somaliensis*. A: flowering shoot. B: inflorescence and flower. C: flower with front petal and three stamens removed. D: gynoecium. E: stamen. F: dehisced capsule. G: seed, dorsal view. H: seed, ventral view. − From Faden & Faden 77/576 (Kenya). Drawn by A. R. Tangerini.

cincinni, terminal and axillary; bracteoles persistent or soon falling. Flowers pedicellate, regular to slightly zygomorphic, bisexual and male; sepals free, subequal, sepaline; petals free, equal, not clawed; sta-

mens 3, antesepalous, sometimes one staminodial, filaments bearded or glabrous; staminodes 3, antepetalous, rarely all lacking, filaments bearded or glabrous, antherodes 3-lobed or hastate; ovary 3-celled. Capsules 3-celled, 3-valved, cells 1−many-

Fig. 53. *Murdannia simplex*. A: plant base, with base of flowering shoot. B: inflorescence. C: bisexual flower, side view. D: bisexual flower, front view, showing style curving away from the stamens. E: dehisced capsule. F: seed, dorsal view. G: seed, ventral view. – From Faden & al. 74/64 (Ghana). Drawn by A. R. Tangerini.

seeded. Seeds uni- or biseriate, hilum punctiform to linear, embryotega lateral to dorsal.

Pantropical and warm temperate genus of c. 50 species, with greatest diversity in tropical Asia.

M. simplex (Vahl) Brenan (1952); *Commelina simplex* Vahl (1805). Fig. 53.

Aneilema sinicum Ker-Gawl. (1822).

Tufted to shortly rhizomatous perennial with erect to ascending or rarely decumbent shoots mostly 30−70 cm long, unbranched or branched near the base; roots uniformly thickened, not tuberous. Leaves spirally arranged or distichous, basal and cauline, widely spaced and strongly reduced upwards, the uppermost bract-like, on the flowering shoot, sheaths to 4 cm long, pilose to nearly glabrous, ciliate at the apex, blade linear to linear-oblong or lanceolate-oblong, (2−)4.5−45 × 0.4−1.7 cm, glabrous to pilose. Inflorescences usually compound, composed of 2 or more thyrses, each thyrse composed of 1−3(−5) elongate, many-flowered cincinni; bracteoles elliptic, 3.5−6 mm long, soon falling. Flowers secund, bisexual and male; pedicels 4.5−8 mm long, erect in fruit, glabrous; sepals 4.5−7 mm long, glabrous; petals ovate-orbicular to ovate-elliptic or obovate-elliptic, c. 5.5−11 mm long, blue or mauve to lavender; stamens 2, filaments bearded; staminodes 4, 3 antepetalous with glabrous filaments and 3-lobed antherodes, 1 antesepalous with a bearded filament and unlobed antherode (or antherode lacking);. Capsules ovoid to broadly ellipsoid, apiculate, (3.5−)4−7.5 × 2.5−4 mm, glabrous, cells 2-seeded. Seeds uni-seriate, transversely elliptic to elliptic or ovate, (1.1−)1.5−1.8 × 1.2−1.7 mm, testa brown, smooth to faintly ribbed, alveolate, or pitted, usually with whitish pustules.

Habitat unknown in Somalia, elsewhere growing in grassland, woodland, rocky places and marshes. S3 ("Foolduqobe−Badade"); tropical Africa, Madagascar, tropical Asia. Kazmi 5211.

3. ANTHERICOPSIS Engl. (1895)
Gillettia Rendle (1896).

Geophytes with tubers at the ends of thin, wiry roots. Leaves all or mainly in a basal rosette, spirally arranged, blade sessile. Inflorescences sometimes subsessile at first flowering, eventually long-pedunculate, scapose or subscapose from the rosette or axillary from a leaf on the peduncle, composed of 1 or 2 sessile, contracted, bracteolate cincinni at the summit of the peduncle, when 2, opposite, but not fused. Flowers pedicellate, regular, bisexual; sepals free, equal, sepaline; petals free, equal, not clawed; stamens 3, antesepalous, filaments glabrous, anthers basifixed, dehiscence introrse; staminodes 3, antepetalous, filaments glabrous, antherodes small unlobed; ovary 3-celled. Capsules 3-celled, 3-valved, cells c. 8−12-seeded. Seeds uniseriate, hilum linear, embryotega semilateral.

A single species restricted to eastern Africa.

A. sepalosa (C.B. Cl.) Engl. (1897); *Aneilema sepalosum* C.B. Cl. (1881); *Gillettia sepalosa* (C.B.

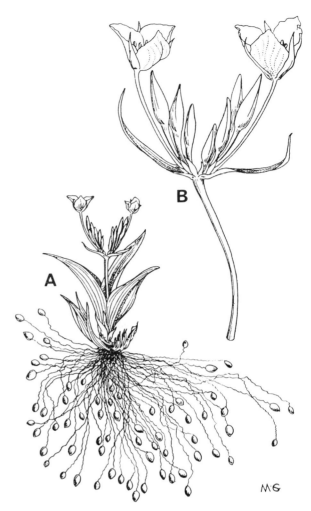

Fig. 54. *Anthericopsis sepalosa.* A: habit, × 0.5. B: inflorescence, × 1. − From J. Linn. Soc. Lond. Bot. 59: 364 (1966). Drawn by M. Grierson.

Cl.) Rendle (1896). Fig. 54.

A. fischeri Engl. (1895).

A. tradescantioides Chiov. (1951).

Completely glabrous perennial; tubers enclosed in a hard, crustaceous layer, ovoid to ellipsoid, 0.7−2 × 0.4−1 cm. Leaves ovate to lanceolate-oblong or oblong-elliptic, 4−31 × 1.2−3.5(−5.5) cm, apex acute to acuminate. Peduncle 6−20(−38) cm long in fruit; bracts and bracteoles herbaceous, 0.7−5 cm long. Flowers c. 2.5 cm wide; pedicels 1−4(−5.5) cm long; sepals linear-lanceolate to lanceolate or lanceolate-elliptic, (10−)15−25 mm long; petals elliptic, c. 15−25 × 13−18 mm, white to pale pink or mauve or rarely bluish; stamens with filaments 3−6 mm long, anthers linear-oblong, 3−5 mm long, yellow; staminodes with filaments c. 2 mm long, antherodes c. 1 mm long; ovary 2.5−4 mm long, style nearly straight, 3−4.5 mm long. Capsules oblong-lanceolate to linear-oblong, 20−30(−35) × 2−3(−4) mm, brown. Seeds transversely oblong, c. 1 mm long, 1.5−3 mm wide, testa smooth, brown to grey, hilum raised within an oblong pit.

Low vegetation on sandy soil; c. 200–400 m. S1; Ethiopia, Kenya, Tanzania, Mozambique, Malawi, Zambia, Zaire. Paoli 1253; Tardelli 150.

Plants may flower before the leaves are fully expanded, thus the leaves may appear ovate in some specimens. The mature leaves are lanceolate-oblong.

4. ANEILEMA R.Br. (1810)
Faden in Smithsonian Contrib. Bot. 76 (1991).

Perennials and annuals. Leaves spirally arranged or distichous, blade petiolate or sessile. Inflorescences thyrses, rarely reduced to a single cincinnus, terminal and axillary, rarely all axillary; bracteoles persistent, commonly perfoliate. Flowers pedicellate, zygomorphic, bisexual and male, rarely female; sepals free, usually subequal, sepaline; petals free, unequal, upper 2 clawed, lower (outer) petal usually different in size, shape and colour; staminodes 3(–2), posterior, antherodes bilobed; stamens 3, anterior, the medial usually differentiated, lateral filaments sometimes bearded, medial glabrous; ovary 2- to 3-celled. Capsules usually dehiscent, 2- to 3-celled, dorsal cell usually 1-seeded or empty, ventral cells 1–6-seeded. Seeds uniseriate, hilum linear, embryotega lateral.

Pantropical genus of c. 64 species, most numerous in Africa.

A. rendlei C.B. Cl. was recorded from Somalia on the basis of its type, Donaldson-Smith s.n. from "Somaliland", but this actually originates from Ethiopia.

1. Inflorescence consisting of a solitary contracted cincinnus partially enclosed in a pair of leafy bracts ....7. *A. lamuense*
   – Inflorescence thyrsiform, composed of (1–)2– many elongate cincinni 2
2. Fruiting pedicels usually erect or slightly further recurved; lower petal reduced and of different colour from paired petals; ventral capsule cells 2–4-seeded; shoots often disarticulating at the nodes when drying 3
   – Fruiting pedicels recurved 120–270°; lower petal usually large and of same colour as paired petals; ventral capsule cells 1–2-seeded; shoots not disarticulating at the nodes 6
3. Inflorescence composed of 2–11(–13) cincinni; cincinnus peduncles (1.2–)1.5–5.5(–8.5) mm long ....4. *A. pusillum*
   – Inflorescence composed of (8–)15–50 cincinni; cincinnus peduncles (2–)5–27 mm long 4
4. Bracteoles not perfoliate; capsules (5–)6–9 mm long, apex usually rounded to truncate; seeds 2–3(–4) per ventral cell, 1.6–2.7 mm long, white farinose in the depressions; roots with distal tubers ....1. *A. somaliense*

– Bracteoles usually perfoliate; capsules 12–18 mm long, apex ± rostrate; seeds 2 per ventral cell, 3.75– 5.4 mm long, not farinose; roots not tuberous 5
5. Cincinnus peduncles (11–)15–27 mm long; bracteoles attached 8–23 mm apart; blade lanceolate to ovate, 1–2(–4) cm wide ...... 2. *A. obbiadense*
   – Cincinnus peduncles 10–20 mm long; bracteoles attached 1.5–7.5 mm apart; blade usually linear-lanceolate and 0.35–0.8 cm wide, rarely lanceolate to lanceolate-elliptic and 0.8–1.9 cm wide ....3. *A. longicapsa*
6. Inflorescence composed of (8–)10–21 cincinni; sepals with marginal glands; capsules 3.4–4 x c. 2 mm; ventral cells 2-seeded with seeds ovate to subtriangular, 1.3–1.7 × 1.35–1.4 mm.........
   .... 5. *A. benadirense*
   – Inflorescence composed of 1–4 cincinni; sepals lacking marginal glands; capsules 4–4.5 x 3.4–4 mm; cells all 1-seeded, the ventral ones with seeds transversely elliptic, 2.7–3.2 × c. 2 mm.........
   .... 6. *A. trispermum*

1. **A. somaliense** C.B. Cl. (1901).
   Perennial, sometimes shortly rhizomatous; roots with distal tubers. Shoots annual, disarticulating at the base and nodes, 7.5–40 cm tall. Leaves spirally arranged, blade sessile, lanceolate-elliptic to elliptic or ovate (1.5–)2.5–7.5(–13) × (1–)2–3.5(–5.5) cm. Inflorescences dense thyrses (2–)3–9.5 × (1.5–) 2–4.5 cm, composed of (8–)17–40(–50) cincinni; bracteoles attached 1–3.5(–4) mm apart, cup-shaped, amplexicaul, not perfoliate. Flowers bisexual, female and male, (7.5–)10–12.5(–17) mm wide; sepals marked with red, puberulous; paired petals 4.8–9 x 3.2–7 mm, white to very pale lilac or pale blue; lower petal green with whitish margins; lateral stamens with filaments sigmoid, glabrous. Capsules oblong-elliptic, 2- to 3-celled, (5–)6–9 × 3.2–4 (–4.8) mm; dorsal cell 0–1-seeded, ventral cells 2–3 (–4)-seeded. Seeds mostly trapezoidal, rectangular or triangular, 1.6–2.7 × (1.2–)1.3–1.6(–2) mm, testa greyish brown to tan, interruptedly furrowed, farinose in the depressions.

   Acacia-Commiphora bushland on red sandy soil over limestone; c. 350 m. S1; E and S Ethiopia, N Kenya. Paoli 1085; Thulin, Hedrén & Dahir 7662.

   Plants from Somalia are unusual in having mainly elliptic to broadly elliptic leaf-blades.

2. **A. obbiadense** Chiov. (1928); type: C1, "Tobungàb", Puccioni & Stefanini 592 (FT holo.). Plate 4 C.

   Arjeg baar, baar (Som.).

   Rhizomatous perennial; rhizome with narrow constrictions; roots spreading horizontally just below the surface, 3–4 mm thick at base, then cord-like, supple, sand-covered. Shoots annual, disarticulating at the nodes and base, prostrate to erect, unbranched or sparsely branched, to 30(–80) cm long (including the inflorescence). Leaves spirally arranged, blade sessile, lanceolate to ovate 2–5(–6.5) × 1–2(–4) cm, apex acute to obtuse, base rounded to broadly cuneate. Inflorescences terminal, lax to moderately lax broadly ovoid to spheroidal thyrses, (4.5–)10–30 × (5.5–)12–27 cm, composed of 11–24 ascending to declinate cincinni to 17 cm long and 9-flowered; cincinnus peduncles (11–)15–27 mm long; bracteoles attached 8–23 mm apart, cup-shaped, perfoliate. Flowers bisexual and male, open in the morning, (11–)12–18(–23) mm wide; pedicels 4–7(–11 in fruit) mm long, recurved c. 180–270° in fruit; sepals 3–6.5 mm long; paired petals 7–11 × 6–8.5 mm, lilac or white, apex hooded. Capsules stipitate, 3-celled, ± fusiform, 13–17 × c. 2–4 mm, falcate, rostrate; dorsal cell 1-seeded, ventral cells (0–)1–2-seeded with seeds transversely elliptic, 3.5–5.4 × 1.6–2.2 mm, testa grey to grey-brown, finely pitted, not farinose.

Open, undulating coastal plains and stabilized sand dunes, in shallow or deep, white or pale tan sand, sometimes over limestone, usually in full sun, sometimes at the edges of or under bushes; 15–300 m. C1; S2; not known elsewhere. Gillett & al. 22229; Thulin, Hedrén & Dahir 7214; Faden & Kuchar 88/152.

The roots are used to weave containers that hold water. This species seems to be used only in the coastal plain where the otherwise preferred plants, *Asparagus africanus* (argeeg) and *Euphorbia longispina* (qabo) are lacking. The large infructescences appear to act as tumbleweeds.

Faden & Kuchar 88/153 was found growing in the shade of shrubs and has much longer shoots (to 80 cm) and larger leaves (6.5 × 4 cm) than any other collection.

3. **A. longicapsa** Faden (1991); type: C2, 17 km S of "Mugakori", Gillett & Beckett 23289 (K holo., EA US iso.).

A. obbiadense var. *angustifolium* Chiov. in Lavori R. Istit. Bot. Catania 1: 10 (1928); type: C1, between "Scermàrca Hassan" and "Tobungàb", Puccioni & Stefanini 579 (FT holo.).

Baar (Som.).

Rhizomatous perennial; rhizome lacking narrow constrictions; roots spreading in all directions, cord-like, stiff and wiry, dark brown, not sand-covered. Shoots annual, disarticulating at the nodes and base, usually erect to ascending, un-branched or sparsely branched, to 50(–100) cm tall. Leaves spirally arranged, blade sessile, linear-lanceolate (to lanceolate-elliptic), 2–9 × 0.35–1.9(–2.5?) cm, apex acute to acuminate, base cuneate to rounded. Inflorescences terminal, moderately dense to moderately lax ovoid thyrses, 2.5–7.5 × 2.5–6.5 (–9) cm, composed of (11–)13–38, ascending to patent cincinni to 5 cm long and 8-flowered; cincinnus peduncles (6–)10–20 mm long; bracteoles spaced 1.5–7.5 mm apart, cup-shaped, perfoliate. Flowers bisexual and male,

open in the morning, 1.5–1.8 cm wide; pedicels 3.5–6.5(–9 in fruit) mm long, recurved (90°–)120°–225° (mostly c. 180°) in fruit; sepals 3–5 mm long; paired petals 7–8.5 × 6.5–8 mm, lilac to blue, purple or violet, apex hooded. Capsules stipitate, 2(–3)-celled, (12–)14–17 × c. 2.5–3 mm, ± fusiform, straight to slightly curved, rarely falcate, rostrate; dorsal cell usually empty (rarely 1-seeded), ventral cells 1–2-seeded with seeds transversely elliptic to transversely oblong-elliptic, 3.8–4.7 × 1.4–1.8 mm; testa light brown to orange-brown or red-brown, finely pitted, not farinose.

Stable sandhills and level to undulating plains with *Acacia-Commiphora* or *Cordeauxia* bushland, shrubland or dwarf-shrub grassland, growing in red or orange sand or red silty soil, sometimes over limestone, often under bushes; 110–420(–450?) m. C1, 2; not known elsewhere. Faden & Kuchar 88/224; Gillett & al. 22609; Thulin & Dahir 6461.

The flowers, capsules and seeds of this and *A. obbiadense* are very similar. *A. longicapsa* has generally shorter, less recurved pedicels and its capsules generally lack a seed in the dorsal cell.

Plants of Thulin 5667 (121 km SW of "Dusa Mareb" along road to "Belet Huen") have unusually broad leaves (to 2.5 cm wide) and thus approach *A. bracteolatum* Faden, which is known only from the Ogaden, Ethiopia. This collection also comes from an unusual habitat ("rocky hillslope, probably sandstone") and from a higher elevation (450 m) than any collection of typical *A. longicapsa*. Its reproductive features agree with this species, however, and it is tentatively placed here.

4. **A. pusillum** Chiov. (1916); type: S2, "Dafet" between "Uanle Uen" and "Ilduc Uen", Paoli 1277 (FT holo.).

Perennial, usually rhizomatous; roots tuberous, fusiform or uniformly thickened, to 5 cm long and 7 mm thick. Shoots annual, erect, unbranched or sparsely branched, 5–15(–18) cm long. Leaves spirally arranged or distichous, all cauline to mostly basal, blade sessile, moderately to strongly succulent, flat to conduplicate, linear-lanceolate or ovate-elliptic, 2–9(–12) × 0.1–2.5 cm, apex acute to acuminate, base broadly cuneate to rounded. Inflorescences moderately lax to moderately dense narrowly ovoid terminal thyrses, 1.5–6.5 × (0.7–)1–3.5 cm, composed of (2–)6–10(–13) ascending to erect or declinate cincinni to 3 cm long and 8-flowered; cincinnus peduncles (1.2–)1.5–5.5(–8.5) mm long; bracteoles attached (0.6–)1–3.6(–6) mm apart, cup-shaped, perfoliate. Flowers bisexual and male, open in the afternoon, 6–9 mm wide; pedicels (1.5–)3–6.5 mm long, erect or recurved 120–270° in fruit; sepals 2–4 mm long; paired petals 3–5.4 × 2.5–4.5 mm, pinkish red, lilac, blue or violet. Capsules sessile, 3-celled, elliptic to oblong-elliptic, 2.4–5 × 1.5–2.5 mm, apex acute to emarginate; dorsal cell (0–)1-seeded, ventral cells (0−)2-seeded with seeds subdeltate to ovate, 1−1.9 × 0.75−1.3 mm, testa tan to orange-tan, sulcate, farinose about the hilum or not.

1. Leaves often mainly basal at flowering time, blade lanceolate to ovate-lanceolate, the broadest 1−2.5 cm wide .... subsp. *thulinii*
− Leaves all or mostly cauline at flowering time, blade linear to linear-lanceolate, 0.1−0.8 cm wide 2
2. Plants rhizomatous; roots ± uniformly thickened, not fusiform; plants usually growing on gypsum ....subsp. *gypsophilum*
− Plants not rhizomatous (or, if so, then other characters not as above); roots fusiform; plants growing on various substrates but rarely on gypsum 3
3. Blade 0.1−0.3(−0.6) cm wide; flowers coral red ....subsp. *pusillum*
− Blade 0.3−0.8 cm wide; flowers pale lilac .... subsp. *variabile*

subsp. **thulinii** Faden in Smithsonian Contrib. Bot. 6: 89 (1991); type: C2, 4 km N of Bulo Burti, c. 3°52'N, 45°34'E, Thulin & Warfa 5320 (US holo., UPS iso.).

Limestone pavement with silt pockets, hillslope in limestone gravel, *Acacia-Commiphora* bushland; 150−270 m. C2; SE Ethiopia. Thulin & Warfa 4570; Kuchar 15578, 17457.

This is a very distinct taxon and might be recognizable at the species level. In cultivation, however, it tends to become smaller and to lose its distinctiveness.

subsp. **gypsophilum** Faden in Smithsonian Contrib. Bot. 76: 88 (1991); type: N2, "Bihen", Glover & Gilliland 1031 (K holo., EA iso.).

Gypsum hills and limestone gorges and pediments; 200−710 m. N2, 3; C1; not known elsewhere. Beckett 1022; Thulin & Warfa 5423, 5395A.

subsp. **pusillum.**

Acacia-Commiphora bushland on red sand or silt/sand over limestone; 135−340 m; C1; S2; not known elsewhere. Gillett & al. 22597; Thulin & Dahir 6578.

The red flowers in subsp. *pusillum* are unique in the genus. Thulin & Dahir 6578 has much broader leaves than the two other collections and, in the absence of flower colour, it could readily have been assigned to one of the other subspecies.

subsp. **variabile** Faden in Smithsonian Contrib. Bot. 76: 88 (1991).

Open low *Acacia-Commiphora* bushland, *Dichrostachys-Acacia* bushland over limestone or gypsum plains; 10−150 m. C1; S2; E Ethiopia, NE Kenya. Gillett & al. 22399; Thulin & Warfa 4659; Thulin, Hedrén & Dahir 7191.

5. **A. benadirense** Chiov. (1936); type: S2, S of "Merca", Ciferri 73 (PAV holo., FT, K iso.).

Probable perennial with shoots erect to ascending, to 22 cm long (or more). Leaves with blade narrowly lanceolate-elliptic to lanceolate or ovate, 3.5−8 × 1.5−3.2 cm. Thyrses moderately dense, ovoid to ovoid-ellipsoid, (2.5−)3−3.5 × 2−4(−5.5) cm, composed of (8−)10−21 often subverticillate cincinni; bracteoles nearly symmetrically cup-shaped, perfoliate, herbaceous, with a prominent subapical gland and smaller marginal glands. Flowers bisexual and male; pedicels (5−)6−7.5(−9) mm long, recurved in fruit 180−270°, puberulous distally; sepals glandular near the apex and along the margins; lateral staminodes with bilobed antherodes, medial staminode vestigial. Capsules substipitate, 3-celled, 2-valved, probably obovate-elliptic to obovate, 3.4−4 × c. 2 mm, apex emarginate; dorsal valve deciduous (tardily?); dorsal cell 1-seeded, ventral cells 2-seeded. Seeds moderately dimorphic; seed from dorsal cell hemispherical, 1.8−2 mm long, 1.3−1.4 mm wide, 1−1.25 mm thick, testa ± smooth, yellow-buff or orange-buff; seeds from ventral cell ovate to subtriangular 1.3−1.6(−1.7) × 1.35−1.4 × 0.9−1 mm, testa scrobiculate, orange-buff, white-farinose in the depressions.

Dunes and littoral; up to c. 100 m. S2; not known elsewhere. Hedberg & Warfa 88.

The flower colour was described as "greenish-white or yellowish-white" by Chiovenda, but this seems unlikely.

6. **A. trispermum** Faden (1995); type: C2, 12 km NE of Ceel Baraf, then 2.5 km N on cutline, Kuchar 17650 (US holo.).

A. petersii (Hassk.) C.B. Cl. subsp. *pallidiflorum* sensu Faden in Smithsonian Contrib. Bot. 76: 97 (1991) as to specimen from Somalia.

Baar yar (Som.).

Much-branched annual to c. 60 cm with linear-lanceolate to narrowly lanceolate-elliptic leaves 4−9 × 0.4−1.9 cm. Thyrses lax, ovoid, to 3.5 × 4 cm, composed of 1−4 alternate cincinni; bracteoles asymmetrically cup-shaped, usually perfoliate. Pedicels 4−8 mm long, ± uniformly recurved in fruit c. 120−180°, puberulous at least apically; sepals with bilobed glands near the apex, without marginal glands, puberulous; petals pale blue; staminodes 3, apparently all with bilobed antherodes. Capsules stipitate, 3-celled, 2-valved, obovate to subquadrate, 4−4.5 × 3.4−4 mm, carinate dorsally, apex emarginate; dorsal valve deciduous; cells all 1-seeded. Seeds strongly dimorphic, transversely elliptic in outline; seed from dorsal cell hemispherical, c. 2.7 mm long, 2 mm wide, 1.7 mm thick, smooth, orange-buff; seeds from ventral cells 2.7−3.2 × c. 2 × 1.3−1.5 mm, testa scrobiculate to transversely ribbed, whole surface covered by a blackish brown to whitish, matted, farinose layer that can be scraped off.

Sandhills and dunes with *Commiphora* or *Acacia*

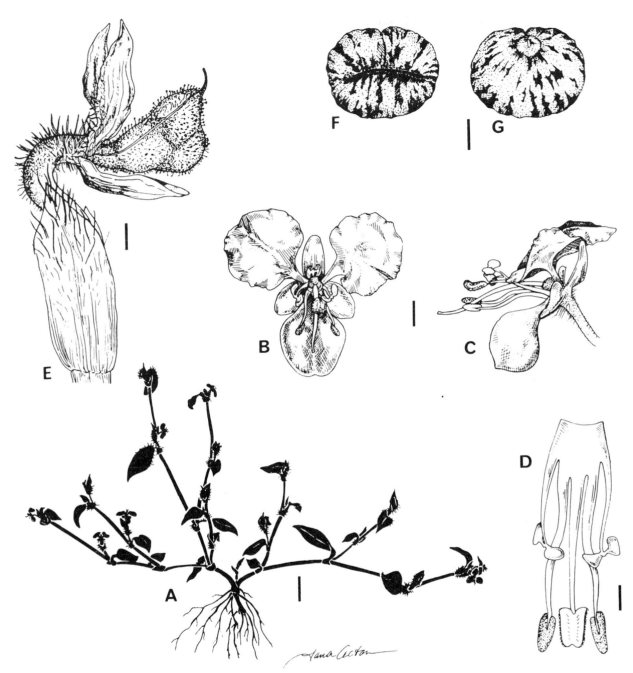

Fig. 55. *Aneilema lamuense*. A: habit. B: bisexual flower, front view. C: bisexual flower, side view. D: stamens and staminodes from bisexual flower, medial staminode removed. E: capsule attached to cincinnus, side view. F: seed from dorsal cell, ventral view. G: seed from dorsal cell, dorsal view. — Modified from Smithsonian Contr. Bot. 76: 135 (1991).

bushland; 150–170 m. C2; S2; not known elsewhere. Thulin & Warfa 5280; Kuchar 17633.

A capsule with three one-seeded cells is a unique feature in the genus.

7. **A. lamuense** Faden (1991). Fig. 55.

Much-branched annual with mostly erect to ascending shoots to c. 20(–40) cm. Leaves spirally arranged (main shoot) or distichous (lateral shoots), blade sessile or shortly petiolate, ovate-elliptic to lanceolate-elliptic or lanceolate, (1–)2–4.5(–5.8) × 1–2(–2.5) cm. Inflorescences terminal on main and short axillary shoots, consisting of a short solitary cincinnus partially enclosed in a pair of herbaceous bracts; cincinnus axis completely covered by overlapping cup-shaped eglandular bracteoles. Flowers bisexual and male, 10.5–14(–15.5) mm wide; pedicels 7–10 mm long, strongly recurved (c. 180°)

only near the apex, pilose-puberulous near the apex, otherwise glabrous; sepals eglandular; paired petals 6–9 × 5–7(–8) mm, lilac, lower petal boat- or cup-shaped, whitish; staminodes all with bilobed antherodes, the lobes yellow with a purplish base; stamen filaments fused basally. Capsules substipitate, 3-celled, 2-valved, obovate to obovate-elliptic, 3.6–4.3(–5) × 1.95–2.1(–2.55) mm, apex emar-ginate; dorsal valve deciduous; dorsal cell 1-seeded, ventral cells 2-seeded. Seeds moderately dimorphic; seed from dorsal cell broadly elliptic, 1.65–2.15 mm long, 1.35–1.55 mm wide, c. 1 mm thick, testa nearly smooth, orange-buff, heavily spotted and striped with dark brown; seeds from ventral cells ovate to sub-triangular, (1.35–)1.45–1.7(–1.8) × 1.4–1.75 × 0.9–1.2 mm, testa scrobiculate or sulcate-scrobiculate, similar in colour to the seed from the dorsal cell, sparsely white-farinose.

Dune area at the coast. S3; coastal Kenya. Kilian 2141 (Lobin 6993).

First record for Somalia. The single collection seen is from 5 km south of Moofe Maam at 0°08'N, 42°46'E.

5. COMMELINA L. (1753)

Annuals or perennials with thin or tuberous roots. Leaves distichous or spirally arranged, blade sessile or petiolate, base usually oblique (sometimes symmetric). Inflorescences terminal and leaf-opposed, composed of 1–2 cymes enclosed in spathes, upper cyme lacking, vestigial or producing 1–several, usually male (occasionally some bisexual) flowers, lower cyme usually several-flowered; spathes with margins free or fused basally, often filled with a mucilaginous liquid. Flowers bisexual and male, zygomorphic; sepals sepaline, free or the anterior 2 connate; petals free, posterior 2 large, clawed, usually blue (sometimes lavender, yellow, peach, apricot or white), anterior petal usually much smaller, sometimes different in colour; filaments all glabrous; stamens 3, anterior, the medial different in form and size from the others; staminodes (2–)3, shorter than the stamens, antherodes commonly 4–6-lobed; ovary 2- to 3-celled, dorsal cell 1-ovulate or obsolete; ventral cells 1–2-ovulate. Capsules 2–3-celled (occasionally 1-celled), 1–5-seeded. Seeds with a linear hilum and lateral embryotega.

About 170 species, almost cosmopolitan, mainly tropical, most diverse in Africa.

In addition to the species treated here *C. diffusa* Burm. f., *C. imberbis* Ehrenb. ex Hassk. and *C. latifolia* Hochst. ex A. Rich. were recorded from Somalia in Cuf. Enum. (1971). The record of *C. imberbis* was based on Gorini 206, which is *C. petersii* and Gorini 207, which is *C.* sp. = Beckett 1478. The records of *C. diffusa* and *C. latifolia* (the latter based on Paoli 303, not seen) have not been

substantiated and therefore these species are also omitted here.

Measurements of spathes in this account refer to the folded spathe.

1. Spathes with margins completely free 2
 – Spathes with margins fused basally 4
2. Flowers blue ....1. *C. stefaniniana*
 – Flowers yellow or buff-orange 3
3. Perennials; flowers yellow ....2. *C. africana*
 – Annuals; flowers buff-orange ....3. *C. arenicola*
4. Spathes solitary, not becoming clustered 5
 – Spathes clustered 10
5. Capsules usually with only the dorsal cell containing a seed (occasionally 1 or more seeds from ventral cells also present); dorsal cell usually with ridges or warts 6
 – Capsules usually with 1–2 seeds in ventral cells; dorsal cell, when present, usually smooth or striate 7
6. Shoots trailing and rooting; plants usually without a definite base; leaves elliptic to linear-lanceolate, pubescent above, usually not inrolled; flowers blue ....7. *C. forskaolii*
 – Shoots ascending to declined; plants with a definite base; leaves linear-lanceolate, usually glabrous above, margin usually inrolled; flowers yellow .... 8. *C.* sp. = Gillett & al. 22610
7. Spathes subsessile (peduncle less than 1 cm long), funnel-shaped, lower margin completely fused 8
 – Spathes distinctly pedunculate (peduncles mostly 1.5–3.5 cm long), not funnel-shaped, lower margin partially fused 9
8. Leaf-blade usually petiolate, pubescent, base oblique, cuneate; sheaths pubescent with long, white or red hairs at the summit; capsules often 5-seeded; roots fibrous; lower petal lanceolate to ovate; cleistogamous flowers sometimes present on short, subterranean shoots from the plant base .... 6. *C. benghalensis*
 – Leaf-blade sessile, glabrous above, base symmet-rical, rounded to cordate or sagittate; sheaths glabrous; capsules 3-seeded; roots tuberous; lower petal linear-oblong; cleistogamous flowers lacking ....9. *C. somalensis*
9. Seeds deeply transversely furrowed, with a single row of tubercles along the ridges, not farinose ....4. *C. petersii*
 – Seeds smooth to alveolate or faintly ribbed, not tuberculate, farinose ... 5. *C.* sp. = Beckett 1478
10. Sheaths with long, white or red hairs at the sum-mit; ventral capsule cells often 2-seeded; leaf-blade lanceolate to ovate; lower petal lanceolate to ovate; cleistogamous flowers sometimes pre-sent on short, subterranean shoots from the plant base .... 6. *C. benghalensis*
 – Sheaths glabrous or with short, white hairs at the summit; ventral capsule cells 1-seeded or empty; blade of lower or all leaves linear to linear-

lanceolate; lower petal linear to linear-lanceolate; cleistogamous flowers lacking 11
11. Upper leaves with bases cordate to sagittate, partially surrounding the spathes, margin strongly undulate .... 9. *C. somalensis*
— Upper leaves with bases cuneate, not surrounding the spathes; margin not or scarcely undulate .. 12
12. Plants appearing woody and shrubby, leafy only distally; spathes semilunate, glabrous; seeds of ventral cell deeply pitted .... 10. *C. frutescens*
— Plants distinctly herbaceous, leafy throughout; spathes ± triangular, pubescent; seeds of ventral cells smooth 13
13. Spathes weakly to not at all falcate; flowers blue ....11. *C. erecta*
— Spathes strongly falcate; flowers usually mauve to lilac (rarely blue or white) ... 12. *C. albescens*

1. C. stefaniniana Chiov. (1916); types: S1, "El Ure", Paoli 1079 (FT syn.), "El Ualac", Paoli 108 bis (FT syn.) & "Baidoa", Paoli 1117 (FT syn.). Fig. 56.

Baar, bacow (Som.).

Perennial 20—30 cm tall, much branched above the simple base; roots tuberous, dark brown, to c. 5 mm thick; shoots annual, erect to ascending, sometimes declinate but not rooting; internodes to 15 cm long, puberulous. Leaves distichous, sheaths to 1 cm long, sparsely to densely puberulous or strigose, blade sessile or lower ones subpetiolate, linear-lanceolate to lanceolate-elliptic, ovate or ovate-lanceolate, 2.5—7.5 x (0.6—)1.2—4 cm, apex acute to acuminate, base oblique, rounded or (in at least the distal leaves) cordate-amplexicaul, margins undulate to nearly flat, ciliolate, both surfaces puberulous, or the upper glabrous, the upper often with c. 3 pairs of maroon spots. Spathes solitary, on puberulous peduncles 1.5—4.2 cm long, 1.6—4.5 x 1—1.9 cm, falcate or not, apex acute to obtuse, base deeply cordate to sagittate, margins free, ciliate at least basally, surfaces puberulous to glabrous outside, usually sparsely strigose inside; upper cincinnus usually shortly exserted and 1-flowered, lower cincinnus 2—5-flowered. Flowers bisexual and male, 1.7—2.6 cm wide; sepals usually sparsely strigose, sometimes glabrous, paired sepals fused into a cup; paired petals blue, lower petal linear, pale lilac; medial stamen with anther saddle-shaped, connective violet, with sterile, yellow, basal lobes, pollen orange. Capsules 3-celled, 2-valved, obovoid, 5—6 x 2.5—3 mm, dorsal valve deciduous, reddish brown, striate, ventral valve stramineous, sometimes spotted and streaked with maroon; dorsal cell (0—)1-seeded, ventral cells (0—)1-seeded, the basal seed aborting in each cell. Seeds of ventral cell elliptic, 3—4.2 x 1.6—1.7 mm, testa reticulate-foveate, dark brown with lighter brown material along the ridges between the pits, white-farinose in the depressions and on the ventral surface.

Pebbly silt or silt/clay plains over limestone or at base of limestone hills with scrub or low shrubland, dried up pond, old gardens; 150—1675 m. N1, 3; C2; S1; Ethiopia, N Kenya. Bally 11825; Kuchar 16984; Popov 1079.

The habitats in N1, from which most collections and all of the higher elevation plants have come, are rarely stated. Thus the plant's ecology is not fully known. It is noteworthy that this species has not been recorded from sandy soils, although in cultivation it grows very well in them.

2. C. africana L. (1753).

Perennial with a distinct base; roots thick, cord-like, usually not tuberous; shoots prostrate to ascending, sparsely to densely branched, to c. 50 cm long. Leaves distichous, sheaths 1—2.7 cm long, with a line of pubescence along the fused edge, ciliate at the apex with colourless hairs, blade lanceolate-elliptic to elliptic, 2.5—7 x 0.7—2 cm, apex acute, base oblique, cuneate to rounded, margins often undulate, surfaces glabrous to puberulous. Spathes solitary, on puberulous peduncles 1—2 cm long, 1.4—3.8 x 0.45—1.2 cm, strongly to not at all falcate, apex acute to acuminate, base cordate to truncate, margins free, ciliate to ciliolate basally, outside glabrous to puberulous, inside pubescent; upper cincinnus lacking or well-developed, long-exserted, 1-flowered, lower cincinnus c. 3—4-flowered; sepals glabrous to pilose, paired sepals free or fused basally; paired petals yellow, lower petal ovate, yellow to yellowish white; staminodes equal or medial reduced; anthers all yellow; ovary strigose ventrally. Capsules oblong-elliptic, 5—7 x 2.5—4 mm, ventrally strigose, dorsal cell 1-seeded, indehiscent, ventral cells 1(—2)-seeded, usually the basal seeds aborting. Seeds of ventral cell transversely oblong, 2.1—2.7 x 1.4—1.8 mm, testa brown, foveolate to foveolate-reticulate, sometimes with abundant, fine, lighter brown granules on the higher parts.

Sheltered places and under shade of bushes; 1400—1555. N1, 2; Djibouti, Ethiopia, widespread in tropical and southern Africa, and in Arabia. Bally 11914; Collenette 289; Thulin, Dahir & Hassan 9132.

Rarely collected in Somalia. The specimens would be assignable to var. *africana* except Collenette 289 which could be placed in var. *villosior* (C.B. Cl.) Brenan, a form with pubescent leaves that may not be worth recognizing.

3. C. arenicola Faden (1995); type: S2, 63 km from Afgooye roundabout on Marka road, Faden & Kuchar 88/111 (US holo.).

C. africana var. *circinata* Chiov., as "*circinnata*", in Atti Ist. Bot. Univ. Pavia, ser. 4, 7: 153 (1936); type: S2, S of "Merca", Ciferri 88 (FT holo., K iso.).

Annual with tufted, much-branched, ascending to prostrate, rarely decumbent shoots to 30(—46) cm long, rooting only near the base; roots thin, fibrous. Leaves spiral or distichous, sheaths 0.5—1 cm long,

Fig. 56. *Commelina stefaniniana*. A: habit. B: spathe and bisexual flower, lateral view. C: bisexual flower, front/side view. D: capsule, dorsal view. E: capsule, lateral view. F: capsule, dorsal view, dorsal valve removed, showing abortive seeds in basal ventral cell. G: seed of ventral cell, ventral view. H: seed of ventral cell, dorsal view. — From Faden & Kuchar 88/232

glabrous or with a line of pubescence along the fused edge, ciliate at the apex, blade sessile, linear to linear-lanceolate or elliptic, usually conduplicate, sometimes strongly recurved, (0.5−)1−5 × 0.4−0.6 cm, apex acute to acuminate, base ± cuneate, surfaces glabrous or pubescent. Spathes with peduncles 0.4−1.4 cm long, solitary, semilunate, not to slightly falcate, 0.9−2 × 0.45−1 cm, apex acute to rounded, base ± hastate, surfaces puberulous to glabrous, margins free, glabrous or ciliate; upper cincinnus present, 1-flowered or abortive, peduncle included (rarely shortly exserted), lower cincinnus 1−4-flowered. Flowers bisexual and male, c. 1 cm wide, petals pale buff, buff-orange or apricot pink, paired petals 5−5.5 × 5−6 mm, staminodes 3, equal. Capsules 5.5−7 × 3−4 mm, dorsal cell indehiscent, 0−1-seeded, ventral cells 1−2-seeded. Seeds of ventral cells transversely oblong-elliptic to broadly ovate, dorsiventrally compressed, 2.7−4.2 × 1.5−2 mm, testa medium to dark brown, foveate to foveate-reticulate with transversely elongate and round pits, each pit filled and bordered with paler farinaceous material, this material often also forming a fine reticulum or partial reticulum on the ridges.

Dunes and littoral, *Acacia tortilis-Combretum* woodland, and cassava cultivation; c. 0−80 m. S2; not known elsewhere. Lavranos & Carter 23182; Raimondo 13/82, 10/86.

The colour of the flowers, testa pattern, and annual habit demonstrate that this species is not related to *C. africana*. The small curled leaves on Ciferri 88 and 88bis and their very compact growth are probably the result of exposure to salt-laden sea breezes.

4. C. petersii Hassk. (1864).
Baar (Som.).

Perennial with thin roots; shoots erect to scrambling, usually sparsely branched, 50−100 cm or more; internodes glabrous or subglabrous. Leaves mainly distichous, sheaths 1−2.5 cm long, glabrous to puberulous, glabrous or ciliolate at the apex, blade of uppermost leaves sessile with a round to truncate base, of lower leaves usually petiolate with a cuneate base, linear-lanceolate to lanceolate or ovate, 4−13.5 × 0.6−3 cm, apex acuminate to acute, margins flat, scabrous, surfaces puberulous. Spathes solitary on puberulous 1−3(−4) cm long peduncles, 2−3 × 0.8−1.4(−1.7) cm, usually distinctly falcate, especially near the apex, densely puberulous and "sticky" to touch, apex usually acuminate, base cordate to truncate, margins fused basally, glabrous; upper cincinnus usually 1-flowered and long-exserted, lower c. 4−5-flowered. Flowers bisexual and male, 1.5−2 cm wide; sepals free; all petals blue, the lower one ovate-lanceolate; anthers entirely yellow. Capsules ± stipitate, oblong, constricted medially, 7.5−9 × (3−)4−5 mm, dorsal cell striate, indehiscent, 1-seeded, ventral cells smooth, dehiscent, (1−)2-seeded. Seeds of ventral cells transversely oblong-

ellipsoid to reniform, 3.3−4.5 × 2.3−2.7 mm, testa dark brown to black or grey, sometimes with contrasting blotches, with 4−6 transverse furrows, each intervening ridge with a row of fine tubercles, farinose granules lacking.

Thickets at base of dune with *Thespesia* and *Acacia* species on sand/silt interface, stabilized dunes with *Acacia senegal* thickets, roadside thorn fence; 65−90 m. S2, 3; Eritrea, Ethiopia, Kenya westwards to Nigeria and southwards to South Africa, Botswana and Namibia, also India and Sri Lanka. Faden & Kuchar 88/269A; Faden & Kuchar 88/115; Paoli 185.

5. C. sp. = Beckett 1478.
Baar, baar maroodi (Som.).

Perennial with thin roots; shoots erect or scrambling, to 2 m. Leaves distichous, sheaths 1.2−3 cm long, puberulous or with a line of pubescence, glabrous to ciliolate at the apex, blade lanceolate-elliptic to lanceolate or ovate, 3.5−13 × 0.7−3.5 cm, apex acuminate, base cordate to rounded or amplexicaul, upper surface scabrous, lower puberulous to glabrescent. Spathes solitary on at least apically puberulous 1.5−5 cm long peduncles, not at all to distinctly falcate, 1.5−3 × 0.8−1.5 cm, puberulous, apex acute to acuminate, margins fused basally; upper cincinnus usually present, 1-flowered, lower cincinnus c. 4-flowered. Flowers bisexual and male; sepals free, all petals blue; lateral stamens with blue anthers, medial stamen anther yellow, sometimes tinged with blue. Capsules stipitate, oblong, constricted in the middle, 6−10 × 4−5.5 mm, dorsal cell striate, indehiscent, 1-seeded, ventral cells (0−)2-seeded. Seeds of ventral cells transversely oblong to reniform, 3−4.3 × 1.5−2.1 mm, testa brown, smooth to alveolate or faintly transversely ribbed, farinose.

Irrigated land near spring, and roadside thorn fence; 60−375 m. S1−3; Kenya, Tanzania. Faden & Kuchar 88/269; Gerrard in SRS 311/2; Mangano s.n.

This plant is *Commelina* sp. D of Upland Kenya Wild Flowers. It closely resembles *C. imberbis* Ehrenb. ex Hassk. of Ethiopia and inland Kenya and is best recognized by its nearly smooth seeds.

6. C. benghalensis L. (1753).
Baar, baar addoi (Som.).

Much-branched annual to c. 30 cm tall, with erect to ascending or decumbent shoots; roots fibrous, thin; cleistogamous flowers sometimes present on short subterranean shoots from the plant base. Leaves spirally arranged or distichous, sheaths 1−2.3 cm long, puberulous or pilose, long-ciliate with rusty or white hairs at the summit, the hairs extended onto the petiole or blade base, blade sessile or petiolate, lanceolate to lancolate-elliptic, ovate-elliptic or ovate, 1.5−7.5 × 0.7−3.5 cm, apex acute to rounded, sometimes mucronulate, base oblique, cuneate, surfaces puberulous or hirsute-puberulous, margins flat or undulate, ciliate basally, scabrous or ciliolate above. Spathes funnel-shaped, subsessile (peduncle to 5 mm long, puberulous), solitary or clustered, 1–1.5(–2.3) × 0.7–1.1 cm, pubescent, falcate or not, pubescent, apex acute, base truncate, upper cincinnus usually well-developed, 1-flowered, shortly exserted, lower cincinnus c. 3-flowered. Flowers bisexual and male, c. 1–1.3 cm wide; sepals free, glabrous; petals blue, lower petal lanceolate-elliptic to ovate; lateral stamens with anthers blue, pollen white, medial stamen with anther yellow, pollen yellow. Capsules oblong-subquadrate to·oblong-elliptic, 3–4.5 × 2–4 mm, cells smooth, the dorsal indehiscent with seed not fused to capsule wall, the ventral (1–)2-seeded. Seeds of ventral cells broadly ovate, 1.6–2.5 × 1.3–1.8 mm, testa reticulate to shallowly and irregularly foveolate-reticulate, brown or grey-brown, farinose.

Rich soil in *Buxus* scrub, grassland, and roadsides on stable, pale orange sand dunes with thick bush dominated by *Acacia senegal* and *Opuntia*; c. 50–1830 m. N1, 2; S1, 2; Djibouti, Ethiopia, Kenya, Arabia, widespread in the Old World tropics, naturalized in tropical and temperate America. Bally 11800; Boaler B50; Faden & Kuchar 88/121.

This species is very variable in Africa, but all of the collections from Somalia seem to belong to the common annual form. *C. benghalensis* is commonly a weed in cultivation in Africa, but it has yet to be recorded as such from Somalia.

7. C. forskaolii Vahl (1805), as "*forskalei*".

Baar, baar biot, horsa-had (Som.).

Perennial (or annual?), sometimes with a definite base and tuberous roots or, more commonly, without a base, and with fibrous roots only; shoots ascending to decumbent and rooting at the nodes, often forming cleistogamous flowers on short, subterranean shoots from the rooted nodes. Leaves distichous, sheaths c. 0.5–1.3 cm long, puberulous, sometimes with a line of longer hairs along the fused edge, ciliate at the apex, blade elliptic to linear-lanceolate, flat to conduplicate, 1.5–6(–7) × 0.4–1.2(–2) cm, apex acute to rounded, often mucronulate and recurved, base oblique, cuneate, surfaces puberulous or pilose-puberulous or the upper sometimes subglabrous, margins strongly undulate. Spathes solitary on 0.4–1.3 cm long peduncles with a line of pubescence, not falcate, 0.8–1.5 × 0.45–0.8 cm, hirsute-puberulous, margins fused basally, glabrous, often violet; upper cincinnus usually well-developed, exserted, 1-flowered, lower 2–3-flowered. Flowers bisexual and male, c. 1.5 cm wide; paired sepals shortly fused basally or ± free; petals dark blue, the lower one very reduced and lanceolate or subulate; lateral stamens with lyrate, winged filaments, anthers blue, medial stamen with larger saddle-shaped, blue or blue-violet connective with yellow, sterile basal lobes. Capsules obovoid, 6–6.5 × 4–4.5 mm when seeds of ventral cells developed, 3.5–4 × 2–2.5 mm when only dorsal cell developed, dorsal cell indehiscent, keeled

on the back and with low, longitudinal, crenate ridges or rows of tubercles, 1-seeded, ventral cells usually abortive, occasionally the apical seed developed. Seeds of ventral cells, when present, transversely elliptic to broadly ovate, c. 2.5–3 × 2 mm, testa brown, smooth, farinose.

Sand dunes, orange sandplains, clay plains, *Acacia* bushland, open bushland, open grassy plains, roadsides, weed in irrigated land; c. 0–1500 m. N1; C1, 2; S1–3; Kenya and Ethiopia southwards to South Africa and Namibia, westwards to Senegal, also in Arabia, Socotra, Madagascar, and India. Faden & Kuchar 88/107; Gillett & al. 24938; Thulin, Dahir & Hedrén 7298.

The habitats at the higher elevations near Hargeisa are not stated and possibly are different from those recorded above. The species is used in Somalia for the treatment of wounds.

8. C. sp. = Gillett & al. 22610.

Baar (Som.).

Tufted perennial herb, sometimes shortly rhizomatous, to c. 25 cm tall; roots fibrous; shoots probably annual, unbranched to densely branched, 5–30 cm long, ascending to declinate, not rooting, internodes to 8 cm long, puberulous to glabrescent. Leaves spiral or distichous, sheaths to 1.3 cm long, sometimes flushed with purple, puberulous, ciliate at the apex, blade sessile, linear-lanceolate, 3–10.5 × 0.3–0.6 cm, flat to completely involute, apex acuminate, base cuneate, upper surface usually glabrous (rarely with a few, long, patent hairs), lower puberulous, margins sometimes ciliate at the base, otherwise papillose. Spathes with 0.4–0.8 cm long puberulous peduncles, solitary, 0.9–1.5 cm × 0.3–0.7 cm, surfaces pilose-puberulous to puberulous, apex acuminate, margins fused basally, sometimes purple, glabrous, upper cincinnus lacking, lower 2–3-flowered. Flowers bisexual and male; paired sepals shortly to longly fused, paired petals c. 8 mm long, yellow, lower petal oblanceolate; staminodes 3, antherodes well-developed; lateral stamens with filaments apparently winged, medial stamen with anther saddle-shaped, sometimes with sterile basal lobes, connective apparently yellow with small dark spots. Capsules with only the seed of the dorsal cell developing, the ventral seeds/ovules aborting. Seed enclosed in the indehiscent dorsal cell and shed with the dorsal capsule valve, dorsal capsule valve c. 4–5 × 2 mm, keeled on the back and with 2–3 regular or irregular longitudinal ridges of spines on each side or with dense, scattered, wart-like projections, or smooth and lacking ridges and warts.

Stable orange sandhills with open bushland, degraded bushland with *Acacia* and *Senna ellisae*, open *Acacia-Commiphora* bushland on red sand over limestone, bushland dominated by *Caesalpinia erianthera* and *Boswellia microphylla*, dwarf-shrub grassland on sandplain; 145–375(–885?) m. N1?;

C2; not known elsewhere. Kuchar 17558; Kuchar 16959; Thulin & Dahir 6553.

This species is very similar to *C. forskaolii* in most characters, differing chiefly in its sometimes distinctly rhizomatous habit, tufted shoots that do not root, long, very narrow leaves (a variable character in *C. forskaolii*) that are usually glabrous above, and yellow flowers. Aside from the flower colour, however, there are no other definite reproductive differences between these species, as can be determined from the limited amount of material (five or six collections) available of this species.

Bally 11829 (K) from "Salawet", foot of "Sheik Pass" at 885 m in N1, if correctly placed here, would greatly extend the geographic and altitudinal ranges of this species. The habitat described as "near stream in grass" is also different from other collections. The specimen consists of four fragments, two with spathes of *C. albescens*, one with a spathe of either *C. forskaolii* or *C.* sp. = Gillett & al. 22610, and a sterile base with tufted shoots that cannot be identified with certainty. It is only the label data "fls. yellow" that suggests that the one fragment might belong here.

9. **C. somalensis** Chiov. (1916); types: S1, "El Ualàc", Paoli 1086 (FT syn.) & "Iscia Baidoa", Stefanini 1211 (FT syn.).

C. baidoensis Chiov. (1951); type: S1, "Baidoa", Corradi 2146 (FT holo.).

Perennial with tuberous roots to 5 mm thick clustered at the base; shoots apparently annual, much branched distally, ascending, 8−40 cm tall; internodes glabrous to sparsely puberulous, to 12.5 cm long. Leaves apparently spirally arranged; sheaths 0.3−1.3 cm long, mostly very short and funnel-shaped, glabrous to sparsely puberulous, not ciliate at the summit; blade sessile, the lower linear-lanceolate, 4−17 × c. 1 cm, broadest at the base, apex acuminate, base rounded, margins strongly undulate, especially towards the base; surfaces glabrous, or lower sparsely puberulous; blade of upper, spathe-subtending leaves narrowly lanceolate to ovate, conduplicate, 2.5−7.5 × 0.6−2 cm, base cordate to sagittate, apex, margins and surfaces similar to those of the lower leaves. Spathes solitary and leaf-opposed, or somewhat clustered due to branching and shortened internodes, subsessile, the base hidden by the subtending leaf base, funnel-shaped, 1−2 × 0.6−1.5 cm, glabrous, apex slightly to not at all recurved, base longly fused, upper cincinnus 1−4-flowered, its peduncle not exserted from spathe, lower cincinnus c. 5-flowered. Flowers bisexual and male, sepals 2−3 mm long, the lower 2 fused except at apex; paired petals c. 7−8 mm long, purple; lower petal linear-oblong; staminodes equal, filaments 2.5−4 mm long, antherodes 6-lobed; lateral stamens with filaments 6−7 mm long, anthers ovate to elliptic, 0.8−1 mm long; middle stamen with filament 4−6 mm long, anther saddle-shaped, c. 1.5

mm long; style c. 7 mm long. Capsules stipitate, 3-celled, 2-valved, laterally compressed, obovoid, c. 7.5 mm long; cells 1-seeded, dorsal indehiscent, ventral dehiscent. Seeds of ventral cells dorsiventrally compressed, ± reniform, 3.5−3.9 × 2−2.1 mm, testa brown, foveate, with whitish ± farinaceous material in the pits, hilum curved.

Bushland on limy soil or limestone pavement with silt pockets; 240 m. C1, 2; S1; not known elsewhere. Kuchar 17456.

10. **C. frutescens** Faden (1995); type: C2, 5 km SSE of Qodqod, Faden & Kuchar 88/229 (US holo.).

Baar (Som.).

Perennial with tufted, erect to ascending or somewhat clambering much branched wiry shoots to 1.2 m tall, appearing woody when dry, shoot bases swollen; roots cord-like, not tuberous; internodes to 7 cm long, glabrous. Leaves borne mostly distally, being shed from the lower nodes, sheaths to 2.7 cm long, glabrous or with a few minute hairs at the summit, blade sessile, linear-lanceolate, 3−8.5 × 0.2−1.1 cm, glabrous or sparsely puberulous with hook-hairs along the midrib beneath. Spathes clustered, subsessile, semilunar, (1.3−)2−3 × (0.7−)1.3−2.5 cm, glabrous, apple green, turning brown and papery, strongly falcate, base longly fused, upper cincinnus lacking, lower 1−2-flowered, bracteoles present. Flowers palest yellow; paired sepals fused; lower petal linear-lanceolate; staminodes 3; medial stamen anther saddle-shaped, larger than the lateral anthers. Capsules broadly elliptic, c. 5 × 4.5 mm, apparently with 3 1-seeded cells, dorsal cell indehiscent, c. 3 mm long, with bumps and pits, ventral cells dehiscent. Seeds of ventral cells hemispherical, elliptic in outline, c. 3.5 × 2.5 mm, testa brown, irregularly and deeply pitted dorsally, with pale tan material in and around the pits, not farinose.

Sand plains with bushland, often dominated by *Cordeauxia edulis*, dwarf shrubland, and dwarf-shrub grassland; 270−395 m. C2; not known elsewhere. Faden & Kuchar 88/229; Kuchar 16955; Thulin & Dahir 6495.

The woody appearance and large, clustered, glabrous spathes make this a very distinctive species. Few seeds have been seen, but it appears that all three cells in a capsule rarely if ever contain a seed. Only one collection (Kuchar 16833) has the flower colour noted, so this should be considered uncertain. The large, papery spathes may serve for wind dispersal. Plants have been observed to grow in shrubs or among spiny *Euphorbia* species, where they are protected from large herbivores.

11. **C. erecta** L. (1753).

C. venusta C.B. Cl. (1901); type: N1, "Golis Range", Cole s.n. (K holo.).

Baar (Som.).

Rhizomatous perennial usually 20−50 cm tall; roots

93

fibrous, thin or cord-like, not tuberous, shoots erect to ascending, or sometimes prostrate, not rooting. Leaves distichous, sheaths to 2.5 cm long, often auriculate at the apex, puberulous, blade usually sessile, linear-lanceolate to lanceolate-elliptic, elliptic or ovate-elliptic, 5−15 × (0.6−)1−3.7 cm, apex acuminate, base oblique, cuneate, surfaces pubescent to subglabrous. Spathes usually clustered, subsessile, 1.5−2.5 × 1−2 cm, usually not or somewhat falcate, occasionally strongly falcate, glabrous to hirsute-puberulous, upper cincinnus lacking, lower several-flowered. Flowers bisexual and male; paired sepals fused into a cup; paired petals blue, lower petal minute, colourless. Capsules c. 2.5−5 × 2.5−6 mm, broader than long when all seeds developed, 3-celled, cells 1-seeded, dorsal cell warty, indehiscent, ventral cells smooth, dehiscent. Seeds of the ventral cell subspherical to transversely ellipsoid, c. 2.5−4 × 2−2.5 mm, testa brown or blackish, smooth, with soft, usually pale patches on the sides or with a whitish torus 3/4 around the circumference of the seed (interrupted by the embryotega), sparsely farinose.

Swampy spot in *Acacia-Terminalia* woodland on pale grey quartz sand, and yellowish white sand with a large, shallow pool, undoubtedly also in drier habitats; c. 35−1500 m. N1; C1; S3; Ethiopia, Arabia, Kenya, widespread in tropical Africa and tropical and temperate America, sometimes treated as pantropical. Gillett & al. 24988, 25295; Wieland 4709 (mixture with *C. albescens*).

It is unclear how widespread this species is in Somalia, because some of the collections may not have been distinguished from the wide-ranging *C. albescens*. Gillett & al. 24988 and 25295, the only collections from S3, are unusually large plants growing in moist situations.

Placing *C. venusta*, known only from the type from the "Golis Range", under *C. erecta* is uncertain. Its spathes are large and strongly falcate, and the very carefully pressed flower is c. 2.5 cm wide, with the limbs of the paired petals c. 20 × 15−20 mm. The petals appear blue. In the absence of further material I am reluctant to accept this as a species and prefer to keep it under the exceedingly variable *C. erecta*.

12. **C. albescens** Hassk. (1867).

? *C. albescens* var. *hirsutissima* Chiov., Fl. Somala 1: 317 (1929); types: S1, "Baidoa", Puccioni & Stefanini 313, 322 (FT syn., not seen).

Baar, hubnaleh (Som.).

Shortly rhizomatous perennial; roots fibrous, thin or moderately thick; shoots tufted, sparsely to densely branched, erect to ascending or prostrate, to c. 35 cm long; internodes sometimes purple; foliage usually somewhat glaucous. Leaves with sheaths 0.8−4.5 cm long, usually not auriculate at the summit, sometimes overlapping basally, sometimes purple-veined, puberulous, ciliate at the apex, blade sessile, linear, 4−12(−16) × 0.2−1 cm, apex acuminate, base oblique, cuneate, surfaces puberulous to pilose or subglabrous. Spathes subsessile, clustered, rarely solitary (in young shoots?), c. 1.2−1.5(−2.5) × 0.8−1(−1.5) cm, usually strongly falcate or recurved, whitish basally, puberulous or hirsute-puberulous, apex acuminate; upper cincinnus lacking, lower 2−3-flowered. Flowers bisexual; paired sepals fused into a cup; paired petals usually lilac or mauve, rarely white (or blue?), lower petal linear-lanceolate, minute, colourless. Capsules c. 3−4 × 2.5−5 mm, broader than long when all seeds developed, 3-celled, cells 1-seeded, dorsal cell verrucose to spinulose, indehiscent, ventral cells smooth, dehiscent. Seeds of ventral cells ovate-subquadrate, c. 2.5−3 × 2−2.5 mm, testa dark brown, smooth, farinose, with varying amounts of soft, tan or whitish material arranged in a torus or partial torus around the seed.

Bushland and grassland, often in rocky ground, sometimes on alluvial plains, rarely on dunes, silty or sandy soil; 90−1645 m. N1−3; C1, 2; S1−3; Djibouti, Sudan, Ethiopia, Kenya, N Tanzania, and in Arabia. Beckett 1013; Gillett & Heemstra 23641; Gillett & Hemming 24810/A.

This species is only moderately separable from *C. erecta*. Its flowers are usually lilac or mauve, but blue flowers are sometimes reported. The more falcate spathes and generally glaucous foliage in *C. albescens* help to distinguish it from *C. erecta*.

163. TYPHACEAE

by M. Thulin

Cuf. Enum.: 1195−1197 (1968); Fl. Trop. E. Afr. (1971).

Rhizomatous aquatic herbs with erect unbranched stems. Leaves linear, in 2 rows, with a long open sheathing base. Inflorescence spike-like, the upper part male, the lower part female, each with a caducous bract at the base, sometimes interrupted by similar bracts. Male flowers of 1−7 stamens with filaments united below; anthers 2-thecous, basifixed. Female flowers sterile and fertile together, with whorls of long straight hairs; ovary superior, stipitate, 1-celled, 1-ovulate; style long, filiform, with flattened stigma; pistil of sterile flowers swollen and club-shaped with a reduced style. Fruit a 1-seeded follicle.

Family with a single genus only.

TYPHA L. (1753)

Description as for the family.

Some 15 species, almost cosmopolitan, in and around fresh or brackish water.

T. domingensis Pers. (1807). Fig. 57.

T. australis Schumach. (1827); *T. angustifolia* L. subsp. *australis* (Schumach.) Kronfeld in Verh. Zool.-Bot. Gesell. Wien 39: 156 (1889); *T. domingensis* var. *australis* (Schumach.) J.B. Gèze in Bull. Soc. Bot. Fr. 58: 459 (1911).

Alool, alool madow (Som.-N); dacar (Som.-S).

Stems to 3 m tall or more. Leaves green or yellowish-green, the sheaths not auriculated or only the uppermost ones somewhat so; blade 5—15 mm wide. Male and female inflorescences usually separated by 1—5.5 cm; male inflorescences usually 15—35 cm long; female inflorescences 12—40 cm long, bright reddish-brown. Stigma linear, exserted.

Swamps, margins of pools and streams; 30—1800 m. N1—3; C2; S1—3; pantropical. Gillett 4341; Bally 11197; Warfa 77.

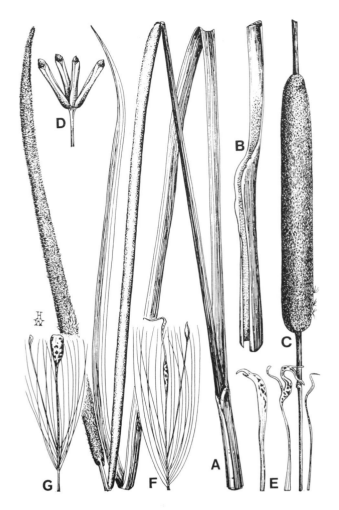

Fig. 57. *Typha domingensis*. A: part of shoot showing mature male and immature female inflorescences, × 0.35. B: leaf sheath, × 0.35. C: mature female inflorescence, × 0.35. D: male flower, × 5. E: male bracteoles, × 9. F: female flower with ovary and bracteole, × 5. G: sterile female flower with club-shaped pistil and bracteole, × 5. — Modified from Fl. Trop. E. Afr. (1971). Drawn by H. Wood.

164. JUNCACEAE

by K. A. Lye

Fl. Trop. E. Afr. (1966); Cuf. Enum.: 1522—1525 (1971); Lye in Lidia 3: 133—140 (1994).

Annual or usually rhizomatic perennial herbs, or rarely subwoody. Leaves mostly basal, grass-like or cylindric, rarely without blades, glabrous or hairy; sheaths open or closed, sometimes with long hairs at the orifice. Inflorescence a terminal panicle that may appear lateral when the main inflorescence bract is cylindric and continuous in the direction of the stem; flowers solitary or in heads; major inflorescence-bracts foliaceous, filiform or cylindric; flower-bracts and bracteoles often present. Flowers regular, small, usually bisexual, greenish or light to dark reddish-brown, more rarely black, white or yellowish; tepals 6. Stamens 3 or 6, shorter than the tepals; filaments linear or triangular; anthers 2-thecous, basifixed, introrse, opening by a longitudinal slit; pollen in tetrads. Ovary superior, 1- or 3-celled; ovules 3—many; style with 3 branches. Fruit a loculicidal capsule. Seeds 3 to many, sometimes with an outgrowth (elaiosome); embryo small, straight, embedded in the endosperm.

Family of eight genera and about 310 species, widely distributed, but with most species in the temperate zones.

Fig. 58. A−C: *Juncus rigidus*, from Snogerup 1993. A: habit, × 0.5. B: pair of flowers with young capsules. C: flower with exserted style. − D: *J. oxycarpus*, habit, × 0.5, from Mooney 5058. − E: *J. punctorius*, habit, × 0.5, from Wickens 1723. F: open flower of *Juncus*. G: floral diagram of *Juncus*. − H: capsule of *Juncus*, opened. − Drawn by G. M. Lye.

JUNCUS L. (1753)

Annual or perennial glabrous herbs, tufted or with elongate rhizomes; stems usually terete. Leaves grass-like, channelled, cylindric or flat, but sometimes reduced to the sheaths only, some species with transverse septa appearing in dry leaves; sheaths usually open, sometimes with auricles. Inflorescence an open or congested terminal or pseudo-lateral panicle. Flowers bisexual; bracteoles sometimes present; tepals 6, free, acute or obtuse. Stames 3 or 6. Ovary sessile. Capsule 1- or 3-celled, or incompletely 3-septate. Seeds many.

About 220 species widely distributed in temperate regions, also reaching the arctic; in the tropics mostly at high altitudes.

1. Plant perennial 2
– Plant annual 4
2. Leaf-blades, if present, and lower bract pungent
   ....1. *J. rigidus*
– Leaf-blades and bracts not pungent 3
3. Leaf-blades 3–6, widely spaced along the stem
   ....2. *J. oxycarpus*
– Leaf-blade 1 only, from the middle or upper part of
   the stem .... 3. *J. punctorius*
4. Flowers borne singly; tepals 2.5–3 mm long;
   capsule almost globose .... 4. *J. sphaerocarpus*
– Flowers in clusters; tepals 4–5 mm long; capsule
   ellipsoid ....5. *J. hybridus*

1. **J. rigidus** Desf. (1798). Fig. 58 A–C.

J. maritimus Lam. var. *somalensis* Chiov. in Lavori R. Ist. Bot. Catania 1: 9 (1928); type: N2, "Gombeia", Puccioni & Stefanini 928 (FT lecto.).

Robust perennial with a thick horizontal rhizome with few–many culms often set in dense rows; stems 50–120 cm long and 2–5 mm thick. Leaves 3–6, basal, reduced to hardened brownish sheaths or 1 or more with ± well developed leaf-blades. Inflorescence much-branched, usually lax, 5–30 cm long and often only 2–7 cm wide, consisting of 1–few sessile to subsessile and 1–many stalked panicles; major involucral bract 3–15 cm long, terete, pungent, erect and continuing in the direction of the stem. Tepals 2–3 mm long, equal or the outer somewhat longer, elliptic-lanceolate, the outer obtuse to acute, the inner obtuse, whitish to light reddish brown with wide transparent border. Stamens 6; anthers 1.5–2 mm long, more than 2.5 times as long as the filaments. Style c. 2 mm long including c. 1 mm long stigmas. Capsule c. 4–5 × 1.5 mm, much longer than the tepals, lanceolate-acute, yellowish-brown, glossy, ending in a c. 0.2 mm long mucro. Seeds 1–1.5 mm long, obliquely lanceolate, reddish-brown with prominent pale basal and apical appendages.

In or near pools and water courses, or in salt marshes, often in gypseous soils; near sea level to 900 m. N1–3; widespread from Sicilia and North Africa

eastwards to Iran, Afghanistan and Pakistan, also in southern Africa. Beckett 1224; Gillett & Watson 23872; Hemming & Watson 3148.

J. rigidus has often been confused with *J. maritimus* Lam., a related species with a more obtuse capsule only slightly longer than the tepals.

2. **J. oxycarpus** E. Mey. ex Kunth (1841). Fig. 58 D.

J. quartinianus A. Rich. (1850).

J. oxycarpus subsp. *sparganioides* Weim. in Svensk Bot. Tidskr. 40: 166 (1946).

Tufted perennial, sometimes decumbent and rooting and branching at the nodes; stems 10–30 cm long. Leaves 3–5 per stem; blades 5–25 cm long and 1–3 mm thick, cylindric, transversely septate; sheaths green or greyish brown, wide. Inflorescence of one sessile globose head of flowers usually subtended by 1–5 stalked heads, and sometimes with additional stalked heads from the base of some stalked primary heads, giving a total of up to 10–20 heads; heads 8–15 mm in diam., often consisting of 20 or more flowers. Tepals equal, 3.5–4.5 mm long, acute, green to light brown when young, turning dark reddish brown when mature. Capsule 2.5–3.5 mm long, oblong with apiculate apex, rather shiny, usually light brown below and dark reddish brown to almost black above. Seeds 0.35–0.45 mm long, ovoid, reticulate.

Swampy ground near water holes or along streams; 1300–1500 m. N2; Ethiopia, East Africa and southwards to the Cape. Newbould 912; Thulin, Dahir & Hassan 9072.

3. **J. punctorius** L. f. (1782). Fig. 58 E.

J. schimperi Hochst. ex A. Rich. (1850).

Robust perennial with horizontal rhizome with closely set or somewhat spaced stems; stems 30–80 cm long and 1–5 mm thick. Lower 2 leaves with wide green to light brown sheaths without blades, but sometimes ending in a short filiform limb; upper sheath ending in a 5–50 cm long cylindric prominently transversely septate blade. Inflorescence consisting of few–numerous hemispherical or globose dense heads of flowers; major peduncles to 7 cm long; smaller heads sometimes crowded into larger composite heads. Tepals equal, 2–3 mm long, acute, light to dark reddish-brown, often with pale margins. Capsule c. 2.5 mm long, ovoid with apiculate apex, shiny, light to dark reddish-brown. Seeds c. 0.3 mm long, ovoid, striate.

Wet soil along streams and in swamps; probably 1500–2000 m. N1; Sudan, Ethiopia, South Africa, North Africa and south-western Asia. Lort Phillips s.n.; Drake-Brockman 309, 310, 468.

4. **J. sphaerocarpus** Nees (1818).

Slender annual with a small root system and many crowded leafy stems; stems 5–20 cm long and 0.3–0.8 mm thick, glabrous, terete to obtusely angular, with 2−4 leaves. Leaves to about 15 cm long and 0.5−1.5 mm wide, flat, gradually tapering to an acute tip. Inflorescence lax, occupying at least 1/2 the total height of the plant; flowers borne singly on a short or long pedicel. Tepals narrowly ovate with a green midrib (sometimes turning brownish with age) and a prominent wide translucent margin; outer tepals c. 3 mm long, prominently acuminate and much longer than the capsule; inner tepals c. 2.5 mm long, acute, usually longer than the capsule. Stamens with linear yellow c. 0.4 mm long anthers. Capsule 2−2.5 × c. 2 mm, almost globose, greenish to brown. Seeds c. 0.3 mm long, smooth to inconspicuously striate.

On clay along stream in evergreen bushland with *Buxus* and *Cadia* on limestone; c. 1300 m. N2; also in Ethiopia, North Africa, Europe, Asia and North America. Thulin, Dahir & Hassan 9103.

First record for Somalia.

5. **J. hybridus** Brot. (1804).

Robust annual with a small root system and many crowded leafy stems; stems 5−20 cm long and 0.5−1.5 mm thick, erect, glabrous, terete or subterete with 3−4 leaves. Leaves to c. 10 cm long and 1 mm wide, flat to inrolled or canaliculate, gradually tapering to an acute tip. Inflorescence lax but with flowers in clusters, to c. 12 cm long and 5 cm wide, from the upper part of the stem only. Bracteoles 2, 1.5−2 mm long, ovate, usually uncoloured. Tepals ovate-lanceolate with a green to brown midrib and a prominent wide translucent margin; outer tepals c. 5 mm long, acuminate and much longer than the capsule; inner tepals c. 4 mm long, obtuse to acute, only slightly longer than the capsule. Capsule 3.5−4 mm long, subprismatic to ellipsoid, obtuse to subacute. Seeds c. 0.3 mm long, smooth or minutely reticulate.

On clay along stream in evergreen bushland with *Buxus* and *Cadia* on limestone; c. 1300 m. N2; Mediterranean area, SW Asia and possibly South Africa, Australia and North America. Thulin, Dahir & Hassan 9104, 9107.

First record for Somalia.

165. CYPERACEAE

by K. A. Lye

Cuf. Enum.: 1414−1495 (1970−71); Haines & Lye, Sedges and rushes of E. Afr. (1983).

Annual or perennial grass-like herbs, rarely tree-like (West African *Microdracoides*); perennial species with short or long rhizomes or with stolons; stems usually solid, triangular, flattened or terete, more rarely 4- or multi-angular, sometimes septate. Leaves often 3-ranked, usually with a closed sheath and a linear blade, but the blade sometimes absent or reduced to a minute limb (e.g. *Eleocharis*). Inflorescence an open or congested anthela (umbel-like structure) or panicle of spikelets, often surrounded by conspicuous leafy bracts, but sometimes reduced to a solitary spikelet. Flowers inconspicuous, unisexual or bisexual, each subtended by a glume (bract) arranged in 1−200-flowered spikelets. Perianth of 3−6 (rarely more) hairs, bristles or scales, but often absent. Stamens 1−3, very rarely 4−6. Ovary solitary, of 2−3 joined carpels, rarely more, superior, 1-celled; style with 2 or 3 branches, rarely more or unbranched. Fruit a small nut (often termed nutlet or achene), sessile or seated on a disc, free but in *Carex* enclosed by a modified prophyll (utricle). The solitary seed with a thin testa not adhering to the fruit-wall.

About 90 genera and 4000 species, especially in the tropics and subtropics, but the largest genus *Carex* more common in temperate regions.

Scleria racemosa Poir. was recorded from southern Somalia in Cuf. Enum.: 1488 (1971) on the basis of a statement by Schumann in Engler, Pflanzenw. Ost-Afr. (1895). As the record has not been substantiated it is omitted here.

1.	Flowers unisexual, i.e. with either ovary or stamen, but not both; female flowers enclosed by a sac-like modified prophyll (utricle) .... 8. *Carex*	
−	Flowers bisexual; female flowers not enclosed by utricle 2	
2.	Flowers with perianth-segments 3	
−	Flowers without perianth-segments 7	
3.	Inflorescence a solitary spike or spikelet 4	
−	Inflorescence of 2−many spikelets 5	
4.	Inflorescence with a stem-like bract continuing in the direction of the stem .... 2. *Schoenoplectus*	
−	Inflorescence bracts scale-like .... 3. *Eleocharis*	
5.	Spikelets few-flowered, producing 1−5 nutlets only .... 7. *Schoenus*	
−	Spikelets many-flowered, producing many nutlets 6	

6. Leaf-blades and leaf-sheaths hairy, but sometimes only on leaf-tips and near throat of leaf-sheaths .... 1. *Fuirena*
− Leaf-blades and leaf-sheaths glabrous or scabrid, or leaf-blades lacking .... 2. *Schoenoplectus*
7. Glumes distichous, i.e. in 2 rows; spikelets often flattened .... 6. *Cyperus*
− Glumes spirally arranged or in 5 rows; spikelets rarely flattened 8
8. Inflorescence a solitary spike or spikelet 9
− Inflorescence of 2−many spikelets at least on some stems 10
9. Leaf-blade reduced to a short triangular limb or absent .... 3. *Eleocharis*
− Leaf-blades conspicuous, but often filiform .... 6. *Bulbostylis*
10. With 3 or more empty glumes at the base of the spikelet .... 7. *Schoenus*
− With only 1−2 (rarely 0) empty glumes at the base of the spikelet 11
11. Inflorescence a panicle; leaves and leaf-sheaths usually hairy .... 1. *Fuirena*
− Inflorescence a lax or congested anthela, often globose 12
12. Major inflorescence-bract stem-like and continuing in the same direction as the stem, the inflorescence therefore appearing lateral .... 2. *Schoenoplectus*
− Major inflorescence-bract(s) leafy or filiform, erect or spreading 13
13. Nutlet with style-base persistent as a distinct knob .... 5. *Bulbostylis*
− Nutlet without persistent style-base 14
14. Leaves flat, often glabrous, without long hairs at opening of leaf-sheaths .... 4. *Fimbristylis*
− Leaves setaceous or narrowly canaliculate, often hairy, with long hairs at opening of leaf-sheaths .... 5. *Bulbostylis*

1. FUIRENA Rottb. (1773)

Annuals or perennials, often with a horizontal rhizome; stems triangular, 5-angular or almost terete, hairy or glabrous, with nodes and leaves throughout their length. Leaf-blades usually flat and hairy at least along margin and at apex; sheaths closed; ligule prominent, tubular. Inflorescence a paniculate corymb consisting of 1−several clusters of spikelets or compound corymbs of spikelets; lower inflorescence-bracts similar to the upper leaves. Spikelets ovate to elongate, producing numerous bisexual flowers; the glumes spirally arranged or (more rarely) 5-ranked making the spikelets angular. Glumes reddish brown to greyish black, usually strongly hairy, and with the midrib excurrent into a straight or recurved mucro or awn. Perianth-segments bristle-like, scale-like, or with a blade on a stalk, 3 or 6, or absent; the 2 whorls of 3 often of different type. Stamens 2−3. Style with 3 long hairy stigmas. Nutlet obovate, 3-angular, smooth or irregularly wrinkled, falling off enclosed in the hypogynous scales or bristles when these are present.

About 30 species in all tropical and warm temperate regions, but with more species in Africa and America.

1. Perennial plant with a horizontal rhizome; achene c. 1.5 mm long .... 1. *F. boreocoerulescens*
− Annual tussocky plant; achene 0.6−1.2 mm long 2
2. Achene 0.6−0.7 mm long, smooth; inner perianth-segments subcordate and swollen to almost ball-like at the apex .... 2. *F. somaliensis*
− Achene 0.8−1.2 mm long with longitudinally raised ribs and narrow rectangular surface cells between the ribs; inner perianth-segments very slender, anchor- or hammer-like at apex.......... .... 3. *F. striatella*

1. **F. boreocoerulescens** Lye (1996); type: N2, Markat, 48°30'E, 10°59'N, Newbould 883 (K holo.). Fig. 59 A−F.

Slender perennial with a horizontally creeping rhizome 2−3 mm thick with stems at about 5 mm intervals, rarely crowded; stems 20−40 cm long and 1−2 mm thick, triangular, minutely pubescent with mostly appressed hairs. Leaves 4−6 and spaced all along the stem, the lowermost with a short triangular limb only, other with well developed blades; the largest blade 10−20 cm long and 3−5 mm wide, minutely hairy on margin at least above, becoming glabrescent with age; sheaths glabrous; ligule 1−2 mm long, cylindrical, glabrous. Inflorescence a terminal cluster of 3−10 crowded spikelets subtended by 2−3 leaf-like involucral bracts 0.5−3 cm long. Spikelets 6−10 × 4−5 mm, ovate-elliptic with obtuse apex and with 40−60 densely imbricate glumes. Glumes 2−2.5 mm long (excluding the 0.7−1 mm long mucro) and 1.2−1.5 mm wide, rectangular, light reddish brown below, greyish above, but often with reddish brown lines, minutely hairy both on surface and margin; midrib 3-nerved, slender, but thickened at apex and excurrent into a thick brown straight hairy mucro. Perianth-segments 6; the 3 outer light reddish brown, hairy, filiform or somewhat flattened in the lower middle part, 1−2 mm long and of very unequal length; the 3 inner (or at least 1) with a well developed obovate stalked light reddish brown blade irregularly thickened and crisped above; midrib prominent, usually excurrent into a c. 0.5 mm long filiform hairy awn; the margin short-hairy or more commonly with hairy lobes or thickenings. Achene 1−1.2 × c. 0.7 mm, light reddish brown, triangular-ovate with cuneate base and tip; apex glabrous or slightly hairy; the surface cells prominent, isodiametric to rectangular with darker

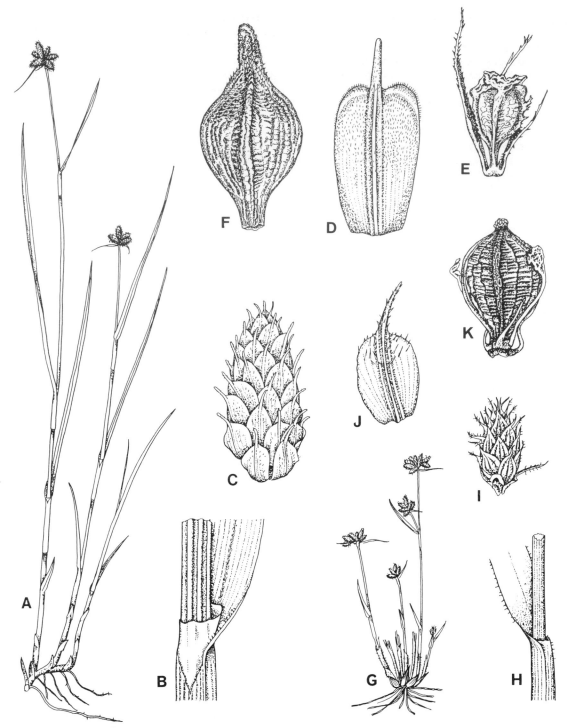

Fig. 59. A−F: *Fuirena boreocoerulescens*, from Newbould 883. A: habit, × 0.5. B: part of stem with base of leaf-blade with ligule, × 5. C: spikelet, × 5. D: glume, × 15. E: perianth, × 25. F: nutlet, × 45. − G−K: *F. striatella*, from Newbould 861. G: habit, × 0.5. H: part of stem with base of leaf-blade with ligule, × 5. I: spikelet, × 5. J: glume, × 15. K: nutlet with perianth, × 40. − Drawn by G. M. Lye.

cell walls.

In open grassy areas on clay; 1350−1500 m. N2; not known elsewhere. Thulin, Dahir & Hassan 9085.

2. **F. somaliensis** Lye (1996); type: S2, "Sablale", Warfa & Warsame 1093 (UPS holo., FT K MOG iso.). Fig. 60 A−D.

Tussocky annual with a small root system; stems

few, clustered, 15−25 cm long and 1−3 mm thick, angular, hairy with short spreading whitish hairs above, glabrous below. Leaves c. 4 per stem, the lower without a blade, the upper with up to 10 cm long and 4−7 mm wide flat densely hairy blades; ligules and sheaths densely hairy but sheaths sometimes glabrescent below. Inflorescence with 3−many spikelets in sessile or stalked clusters from the 2−3 upper leaf-sheaths. Spikelets 4−9 × 3−4 mm, ovate-lanceolate, variegated green and brown, 40−80-flowered with numerous spirally arranged glumes. Glumes c. 3 mm long including a prominent 1−1.5 mm long green strongly hairy mucro, midrib 3-nerved, often greenish with other part of glume dark grey with light brown patches, the entire glume densely hairy, but the hairs on margin and midrib prominently longer, up to 1 mm long. Outer perianth-segments 0.5−0.8 mm long, filiform; the inner c. 1.2 mm long with a subcordate light reddish brown blade that is very pale and swollen and almost ball-like at the apex; lateral nerves not very prominent. Achene 0.6−0.7 × 0.5-0.6 mm, sharply triangular with a prominent long stipitate base, grey with a greenish tinge, smooth.

In alluvial soil; c. 50 m. S2; not known elsewhere.

3. F. striatella Lye (1996); type: N2, Ragad, 48°30'E, 10°57'N, Newbould 861 (K holo.). Fig. 59 G−K.

Tussocky annual; stems few−many, crowded, 2−10 cm long and c. 1 mm thick, often almost terete with longitudinal ridges, glabrous or with a few stiff hairs. Leaves 3−5, the lowermost without or with very short blade; sheaths light reddish brown, brownish or green, densely hairy with short spreading hairs but uppermost sheaths usually glabrescent; ligule c. 1 mm tall, densely short-hairy to glabrous; blades probably to 20 cm or more long and 2−5 mm wide, rather thick, hairy on margin and sometimes on upper surface. Inflorescence a terminal cluster of 3−10 crowded sessile or subsessile spikelets, sometimes with an additional stalked spikelet-cluster from the uppermost sheath of the stem. Spikelets 4−5 × 2−3 mm, ovate with obtuse apex, with 20−40 densely imbricate glumes. Glumes c. 1.5 mm long (excluding the up to 1 mm long mucro), ovate-elliptic, reddish brown to greyish, hairy particularly near margin and on the excurrent greenish midrib. Perianth-segments usually 6, whitish; the 3 outer filiform, often curved and only 0.2−0.3 mm long; the 3 inner c. 0.8 mm long, filiform below, but ending in a minute irregular or anchor-like blade only c. 0.2 mm wide. Achene 0.8−1 × c. 0.5 mm, obovate, obtusely triangular, light reddish brown with longitudinally raised ribs and narrow rectangular surface cells between the ribs.

In clay flush with heavily grazed vegetation; 1350−1400 m. N2; not known elsewhere.

The nearest relative of this appears to be *F. wallichiana* Kunth in India.

2. SCHOENOPLECTUS Palla (1888)
Raynal in Adansonia sér. 2, 15: 537−542 and 16: 119−155 (1976).

Small annuals or large perennials; stem terete or triangular, with or without a node above the base. Leaves mostly basal; blades present or absent; main inflorescence-bract leafy or more commonly stem-like and continuing in the direction of the stem, with or without transverse septa. Inflorescence usually a dense apparently lateral cluster of few−many spikelets, more rarely a lax terminal anthela. Spikelets ovate to lanceolate, usually with numerous bisexual flowers subtended by spirally arranged glumes, rarely of 1−10 flowers only. Glumes ovate to broadly cordate, glabrous or minutely scabrid, but often with longer hairs along margin. Perianth of usually 6 needle-like bristles with or without recurved barbs, or of 4−5 flattened plumose segments (in *S. subulatus* only). Stamens usually 3. Style with 2−3 stigmas. Nutlet obovate to nearly orbicular, triangular or biconvex, smooth or transversely wrinkled. Annual species often with basal cleistogamous flowers hidden in the lower leaf-sheaths and with very long styles with the stigmas protruding from the opening of the sheaths, and with nutlets larger than those produced in aerial spikelets.

Genus of about 50 species in all tropical and temperate parts of the world, sometimes included in *Scirpus* L.

1. Perennial with triangular stems bearing leaf-blades high up the culm; involucral bracts many, green, leaf-like .... 1. *S. maritimus*
− Perennial or annual, often entirely without leaf-blades, and never with long green leafy involucral bracts 2
2. Tall perennial; perianth-segments plumose; inflorescence a lax anthela of many spikelets.......... .... 2. *S. subulatus*
− Perennial or annual; perianth-segments needle-like or absent; inflorescence a lax anthela, a solitary spikelet or a dense head of few−many crowded spikelets 3
3. Inflorescence a dense lateral cluster of few−many sessile spikelets 5
− Inflorescence an anthela with at least some spikelet-clusters stalked 4
4. Perennial; stem 2−8 mm thick . 3. *S. corymbosus*
− Annual; stem 0.4−2 mm thick ... 7. *S. lateriflorus*
5. Thickest stem more than 2 mm thick; erect inflorescence-bract with distinct transverse septa .... 4. *S. articulatus*
− Thickest stem 0.5−2 mm thick, with or without transverse ribs 6
6. Glumes 2−2.5 mm long; nutlet wavy on angles .... 6. *S. roylei*
− Glumes 2.5−3.5 mm long; nutlet not wavy or wrinkled on angles (but prominently so on the faces) 7

Fig. 60. A—D: *Fuirena somaliensis*, from Warfa & Warsame 1093. A: habit, × 0.5. B: spikelet, × 5. C: perianth, × 35. D: nutlet, × 35. — E—G: *Schoenoplectus maritimus*, from Lobin 6914. E: habit, × 0.5. F: spikelet, × 3. G: flower with 6 setae, 3 flattened filaments and an ovary with 3-branched style, × 5. — Drawn by G. M. Lye.

7. Glumes strongly concave at apex and with a golden tinge; inflorescence-bract with distinct transverse septa .... 5. *S. senegalensis*
  – Glumes not concave at apex and with or without a golden tinge; inflorescence-bract without transverse ribs 8
8. Glumes in distinct rows; achene 1–1.1 mm wide with very narrow and sharp transverse frills .... 8. *S. junceus*
  – Glumes not in distinct rows; achene c. 0.8 mm wide with obtuse or sharp less narrow transverse frills .... 7. *S. lateriflorus*

1. **S. maritimus** (L.) Lye (1971); *Scirpus maritimus* L. (1753); *Bolboschoenus maritimus* (L.) Palla (1904). Fig. 60 E–G.

Perennial with long hardened stolons ending in tubers or stems; stems 40–120 cm long and 2–5 mm thick, sharply triangular, pith-filled or hollow. Leaf-blades 25–40 cm long and 3–10 mm wide, flat; sheaths greenish. Inflorescence with clusters of spikelets on branches of very unequal length (usually 0.5–3 cm long); total number of spikelets usually 5–40; largest involucral bracts usually 8–15 cm long, erect or spreading, similar to the leaf-blades. Spikelets 10–50 mm long and 3–5 mm wide, ovate to elongate and often somewhat curved, golden brown to reddish brown. Glumes 5–6 mm long, ovate with midrib excurrent; margin and surface with minute hairs. Perianth-bristles 6. Style with 3 branches. Nutlet 2–2.6 × 1.4–1.8 mm, 3-angular, dark brown and smooth as mature.

Saline soil in pools and seasonally wet swamps and grasslands; near sea-level to 100 m. S2, 3; also in Djibouti, pantropical as well as in temperate regions. Paoli 210; Moggi & Bavazzano 270; Friis & al. 4604.

2. **S. subulatus** (Vahl) Lye (1971); *Scirpus subulatus* Vahl (1805).
Scirpus littoralis Schrad. (1806).

Tall perennial with horizontal stolons; stems 30–450 cm long and 5–12 mm thick below, terete to triangular, pith-filled. Leaf-blades short or up to 70 cm long; sheaths to 70 cm long and 3.5 cm wide. Inflorescence a lax anthela with spikelets or groups of spikelets on branches of very unequal length, nearly all spikelets stalked; major involucral bract 4–6 cm long, erect and stiff, leafy to stem-like. Spikelets 6–15 × 2–5 mm, ovoid to oblong, pale to medium brown. Glumes 3–4 mm long, very broad and rounded at apex, strongly concave, pale with darker brown lines or patches; margin with a prominent pallid border and often numerous hairs, especially near the apex; midrib distinct, extended into a short hairy or scabrid whitish apex. Perianth-segments 4–6, plumose, shorter or longer than the nutlet. Style usually with 2 stigmas. Nutlet 1.5–2(–3) mm long, obovate, biconvex, almost smooth, dark brown when mature.

Temporary pools and river-margins; 60–900 m. N2, 3; tropical and temperate regions of the Old World. Puccioni & Stefanini 670; Hemming 1575; Thulin 9230.

3. **S. corymbosus** (Roem. & Schult.) Raynal (1976); *Isolepis corymbosa* Roem. & Schult. (1817).
Scirpus inclinatus (Del.) Asch. and Schweinf. ex Boiss. (1882).

var. **brachyceras** (Hochst. ex A. Rich.) Lye in Nord. J. Bot. 3: 242 (1983); *Scirpus brachyceras* Hochst. ex A. Rich. (1850).

Robust tussocky perennial with a short thick and woody rhizome; stems 50–300 cm long and 2–8 mm thick, terete and pith-filled. Leaf-sheaths often splitting to give a fine filigree pattern across the split, the upper ending in short lobes; blades absent. Inflorescence a lax anthela with clusters of spikelets on very unequal branches; largest branches 2–12 cm long; major inflorescence bract usually 1–2 cm long, sometimes stem-like or flattened and boat-shaped. Spikelets 3–8 × 1.5–2.5 mm, ovoid, acute, light to dark brown. Glumes 2–4 mm long, ovate, dark reddish brown or greyish with reddish brown lines or patches; margin glabrous or hairy near the apex, without a pale border; midrib scabrid and excurrent into a short mucro. Perianth-segments absent. Style usually with 3 stigmas, rarely 2 or 4, often splitting irregularly. Nutlet 1.2–2 mm long, broadly ovate, almost smooth, dark brown or blackish when mature.

Swamps, pools and lake-margins, c. 1000–1500 m. N2; tropical and South Africa, Madagascar. Glover & Gilliland 921.

4. **S. articulatus** (L.) Palla (1889); *Scirpus articulatus* L. (1753). Fig. 61 A.

Robust tufted annual with few–many crowded stems; stems usually 4–30 cm long (excluding the stem-like bract above the inflorescence) or 12–70 cm long when including the bract, terete. Leaf-sheaths often much wider than the stem; blades absent. Inflorescence of few–numerous crowded, apparently lateral spikelets in a dense head; main inflorescence-bract stem-like, 6–40 cm long and 1–8 mm wide, with conspicuous transverse septa, much longer than the actual stem which ends at the inflorescence. Spikelets 6–18 × 4-10 mm, ovoid, acute, often variegated grey, greenish and reddish brown. Glumes 4–6.5 mm long, ovate-triangular, golden to reddish brown with a greenish slightly excurrent midrib. Perianth absent. Style with 3 stigmas. Nutlet 1.8–2 mm long, somewhat triangular, smooth, almost black when mature. Some basal flowers usually present and producing more globular nutlets, 4–5 × 3–4 mm.

Irrigation channel; below 100 m. S2; Old World tropics. Bavazzano 316; Lobin 6911.

5. S. senegalensis (Steud.) Raynal (1976). Fig. 61 B, C.

Scirpus praelongatus sensu Cuf. Enum.: 1472 (1970), non Poir.

Tufted annual with few—many crowded stems; stems 1—30 cm long below the inflorescence, or total length 6—60 cm when including the bract, 0.3—1.6 mm thick, terete to somewhat angular, hollow with transverse septa. Leaf-sheaths pale to light reddish, without blades. Inflorescence of 1—25 sessile spikelets in an apparently lateral globose cluster; major inflorescence-bract stem-like or flattened, 5—30 cm long, with some transverse septa, much longer than the stem. Spikelets 3—9 × 2—4 mm, ovoid, obtuse, golden brown to bronze-coloured. Glumes 2.5—3.2 mm long, broadly ovate, very concave, pale golden below, darker above, often ending in a short indistinct mucro; midrib distinct only near apex. Perianth absent. Style with 3 branches. Nutlet 1.2—1.5 mm long, obovate, triangular with concave sides, strongly transversely wrinkled on the 3 sides, dark brown when mature. Basal flowers sometimes present and producing 2—2.5 mm long nutlets.

Seasonally wet soils in swamps, pools, ditches and on lake-shores; c. 20—200 m. S1, 3; tropical and South Africa, Egypt, India. Paoli 654, 735; Lobin 7057.

6. S. roylei (Nees) Ovczinn & Czukav. (1963); *Scirpus roylei* (Nees) Parker (1929).

Tufted annual with few—many crowded stems; stems 2—8 cm long below the inflorescence, or total length 5—25 cm including the bract, 0.5—0.9 mm thick, terete; transverse septa present but not prominent. Leaf-sheaths pale to light reddish brown, without blades. Inflorescence of 2—10 sessile spikelets in an apparently lateral globose cluster; major inflorescence-bract stem-like, 6—17 cm long and 0.7—1 mm thick, somewhat flattened, much longer than the stem. Spikelets 4—6 × 2—3 mm, ovoid, variegated pale reddish brown to straw-coloured or with a golden tinge. Glumes 2—2.5 mm long, ovate and slightly concave, light reddish brown. Perianth absent. Style with 3 branches. Nutlet 0.8—1.2 mm long, obovate, triangular, transversely wavy even on the angles, dark brown to blackish when mature. Basal flowers sometimes present and producing 2—2.5 mm long nutlets.

Somewhat saline, seasonally wet swamp; below 50 m. S3; tropical Africa, India. Moggi 190.

7. S. lateriflorus (Gmel.) Lye (1971). Fig. 61 D, E.

Scirpus supinus sensu Cuf. Enum.: 1473 (1970), non L.

Tufted annual with few—many crowded stems; stems 4—40 cm long, or total length 8—60 cm when including the bract, 0.4—2 mm thick, triangular to almost terete. Leaf-sheaths ending in minute lobes or well developed blades 10—30 cm long. Inflorescence lax with clusters of spikelets on branches of unequal length, or contracted to a pseudo-lateral globose cluster; major involucral bract 3—12 cm long, stem-like, shorter than the stem. Spikelets 4—10 mm × 2—2.5 mm, ovoid, acute, variegated grey, green and brown. Glumes c. 3 mm long, ovate, reddish brown or pale with reddish brown lines or patches; margin shortly hairy or glabrous; midrib green, excurrent in a short mucro. Perianth absent. Style with 3 branches. Nutlet 1—1.3 × c. 1 mm, broadly ovoid, triangular, transversely wrinkled, blackish when mature. Basal flowers often present and producing c. 1.5 mm long and wide almost smooth nutlets.

Seasonally wet habitats; sea-level to 500 m. S1; Old World tropics. Lobin 6853.

8. S. junceus (Willd.) Raynal (1976); *Schoenus junceus* Willd. (1794); *Scirpus aureiglumis* Hooper (1972). Fig. 61 F, G.

Tufted leafless glabrous annual with shallow root-system; stems few to numerous, 4—25 cm long, or total length 10—40 cm when including the bract, 0.6—1.5 mm thick, terete with rounded ridges, pith-filled without transverse septa. Leaf-blade absent; leaf-sheath pale, the upper with distinct translucent margin, ending in a 5 mm long linear lobe. Inflorescence consisting of 1—12 sessile apparently lateral spikelets in a dense head-like cluster (rarely with a few sessile spikelets on c. 5 mm long stalk). Main inflorescence-bract leafy to stem-like, 6—25 cm long and 0.5—1.5 mm wide, flattened with longitudinal ridges, usually about as long as the stem. Spikelets 5—9 mm long and 3—4 mm wide, greyish golden, angular with the acute spreading glumes. Glumes 2.5—3 mm long, ovate, concave, arranged in longitudinal rows, golden yellow (but young glumes often uncoloured) occasionally with golden-brown patches or borders; midrib green, 1-nerved, excurrent in a short mucro. Perianth absent. Stamens 3. Style-branches 3. Nutlet c. 1 mm long and 0.8 mm wide, triangular with very distinct transverse frills, as mature dark brown to black. Basal cleistogamous flowers sometimes present, with c. 10 mm long style and nutlet 2—2.5 mm long and c. 1.5 mm wide, transversely wavy and dark brown when mature.

In seasonally wet depressions; from near sea-level to c. 500 m. S3; Old World tropics. Lobin 7039.

3. ELEOCHARIS R.Br. (1810)

Small to large annual or perennial glabrous blade-less herbs; stems usually terete or triangular, hollow or pith-filled, sometimes with transverse septa. Leaf-sheaths usually pale brownish, but often purple at the base, tubular, truncate above and ending in a short triangular or lanceolate lobe. Inflorescence a single terminal ovate to lanceolate, rarely almost globose, spikelet; leafy involucral bracts absent. Lower 1—2

Fig. 61. A: *Schoenoplectus articulatus*, habit, × 0.5, from Lobin 6911. — B, C: *S. senegalensis*, from Lobin 7057. B: habit × 0.5. C: spikelet, × 5. — D, E: *S. lateriflorus*, from Lobin 6853. D: habit, × 0.5. E: spikelet, × 5. — F, G: *S. junceus*, from Lobin 7039. F: habit, × 0.5. G: spikelet, × 5. — Drawn by G. M. Lye.

glumes of spikelet sometimes much stiffer than the fertile glumes. Fertile glumes spirally arranged and imbricate, usually numerous, pale grey to dark brown.

Perianth of 3–9 bristles shorter or longer than the nutlet, or absent. Stamens 1–3. Style with 2–3 branches. Nutlet obovate or urn-shaped, whitish, brown or black, smooth, reticulate, pitted or longitudinally grooved with transverse ridges, with the style base persistent as a conical or flattened appendage.

About 180 species in all tropical and temperate parts of the world, but with most species in America.

1. Glumes c. 1 mm long; nutlet c. 0.5 mm long
   ....3. *E. atropurpurea*
 - Glumes 2−3 mm long; nutlet 0.8−1 mm long ... 2
2. Inflorescence ovate-lanceolate with acute apex, 2−2.5 mm wide; appendage of nutlet as long as wide ....1. *E. intricata*
 - Inflorescence nearly globose with rounded apex, 2.5−3.5 mm wide; appendage of nutlet wider than long ....2. *E. geniculata*

1. **E. intricata** Kük. (1914). Fig. 62 G.

Fairly slender perennial or perhaps sometimes annual forming dense swards with numerous crowded stems from the end of stolons and more scattered stems from horizontally spreading c. 1 mm thick stolons at soil level or slightly above; stems 3−20 cm long and 0.5−1.2 mm thick, terete to angular. Sheath grey above, brown or reddish brown below, the tip triangular, membranous. Spikelet ovate to lanceolate with acute tip, usually 4−5 × 2−2.5 mm, 10−30-flowered. Glumes 2.2−2.5 mm long, broadly ovate, grey to reddish brown, with a green thick midrib ending at or below the rounded apex; margin with pale irregular border. Perianth-bristles 6−7, of which at least 3 longer than the nutlet, light reddish brown with numerous recurved bristles. Style with 2 long branches. Nutlet 0.8−1 mm long (excluding the appendage), dark reddish brown and glossy, with a 0.3 mm long triangular pale appendage.

In marshy ground or by springs in limestone hills; 1350−1500 m. N1, 2; scattered in Africa but rare. Wood 72/140; Newbould 911; Thulin, Dahir & Hassan 9087.

2. **E. geniculata** (L.) Roem. & Schult. (1817). Fig. 62 A−C.

Tufted annual with crowded stems; stems 5−40 cm long and 0.6−1.2 mm thick, terete or angular. Sheath grey above, purple below, ending in a short triangular acute lobe with thinner hyaline margin. Spikelet shortly ovate to almost globose, usually 3−4 × 2.5−3.5 mm, 20−50-flowered. Glumes 2−3 mm long, greyish below, brown above, rounded at the apex, margin usually somewhat frayed and without a distinct hyaline border. Perianth-bristles 5−7, glabrous, longer than the nutlet. Stamens 1−3. Style with 2 stigmas. Nutlet c. 0.8 × 0.6 mm, obovate, biconvex, blackish-purple, smooth and shiny; appendage small, low-conical, pallid.

Swamps and shallow water; near sea-level to 1000 m. N1−3; C2; pantropical, extending to South Africa. Gillett 4339; Bally & Melville 16072; Thulin & Warfa 5620.

3. **E. atropurpurea** (Retz.) Presl (1828). Fig. 62 D−F.

Dwarf tufted annual; stems 1−12 cm long and 0.2−0.3 mm thick, angular. Sheaths usually 2, at least the basal purplish, the upper ending in a short triangular lobe. Spikelets 2−4 × 1−2 mm, ovate, usually 10−20-flowered. Glumes c. 1 mm long, pale to dark reddish brown with a green midrib ending in or near the obtuse tip, with or without a narrow pale margin. Perianth-bristles usually present, shorter than the nutlet, 4−5, with recurved spine-like teeth. Stamens 2. Style with 2 stigmas. Nutlet c. 0.5 × 0.4 mm, broadly obovate, biconvex, smooth and glossy, as mature almost black; appendage greyish, minute, flattened.

Seasonally wet swamps or grassland; below 100 m. S3; pantropical. Lobin 7048 & Kilian 2196.

4. **FIMBRISTYLIS** Vahl (1805)

Tussocky annual or perennial herbs, or stems more distant from rhizomes or stolons; stems terete, angular or flattened, glabrous or scabrid. Leaves mostly basal; sheaths glabrous or short-hairy, without long flexuose whitish hairs at opening; blades usually well developed, flat, linear, to c. 40 cm long and most often 1−3 mm wide, glabrous or short-hairy. Inflorescence a lax anthela of 1 sessile spikelet and 1−numerous stalked spikelets or new groups of sessile and stalked spikelets, more rarely spikelets congested, or inflorescence a terminal spikelet only; involucral bracts usually leafy and spreading, shorter or longer than the inflorescence. Spikelets ovate to elongate, with numerous bisexual flowers subtended by spirally arranged glumes. Glumes ovate to lanceolate or almost orbicular, glabrous or short-hairy, sometimes with longer hairs on margin; midrib ending below apex or excurrent in a short or long mucro or awn. Perianth absent. Stamens 1−3. Style often strongly flattened and fimbriate, with 2 or 3 stigmas; base of style widened but falling with the rest of the style, very rarely with slender processes descending from the style-base over the nutlet. Nutlet orbicular or obovate to almost cylindrical, triangular or biconvex, smooth, tuberculate or reticulate with hexagonal or linear surface cells that are never longer than wide when seen from base towards apex.

About 150 species in all tropical and subtropical parts of the world, but especially numerous in Asia and Australia.

1. Stigmas 3; nutlet triangular to obovate; stem flattened ....1. *F. complanata*
 - Stigmas 2; nutlet biconvex or flattened; stem angular or terete 2
2. Upper part of glumes grey with densely set minute hairs ....3. *F. ferruginea*
 - Glumes glabrous on upper surface 3

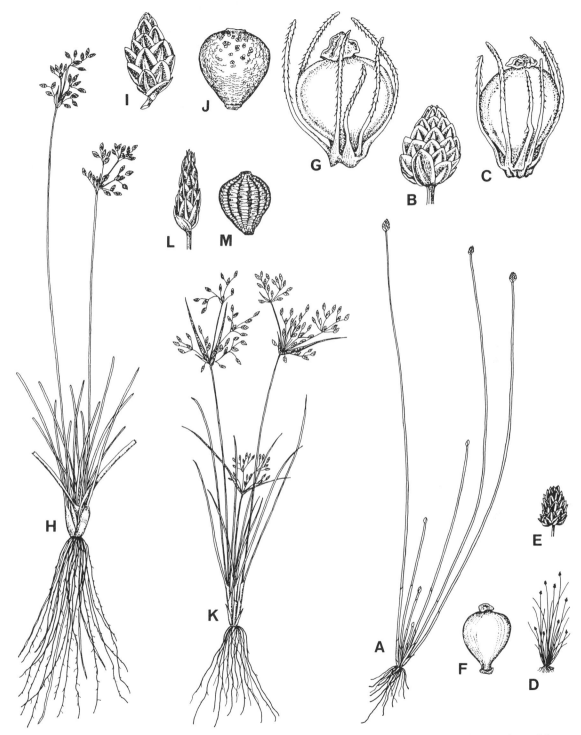

Fig. 62. A−C: *Eleocharis geniculata*, from Thulin & Warfa 5620. A: habit, × 0.5. B: spikelet, × 5. C: nutlet, × 25. − D−F: *E. atropurpurea*, from Lobin 7048. D: habit, × 0.5. E: spikelet, × 5. F: nutlet, × 25. − G: *E. intricata*, nutlet, × 25, from Wood 72/140. − H−J: *Fimbristylis cymosa* subsp. *spathacea*, from Bizi 128. H: habit, × 0.5. I: spikelet, × 5. J: nutlet, × 25. − K−M: *F. bisumbellata*, from Lobin 6897. K. habit, × 0.5. L: spikelet, × 5. M: nutlet, × 25. − Drawn by G. M. Lye.

3. Spikelets clustered; nutlet not longitudinally striate .... 2. *F. cymosa*

− Spikelets never clustered; nutlet longitudinally striate .... 4. *F. bisumbellata*

1. F. complanata (Retz.) Link (1827).

Tufted perennial with short woody rhizome; stems 25—180 cm long and 1—2 mm thick, flat to almost winged, glabrous or scabrid on margins just below the inflorescence. Leaves with compressed sheaths and well-developed blades, the largest 5—20 cm long and 1.5—2 mm wide, glabrous but minutely scabrid on margin; ligule a rim of densely set short hairs. Inflorescence a lax anthela with a sessile spikelet subtended by stalked spikelets and new groups of sessile and stalked spikelets; major involucral bracts usually shorter than the inflorescence. Spikelets 5—8 mm long, but to 12 mm long when fruiting, c. 2 mm wide, ovoid to lanceolate, acute, brown, 5—15-flowered. Glumes 2.5—3 mm long, ovate, brown to dark brown with pale margin; midrib keeled, ending in an acute apex or shortly excurrent. Stamens 3. Style with 3 stigmas. Nutlet 0.7—0.9 mm long, obovate, obscurely triangular, with cells in longitudinal rows and with scattered tubercles.

Swamps, along streams, and near water holes; 1350—1500 m. N2; pantropical. Newbould 908; Thulin, Dahir & Hassan 9077, 9084.

2. F. cymosa R.Br. (1810).

subsp. **spathacea** (Roth) Koyama in Micronesia 1: 83 (1964). Fig. 62 H—J.

Densely tufted perennial with a short erect rhizome and numerous flat obtuse basal leaves 2—20 cm long and 0.7—2 mm wide ending abruptly in a short mucro. Flowering stems 3—50 cm long, angular, glabrous. Inflorescence a compound anthela with numerous small pedunculate clusters of sessile spikelets. Spikelets 3—5 × 1.5—2.5 mm, pale brown. Glumes c. 1.5 mm long, pale brown with a narrow green midrib ending below the rounded or emarginate slightly fimbriate apex; margin pale, wide and prominent. Ovary with 2-branched style. Nutlet 0.7—0.8 mm long, obovate, smooth or rough with isodiametric cells in longitudinal rows, as mature dark brown to almost black.

On sandy beaches, grasslands, and salt marshes; from near sea-level to c. 300 m. N1—3; S2, 3; pantropical. Moggi & Bavazzano 1903; Gillett 23057; Lavranos & Carter 23502.

3. F. ferruginea (L.) Vahl (1805).

Tufted perennial with a short rhizome; stems 50—100 cm long and 1—3 mm thick, somewhat compressed, glabrous or with scattered spine-like or obtuse teeth. Inflorescence a lax or congested anthela of sessile and stalked spikelets. Spikelets 8—18 × 3—5 mm, oblong-ovoid, 20—40-flowered. Glumes 3—4 mm long, broadly ovate, light to dark reddish brown, densely short-hairy in upper half; the midrib slightly extended. Stamens 3. Style with 2 long stigmas. Nutlet 1—1.7 mm long including the short gynophore, biconvex, almost smooth with isodiametric surface-cells.

subsp. **ferruginea.**

Basal sheaths coriaceous, shining brown. Leaf-blades up to 10 cm long and 0.5—1.5 mm wide. Involucral bracts usually shorter than the inflorescence. Spike- lets acute, 3—4 mm wide. Nutlet 1—1.4 mm long, obovate.

On sea-shores and in other saline habitats; sea-level to 50 m. N2; S3; pantropical, but uncommon in Africa. Paoli 174; Moggi & Bavazzano 1709; Newbould 1020A.

subsp. **sieberiana** (Kunth) Lye in Nord. J. Bot. 2: 335 (1982); *F. sieberiana* Kunth (1837); *F. mauritiana* Tausch (1824).

Basal sheaths herbaceous, paler. Leaf-blades on fertile stems to 30 cm long and 1.5—2 mm wide. Lower involucral bract often longer than inflorescence. Spikelets more obtuse and 3—5 mm wide. Nutlet 1.3—1.7 mm long, obovate to orbicular.

Wet grassland, stream beds and swamps on saline or alkaline soils; sea-level to 1350 m. N1—3; S1, 3; also Djibouti and Socotra, tropics and subtropics of the Old World. Gillett 4338; Glover & Gilliland 132; Thulin, Dahir & Hassan 9110.

4. F. bisumbellata (Forssk.) Bub. (1850). Fig. 62 K—M.

Tufted annual sometimes forming very dense tussocks; stems usally 8—35 cm long and 0.5—1 mm thick, triangular, glabrous. Leaf-sheaths brownish or straw-coloured; ligule a dense rim of short hairs; blades often 2—5 cm long and 1—2 mm wide, scabrid on margin and major ribs. Inflorescence a lax anthela of one sessile spikelet and few—many stalked spikelets or new groups of sessile and stalked spikelets; peduncles slender; major involucral bract leaf-like, shorter or longer than the inflorescence. Spikelets 3—8 × c. 1.5 mm, angular, acute. Glumes 1.3—1.8 mm long, somewhat concave, brown with a green 3-nerved midrib excurrent in a short mucro; margin often shortly ciliate. Stamen 1. Style with 2 stigmas. Nutlet 0.6—0.7 × 0.4—0.5 mm, obovate, biconvex, conspicuously trabeculate with 5—7 longitudinal rows of cells on each side, tubercles absent.

Sandy seasonally wet places, especially on river banks; sea-level to 150 m. S2, 3; Old World tropics from West Africa and the Mediterranean to Australia. Raimondo 8; Lobin 6060, 6897.

5. BULBOSTYLIS Kunth (1837), nom. cons.

Annual or perennial herbs of small to medium size; stems erect or somewhat curved, triangular, polyangular or almost terete, glabrous, scabrid or hairy. Leaves near the base only; blades well-developed, flat or filiform or rarely reduced to sheaths, usually with long flexuose hairs at the opening of the sheaths. Inflorescence a lax anthela, a head of congested spikelets or reduced to a single terminal spikelet; involucral bracts leafy or glume-like, often shorter than the inflorescence. Spikelets with few—numerous spirally arranged glumes, all scales fertile or 1—2 lowermost without flowers. Glumes pale, reddish brown to almost black, often with green midrib, ovate to orbicular with obtuse or acute apex, sometimes the midrib extended into an awn, glabrous or more commonly scabrid or short-hairy. Perianth absent. Stamens 1—3. Style with 2 or (commonly) 3 stigmas. Nutlet obovate to obcordate, obtusely or sharply triangular, rarely biconvex, pale, grey, reddish brown to almost black; surface smooth, transversely wrinkled, papillose, reticulate or longitudinally striate; the style base swollen and often persisting on the mature nutlet as a small knob.

About 150 species in all tropical regions, but especially common in tropical Africa, and growing in relatively drier habitats than most other members of *Cyperaceae*.

1. Annuals; nutlet smooth or minutely reticulate .. 2
— Annuals or perennials; nutlet transversely wrinkled 3
2. Glumes 2.5—3 mm long .... 4. *B. pallescens*
— Glumes 1.5—2 mm long .... 5. *B. barbata*
3. Perennial with a short rhizome; glumes 4—6 mm long ....1. *B. craspedota*
— Annuals with a slender root-system; glumes 2—5 mm long 4
4. Nutlet with prominent papillae or tubercles on the 3 angles .... 3b. *B. hispidula* subsp. *pyriformis*
— Nutlet without prominent papillae or tubercles on the angles 5
5. Glumes 4—5 mm long..................................
.... 3c. *B. hispidula* subsp. *macroglumis*
— Glumes 2—3 mm long 6
6. Basal leaf-sheaths reddish..........................
.... 2c. *B. somaliensis* subsp. *microcarpa*
— Basal leaf-sheaths pallid, straw-coloured or brown
....7
7. Nutlet prominently triangular with 5—6 transverse undulations on each side..........................
....................3a. *B. hispidula* subsp. *hispidula*
— Nutlet obtusely triangular to almost globose with 7—10 transverse undulations on each side 8
8. Nutlet almost globose................................
...............2a. *B. somaliensis* subsp. *somaliensis*
— Nutlet obovate, obtusely triangular..................
...................2b. *B. somaliensis* subsp. *confusa*

1. **B. craspedota** Chiov. (1932); types: S3, "Baddada", Senni 382 (FT syn.) and "Licchitore", Senni 214 (FT syn.).
Tufted perennial with woody horizontal rhizome and crowded shoots at the growing end; stems 20—70 cm long and 0.5—1.5 mm thick, angular, glabrous except for minute spine-like hairs below the inflorescence. Leaf-sheaths light brown, glabrous except for long flexuose hairs at their openings; blades 5—15 cm long and 0.5—1 mm wide, flat or canaliculate, almost glabrous. Inflorescence of 2—12 spikelets, either all sessile forming a head or with 1—3 additional stalked spikelets or spikelet-clusters; involucral bracts inconspicuous. Spikelets 6—10 × 2—4 mm, ovoid. Glumes 4—6 mm long, reddish brown with a green midrib, minutely hairy, ovate with acute apex; margin ciliate. Nutlet 1.4—1.8 mm long, obovate, transversely wrinkled; style base persistent.

Dry grassland; near sea-level to 100 m. S3; northern Kenya.

2. **B. somaliensis** Lye (1996); type: C1, 17—18 km N of Hobyo on road to Jirriiban, Thulin & Dahir 6653 (UPS holo., K iso.).

a. subsp. **somaliensis**. Fig. 63 I—L.
Slender tussocky annual; stems 5—25 cm long and 0.3—0.6 mm thick, angular, densely short-hairy. Leaf-sheaths pale to light reddish brown with green central part; blades 2—6 cm long and 0.2—0.5 mm wide, flat or inrolled, densely hairy on midrib and margins. Inflorescence a lax anthela up to 2 cm wide and long consisting of one sessile spikelet and 1—4 stalked spikelets; major involucral bract 3—10 mm long, with a leaf-like upper part. Spikelets 4—7 × 2—2.5 mm, ovate to lanceolate with acute tip, angular, 10—15-flowered. Glumes 2—2.5 mm long, light to medium reddish brown but with a prominent green midrib excurrent into a short mucro, densely short-hairy. Nutlet 0.7—0.8 × 0.6—0.7 mm, obovate to almost globose, not triangular, pale to light reddish brown with c. 10 transverse undulations.

Sand over flat, open limestone rocks; below 100 m. C1; not known elsewhere.

b. subsp. **confusa** Lye (1996); type: S1, 3 km SW of Diinsoor, 2°24'N, 42°58'E, Thulin, Hedrén & Dahir 7615 (UPS holo.).
Slender annual; stems 5—20 cm long and 0.3—0.6 mm thick, angular, hairy or almost glabrous. Leaf-sheaths pale, densely hairy to almost glabrous with long hairs at the openings; blades usually less than 15 cm long and frequently only 1—2 cm long and 0.2—0.5 mm wide, flat or canaliculate, usually densely hairy. Inflorescence a simple or compound lax anthela of one sessile spikelet subtended by 2—many stalked spikelets or new groups of sessile and stalked spikelets, rarely all spikelets almost sessile. Involucral bracts usually shorter than the inflorescence. Spikelets 4—15 × 2—3 mm, ovate to elongate. Glumes 2—2.5 mm long, light to dark reddish brown, but often with paler midrib and margin, minutely hairy both on surface and margin. Nutlet 0.7—0.8 mm long, pale, grey or light brown, obovate, obtusely triangular with 7—10 rounded transverse wrinkles on each of the 3 sides; angles smooth; swollen style-base usually not persistent on the mature nutlet.

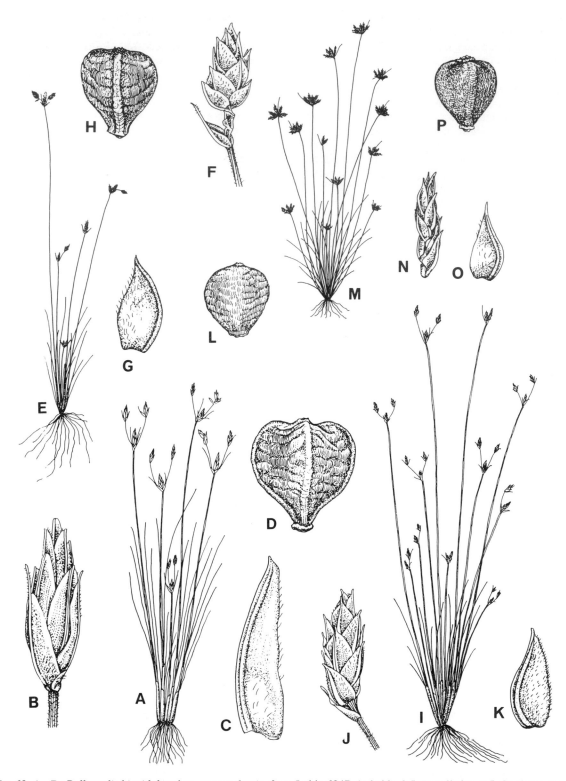

Fig. 63. A−D: *Bulbostylis hispidula* subsp. *macroglumis*, from Lobin 6847. A: habit, 0.5. B: spikelet, × 5. C: glume, × 10. D: nutlet, × 25. − E−H: *B. hispidula* subsp. *hispidula,* from Lobin 6972. E: habit, × 0.5. F: spikelet, × 5. G: glume, × 10. H: nutlet, × 25. − I−L: *B. somaliensis* subsp. *somaliensis,* from Thulin & Dahir 6653. I: habit, × 0.5. J: spikelet, × 5. K: glume, × 10. L: nutlet, × 25. − M−P: *B. barbata*, from Thulin & al. 7607. M: habit, × 0.5. N: spikelet, × 5. O: glume, × 10. P: nutlet, × 25. − Drawn by G. M. Lye.

Open grassland, often on sand or on shallow soils over rocks; sea-level to 350 m. S1—3; not known elsewhere. Moggi & Bavazzano 75; Thulin & Mohamed 6783; Thulin, Hedrén & Dahir 7510.

c. subsp. **microcarpa** (Chiov.) Lye (1996); *Fimbristylis cioniana* Savi var. *microcarpa* Chiov., Result. Sci. Miss. Stefanini-Paoli: 180 (1916); type: S1, between "Goriei" and "El-Magu", Paoli 629 (FT holo.).

Slender tussocky annual. Stems 10—25 cm long and 0.3—0.5 mm thick, densely short hairy. Leaves with prominent red basal sheaths contrasting brightly with the long white hairs from the orifices of the leaf-sheaths; leaf-blades 2—10 cm long and 0.3—0.5 mm wide, flat or incurved, densely short hairy. Inflorescence a lax anthela of 1 sessile spikelet subtended by 1—8 solitary spikelets or groups of spikelets. Spikelets 4—6 x 1.5—2.5 mm, light brown, ovate to lanceolate. Glumes 1.7—2 mm long, light reddish brown, but often darker near the midrib above and paler near margin, but without prominent pale marginal border, densely short-hairy both on surface and margin. Nutlet 0.6—0.7 x 0.4—0.5 mm, obovate, white to light reddish brown, with 10—15 transverse wrinkles on each of the 3 sides; the angular ribs papillose.

Sandy seasonally damp soil; c. 150 m. S1; not known elsewhere.

3. **B. hispidula** (Vahl) R. Haines (1983); *Fimbristylis hispidula* (Vahl) Kunth (1837).

a. subsp. **hispidula.** Fig. 63 E—H.

Slender annual. Stems 5—20 cm long and 0.3—0.5 mm thick, angular, hairy to glabrescent. Leaf-sheaths straw-coloured to brownish, prominently nerved, densely short-hairy with longer hairs at the orifices; blades to c. 10 cm long and 0.5 mm wide, flat or somewhat inrolled, hairy. Inflorescence a solitary terminal spikelet or consisting of 1 sessile spikelet subtended by 1—3 stalked spikelets; involucral bracts usually shorter than the inflorescence. Spikelets 4—6 x c. 3 mm, ovate with obtuse or subacute apex, 10—15-flowered. Glumes 2.2—2.5 mm long, ovate-triangular, medium to dark reddish brown, but frequently pale towards margin and apex, densely short-hairy both on surface and margin; midrib green, minutely hairy and excurrent into a short mucro. Nutlet c. 0.8 mm long, prominently triangular both in outline and in section, light reddish brown, with 5—6 prominent transverse undulations on each of the 3 sides; swollen style-base usually not persistent on the mature nutlet.

Wooded sand-dunes; below 50 m. S2; widespread in tropical Africa, but the Somali plants have smaller glumes than plants from elsewhere. Lobin 6972 & Kilian 2120.

b. subsp. **pyriformis** (Lye) R. Haines in Sedges and rushes of E. Afr., Appendix 3: 1 (1983).

Slender tufted annual with numerous stems and slender roots; stems 5—20 cm high and 0.2—0.5 mm thick, deeply ridged and densely set with short hairs; base of stem covered by very pale translucent old leaf-sheaths. Leaves 5—12 cm long and c. 0.2 mm wide, canaliculate and densely set with short hairs (0.1—0.2 mm long), but orifice and leaf-sheath with 2—3 mm long slender hairs; leaf-sheath very pale, transparent. Inflorescence a simple umbel-like anthela or spikelet solitary; major involucral bracts leafy, 5—15 mm long, densely hairy. Spikelets 4—8 x 2—3 mm. Glumes c. 3 mm long, reddish brown with a very distinct green midrib (distinctly protruding in the lower glumes, slightly or not protruding in the upper ones); short hairs present on margin, surface and midrib. Nutlet c. 1.3 x 1 mm, with a very distinctly cuneate base, white or pale brown; surface transversely undulate except for the 3 protruding tuberculate ribs; swollen base of the style either deciduous or persistent as a distinct brown knob.

Probably in open woodland or grassland, c. 500—1500 m. N1; East Africa. Lort Phillips s.n.

c. subsp. **macroglumis** Lye (1996); type: S1, Buur Heybo, 44°17'E, 3°00'N, Lobin 6847 (BONN holo.). Fig. 63 A—D.

Densely tussocky annual with many crowded stems; stems 5—20 cm long and 0.3—0.6 mm thick, angular and densely pubescent with 0.3—0.7 mm long hairs. Leaves often 3—4 per stem with new leafy shoots produced intravaginally; sheath pale to light reddish brown, pubescent with long flexuose hairs at their orifices; blades 3—10 cm long and 0.2—0.4 mm wide, flat, pubescent. Inflorescence a simple anthela of 1 sessile spikelet subtended by 1—3 stalked spikelets on up to 2 cm long peduncles, rarely reduced to a solitary spikelet; involucral bracts erect or spreading, leaf-like and to 2 cm long. Spikelets 5—9 x 3—4 mm, ovate. Glumes 4—5 mm long, medium to dark reddish brown, but with a wide pale margin, minutely pubescent and with a slightly scabrid raised pale midrib ending in or slightly below the apex, but the 2—3 lowermost scales with the midrib excurrent into a prominent up to 2 mm long awn. Nutlet 1—1.2 x c. 1 mm, grey to reddish brown, obtriangular in outline, triangular in section; surface with 6—10 prominent transverse undulations; swollen triangular style-base not persistent on the mature nutlet.

In cracks in granite outcrops; 450 m. S1; not known elsewhere.

This is very similar to subsp. *brachyphylla* (Cherm.) R. Haines, but is an annual plant with a shorter stem and a smaller inflorescence.

4. **B. pallescens** (Lye) R. Haines (1983); *Abildgaardia pallescens* Lye (1982).

Slender tufted annual; stems 5—12 cm long and

0.3−0.5 mm thick, scabrid. Leaf-blades 1−5 cm long and 0.3−0.7 mm wide, densely scabrid on margin and ribs; leaf-sheaths straw-coloured to light brown, densely scabrid. Inflorescence a solitary terminal cluster (5−10 mm wide), of 2−5 sessile spikelets; major inflorescence-bract filiform, 5−10 mm long. Spikelets 3−6 × 2−4 mm, ovoid, greyish. Glumes 2.6−3 mm long, ovate, pubescent, greyish or straw-coloured, but often reddish-brown below; midrib green, excurrent in a distinct mucro. Nutlet 0.7−0.8 × 0.4−0.5 mm, triangular, light brown with the base of the style persisting as a darker knob, indistinctly reticulate with rectangular surface-cells.

Open sandy soil in shrubland; 20−100 m. S3; Kenya. Moggi & Bavazzano 1442.

5. **B. barbata** (Rottb.) C.B. Cl. (1893); *Scirpus barbatus* Rottb. (1773). Fig. 63 M−P.

Slender tufted annual; stems 5−25 cm long and 0.2−0.4 mm thick, glabrous. Leaf-blades 1−10 cm long and 0.2−0.5 mm wide, scabrid on margin at least above; sheath with long slender hairs at its opening. Inflorescence a solitary terminal head of few−numerous spikelets, 3−15 mm in diam. Spikelets 3−8 × 1−1.5 mm, ovoid-lanceolate. Glumes 1.5−2 mm long, ovate, reddish brown with paler margin and usually green slightly excurrent scabrid midrib, glabrous or sparsely short-hairy. Stamen usually 1 only. Nutlet 0.5−0.7 mm long, obovate, triangular, light brown, smooth with isodiametric surface-cells; the style base persisting on the mature nutlet as a small knob.

Sandy soils; from near sea-level to 300 m. C2; S1, 3; widespread in the Old World tropics. Moggi & Bavazzano 1620; Thulin & Mohamed 6782; Thulin, Hedrén & Dahir 7607.

6. CYPERUS L. (1753)

Perennial or (more rarely) annual herbs, tufted or with creeping rhizomes or stolons, sometimes producing tubers or bulbs; stems trigonous or (more rarely) subterete or 6-angular, usually leafy only at the base, rarely halfway up, but some species lacking basal leaves altogether; stem-base sometimes swollen and succulent. Leaves tristichous, linear and grass-like, the lower ones often scale-like, covering the base of the stem and the rhizome, rarely all reduced to their sheaths. Stem- and leaf-anatomy of eucyperus-type (in species 1−10) or of chlorocyperus-type (in species 11−86). Inflorescence terminal, often a subumbel-like open anthela or congested into a dense head (capitate), usually consisting of numerous spikelets set in distinct spikes or in digitate clusters; involucral bracts usually similar to the leaves, the base of each branch (ray) enclosed in a tubular, two-keeled prophyll. Spikelets ± compressed, linear or ovate, 1−50 flowered; axis (rhachilla) often winged

by the decurrent base of the glumes, persistent or caducous (then spikelets falling off as a whole). Glumes distichous, white, grey, green, brown, reddish brown or blackish, often variegated, with 1−5 ± distinct nerves on each side of the midrib; midrib often green and of another colour than the other parts of the glume, sometimes excurrent in a straight or recurved mucro. Flowers bisexual. Hypogynous bristles or scales absent. Stamens 3, 2 or 1; the connective often produced into an apical appendage. Style 3- or 2-fid, rarely almost undivided. Nutlet sessile, trigonous or lenticular, usually obovoid or ellipsoid; surface often tuberculate or papillose.

Large genus of about 650 species; it is taken here in a wide sense including *Anosporum* (species 10), *Kyllinga* (species 70−81), *Mariscus* (species 49−51, 53−69), *Pycreus* (species 83−86), *Queenslandiella* (species 82) and *Sorostachys* (species 8−9).

1. Inflorescence of one or more spikes having a distinct rhachis (axis); at least some of the spikes stalked, but sometimes the stalks only 2−5 mm long 2
− Inflorescence a solitary usually dense head (very rarely with 1−4 additional heads), or inflorescence more open but then spikelets arranged in digitate clusters and not in spikes 31
2. Plants without leaf-blades; stem with transverse rings; plants perennial ....15. *C. articulatus*
− Plants with the largest leaf-blades more than 3 cm long or, if shorter, then annual plants; stem without prominent transverse rings; plants perennial or annual 3
3. Style with 2 branches; nutlet flattened 4
− Style with 3 branches; nutlet triangular 7
4. Plants annual with a minute root-system 5
− Plants perennial with woody rhizome or stolons 6
5. Glumes with midrib excurrent in prominent mucro .... 82. *C. hyalinus*
− Glumes obovate and obtuse 83. *C. macrostachyos*
6. Plant with woody rhizome; in marshes......... ....13. *C. alopecuroides*
− Plant with slender stolons ending in bulbs; on sand-dunes .... 22. *C. afrodunensis*
7. Plants annual with a minute root-system; glumes 1.3−1.7 mm long 8
− Plants perennial with stolons, rhizome or slightly woody or hardened stem-bases 9
8. Glumes elliptic with the midrib excurrent into a long mucro ....30. *C. squarrosus*
− Glumes obovate to rounded with the midrib excurrent into a very short mucro ....26. *C. iria*
9. Inflorescence a single spike without leafy involucral bracts ....24. *C. bulbosus*
− Inflorescence of few−many spikes in a simple or compound anthela or, if reduced to a single spike, then at least some spikes with leafy involucral bracts 10

10. Basal sheaths semisucculent producing bulbous-swollen bases; spikelets falling off entire when mature 11
— Basal sheaths not succulent; spikelets sometimes with rhachilla persistent after the lower glumes are shed 14
11. Glumes 2.5−3 mm long, greyish white to greenish ....62. *C. phillipsiae*
— Glumes 3−4 mm long, light to dark reddish brown 12
12. Leaf-blades 0.2−0.6 mm wide, filiform .... 66. *C. chaetophyllus*
— Leaf-blades 1−6 mm wide, flat 13
13. Glumes with midrib excurrent into a short recurved mucro ....63. *C. vestitus*
— Glumes with midrib not excurrent.................. .... 65. *C. amauropus*
14. Plants with slender stolons, sometimes ending in bulbs or tubers 15
— Plants without stolons; plant base swollen or hardened or stems coming from woody rhizomes 25
15. With bulbs 16
— Without bulbs, but tubers often present 18
16. Bulbs underground; stems distant from the bulb; spikelets 3−4 mm wide ......22. *C. afrodunensis*
— Bulbs often at ground-level; stems arising directly out of the bulbs; spikelets 1.5−3 mm wide 17
17. Spikelets dark reddish, 1.5−2 mm wide.......... ....24. *C. bulbosus*
— Spikelets tawny, 1.5−3 mm wide................... ....23. *C. grandibulbosus*
18. Spikelets falling off entire when mature.......... .... 49. *C. ferrugineoviridis*
— Spikelets remaining attached to the rhachis whilst the lower mature glumes and nutlets are shed... 19
19. Stolons stout, rhizome-like, not ending in tubers; stem-base not much swollen .... 16. *C. longus*
— Stolons slender, frequently ending in small tubers, or stem-base swollen 20
20. Spikelets with distant obtuse glumes.............. .... 21. *C. dilatatus*
— Spikelets with crowded closely imbricate glumes 21
21. Rhachilla (axis) of spikelet not or hardly winged .... 14. *C. procerus*
— Rhachilla of spikelet distinctly winged 22
22. Spikelets yellowish brown or rusty brown; glumes 2.2−2.6 mm long, with raised nerves almost to the margin .... 20. *C. esculentus*
— Spikelets reddish brown to dark purplish; glumes 2−4.5 mm long, with a wide marginal border without nerves 23
23. Glumes 2−3 mm long with a wide uncoloured margin ....19. *C. maculatus*
— Glumes 2.7−4 mm long with a narrow un-coloured margin or such margin absent........ 24

24. Glumes without a pale marginal border or, if such a border is present, then glumes less than 3.5 mm long ....17. *C. rotundus*
— Glumes with a pale marginal border, 3.5−4 mm long ....18. *C. nubicus*
25. Spikelets falling off entire when mature......... ....50. *C. hemisphaericus*
— Spikelets remaining attached to the rhachis whilst the lower mature glumes and nutlets are shed... 26
26. Spikes less than twice as long as wide, with few spikelets; largest spikelets often 15−40 mm long 27
— Spikes at least twice as long as wide, often crowded; spikelets not exceeding 12 mm when flowering 29
27. Spikelets less than 1 mm wide; glumes 1.7−2.6 mm long .... 48. *C. distans*
— Spikelets 1−1.5 mm wide; glumes 2−3 mm long 28
28. Stem-base not prominently swollen................. .... 16. *C. longus*
— Stem-base prominently swollen.................... ....19. *C. maculatus*
29. Glumes with a rounded keel and inrolled margin .... 13. *C. alopecuroides*
— Glumes with a distinct green keel; margin not inrolled 30
30. Spikes 6−15 mm wide with 30−120 crowded spikelets; glumes 1.2−1.8 mm long... 12. *C. dives*
— Spikes 15−25 mm wide often with only 15−30 rather distant spikelets; glumes 1.8−2.4 mm long ....11. *C. exaltatus*
31. Inflorescence rather open with spikelets in digitate clusters 32
— Inflorescence congested into a solitary, usually dense, occasionally cylindrical head, very rarely with 1−4 additional sessile or stalked heads ... 71
32. Without basal leaf-blades, but inflorescence bracts often leafy 33
— With leaf-blades in lower part of stem 35
33. Involucral bracts leafy, numerous and far over-topping the inflorescence ......1. *C. alternifolius*
— Involucral bracts scale-like or leafy, but only 5−30 mm long 34
34. Major inflorescence-branches all of equal length; stem without transverse rings ....2. *C. prolifer*
— Major inflorescence-branches of unequal length; stem with transverse rings ..... 15. *C. articulatus*
35. Glumes 1.5−2 mm long; style with 2 stigmas; nutlet flattened or rounded 36
— Glumes often larger; style with 3 stigmas; nutlet triangular 38
36. Glumes 1.5−2 mm long, to a large extent whitish with large surface cells; nutlet c. 1.2 mm long .... 86. *C. dwarkensis*
— Glumes 1−1.3 mm long, reddish brown without prominent surface-cells; nutlet 0.5−0.6 mm long 37

37. Glumes conspicuously mucronate with excurrent midrib; nutlet obovate .... 84. *C. pumilus*
– Glumes with midrib ending below the obtuse apex; nutlet broadly obovate to angular in outline .... 85. *C. micropelophilus*
38. Plants annual with a minute root system 39
– Plants perennial with woody rhizome or slender or tough stolons, or stem base swollen 45
39. Leaf-blades prominently spirally twisted.......... .... 58. *C. micromedusaeus*
– Leaf-blades not prominently spirally twisted... 40
40. Spikelets 3–5 mm wide; glumes 3.5–5 mm long ....27. *C. compressus*
– Spikelets 0.8–2 mm wide; glumes 0.5–2 mm long 41
41. Glumes 0.5–0.8 mm long .... 3. *C. difformis*
– Glumes 1–2 mm long 42
42. Glumes truncate to emarginate with 3-nerved midrib ending in a short or long recurved mucro; nutlet 0.5–0.8 mm long 43
– Glumes oblong to ovate, obtuse with the midrib ending in apex; nutlet 0.4–0.6 mm long 44
43. Glumes 1.5–2 mm long including a long re-curved mucro ....29. *C. cuspidatus*
– Glumes 1.3–1.6 mm long with a very short mucro .... 28. *C. amabilis*
44. Glumes straight; connective of the anthers setu-lose .... 5. *C. haspan*
– Glumes spreading; connective of the anthers smooth ....6. *C. tenuispica*
45. Glumes 3.5–9 mm long 46
– Glumes 0.5–3 mm long 48
46. Glumes 7–9 mm long, whitish to pale dirty brown; at low altitudes only, often on sandy sea shores ....31. *C. crassipes*
– Glumes 3.5–5 mm long, grey, brown or reddish 47
47. Spikelets strongly compressed; often inland; base of plant not woody ....27. *C. compressus*
– Spikelets not strongly compressed; only near the sea-shore; base of plant woody ...... 37. *C. frerei*
48. Plant with stolons 49
– Plant without stolons 51
49. Glumes 3–4 mm long; nutlet 1.6–1.8 mm long; stem-base succulent .... 65. *C. amauropus*
– Glumes 0.6–2 mm long; nutlet 0.6–0.9 mm long; stem-base not succulent 50
50. Glumes 0.6–0.8 mm long .... 3. *C. difformis*
– Glumes c. 2 mm long .... 4. *C. commixtus*
51. Stem-base with semisucculent basal sheaths pro-ducing oval to cylindric swollen bases 52
– Stem-base not succulent 59
52. Glumes 1.5–2.5 mm long 53
– Glumes 3–6 mm long 54
53. Stem-base ovate and bulb-like; leaf-blades not twisted; glumes 1.5–1.8 mm long.................. ....25. *C. densibulbosus*

– Stem-base cylindrical; leaf-blades prominently spirally twisted; glumes 2–2.5 mm long....... .... 58. *C. micromedusaeus*
54. Largest spikelet at least 25 mm long; glumes 5–6 mm long ....57. *C. ossicaulis*
– Largest spikelet to 20 mm long; glumes 3–4 mm long 55
55. Leaf-blades flat, the largest 2–4 mm wide..... ....52. *C. rubicundus*
– Leaf-blades filiform, the largest 0.5–1.5 mm wide 56
56. Glumes 3–3.5 mm long; nutlet 0.7 × 0.5 mm ....56. *C. baobab*
– Largest glumes usually 3.5–4 mm long; nutlet 1–1.5 × 0.8–1.2 mm 57
57. Glumes triangular with acute apex and mucronate midrib 58
– Glumes ovate to ligulate with obtuse to retuse apex, but midrib excurrent into a mucro..... .... 55. *C. obbiadensis*
58. Midrib of glume prominently keeled and scabrid in upper half of glume, excurrent into a 0.6–0.8 mm long recurved mucro.......................... ....53. *C. pseudosomaliensis*
– Midrib of glume only slightly keeled near apex, glabrous or with a few teeth, excurrent into a very short mucro .... 54. *C. cunduduensis*
59. Glumes 1.3–2 mm long; leaf-blades rather thin 60
– Glumes 2.3–5 mm long; leaf-blades thicker... 63
60. Stems 5–10 cm tall with a bulb-like base........ ....25. *C. densibulbosus*
– Largest stems 15–80 cm tall 61
61. Nutlet elliptic, 1 × 0.4 mm .... 4. *C. commixtus*
– Nutlet obovate, 0.5–0.6 × 0.3 mm 62
62. Stems 10–40 cm tall; glumes 1.3–1.6 mm long; nutlet obovate to spherical ....5. *C. haspan*
– Stems 50–80 cm tall; glumes 1.5–2 mm long; nutlet obovate-triangular ..7. *C. altomicroglumis*
63. Spikelets 1–1.5 mm wide 64
– Spikelets 2–5 mm wide 65
64. Glumes c. 3.5 mm long; nutlet 1.1–1.3 mm long .... 61. *C. recurvispicatus*
– Glumes 2.3–2.7 mm long; nutlet 0.8–1 mm long .... 36. *C. mudugensis*
65. Glumes 4–5 mm long 66
– Glumes 2.5–4 mm long 67
66. Roots densely covered by tomentum; nutlet 1.8–2 mm long, triangular .... 33. *C. macrorrhizus*
– Roots without tomentum; nutlet c. 1.5 mm long, compressed and curved ...... 39. *C. benadirensis*
67. Spikelets 4–6 × 2 mm .... 60. *C. somalidunensis*
– Largest spikelets more than 6 × 2 mm 68
68. Roots densely covered by tomentum................ ....34. *C. conglomeratus*
– Roots without tomentum 69
69. Glumes 2.5–3 mm long; spikelets 2–3 mm wide; nutlet triangular .... 35. *C. jeminicus*

− Glumes 3−5 mm long; spikelets 3−5 mm wide; nutlet compressed and prominently curved 70
70. Stem smooth; nutlet c. 1.5 mm long..............
.... 39. *C. benadirensis*
− Stem prominently scabrid at least on some ridges; nutlet c. 1.2 mm long .... 40. *C. scabricaulis*
71. Inflorescence bright yellow...........................
.... 44. *C. niveus* var. *flavissimus*
− Inflorescence white or brownish 72
72. One head of spikelets sessile, subtended by 1−4 stalked heads 73
− Head of spikelets solitary 74
73. 1−2 stalked heads present; spikelets 3−5 mm wide .... 41. *C. mogadoxensis*
− 2−4 stalked heads present; spikelets 1−1.5 mm wide .... 51. *C. pluricephalus*
74. Leaf-blades absent or very short; involucral bracts 1−2, often shorter than the inflorescence 75
− Leaf-blades well developed; involucral bracts 2−many, the largest much longer than the inflorescence 76
75. Plant floating; stem strongly triangular; glumes 4.5−5.5 mm long .... 10. *C. colymbetes*
− Plant not floating; stem terete to angular; glumes 2−3.5 mm long .... 47. *C. laevigatus*
76. Style with 3 long stigmas or stigma unbranched or shortly 2-branched; nutlet triangular to rounded 77
− Style with 2 long stigmas; nutlet flattened 111
77. Plants annual with a minute root-system 78
− Plants perennial with rhizome, stolons or a swollen stem-base 82
78. Glumes 0.6−0.8 mm long .... 3. *C. difformis*
− Glumes 1.3−5 mm long 79
79. Glumes 1.3−2 mm long ending in long recurved mucro 80
− Glumes 2.5−5 mm long with or without mucro...
.... 81
80. Spikelets 1.7−2 mm wide; glumes 1.6−2 mm long
....29. *C. cuspidatus*
− Spikelets 2−3 mm wide; glumes 1.4−1.7 mm long .... 30. *C. squarrosus*
81. Stem-base not succulent; spikelets 10−50 mm long; glumes 3.5−5 mm long; nutlet 1.5−1.7 mm long .... 27. *C. compressus*
− Stem-base succulent; spikelets 4−15 mm long; glumes 2.5−4 mm long; nutlet 0.9−1.1 mm long .... 52. *C. rubicundus*
82. Inflorescence light to dark reddish brown, or at least with larger dark coloured patches 83
− Inflorescence white, but may dry to a pale pinkish brown, with or without a few reddish brown spots 102
83. Plants with long stolons or horizontal rhizomes; sea-level to 150 m 84
− Plants growing in small tussocks and without a well-defined rhizome; from sea-level to 2000 m 86

84. Spikelets 4−10 mm long; glumes 3−4 mm long
.... 32. *C. chordorrhizus*
− Spikelets 10−40 mm long; glumes 4.5−8 mm long 85
85. With scale-covered stolons; glumes 6−8 mm long, pale brown to white .... 31. *C. crassipes*
− With horizontal rhizome; glumes 4.5−5 mm long; reddish brown .... 37. *C. frerei*
86. Glumes 1.3−1.5 mm long; nutlet 0.5−0.6 mm long, pear-shaped ....45. *C. meeboldii*
− Glumes 2.3−6 mm long; nutlet 0.7−2 mm long 87
87. Glumes 5−6 mm long; nutlet 2 mm long.......
.... 38. *C. poecilus*
− Glumes 2.3−4 mm long; nutlet 0.7−1.5 mm long 88
88. Nutlet as long as wide, compressed and curved or with 3 very concave faces 89
− Nutlet longer than wide, obscurely to prominently triangular, not compressed 91
89. Stem prominently scabrid; leaf-blades 2−3 mm wide ....40. *C. scabricaulis*
− Stem glabrous; leaf-blades 0.5−2 mm wide ... 90
90. Leaf-blades straight or curled at the apex only; glumes 3−3.5 mm long; nutlet c. 1.2 mm long and wide .... 41. *C. mogadoxensis*
− Basal leaf-blades prominently and repeatedly curled; glumes 2−3 mm long; nutlet 0.9−1 mm long and wide .... 42. *C. medusaeus*
91. Stem-base succulent with fleshy leaf-sheaths forming an ovate or cylindrical pseudobulb ...92
− Stem base not succulent, but rhizome sometimes thickened at the stem-base 98
92. Leaf-blades flat, the largest 2−4 mm wide 93
− Leaf-blades usually filiform or canaliculate, the largest 0.3−1.5 mm wide 95
93. Spikelets 4−6 mm wide, not disarticulating before the glumes and nutlets are shed; nutlet 0.9−1.1 mm long .... 52. *C. rubicundus*
− Spikelets 1.5−4 mm wide, disarticulating as one unit with the glumes and nutlets persistent; nutlet 1.6−1.8 mm long 94
94. Spikelets 4−5 x 1.5−2 mm ...... 64. *C. cruentus*
− Spikelets 5−17 x 2−4 mm .... 65. *C. amauropus*
95. Glumes 3−3.5 mm long; nutlet 0.7 x 0.5 mm ....56. *C. baobab*
− Largest glumes usually 3.5−4 mm long; nutlet 1−1.5 x 0.8−1.2 mm 96
96. Glumes ovate to ligulate with obtuse to retuse apex, but midrib excurrent into a mucro .... 58. *C. obbiadensis*
− Glumes triangular with acute apex and mucronate midrib 97
97. Midrib of glume prominently keeled and scabrid in upper half of glume, excurrent into a 0.6−0.8 mm long recurved mucro..........................
.... 53. *C. pseudosomaliensis*

— Midrib of glume only slightly keeled near apex, glabrous or with a few teeth, excurrent into a very short mucro .... 54. *C. conduduensis*

98. Spikelets c. 1.5 mm wide; nutlet 0.8−1 × 0.4−0.5 mm .... 36. *C. mudugensis*
— Spikelets 2−4 mm wide; nutlet 1−1.5 × 0.8−1.3 mm 99

99. Spikelets 4−6 mm long, 5−9-flowered..........
.... 60. *C. somalidunensis*
— Spikelets 5−30 mm long, 10−50-flowered... 100

100. Nutlet c. 1 mm long, prominently triangular
.... 59. *C. gypsophilus*
— Nutlet 1.2−1.5 mm long, obscurely triangular
.... 101

101. Roots densely covered by tomentum; spikelets compressed to subterete . 34. *C. conglomeratus*
— Roots without tomentum; spikelets strongly compressed .... 35. *C. jeminicus*

102. Plants with long stolons .... 31. *C. crassipes*
— Plants without stolons 103

103. Spikelets 4−25 mm long, 8−30-flowered, the rhachilla persistent after at least some of the lower glumes and nutlets have fallen 104
— Spikelets 2−8 mm long, 1−8-flowered, falling off entire when mature 109

104. Glumes 0.8−1.6 mm long 105
— Glumes 2−6 mm long 107

105. Leafy involucral bracts 5−8, longer than the stem; glumes 1.3−1.6 mm long 46. *C. pygmaeus*
— Leafy involucral bracts 2−3, much shorter than the stem; glumes 0.8−1.5 mm long 106

106. Glumes 1.3−1.5 mm long .... 8. *C. pulchellus*
— Glumes 0.8−1 mm long ...... 9. *C. microglumis*

107. Nutlet obscurely triangular, minutely papillose
.... 34. *C. conglomeratus*
— Nutlet triangular with concave adaxial side, smooth 108

108. Glumes 3−3.5 mm long ......43. *C. somaliensis*
— Glumes 4−6 mm long .... 44. *C. niveus*

109. Leaf-blades 0.5−1 mm wide ... 67. *C. scleropus*
— Leaf-blades 1.5−3.5 mm wide 110

110. Glumes 3.5−4.5 mm long ....68. *C. paolii*
— Glumes 2−2.5 mm long .... 69. *C. dubius*

111. Inflorescence a cluster of 5−10 easily discernable pale spikelets .... 86. *C. dwarkensis*
— Inflorescence a dense globose or cylindrical head of numerous crowded spikelets 112

112. Inflorescence white, cream or greenish white, but sometimes greyish when mature (the colour of dark mature nutlets is seen through the thin glumes) 113
— Inflorescence yellow, golden or greenish 120

113. Long slender stolons present.......................
.... 73. *C. purpureoglandulosus*
— Stolons absent, but rhizome sometimes horizontal 114

114. Keel of glumes winged; wings toothed..........
.... 80. *C. alatus*

— Keel of glumes not winged, but sometimes scabrid or hairy 115

115. Keel of glumes with long cilia; glumes to 2 mm long .... 75. *C. welwitschii*
— Keel of glumes glabrous or scabrid of short spine-like teeth 116

116. Spikelets c. 1 mm long, perfecting 1 nutlet only; unbranched part of style very short.............
.................................76. *C. microstylis*
— Spikelets 3−8 mm long, perfecting 1−5 nutlets; unbranched part of style usually long 117

117. Glumes not prominently keeled; midrib distinct near apex only .... 72. *C. brunneofibrosus*
— Glumes keeled with midrib distinct at base also
.... 118

118. Base of stem without remains of old leaf-sheaths
.... 74. *C. sesquiflorus* subsp. *appendiculatus*
— Base of stem covered by fibrous remains of old leaf-sheaths 119

119. Glumes 4.5−7 mm long ....70. *C. eximius*
— Glumes 3−4 mm long ....71. *C. comosipes*

120. Keel of glumes winged, the wing toothed and hairy .... 81. *C. aureoalatus*
— Keel of glumes not winged, but sometimes toothed, scabrid or hairy 121

121. Plants tufted without a prominent horizontal rhizome; stem-base swollen and covered by the fibrous remains of old leaf-sheaths.............
.....................77. *C. oblongus* subsp. *jubensis*
— Plants with a horizontal creeping rhizome; stem-base without fibrous remains 122

122. Keel of glumes glabrous; involucral bracts 1−4; inflorescence often a solitary ovate or hemispherical head .... 78. *C. erectus*
— Keel toothed; involucral bracts 4−8; inflorescence an irregular head consisting of 1 large and several smaller spikes .... 79. *C. aromaticus*

1. **C. alternifolius** L. (1771).
subsp. **flabelliformis** (Rottb.) Kük., Cyp. E. Afr.: 4 (1936).
Robust perennial with a 2−10 mm thick creeping woody rhizome and several stems usually placed in a straight row; stems 25−150 cm long, 1.5−7 mm thick (but to 12 mm across the leaf-sheaths), rounded with longitudinal ridges, minutely scabrid; the basal part covered with leaf-sheaths and shorter black scales. Leaf-blades absent or the uppermost sheath with a 1−8 cm long somewhat leafy limb (the function of the leaves has been taken over by the leafy involucral bracts). Inflorescence a compound anthela 5−25 cm in diam., subtended by 15−25 leafy involucral bracts; bracts 10−35 cm long and 3−20 mm wide, scabrid on margin and major ribs, spirally arranged along a 1−5 cm long axis; largest peduncle 3−13 cm long; each spikelet-cluster digitate consisting of 3−20 spikelets, if more than 10 then the cluster is almost capitate. Spikelets 2−10 × 1−2.5 mm, ligth to medium brown,

lanceolate, much compressed, 15—30-flowered. Glumes 1—2 mm long, straw-coloured or golden to reddish brown with 3-keeled midrib excurrent in a short mucro or acute tip. Style 3-branched. Nutlet 0.8—0.9 × 0.4—0.5 mm, triangular, lanceolate to oblong, yellow as young turning brownish when mature, minutely papillose.

In swamps, wet grasslands, and on stream-banks, also cultivated as a garden plant; 1000—1300 m. N1; widespread in tropical Africa. Drake-Brockman 466; Burne 28; Wood 71/40.

2. C. prolifer Lam. (1791).

Fairly robust perennial with crowded stems on a thick creeping rhizome; stems 25—110 cm long and 2—7 mm thick, terete or triangular, glabrous. Leaf-blades absent; sheaths reddish brown to dark purple, glabrous; the apex ending in a short mucro. Inflorescence an open umbel-like anthela 10—20 cm in diam.; major involucral bracts 0.5—3 cm long, leafy; major inflorescence-branches 50—100 per stem, all of about equal length, 4—10 cm long, carrying one digitately arranged cluster of spikelets or one sessile and 1—several stalked groups of spikelets; some inflorescences with 8—25 cm long secondary stems, each carrying a few basal sheaths and a new smaller inflorescence. Spikelets 3—12 × 1—1.3 mm, linear, 7—25-flowered. Glumes 1.2—1.5 mm long, light reddish brown with paler margin; midrib ending in the apex or slightly excurrent. Style 3-branched. Nutlet 0.4—0.5 × 0.3—0.4 mm, obovate, triangular, whitish to brown, almost smooth to minutely papillose.

In swamp-edges, at stream-sides, and in seasonally flooded habitats, especially along the coast; from sea-level to 500 m. N1; scattered in east tropical Africa, South Africa and the Madagascar region. Robecchi-Bricchetti 116.

3. C. difformis L. (1756).

Slender to medium-sized annual; stems 6—80 cm long and 0.7—3 mm thick, triangular, glabrous. Largest leaf-blades 5—25 cm long and 2—6 mm wide, flat, smooth or scabrid on margin and midrib; sheaths green to reddish brown, rather wide, the basal without leaf-blades. Inflorescence a solitary congested anthela or with many subumbellately arranged heads giving an anthela 1—8 cm wide, and with 2—3 leafy bracts; largest involucral bract 3—25 cm long and 1—6 mm wide, erect or spreading, much over-topping the inflorescence; each head 5—12 mm in diam. and with 10—60 spikelets in digitate clusters. Spikelets 2—6 × 0.8—1.2 mm, pale yellowish grey to dark brown, 6—30-flowered. Glumes 0.6—0.8 mm long, obovate, yellowish to dark reddish brown with a wide green midrib ending in a short mucro. Style 3-branched. Nutlet 0.6—0.8 × 0.3—0.4 mm, triangular, obovate-elliptic, yellowish brown, minutely papillose.

In seasonally wet grassland and temporary swamps and pools; below 50 m. S3; pantropical, also in southern Europe. Moggi & Bavazzano 1828; Lobin 7040, 7058.

4. C. commixtus Kük. (1931); type: N2, "Ahl Mts", Hildebrandt 873c (K holo., lost?). Fig. 64 D, E.

Slender perennial with a short woody rhizome sometimes producing slender stolons; stems few, 20—30 cm long and 0.7—2 mm thick, obtusely triangular, glabrous. Leaves from the basal 5 cm only, usually 3—4 per stem; sheaths light brown to pale or greenish with uncoloured margin, glabrous; blades 5—40 cm long and 1—2.5 mm wide, flat and rather thin, glabrous or slightly scabrid on midrib and margin. Inflorescence a lax anthela of 1 sessile and 1—4 stalked digitate groups of spikelets (usually 3—6 spikelets per group); involucral bracts 2—3, erect or spreading, leaf-like and the largest 4—15 cm long; peduncles to 4 cm long, 0.2—0.3 mm thick, almost terete, glabrous. Spikelets 5—10 × c. 2 mm, linear or linear-lanceolate, flattened with acute or somewhat obtuse apex, 15—25-flowered. Glumes c. 2 mm long, triangular-ovate, light to dark reddish brown with somewhat paler margin and 3 slender lateral nerves on each side of the greenish midrib which is thickened above and ending in the subacute apex. Stamens 3. Style c. 2 mm long; branches 3, 1 mm long. Nutlet c. 1 × 0.4 mm, elliptic, triangular with flat sides, light reddish brown and minutely papillose.

In small lakes or pools with very variable water level, sometimes inundated at high water level; 500—1000 m. N2, 3; not known elsewhere. Glover & Gilliland 141; Hemming & Watson 3059.

The type at K seems to have disappeared, but the two collections cited fit the original description very well.

5. C. haspan L. (1753).

Annual or perennial herb without or with a very short rhizome; stems usually crowded, 5—40 cm long and 0.8—2.5 mm thick, triangular, green, glabrous. Leaf-blades present at least on some shoots, up to 20 cm long and 1—4 mm wide; leaf-sheaths light to dark reddish brown, glabrous, often with undulate margin and orifice. Inflorescence a lax anthela 2—15 cm in diam.; major involucral bract leafy, 1.5—7 cm long, usually shorter than the inflorescence; largest inflorescence-branches 1—7 cm long, carrying 1 sessile and a subumbel-like group of digitately arranged spikelets. Spikelets 3—12 × 1—1.5 mm, linear-lanceolate, light reddish brown. Glumes 1.3—1.6 mm long, reddish brown with a paler midrib which is excurrent in a short straight mucro. Style 3-branched. Nutlet 0.5—0.6 × 0.3—0.4 mm, obovate or subspherical, greyish brown, irregularly papillose.

In seasonally wet habitats, sometimes as a weed in rice-fields; below 100 m. S2, 3; widespread in the

Fig. 64. A–C: *Cyperus altomicroglumis*, from Vatova 46. A: habit, × 0.5. B: 3 spikelets, × 5. C: nutlet, × 30. – D, E: *C. commixtus*, from Glover & Gilliland 141. D: habit, × 0.5. E: spikelet, × 5. – F–H: *C. pulchellus*, from Lobin 6852. F: habit, × 0.5. G: spikelet, × 5. H: nutlet, × 30. – Drawn by G. M. Lye.

Old World tropics. Senni 373; Ciferri s.n.

The specimens above have not been seen by the author, but were cited as this species in Cuf. Enum.: 1427 (1970). The record therefore needs confirmation.

6. C. tenuispica Steudel (1855).

Slender annual with few—many stems and a small root system; stems 5−20 cm long and 0.5−1.5 mm thick, triangular or 6-angular, green, glabrous. Leaf-blades present, usually short, up to 10 cm long and 0.5−3 mm wide; leaf-sheaths light reddish brown to purple, glabrous. Inflorescence an open anthela 3−10 cm in diam.; major involucral bract leafy, 2−10 cm long, usually longer than the inflorescence; largest inflorescence-branches 0.5−5 cm long, carrying 1 sessile and several stalked groups of digitately arranged spikelets. Spikelets 2−12 × 0.8−1.3 mm, linear-lanceolate, green or reddish brown. Glumes 1−1.3 mm long, truncate, reddish brown with a paler midrib excurrent in a short recurved mucro. Stamens 2. Style 3-branched. Nutlet 0.4−0.5 × 0.25−0.3 mm, obovate, whitish or pale brown with large rectangular surface cells with raised cell-walls.

In seasonally wet habitats, near swamps, streams, and in rice-fields; below 100 m. S2, 3; widespread in the Old World tropics. Senni 81; Ciferri s.n.

The specimens above have not been seen by the author, but were cited as this species in Cuf. Enum.: 1439 (1970). The record therefore needs confirmation.

7. C. altomicroglumis Lye (1996); type: S1, "Baidoa", Vatova 46 (FT holo.). Fig. 64 A−C.

Tall fairly robust perennial with a short rhizome producing many crowded stems. Stems 50−80 cm long and 2−3 mm thick, triangular at least below the inflorescence, glabrous. Leaves many; the sheaths grey to straw-coloured or (the lowermost) brownish; the two uppermost usually developing 5−20 cm long and 2−4 mm wide fairly thick blades. Inflorescence lax, 3−9 cm in diam., consisting of 1−several sessile or subsessile clusters of spikelets subtended by new groups of sessile and stalked digitately arranged clusters of spikelets on up to 4 cm long major branches (rays); involucral bracts leafy, spreading and up to 20 cm long. Spikelets 4−12 × 2−3 mm, linear-lanceolate, prominently compressed, variegated grey and reddish brown, 15−25-flowered. Glumes 1.5−2 mm long, ovate, light to dark reddish brown with a prominent wide uncoloured margin and a wide green or straw-coloured midrib ending the acute apex or slightly excurrent; the surface cells of the glumes rectangular and prominent (except where strongly coloured). Stamens 2−3. Style with 3 long stigmas. Nutlet 0.5−0.6 × 0.3 mm, obovate-triangular, light greyish, minutely papillose.

In damp ground near a spring; c. 200 m. S1; not known elsewhere. Puccioni & Stefanini 352; Beckett 1483.

8. C. pulchellus R.Br. (1810). Fig. 64 F−H.

Slender perennial with swollen stem-bases covered by fibrous remains of old sheaths, often with many crowded stem-bases giving a rhizome-like structure; stems 6−40 cm long and 0.4−1 mm thick, glabrous or slightly scabrid, triangular, covered by reddish brown to blackish sheaths. Leaf-blades 2−15 cm long and 0.5−2 mm wide, flat or V-shaped, scabrid on margin and midrib; sheaths green to light reddish brown, the basal sheaths darker. Inflorescence a congested anthela with sessile spikelets only; major involucral bract leafy, 2−10 cm long and 1−1.5 mm wide, reflexed or spreading; anthela 0.7−1.5 cm in diam., consisting of 15−60 crowded spikelets. Spikelets 4−8 × 1−2.5 mm, greyish white with a light pinkish brown (cinnamon) tinge, 10−20-flowered. Glumes 1.3−1.5 mm long, lanceolate, greyish white with a cinnamon tinge; midrib indistinct; margin often incurved. Style 3-branched. Nutlet 0.6−0.8 × 0.2−0.3 mm, obovate, flattened triangular, grey to pale brown, minutely papillose.

In seasonally wet habitats; 300 m. S1; widely distributed in the Old World tropics. Lobin 6852 & Kilian 1999.

9. C. microglumis D. A. Simpson (1990); type: C1, 6 km W of Saddex Higle, Beckett 217 (K holo.).

Slender perennial with swollen stem-bases covered by fibrous remains of old sheaths; stems 2−15 cm long and 0.2−0.5 mm thick, terete to triangular, glabrous. Leaf-blades to 10 cm long and 0.4−0.6 mm wide, flat or channelled, scabrid on margin. Inflorescence a congested dense anthela with up to 30 spikelets, 4−10 mm in diam.; involucral bracts 3−4, spreading or reflexed. Spikelets 2−6 × c. 1 mm, linear-elliptic, 18−25-flowered. Glumes 0.7−1 × 0.4−0.5 mm, ovate, whitish to light brown, indistinctly nerved; midrib excurrent into a short thickened mucro. Stamen 1. Style 3-branched. Nutlet obovate to subglobose, 0.3−0.4 × 0.2−0.4 mm, obscurely trigonous, dark greyish brown, minutely papillose.

Seasonally wet sandy areas; 160−315 m. C1; not known elsewhere. Beckett 227.

10. C. colymbetes Kotschy & Peyr. (1867).

Fairly robust perennial with an erect or creeping subwoody rhizome, from which new stems develop at irregular intervals; stems 30−70 cm long and 3−5 mm thick, sharply triangular to winged, green, glabrous. Leaf-blades absent; sheaths very wide, reddish brown to purple, ending in a thin ligule and a thick triangular apex. Inflorescence a congested pale and almost whitish anthela with sessile spikelets only; solitary involucral bract leafy to stem-like, 8−12 mm long and continuing in the direction of the stem; anthela 1−2.5 cm in diam., consisting of 6−15 crowded spikelets. Spikelets 8−12 × 5−7 mm, ovate, light green to pale reddish brown, 12−30-flowered.

Glumes 4.5—5.5 mm long, ovate, reddish brown with large surface-cells, 3—9-nerved (but nerves often indistinct); midrib thicker and often scabrid at apex, not excurrent. Stamens 3. Style 3-branched, or rarely 2-branched. Nutlet 4—5 × 1.2—1.4 mm including a 1—2 mm long narrow beak and a 0.8—1.2 mm long basal tissue; the nutlet brown, surrounding sterile tissue yellow.

In standing water, often floating in swamps and irrigation canals; below 100 m. S2; scattered in tropical Africa and Madagascar. Bavazzano s.n.

11. C. exaltatus Retz. (1789).

Very robust perennial with crowded stems on a short c. 1 cm thick woody rhizome, the scales of the rhizome breaking up into fibrous remains; stems 40—150 cm long and 3—10 mm thick, triangular, smooth, the base slightly swollen. Basal leaves many, up to 80 cm long and 8—20 mm wide, flat, scabrid at least on margins and major ribs; leaf-sheaths green to purple. Inflorescence a 10—30 cm long and 10—40 cm wide open anthela consisting of 4—6 subsessile and stalked spikes and 5—15 stalked groups of spikes on 5—20 cm long inflorescence-branches (rays); primary involucral bracts leafy, the largest 30—60 cm long and 8—20 mm wide, erect or spreading. Spikes 2—5 × 1.5—2.5 cm when mature, with 15—60 spreading rather distant spikelets. Spikelets 6—15 × c. 1.5 mm, linear but sometimes slightly curved, flattened, brown, 12—30-flowered. Glumes 1.8—2.4 mm long, ovate-elliptic, reddish brown to golden with or without 2—3 nerves on each side of the excurrent green midrib. Stamens 3. Style 3-branched. Nutlet 0.8—1 × 0.5—0.6 mm, triangular, elliptic, greyish, almost smooth.

In swamps, river beds or in open water; below 200 m. S1—3; widespread in tropical and subtropical parts of the Old World. Moggi & Bavazzano 1666; Tardelli 259; Lobin 6921.

12. C. dives Del. (1813).

Robust perennial with a few stems from a short woody rhizome; roots often reddish; stems 50—150 cm long and 5—15 mm thick, triangular, smooth, the basal part covered by rather thick leaf-sheaths. Basal leaves many, up to 80 cm long and 1.5—3.5 cm wide, flat, scabrid at least on margin and major ribs; leaf sheaths purple at least below. Inflorescence a 10—30 cm long and 15—30 cm wide open anthela consisting of a few sessile or shortly stalked spikes and 5—15 clusters of spikes on 2—20 cm long inflorescence-branches (rays); primary involucral bracts leafy, the largest 20—80 cm long and 1.5—2.5 cm wide, erect or spreading. Spikes 1—3 × 0.6—1.5 cm, with 30—120 spreading and very crowded spikelets. Spikelets 3—5 (rarely to 10) mm long and 1—1.5 mm wide, linear-lanceolate, flattened, brown or golden, 6—20-flowered. Glumes 1.2—1.8 mm long, ovate, golden with darker reddish brown margin and an excurrent

green midrib; lateral nerves indistinct. Stamens 3. Style 3-branched. Nutlet 0.6—0.8 × 0.4—0.5 mm, triangular, elliptic, greyish and almost smooth.

In swamps, on river-banks or in open water; below 100 m. S2; widespread in Africa (including Macaronesia and Madagascar) and India. Bavazzano 281.

13. C. alopecuroides Rottb. (1773); *Juncellus alopecuroides* (Rottb.) C. B. Cl. (1893).

Fairly robust tussocky perennial with 50—150 cm long triangular stems. Basal leaves crowded, 5—15 mm wide, strongly scabrid on margin; sheath reddish brown to blackish. Inflorescence an umbel-like anthela 10—30 cm long and wide, consisting of 1 sessile and few—many stalked clusters of spikes; largest peduncles with tertiary clusters of spikes. Spikes 1—4 × 0.4—1.2 cm, consisting of numerous crowded spikelets. Spikelets 2.5—8 × 1.5—2.5 mm, lanceolate, pale to golden (or more rarely reddish) brown, 10—15-flowered. Glumes 1—1.7 mm long, ovate, rounded on the back, excurrent in a short mucro; margin in the lower half inrolled to enclose the nutlet. Style with 2 long stigmas. Nutlet 0.7—0.9 × 0.5—0.6 mm, usually flattened, smooth or minutely reticulate.

In swamps and river beds; sea-level to 500 m. S2, 3; widespread in the tropics of the Old World, probably introduced into America. Warfa 116; Tardelli 480; Lobin 6666.

14. C. procerus Rottb. (1773).

Fairly robust perennial with up to 15 cm long very slender stolons covered by distantly spaced blackish or grey scales; stems 50—90 cm long and 2—5 mm thick (but wider across the leaf-sheaths), triangular. Leaves many in the lower part, the largest 20—60 cm long and 3—10 mm wide. Inflorescence a lax or somewhat congested anthela 5—20 cm long and 5—15 cm wide, consisting of 2—4 leafy bracts and 3—6 stalked spikes with or without stalked or sessile secondary spikes from the base of the primary spikes. Spikes 2—4 × 2—5 cm, with 5—20 spreading spikelets. Spikelets 5—30 × 1.5—3.5 mm, linear. Glumes 2.5—3 mm long, ovate, light to dark reddish brown with an uncoloured margin and a rounded apex. Style 3-branched. Nutlet 1.2—1.6 × 0.7—0.9 mm, obovate, triangular, brown.

In seasonally wet grassland and swamps; near sea-level. S2; tropical Africa. Macaluso 67.

The specimen above was cited in Cuf. Enum.: 1435 (1970) as *C. procerus* var. *stenanthus* Kük. This specimen has not been seen by the author and the record needs confirmation.

15. C. articulatus L. (1753).

Robust leafless perennial with solitary stems from the end of stolons; stolons to 10 cm or more long (but frequently with stems at 1—3 cm intervals) and 2—8 mm thick, often woody, clothed with blackish or purple scales; stems 80—160 cm long and 3—12 mm thick below, but only 1—3 mm thick below the inflorescence, rounded, pith-filled with transverse rings at 5—50 mm intervals (these rings show best when dry owing to shrinkage of the pith); lower part of stem covered with 3—5 leaf-sheaths, the very base swollen and woody. Inflorescence a compound terminal lax anthela 4—15 cm in diam. with 1—3 sessile spikelet-clusters and 2—10 pedunculate clusters or umbels of new sessile and stalked clusters; largest peduncles 1—12 cm long; inflorescence-bracts scale-like, only 5—15 mm long. Spikelets 5—55 × 1—2 mm, linear, somewhat flattened, light to dark reddish brown, 20—50-flowered. Glumes 3—4 mm long, ovate, reddish brown (or straw-coloured when very young) with paler midrib; apex obtuse. Stamens 3. Style with 3 long style-branches. Nutlet 1.4—1.7 × 0.4—0.5 mm, narrowly elliptic, triangular, shortly apiculate, greyish yellow as young, reddish brown to dark olive brown when mature; surface smooth.

In stagnant water, often near pools; below 100 m. S2; pantropical. Bavazzano 259.

16. C. longus L. (1753).

C. fenzelianus Steud. var. *badiiformis* Chiov. in Ann. Bot. Roma 13: 376 (1915); *C. longus* var. *pallidus* Boeck. forma *badiiformis* (Chiov.) Kük. in Pflanzenr. IV.20: 101 (1936); type: S1, "El Ualac", Paoli 1095 (FT holo.).

Fairly robust perennial with rather thick horizontal, often curved, scale-covered 3—10 mm thick stolons and only slightly swollen stem-bases; stems 40—90 cm long and 1—4 mm thick, glabrous, triangular above, terete below. Leaf-blades few, withering early, the largest 15—30 cm long and 3—5 mm wide; leaf-sheaths light brown to dark reddish brown. Inflorescence a 5—13 cm long and 2—10 cm wide anthela consisting of 1 sessile and 4—8 stalked spikes on 0.5—10 cm long peduncles, with or without secondary spikes from the base of the primary spikes; inflorescence-bracts 3—7, leafy, erect or spreading, scabrid on margin and midrib, the largest 6—12 cm long and 2—5 mm wide. Spike 1—3 × 1—5 cm with 3—12 spreading spikelets. Spikelets 10—25 × 1—2 mm, linear-lanceolate, reddish brown with 8—20 flowers. Glumes 2—3 mm long, ovate, light to dark reddish brown with a narrow uncoloured margin and a greenish midrib ending below the obtuse tip; nerves very slender. Stamens 3. Style 3-branched. Nutlet 1.3—1.5 × 0.5—0.6 mm, elliptic, triangular, brown; surface almost smooth.

In shallow pools or periodically wet depressions in grassland or bushland; below 500 m. S1, 3; widespread in Africa, southern Europe and western Asia. Ciferri 55, 116.

17. C. rotundus L. (1753).

Gocondho, quunje (Som.).

Medium-sized perennial with a somewhat swollen (sometimes tuber-like) stem-base arising from slender to fairly robust stolons with rather remote scales; stems 25—80 cm long and 1—4 mm thick, glabrous, triangular with many crowded leaves in the basal part. Largest leaf-blades 15—40 cm long and 4—8 mm wide, flat or enrolled, scabrid at least on margin and major ribs; leaf-sheaths green to brown. Inflorescence a 3—15 cm long and 2—12 cm wide anthela consisting of 1(—many) sessile and 1—8 stalked spikes on 0.5—12 cm long peduncles with or without secondary spikes from the base of primary spikes; inflorescence-bracts 1—7, leafy, erect or spreading, the largest 3—20 cm long and 2—9 mm wide. Spikes 1—5 × 1.5—7 cm with 4—15 erect or spreading spikelets. Spikelets 6—70 × 1—2.5(—4) mm, linear-lanceolate, light to dark reddish brown with 8—35 flowers. Glumes 2.7—4.3 mm long, ovate, almost uncoloured or light to dark reddish brown with or without a narrow uncoloured margin and 1—2 nerves on each side of the midrib; midrib green, glabrous or scabrid ending in or below the obtuse apex. Style 3-branched. Nutlet 1.5—1.8 × 0.7—1 mm, obovate, triangular, greyish, minutely papillose.

In seasonally wet grassland, swamps and margins of springs and streams; near sea-level to 500 m. N2, 3; C2; S1—3; pantropical, and also in temperate regions. Beckett 1484; Friis & al. 4670; Lobin 6664.

This is a very variable species, particularly as regards the colour and size of the glumes. The following subspecies apparently occur in Somalia: subsp. *rotundus* with 3.3—4.3 mm long light to dark reddish brown glumes, subsp. *tuberosus* (Rottb.) Kük. with 3.7—4 mm long light brown to uncoloured glumes, and subsp. *merkeri* (C.B. Cl.) Kük. with 2.7—3.2 mm long dark reddish brown to almost blackish glumes. Most collections are, however, too young to be referable to any of the subspecies. The following subspecies is most distinctive:

subsp. **divaricatus** Lye (1996); type: N3, "Dudo" in "Shol" plain, Merla, Azzaroli & Fois s.n. (FT holo.). Fig. 65 A, B.

Robust perennial; stems 30—40 cm long and 1.5—3 mm thick. Leaf-blades to 40 cm long and 4—8 mm wide, flat. Inflorescence a lax anthela to c. 10 cm long and wide, consisting of many sessile as well as stalked spikes or clusters of spikes. Spikes c. 2 × 3—4 cm, consisting of 4—10 spikelets mostly spreading at right angles; rhachis 1—5 mm long. Spikelets 15—22 × 3—4 mm, linear, 20—25-flowered; rhachilla prominently zigzag with rather widely spaced flowers. Glumes 3—3.5 mm long, ovate, reddish brown without pallid margin, with 3—5 slender nerves on each side of the green midrib which is excurrent into a c. 0.2—0.3 mm long mucro. Style 3-branched. Nutlet 1.8 × 0.7—0.8 mm, obovate, prominently triangular, greyish, minutely papillose.

Depression in wadi; 350—400 m. N3; not known elsewhere.

18. C. nubicus C.B. Cl. (1902).

C. rotundus subsp. *retzii* (Nees) Kük. in Pflanzenr. IV.20: 114 (1936).

Robust perennial producing solitary stems from the end of 5−10 cm long and 1−2 mm thick stolons with 10−15 mm long prominent scales; stem 30−60 cm long and 2−5 mm thick, triangular, glabrous. Leaves usually 4−5 from the basal 10−20 cm; the sheaths grey, wide and loose; blades 15−30 cm long and 7−10 mm wide, flat, scabrid on margin and midrib. Inflorescence a 5−15 cm long and 4−15 cm wide lax anthela consisting of 1 sessile spike subtended by 2−10 stalked spikes or new groups of sessile and stalked spikes on up to 12 cm long peduncles; involucral bracts usually 3, leafy. Spikes 1−3 × 2−4 cm with 5−10 spreading spikelets. Spikelets 6−15 × 2−3 mm, linear or linear-lanceolate, spreading. Glumes 3.5−4 mm long, ovate, greyish to medium reddish brown with a pale marginal border and a green midrib excurrent into a short straight or recurved mucro; lateral nerves 3−4. Style 3-branched. Nutlet c. 1 × 0.6 mm, obovate, triangular, as young brownish, as mature dark grey to almost black, minutely papillose.

In pools and river beds; 15−930 m. N1, 2; Sudan. Cole s.n.; Glover & Gilliland 1109; Hemming & Watson 3147.

19. C. maculatus Boeck. (1864).

Slender to robust perennial with up to 15 cm long stolons, but when growing in narrow rock-cracks the stolons are reduced and the basal parts of the plant consist of many densely crowded swollen woody stem-bases; stems 10−70 cm long and 1−3 mm thick, triangular to subterete, glabrous. Largest leaf-blades 4−40 cm long and 1.5−5 mm wide, flat, somewhat bluish green, scabrid on margin and midrib; leaf-sheaths green to light reddish brown. Inflorescence a 1−12 cm long and 1−10 cm wide anthela consisting of 1 sessile and 1−6 stalked spikes, but usually with 1−5 secondary sessile or stalked spikes from the base of the primary spikes, but Somali plants sometimes with 5−15 spikelets per inflorescence only; bracts 2−4, leafy, erect or spreading, the largest 1−30 cm long and 1−5 mm wide. Spikes 1−5 × 1−3 cm, with 3−10 erect or spreading spikelets; rhachis only 2−10 mm long. Spikelets 8−40 × 1−1.5 mm, linear-lanceolate with acute tip, straight or curved, variegated greenish and dark reddish brown, 10−50-flowered. Glumes 2.2−3 mm long, ovate-elliptic, closely overlapping, reddish brown with a wide uncoloured marginal border and greenish midrib ending in the apex, lateral nerves absent. Stamens 3. Style 3-branched. Nutlet 1−1.2 × 0.5−0.6 mm, obovate, triangular, brown.

In sandy habitats near pools and rivers; below 300 m. C2; S2; widespread in tropical Africa. Puccioni & Stefanini 118, 125; Lobin 6664.

20. C. esculentus L. (1753).

Fairly robust stoloniferous perennial; stolons to c. 15 cm long and 0.5−1.5 mm thick, covered with brown to blackish scales and ending in a blackish tuber 3−8 mm in diam.; stems 15−70 cm long and 1−5 mm thick, triangular, glabrous, with 3−many crowded leaves near the base. Largest leaf-blades 10−30 cm long and 3−9 mm wide, flat, scabrid on margin and major ribs; leaf-sheaths green to reddish brown, rarely blackish. Inflorescence a 3−20 cm long and 3−15 cm wide anthela consisting of 1 sessile and 3−10 stalked spikes on 0.5−15 cm long peduncles, often with 1−5 secondary (usually stalked) spikes from the base of some primary spikes; primary inflorescence-bracts 3−9, leafy, erect or spreading, the largest 3−20 cm long and 2−9 mm wide. Spikes 1−3 cm long and wide, with 4−12 spreading spikelets. Spikelets 5−20 × 1.5−2 mm, linear-lanceolate with obtuse tip, brown or rust-coloured, 6−22-flowered. Glumes 2.2−2.6 mm long, ovate-elliptic, reddish brown with an uncoloured marginal border and 3−4 distinct nerves on each side of the midrib; midrib ending in the obtuse apex or slightly excurrent. Stamens 3. Style 3-branched. Nutlet 1.3−1.5 × 0.6−0.7 mm, elliptic, triangular, grey and shiny; surface with minute isodiametric cells.

Weed of cultivations, and also in seasonally wet grassland and swamps; below 100 m. S3; pantropical and subtropical. Hemming & Deshmukh 279.

21. C. dilatatus Schumach. & Thonn. (1827).

C. pseudosphacelatus Chiov. (1915), nom. illeg. non Boeck. (1890); *C. esphacelatus* Kük. (1936); type: S3, "Giumbo", Paoli 221 (FT holo.).

Fairly robust perennial with solitary stems and somewhat swollen stem bases from the end of long slender stolons; stems 40−70 cm long and 1.5−2.5 mm thick, glabrous, triangular or somewhat compressed. Leaves with green or somewhat reddish sheaths and 15−25 cm long and 3.5−5 mm wide blades. Anthela lax with 5−8 up to 10 cm long major branches; involucral bracts 5−7, up to 20 cm long. Spikes with 5−15 spikelets only. Spikelets 10−15 × 1.5−2.5 mm, linear-lanceolate, terete to somewhat flattened, straw-coloured to light reddish brown or somewhat golden, 10−18-flowered. Glumes 3−4 mm long, oblong-lanceolate with acute or subobtuse apex, light reddish brown with green 5-nerved midrib. Stamens 3. Style with 3 long stigmas. Nutlet elliptic-obovate, triangular.

In shrubland, or as a weed in irrigated fields; probably only below 50 m. S3; scattered but rare in tropical Africa. Lobin 7013a; Tozzi 307; Terry 3397.

22. C. afrodunensis Lye (1983).

Fairly robust perennial producing very slender stolons ending in bulbs; stolons to c. 5 cm long and 0.5−0.8 mm thick, brown; scales c. 6−8 mm long,

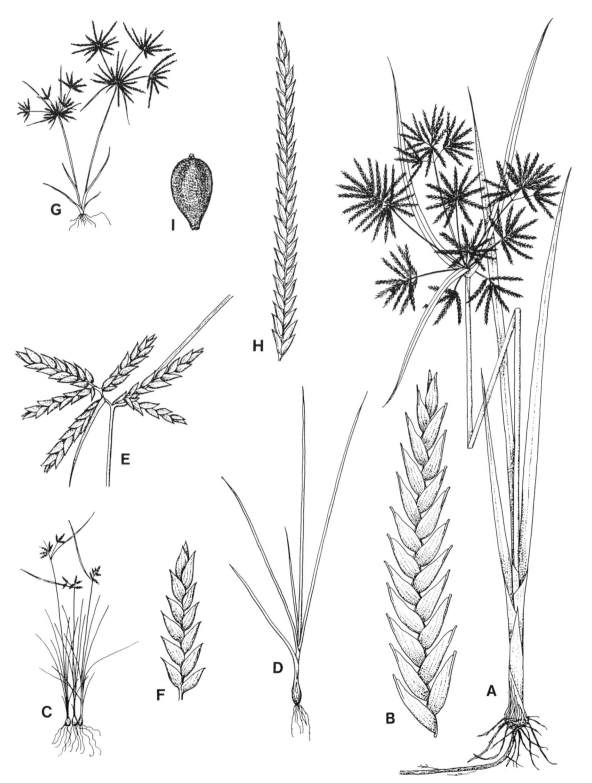

Fig. 65. A, B: *Cyperus rotundus* subsp. *divaricatus*, from Merla & al. s.n. A: habit, × 0.5. B: spikelet, × 5. − C−F: *C. densibulbosus*, from Glover & Gilliland 149. C: habit, × 0.5. D: vegetative shoot, × 1. E: inflorescence, × 3. F: spikelet, × 5. − G−I: *C. amabilis*, from Thulin & Mohamed 6807. G: habit, × 0.5. H: spikelet, × 5. I: nutlet, × 30. − Drawn by G. M. Lye.

pale brown; bulbs 2—2.5 × c. 1 cm in diam., covered by rather thin pale reddish brown scales; stems 10—40 cm long and 1.5—2.5 mm thick, triangular, glabrous. Leaves from the basal 5—10 cm only; leaf-blades 10—20 cm long and 3—8 mm wide, rather thick and fleshy, scabrid at least on margins; leaf-sheaths rather wide, greyish white above, brownish below. Inflorescence an open anthela 2—12 cm wide, consisting of 1—4 sessile or almost sessile spikes and 2—8 stalked spikes on 0.5—5 cm long glabrous peduncles; inflorescence-bracts 5—8, leafy, the largest 10—15 cm long and 3—5 mm wide. Spikes up to 1.5—2.5 × 2—3 cm, consisting of 4—12 spikelets from an up to 10 mm long axis, the lowermost spikelets sometimes with a minute stalk and with an additional spikelet from its base. Spikelets 7—20 × 3—4 mm, pale reddish brown to golden brown, only slightly flattened, 15—30-flowered. Glumes 3—4 mm long, pale reddish brown to brown, many-nerved; midrib only slightly excurrent. Style with 2 very long branches.

On sand dunes and in sandy fields near the seashore, more rarely as a weed in gardens. S2, 3; Kenya. Senni 611 ter; Tardelli & Bavazzano 511, 619; Terry 3347.

23. C. grandibulbosus C.B. Cl. (1902).

C. giolii Chiov. (1915); type: S1, between "Baidoa" and "Bur Acaba", Paoli 1133 (FT holo.).

Mariscus nogalensis Chiov. (1928); type: N3, "Bei Dagoi", Puccioni & Stefanini 882 (FT holo.).

Fairly robust perennial with long slender stolons and a 7—10 mm thick bulb; stems 10—60 cm long and 1—3 mm thick, triangular, growing directly from the bulb. Leaves many from the base, the largest 15—30 cm long and 2.5 mm wide; leaf-sheaths rather wide and fleshy, brown. Inflorescence a congested anthela of few—many crowded spikes, 2—4 cm in diam., more rarely with 1—4 additional stalked spikes; inflorescence-bracts 4—8, leafy, spreading or reflexed, the largest 8—20 cm long. Spikelets 5—20 × 1.5—3 mm, lanceolate, golden yellow to yellowish brown, 8—15-flowered. Glumes 3.8—5 mm long, ovate-lanceolate, yellowish brown to medium reddish brown with a green excurrent midrib (in the uppermost glumes the mucro is at least 0.5 mm long and recurved) and 3—5 prominent lateral nerves on each side of the midrib. Style 3-branched.

In seasonally wet habitats, often in irrigated land; near sea-level to 1250 m. N2, 3; S1, 2; also in East Africa. Glover & Gilliland 974; Eagleton 86; Terry 3411, 3428.

24. C. bulbosus Vahl (1805).

Perennial with a basal bulb covered by brown to blackish scales and with very slender stolons ending in new 3—7 mm thick bulbs from which the stems emerge; stems 8—50 cm long, triangular. Leaves many, 5—25 cm long and 2—6 mm wide. Inflorescence a lax anthela 3—12 cm long and 5—12 cm wide, consisting of 4—6 leafy bracts, 1 sessile spike and 3—6 stalked spikes or reduced to 1 single spike. Spikes 1.5—4 × 1.5—4 cm, consisting of 5—20 spreading spikelets. Spikelets 5—25 × 1.5—3 mm, linear-lanceolate, dark reddish brown to yellowish green, 5—10-flowered. Glumes 3—4 mm long, ovate-lanceolate, dark reddish brown to almost blackish, with a paler slightly excurrent green midrib (even in the uppermost glumes the mucro is c. 0.2 mm long only) and prominent lateral nerves. Style 3-branched. Nutlet c. 0.8 × 0.5 mm, triangular, papillose.

In seasonally wet grassland or open sand; near sea-level to 600 m. N1, 2; C2; S1—3; widespread in Africa, also Arabia to India and Malaysia, and N Australia. Hemming & Deshmukh 306; Hansen & Heemstra 6304; Alstrup & Michelsen 107.

25. C. densibulbosus Lye (1996); type: N2, between "Bihen" and "Las Anod", Glover & Gilliland 149 (K holo.). Fig. 65 C—F.

Small slender perennial with a short woody rhizome with crowded bulbs producing many leafy shoots and few fertile shoots; bulbs 3—5 × 2—3 mm, shortly oval to almost globose (previous years' bulbs may appear cylindrical), consisting of a woody base (and sometimes basal stem) surrounded by numerous pale glabrous leaf-sheaths with prominent blackish nerves; c. five of the sheaths with a leaf-blade. Leaf-blades to 6 cm long and 0.5—1 mm wide, flat but often inrolled, rather thick and without prominent nerves, glabrous or minutely scabrid on margin. Stems 5—10 cm long and 0.2—0.4 mm thick, obtusely angular, glabrous, with blades from the lower 1 cm only. Inflorescence a small anthela consisting of 1—2 sessile spikelets (in depauperate specimens) or an up to 1 cm wide digitate cluster of 3—6 spikelets with or without an additional stalked spikelet or digitate cluster of 2—3 spikelets; peduncle to 8 mm long with a tubular basal prophyll to 3 mm long. Spikelets 3—8 × 1.5—2 mm, linear-lanceolate with acute apex, with 7—15 rather distantly placed flowers. Glumes 1.5—1.8 mm long, ovate, medium to dark reddish brown with usually indistinct paler margin and 3—4 slender paler nerves on each side of the greenish midrib which ends in the acute apex or is very slightly excurrent. Stamens 3. Style 1.5—2 mm long, medium reddish brown with 3 c. 1 mm long branches. Nutlet triangular, immature.

Near pool with greatly changing water-level; c. 600 m. N2; not known elsewhere.

26. C. iria L. (1753).

Tussocky annual with numerous short roots; stems 8—50 cm long and 0.6—3 mm thick, triangular, glabrous, green. Largest leaf-blades 4—25 cm long and 1—5 mm wide, flat, scabrid on margin and major ribs; leaf-sheaths green to reddish brown. Inflorescence a lax anthela 1.5—20 cm long and 1—20 cm wide, with groups of spikes sessile or on 0.5—15 cm long major peduncles (rays); involucral bracts leafy, the largest 5−30 cm long and 1−6 mm wide. Spikes sessile or almost so, rather irregular in shape and length. Spikelets 2−10 × 1.5−2 mm, 5−20-flowered, golden to yellowish green. Glumes 1.3−1.6 mm long, obovate or rounded, golden brown with an uncoloured margin and a greenish slightly excurrent midrib. Style 3-branched. Nutlet 1.2−1.4 × 0.6−0.7 mm, obovate, triangular with a short apiculus, dark brown to nearly black as mature; surface almost smooth.

In seasonally wet habitats, such as temporary pools and in rice-fields; probably below 100 m. ?S3; pantropical at least as a weed.

This was said to occur in southern Somalia in Cuf. Enum. (1970), but no specimen has been seen and the record needs confirmation.

27. C. compressus L. (1753).

Small to fairly robust annual with a single or a few stems; roots brown to reddish; stems 10−60 cm long and 0.7−5 mm thick, triangular, almost glabrous. Leaves 3−30 cm long and 1.5−6 mm wide, flat, scabrid on margins and ribs; leaf-sheaths grey to reddish brown. Inflorescence in small specimens a solitary head of 3−6 spikelets, but more frequently a compound anthela 1−15 cm long and 0.5−25 cm wide, consisting of 1 sessile group of spikelets and 1−10 stalked groups on 0.5−12 cm long peduncles; involucral bracts leafy, erect or spreading, the largest 2−25 cm long. Spikelets 10−50 × 3−5 mm, linear-lanceolate, flattened, greenish grey to reddish brown, 10−60-flowered. Glumes (3−)3.5−5 mm long, ovate-elliptic, grey to brown with or without yellow patches; midrib green and shortly excurrent. Stamens 3. Style 3-branched. Nutlet 1.5−1.7 × 1.1−1.3 mm, obovate, reddish brown and almost smooth.

In seasonally wet habitats; sea-level to 300 m. S1−3; throughout warm parts of Africa, Asia and America. Bavazzano & Tardelli 839; Thulin, Hedrén & Dahir 7509; Hemming & Deshmukh 86−123.

Lobin 6665 differs in having glumes with a more prominently excurrent midrib. It could be considered as a separate variety.

28. C. amabilis Vahl (1805). Fig. 65 G−I.

Slender to fairly robust annual with solitary or crowded stems; stems 5−25 cm long and 0.3−1.3 mm thick, triangular, glabrous. Leaf-blades 2−10 cm long and 0.5−2.5 mm wide, flat or inrolled, scabrid on margin; sheaths reddish brown to purple, rather short. Inflorescence an open anthela 2−8 cm in diam.; major involucral bract leafy, 1−10 cm long and 0.5−2.5 mm wide, usually shorter than the inflorescence; largest inflorescence-branches 1−7 cm long, with 1 sessile and 1−9 stalked groups of digitately arranged spikelets, occasionally some stalked spikelet-clusters with a secondary stalked spikelet-cluster, the tubular prophyll at the base of

each peduncle greenish or light brown. Spikelets 3−17 × 1−1.5 mm (or to 3.5 mm wide with glumes spreading at right angles), linear, reddish brown to golden. Glumes 1.3−1.5 mm long, ovate, 3-nerved, orange to reddish brown with a greenish midrib which is usually very shortly excurrent in a slightly recurved mucro. Stamen 1. Style 3-branched. Nutlet 0.5−0.6 × 0.3−0.4 mm, obovate, triangular, brownish, minutely papillose.

In seasonally wet habitats, often on sandy or silty soil; near sea-level to c. 230 m. S1, 3. Paoli 775; Bavazzano 858; Thulin & Mohamed 6807.

29. C. cuspidatus H.B.K. (1815).

Slender annual with a minute root system; stems 3−14 cm long and 0.2−0.5 mm thick, triangular, glabrous. Leaf-blades flat or inrolled, slightly scabrid near the apex, 1−8 cm long and 0.2−0.8 mm wide; sheaths reddish brown to purple. Inflorescence a single spikelet-cluster of 4−20 (rarely 1−3) spikelets, or of 1 sessile and 1−3 stalked spikelet-clusters; major involucral bracts 2−8 cm long, leafy to filiform; major inflorescence-branches (when present) 0.5−2 cm long. Anthela 2.5−4 cm in diam. when consisting of several spikelet-clusters; each spikelet-cluster 1−2 cm across. Spikelets 4−10 × 1.7−2 mm, linear, squarrose, reddish brown, 8−25-flowered. Glumes 1.6−2 mm long including the 0.5−0.6 mm long excurrent recurved midrib, strongly 3-nerved, truncate, reddish brown with green midrib. Stamens 1−3. Style 3-branched. Nutlet 0.6−0.8 × 0.3−0.4 mm, obovate, strongly triangular, reddish brown with dark grey angles, densely papillose.

In shallow seasonally wet soil often near pools on rock outcrops; below 500 m. S1; pantropical. Paoli 624 bis.

30. C. squarrosus L. (1756).

Fairly slender annual with solitary or crowded stems and a minute root-system; stems 3−20 cm long and 1−3 mm thick, triangular, almost glabrous, not swollen at the base. Leaves 2−10 cm long and 1−4 mm wide, flat, slightly scabrid or glabrous; sheaths green to purple, rather wide. Inflorescence a 1−5 cm long and 1−4 cm wide anthela consisting of 1−2 sessile spikes and 1−7 stalked spikes on 0.5−4 cm long peduncles, rarely reduced to a solitary spike. Spikes 0.5−1.5 × 0.5−1.2 cm, consisting of numerous crowded spikelets. Spikelets 3−7 × 2−3 mm, linear, yellowish green to reddish brown, flattened, squarrose with recurved glume-apices, 6−15-flowered. Glumes 1.4−1.7 mm long, elliptic, yellow to reddish brown with a green strongly excurrent midrib and 3−4 nerves on each side of the midrib. Stamen 1. Style with 3 long branches. Nutlet 0.6−0.7 × 0.2−0.3 mm, narrowly oblong, triangular, dark grey, minutely papillose, disarticulating at its base but held by the persistent glume so that it falls with the spikelet.

In seasonally wet grassland, often on silt near rock pools, streams and roadsides; up to 250 m. S1; pantropical, but rare in America. Thulin & Mohamed 6806.

31. C. crassipes Vahl (1805).
Daleen (Som.).
Robust tussocky perennial with a thick branched rhizome and long stolons; stems 1—35 cm long and 2—5 mm thick, triangular to almost rounded. Leaves many, 10—40 cm long and 3—8 mm wide, flat or folded, rather stiff, scabrid on margin; sheaths green to light reddish brown. Inflorescence a large globose head or sometimes a more open anthela surrounded by 4—8 long leafy bracts; major involucral bract 10—25 cm long and 4—6 mm wide, strongly scabrid on margin, spreading or reflexed. Anthela 4—8 cm in diam., consisting of numerous almost sessile digitate spikelet-clusters, or sometimes the spikelet-clusters on 0.5—2.5 cm long peduncles; each spikelet-cluster with 6—15 spikelets. Spikelets 14—20 × 3—5 mm, linear-lanceolate, pale brown to almost whitish, 10—20-flowered, only slightly compressed. Glumes 6—8 mm long, broadly ovate, concave, light reddish brown with paler margin and numerous narrow nerves; midrib thickened above and slightly excurrent or ending in apex. Stamens 3. Style with 3 long branches. Nutlet 2—2.3 × 1—1.2 mm, obovate, dark brown, minutely papillose.
In dry sandy places along river-banks and on coastal dunes; sea-level to 280 m. C1; S2, 3; along most tropical African coasts. Hemming 3413; Thulin, Hedrén & Dahir 7181, 7252.

32. C. chordorrhizus Chiov. (1926); type: S3, "Chisimaio", Gorini 97 (FT lecto.). Plate 4 D.
Perennial with a long creeping branched stolon to 1 m or more long with shorter lateral sterile shoots from the nodes (these are frequently crowded at the tip of the stolon), and fertile terminal shoots. Leaves 0.5—8 cm long and 0.5—1.5 mm wide, subterete with scabrid margin and frequently with a prominent transverse impression below the apex; leaf-sheaths grey, translucent, but with darker reddish brown nerves, the old sheaths darker and splitting up into fibres. Inflorescence a solitary head or spikelet-cluster on a 1—8 cm long stem, consisting of 2—10 rather flattened reddish brown spikelets 4—10 × 3—4 mm. Glumes 3—4 mm long, reddish brown with a paler margin. Style 3-branched.
On maritime sand-dunes; near sea-level to c. 50 m. C1; S2, 3; Kenya. Hemming 3344; Hansen 6000; Gillett & Hemming 24432.

33. C. macrorrhizus Nees (1834).
Densely tussocky perennial with a c. 5 mm thick horizontal rhizome densely covered by long chestnut brown glossy leaf sheaths not splitting into fibres; roots with tomentum; stems somewhat crowded, 30—80 cm long and 2—3 mm thick, terete below, ob-

tusely triangular below the inflorescence, glabrous. Leaves from the basal 15 cm, often 10—15 from each stem; sheaths straw-coloured above, brown below but with an uncoloured margin, glabrous; blades 10—40 cm long and 1—3 mm wide, semiterete to canaliculate, glabrous or somewhat notched on margin, the apex obtuse. Inflorescence to c. 10 cm long and wide, a globose head of many crowded spikelets or with 1 globose group of spikelets and 1—8 stalked globose or digitate groups of spikelets on 1—6 cm long somewhat compressed peduncles; major involucral bracts 1—4, leaf-like, erect or spreading, the largest to 20 cm long and 2—3 mm wide. Prophylls at base of peduncles 6—12 mm long, tubular, pallid to light reddish brown. Spikelets 8—15 × 3—5 mm, ovate-lanceolate, whitish with acute apex, 15—25-flowered with closely imbricate glumes; axis conspicuously flattened. Glumes 4—5 mm long, ovate-elliptic, whitish but often with a yellow or orange-brown patch at apex and sometimes along margin, with many distinct lateral nerves and usually green midrib ending in the obtuse apex or slightly excurrent. Stamens 3. Style c. 3 mm long, 3-branched almost to the base. Nutlet 1.8—2 mm long, obovate, almost smooth.
Sand dunes between hills and on coastal plain; near sea-level to 150 m. N1, 2; widespread but rare from North Africa to India. Hemming 2380; Thulin, Dahir & Hassan 9018.

34. C. conglomeratus Rottb. (1772).
Slender to very robust tussocky perennial with a short woody rhizome and numerous c. 0.5 mm thick wiry roots that are often covered by a thick greyish tomentum and then c. 2 mm thick; stems 1—40 (rarely to 120) cm long and c. 1—4 mm thick, triangular to almost terete, glabrous. Leaves many, from the basal 3—15 cm only; sheaths rather thin, grey to reddish brown, the lower sometimes splitting up into reddish brown fibres; blades 5—25 cm long and 1—3 mm wide, often rather thick, weakly or strongly curved, sometimes folded as dry. Inflorescence a terminal head of 3—50 crowded spikelets usually 3—5 cm in diam., but often subtended by 1—8 new heads or groups of spikelets on 1—6 cm long peduncles; involucral bracts 1—5, leafy, erect or spreading, the largest 1—25 cm long, usually overtopping the inflorescence. Spikelets 5—30 × 2—4 mm, lanceolate, compressed or subterete, greyish white or variegated grey and reddish brown, 10—50-flowered. Glumes 2.5—4 mm long, ovate, greyish white to brown or reddish brown, many-nerved, but the nerves sometimes indistinct; midrib often green and excurrent into a short straight awn. Style 3-branched. Nutlet c. 1.5 × 0.8—1 mm, obovate, obscurely triangular, blackish, minutely papillose.
In sandy soils, more rarely on rocks, clay-soils and gravel; near sea-level to 800 m. N1—3; C1; S2, 3; widespread in dry and hot parts of Africa and western Asia. Gillett 4738, 4816; Hemming 2101.

35. C. jeminicus Rottb. (1772); *C. conglomeratus* subsp. *jeminicus* (Rottb.) Lye in Lidia 3: 131 (1994).

Slender perennial forming small tussocks not connected by rhizomes; roots slender and without tomentum; stems 5–20 cm long and 0.5–2 mm thick, triangular, glabrous. Leaves basal with prominent pale sheaths; blades 2–10 cm long and c. 1 mm wide, often thick and sometimes curved, canaliculate or folded as dry. Inflorescence a sessile terminal cluster of 5–20 crowded spikelets with or without additional stalked groups of spikelets; involucral bracts erect or spreading, leaf-like, to c. 10 cm long. Spikelets 5–15 × 2–3 mm, linear-lanceolate, strongly compressed, brown or reddish brown, 10–25-flowered. Glumes 2.5–3 mm long, ovate, reddish brown, many-nerved; midrib green and excurrent into a short mucro. Style 3-branched. Nutlet 1.2–1.5 mm long, obovate, obscurely triangular, brown, minutely papillose.

In sand dunes or sandy soils; near sea-level to 500 m. N1, 3; North Africa and south-western Asia. Gillett 4434; Popov 1181; Collenette 200.

It is very doubtful if this is a good species, and it should probably be included in *C. conglomeratus*, from which it differs in its more flattened reddish brown spikelets and the slender roots without tomentum. It is intermediate between *C. mudugensis* and *C. conglomeratus*. The Somali collections of this species are very young and poor.

36. C. mudugensis Simpson (1994); type: C1, 20 km N of Hobyo on road to Jirriiban, Thulin & Dahir 6661 (K holo., UPS iso.). Fig. 66 A–C.

Small tussocky perennial with a very short rhizome; stems crowded, 3–8 cm long and 0.6–0.8 mm thick, angular to almost terete, glabrous, base slightly swollen and covered by a dense layer of old leaf-sheaths. Leaves basal; sheaths brown with a wide pale membranous margin; blades 2–6 cm long and up to 1 mm wide, flat to canaliculate but very thick and sometimes almost semiterete at base, often curved towards the apex, margin minutely scabrid. Inflorescence a sessile terminal cluster of 4–15 crowded spikelets with or without 1–2 additional stalked spikelet groups on up to 8 mm long peduncles; involucral bracts spreading, leaf-like, up to 2.5 cm long. Spikelets 6–14 × c. 1.5 mm, linear, acute, 10–20-flowered. Glumes 2.3–2.7 mm long, ovate with obtuse or subacute apex, reddish brown with a green midrib ending in apex or shortly excurrent, with 2–4 prominent lateral nerves on each side of the midrib. Stamens usually 2. Style with 3 long stigmas. Nutlet 0.8–1 × 0.4–0.5 mm (immature), brown, minutely papillose.

In coastal plain on gently sloping, open limestone rocks; c. 70 m. C1; not known elsewhere.

37. C. frerei C.B. Cl. (1901).

C. chisimajensis Chiov. (1926); type: S3, "Chisimaio", Gorini 59 (FT holo.).

Robust perennial with a thick horizontal, often curving rhizome 3–5 mm thick with stems set at intervals of c. 0.5–2 cm; stems 20–80 cm long and 1–4 mm thick, triangular, smooth or slightly scabrid on angles, with numerous (10–15) crowded leaves at their base. Leaf-blades 15–60 cm long and 2–6 mm wide, flat or folded with a very prominent keeled midrib on lower surface. Inflorescence a congested or lax anthela of numerous crowded whitish to reddish brown spikelets or of 1 sessile group of spikelets and 1–several stalked groups of spikelets. Spikelets 8–40 × 3–5 mm, lanceolate, flattened, 8–25-flowered. Glumes 4.5–6 mm long, ovate, whitish to light or medium reddish brown with prominent lateral nerves and a green midrib excurrent into a short awn. Style 3-branched. Nutlet 1.7–1.9 × 1.2–1.3 mm wide, black, elliptic, prominently triangular with flat sides and a short apiculus; surface smooth and shiny.

In coastal dunes; sea-level to 50 m. S3; Kenya. Tardelli 147.

38. C. poecilus C.B. Cl. (1902); type: N1, "Mandira", Keller 58 (K holo.).

Tussocky perennial with a horizontal rhisome set with 1 fresh stem and few–many old stem-bases often in a straight row; stem 20–35 cm long and 1–2 mm thick, terete to angular, smooth. Leaves from the basal 6 cm only; sheaths loose, pallid above, brown below; blades 5–15 cm long and 2–3 mm wide, flat or inrolled, scabrid on margin. Inflorescence a dense head 2–3 cm in diam., consisting of many crowded spikelets; involucral bracts 3–5, leafy, spreading or reflexed. Spikelets 10–15 × 4–5 mm, ovate, compressed, 10–20-flowered, sometimes pale as young, otherwise brownish. Glumes 5–6 mm long, pale to medium reddish brown with green midrib excurrent into a slightly recurved mucro at least 1 mm long in the uppermost glumes; lateral nerves prominent, 3–4 on each side of the midrib. Style with 3 long stigmas. Nutlet c. 2 × 1.2 mm, obovate, triangular with concave sides.

In dry habitats, often on rocks or among stones; 1500–1600 m. N1; not known elsewhere. Godding 178.

39. C. benadirensis Chiov. (1932); *C. poecilus* var. *evolutus* Kük. in Pflanzenr. IV.20: 283 (1936); type: S3, "Bender Suguma", Paoli 258 (FT holo.).

Uda leef (Som.).

Robust tussocky perennial with swollen hard stem bases crowded on a short woody rhizome and covered by brown and pallid sheaths; roots fairly slender and without tomentum; stems 15–50 cm long and 1–3 mm thick, obtusely triangular, smooth. Leaves from the lower 10 cm only, c. 5–8 per stem but with additional infravaginal leafy shoots; upper sheaths prominently whitish, the lower brown; blades 5–20 cm long and 2–3 mm wide, flat or with the scabrid margin inrolled. Inflorescence a lax anthela c. 4–10 cm in diam. consisting of 1 sessile and 5–10 stalked

Fl. Somalia

Fig. 66. A−C: *Cyperus mudugensis*, from Thulin & Dahir 6661. A: habit, × 0.5. B: spikelet, × 5. C: nutlet, × 30. − D, E: *C. scabricaulis*, from Hansen 6038. D: habit, × 0.5. E: spikelet, × 5. − F−H: *C. medusaeus*, from Beckett 380. F: habit, × 0.5. G: spikelet, × 5. H: nutlet, × 30. − Drawn by G. M. Lye.

digitate clusters of spikelets on up to 5 cm long peduncles; involucral bracts erect or spreading, the largest 5—15 cm long. Spikelets 7—15 × 3—5 mm, ovate with acute apex, prominently compressed, 10—25-flowered. Glumes 3—5 mm long, ovate, reddish brown with several slender lateral nerves; midrib green and prominent, excurrent into a 0.5—1 mm long mucro. Style with 3 long branches. Nutlet c. 1.5 mm long, almost rounded, flat and strongly curved with the adaxial side concave and the abaxial side convex, or with 3 concave sides, grey and minutely papillose.

In sand dunes and bushland; mostly near sea-level. C2; S2, 3; not known elsewhere. Senni 212; Moggi & Bavazzano 102; Elmi & Hansen 4023.

40. C. scabricaulis Lye (1996); type: S2, 5 km SW of Mogadishu, "Gezira Costal Range", Hansen 6038 (K holo., BR C EA M WAG iso.). Fig. 66 D, E.

Medium-sized somewhat tussocky perennial with a few crowded stems with swollen stem bases from a very short woody rhizome; roots without tomentum; stems 10—30 cm long and 1—2 mm thick, terete to obtusely triangular, minutely or prominently scabrid at least on some ridges. Leaves from the lower 1—6 cm only, c. 5—8 per stem; upper sheaths pale with distinct auricles, the lower brown, sometimes splitting up into fibres; blades 5—20 cm long and 2—3 mm wide, flat but strongly folded or inrolled when dry; margin and lower surface densely set with minute prickles or projections in numerous longitudinal rows, but without a midrib, upper surface smooth. Inflorescence a dense head of crowded spikelets 2—2.5 cm in diam. or more commonly consisting of 1 dense head of spikelets with 1—7 additional stalked heads or digitate clusters of spikelets on up to 3 cm long peduncles; involucral bracts usually 3—5, up to c. 10 cm long and 2—3 mm wide, spreading or reflexed, similar to the leaf-blades. Spikelets 7—9 × 3—5 mm, ovate with acute apex, compressed, 15—20-flowered with very densely crowded glumes. Glumes 3—4 mm long, broadly ovate with obtuse apex, medium reddish brown with or without an irregular paler margin and apical part; midrib light brown to greenish, excurrent into a short thick mucro c. 0.3 mm long; lateral nerves slender, 3—4 on each side of the midrib. Style with 3 slender branches. Nutlet c. 1.2 mm long and wide, rounded to squarish, flat and strongly curved with the adaxial side prominently concave and the abaxial side convex, medium reddish brown, smooth or minutely papillose.

In sand dunes near the coast; sea-level to 140 m. S2; not known elsewhere. Gillett, Hemming & Watson 22099B.

41. C. mogadoxensis Chiov. (1940); type: S2, Mogadishu dunes, Senni 1273 (FT holo.).

Fairly slender perennial with 1—5 crowded stems with small swollen stem-bases c. 5 mm thick frequently set in a horizontal row and covered by reddish brown leaf-sheaths sometimes splitting into fibres; roots very slender and without tomentum; stems 3—25 cm long and 0.5—1 mm thick, obtusely triangular to terete, smooth. Leaves from the basal 1—5 cm only, usually 4—6 per stem; blades to c. 10 cm long and 1—2 mm wide, flat or with margins inrolled, very strongly and prominently scabrid (almost serrulate) on margin; upper sheaths conspicuously whitish. Inflorescence a terminal cluster 1—3 cm in diam. consisting of 5—20 densely crowded spikelets, rarely with 1—2 additional clusters on up to 15 mm long peduncles; involucral bracts 2—5, spreading or reflexed, the largest 2—10 cm long. Spikelets 5—10 × 3—5 mm, ovate with acute or subacute tip, prominently flattened, 15—30-flowered with very densely set glumes. Glumes 3—3.5 mm long, ovate and only slightly concave, pale to reddish brown at least in patches; midrib ending at the apex in the lower glumes but excurrent into a short mucro in the upper glumes. Style with 3 long branches. Nutlet c. 1.2 mm long and wide, rounded in outline, prominently triangular in section with all 3 sides very concave, grey and minutely papillose; style base persisting as a minute reddish brown knob.

In sand dunes and coastal bushland; up to c. 50 m. C1; S2; not known elsewhere. Moggi & Tardelli 221; Thulin & Dahir 6645, 6739.

42. C. medusaeus Chiov. (1928); type: C1, between "Obbia" and "Sissib", Puccioni & Stefanini 391 (FT holo.). Fig. 66 F—H.

Tussocky perennial with a prominent woody rhizome to c. 5 mm thick; roots slender, without tomentum; stems solitary or a few together, 5—25 cm long and 0.7—1.5 mm thick, obtusely angular to somewhat compressed or almost terete, glabrous. Leaves numerous, mostly basal and only 2—3 cm up the stem; blade to 20 cm long but mostly prominently coiled up and then often less than 3 cm long, 0.5—2 mm wide, very thick, flat below, more squarish to triangular above, prominently but irregularly scabrid on margin and lower side of midrib; sheath glabrous, straw-coloured to reddish brown with a prominent almost white membranous margin. Inflorescence a terminal head 9—13 mm in diam., consisting of numerous crowded sessile spikelets; involucral bracts 2—3, to 9 cm long, spreading, leaf-like, but often not coiled up or coiled at the apex only. Spikelets 5—7 × 3—4 mm, ovate with obtuse or subacute apex, only slightly compressed, variegated grey-reddish brown, 10—20-flowered; the rhachilla prominently notched. Glumes 2—3 mm long, ovate, concave, reddish brown with a pale marginal border and 3—4 paler narrow nerves on each side of the midrib that is green and prominent in upper half of glume only; midrib ending in the obtuse apex or excurrent in a very short mucro. Style with 3 long branches. Nutlet 0.9—1 mm long and wide, rounded in outline, prominently triangular in section with all 3 sides very concave, grey and minutely papillose.

Sand dunes and coastal grasslands; probably below 50 m only. C1; S2; not known elsewhere. Beckett 380; Bavazzano 1015.

This is very close to *C. mogadoxensis*, but differs in its more coiled leaf-blades, and in its smaller and less compressed spikelets with smaller glumes and nutlets. Bavazzano 1015 from "Uarsciek" (S2) is somewhat intermediate with 2.7–3 mm long glumes.

43. C. somaliensis C.B. Cl. (1895); type: N1, without precise locality, Cole s.n. (K lecto.). Fig. 67 A–C.

Slender tussocky perennial with numerous crowded basal leaves and 1 or a few fertile stems from each tussock; stem 5–20 cm long and 0.5–1 mm thick, triangular, glabrous. Leaves from the lower 2 cm only; base covered by a very thick layer of brown sheaths; blades 2–10 cm long and 0.4–0.8 mm thick, flat (but appears filiform) or inrolled, scabrid on margin. Inflorescence a dense hemispherical head of numerous crowded spikelets; involucral bracts 2–4, leafy, erect, spreading or reflexed, 1–8 cm long. Spikelets 5–7 × 2.5–3.5 mm, ovate-elliptic, compressed, 8–12-flowered. Glumes 3–3.5 mm long, ovate-elliptic, pale brown with 4–7 narrow nerves on each side of the indistinct midrib that ends below the obtuse apex. Style 3-branched. Nutlet c. 1.5 × 1 mm, obovate-triangular with concave adaxial side, black, almost smooth.

In bare ground in open forest, on bare limestone or in gravelly plain; 1200–1400 m. N1, 2; not known elsewhere. Bally 7326; Newbould 757, 921.

44. C. niveus Retz. (1791).

Perennial with crowded stems often growing in a straight line; stem-bases swollen and fused into a horizontal rhizome; roots slender; stems 10–50 cm long and 0.7–2.5 mm thick, triangular to rounded, glabrous, the base covered by hard leaf-sheaths not breaking into fibres. Leaves 5–35 cm long and 0.5–5 mm wide, scabrid on margin and midrib at least near the tip; sheaths hard, light to dark brown. Inflorescence a solitary usually globose head of 5–50 spikelets, 1–4 cm in diam.; involucral bracts leafy, the largest 1–16 cm long. Spikelets 5–17 × 3–7 mm, lanceolate, compressed, 10–30-flowered. Glumes 4–6 mm long, elliptic-lanceolate, white with a pinkish brown tinge and 6–8 conspicuous striations on each side of the rather obscure midrib; apex obtuse. Stamens 3. Style 3-branched. Nutlet 1.3–2 × 1–1.3 mm, obovate in outline, triangular, brown to black, smooth, shortly apiculate.

var. leucocephalus (Kunth) Fosberg in Kew Bull. 31: 835 (1977).

C. obtusiflorus Vahl (1805).

Inflorescence white as young, pale reddish brown when mature.

In dry grassland, on rocky or stony slopes, and on shallow soil over rocks or on coastal dunes; near sea-level to 1800 m. N1, 2; C2; S2, 3; widespread in Africa and western Asia. Gillett 4891; Bally 11106; Tardelli & Bavazzano 510.

var. flavissimus (Schrad.) Lye in Sedges and rushes of E. Afr.: 257 (1983).

Inflorescence bright yellow or orange.

In open forest or on rocky slopes; 1500–1800 m. N2; scattered in southern and eastern Africa. Glover & Gilliland 1139; Popov 1217; Thulin, Dahir & Hassan 9028.

45. C. meeboldii Kük. (1922).

Rather slender perennial with swollen stem-base covered by fibrous remains of old sheaths; root-system much reduced; stems 4–20 cm long and 0.4–1 mm thick, triangular, glabrous or slightly scabrid, swollen at the base. Leaf-blades 3–12 cm long and 0.5–2 mm wide, flat or inrolled, slightly scabrid on margin; sheaths grey to light brown, the oldest disintegrating into blackish fibres. Inflorescence a congested anthela with numerous sessile spikelets; major involucral bract leafy, 4–10 cm long and 1–4 mm wide, much longer than the inflorescence; anthela 0.7–2 cm in diam., consisting of 15–60 crowded spikelets. Spikelets 4–10 × 1–2 mm (or to 2.5 mm wide with glumes spreading), light to dark brown, linear, 15–30-flowered. Glumes 1.3–1.5 × c. 0.5 mm, ovate, reddish brown with 3-nerved midrib which ends in the slightly recurved apex. Stamen usually 1. Style unbranched or with 3 branches. Nutlet 0.5–0.6 × 0.3–0.4 mm, pear-shaped, triangular, grey, minutely papillose.

In seasonally wet habitats in grassland, or on rocks; c. 450 m. S1; East Africa, and in India. Lobin 6848 & Kilian 1995.

46. C. pygmaeus Rottb. (1773); C. michelianus (L.) Link subsp. pygmaeus (Rottb.) Aschers. & Graebn. in Syn. Mitteleur. Fl. 2, 2: 273 (1904).

Tussocky annual with crowded stems; stems 1–15 cm long and 0.5–1.5 mm thick (but wider across the sheaths), triangular, glabrous. Leaf-blades 1–6 cm long and 1–2 mm wide, flat, but as dry often folded and twisted, scabrid on margin and midrib at least near the apex; sheaths reddish to purple. Inflorescence a congested anthela of irregular outline and surrounded by 4–6(–8) long leafy spreading bracts; major involucral bract 3–12 cm long and 1.5–2.5 mm wide; anthela 0.5–1.2 cm in diam., consisting of several crowded hardly discernable spikes, each with many crowded spikelets. Spikelets 2.5–4 × 1–1.5 mm, green, oblong-lanceolate, 8–15-flowered. Glumes 1.3–1.6 mm long, ovate-lanceolate, 3–7-nerved, uncoloured below, light reddish brown with a thick green midrib above. Stamens 1–2. Style 2-branched. Nutlet 1–1.2 × 0.3–0.4 mm, oblong,

Fig. 67. A–C: *Cyperus somaliensis*, from Newbould 757. A: habit, × 0.5. B: spikelet, × 5. C: nutlet, × 30. – D–F: *C. pluricephalus*, from Kuchar 17635. D: habit, × 0.5. E: spikelet, × 5. F: nutlet, × 30. – G–I: *C. baobab*, from Moggi & Bavazzano 344. G: habit, × 0.5. H: spikelet, × 5. I: nutlet, × 30. – Drawn by G. M. Lye.

lenticular with one flat and one rounded side, yellow to apricot, minutely papillose.

In seasonally wet habitats, often in sandy places near pools or watercourses; near sea-level to c. 200 m. N3; S2; widespread in the tropics and subtropics of the Old World. Agrarian 22; Beckett & White 1663.

47. C. laevigatus L. (1771); *Juncellus laevigatus* (L.) C.B. Cl. (1893).

Tufted perennial with crowded stems or stems solitary on a long creeping rhizome; rhizome 1−5 mm thick, light brown to purplish black; stems usually 3−60(−150) cm long and 0.5−2(−6) mm thick, rounded to triangular or angular, glabrous, the basal part covered with short scales and rather loose leaf-sheaths and the stem therefore wider across the sheaths. Leaves to 4 cm long and 0.5−2 mm wide, usually inrolled and almost stem-like, scabrid on margin, blade sometimes absent; sheaths light to dark purple brown, all or the lowest only without leaf-blades. Inflorescence a solitary spikelet or more commonly a lax to crowded head of 2−30 spikelets (rarely to 80 spikelets), 0.5−3.5 cm in diam.; major involucral bract 1−3 cm long, scabrid on margin, slightly flattened but stem-like and continuing in the direction of the stem, the inflorescence therefore apparently lateral. Spikelets 5−20 × 1.5−3 mm, straight or curved, linear to lanceolate, somewhat flattened, pale yellowish grey, rarely variegated dark brown, 15−30-flowered. Glumes 2−3.5 mm long, very closely overlapping, broadly elliptic, rounded on the back that is without keel except near the tip, pale yellowish with reddish brown dots; apex acute, shortly mucronate or frayed. Stamens 3. Style with 2 long branches. Nutlet 1.5−1.7 × 0.8−1 mm, obovate with short apiculus, flat on 1 side, rounded on the other, grey to brown; surface smooth but with distinct rather large isodiametric surface-cells.

Forming a very dense sward on saline shores and in seepage zones, also near hot springs; sea-level to 1200 m. N1, 2; C2; pantropical. Bally 10893; Gillett 4304, 4884.

48. C. distans L.f. (1782).

Tufted perennial with a short thick rhizome and stems usually set in a row, or stems solitary; stems 20−60 cm long and 1.5−5 mm thick, triangular, glabrous, the basal part covered with leaf-sheaths. Leaves 5−30 cm long and 2−8 mm wide, flat, scabrid on margin and midrib; leaf-sheaths grey to dark purple (or black on old stem), rather lax. Inflorescence a compound umbel-like anthela 5−25 cm wide and 3−25 cm long; major branches 5−15, 1−18 cm long; secondary and tertiary branches short or spikes sessile. Spikes 2−4 × 1−4 cm, spikelets rather loosely set and often spreading at an angle of 90°. Spikelets 7−20 × 0.5−1 mm (or 1−2 mm wide with glumes spreading), linear or zigzag when glumes spreading, often breaking at its base with the glumes and nutlets persistent on its rhachis, 10−20-flowered; rhachis with a wide transparent wing on two sides. Glumes 1.7−2.6 mm long, oblong-elliptic, straw-coloured, light to dark reddish brown with a 3−5-nerved often green or paler keel; apex obtuse with a reddish brown or transparent margin into which the keel does not reach; glumes placed rather distant, falling off with the nutlet or persistent until the whole spikelet falls as one unit. Style 3-branched. Nutlet 1.4−1.7 × 0.4−0.5 mm, narrowly elliptic with short apiculus, yellowish and almost smooth when young, grey with a metallic shine and minute papillae in longitudinal rows when mature.

Edges of pools and ditches, also in dry forest, on roadside banks and in cultivations; up to 500 m. S1, 3; pantropical. Moggi & Bavazzano 1286, 1716; Kazmi & al. 733.

49. C. ferrugineoviridis (C.B. Cl.) Kük. (1936).

Robust stoloniferous perennial with a swollen stem base and with or without a short rhizome; stolons to 15 cm long and 0.5−1 mm thick, but to 3 mm thick with persistent fibrous sheaths; stems 25−70 cm long and 3−5 mm thick, triangular, with leaves in lower half. Leaves 15−40 cm long and 5−15 mm wide, flat, with scabrid margin and midrib; sheaths rather conspicuous, green to brownish above, dark brown to purple near the stem base. Inflorescence a lax anthela; major peduncles 5−10, 0.5−25 cm long, carrying single spikes or with 1−3 secondary spikes at the base of some of the main spikes; involucral bracts 6−10, leafy, erect or spreading, the largest 15−40 cm long and 6−12 mm wide. Spikes 15−45 × 20−50 mm, with many spreading spikelets; rhachis strongly winged. Spikelets 10−30 × 0.7−1.5 mm, linear-lanceolate, falling off entire when mature. Glumes 3.5−4.5 mm long, ovate-lanceolate, greenish, golden or reddish brown with translucent border and a greenish midrib (which is sometimes slightly excurrent). Style 3-branched.

In cleared forest, grassland and as a weed in cultivated land; 0−50 m. S2; scattered in eastern Africa. Bizi 29, 33, 35.

50. C. hemisphaericus Boeck. (1859).

Robust tussocky perennial with a short creeping rhizome and numerous crowded basal leaves; stems 15−90 cm long and 2−4 mm thick, triangular. Leaves 10−90 cm long and 5−13 mm wide, flat or folded, rather stiff, scabrid on margin and midrib. Inflorescence an open anthela of rather short and wide spikes, but sometimes rather much congested; involucral bracts 4−8, the largest 5−40 cm long. Spikes 10−40 × 12−30 mm, with numerous spreading spikelets, sometimes with 1−2 secondary spikes from the base of a primary spike; peduncles 0.5−6 cm long. Spikelets 10−15 × 1−2 mm, linear-lanceolate, falling off entire when mature. Glumes 3−4 mm long, ovate-lanceolate, yellowish white with a paler margin. Style 3-branched.

In seasonally wet grassland or savanna; near sea-level to 100 m. S1, 3; widespread in eastern Africa. Senni 329; Paoli 1169.

The Somali collections are very young and poor and the indentification is tentative.

51. C. pluricephalus Lye (1996); type: C2, 12 km NE of Ceel Baraf, then 2.5 km N on cutline, 3°19′N, 45°05′E, Kuchar 17635 (UPS holo.). Fig. 67 D−F.

Robust perennial with a c. 5 mm thick horizontal woody rhizome and crowded stems; roots sometimes with tomentum; stems 60−90 cm long and 1−2 mm thick, terete, but sometimes obtusely triangular above, with very weak longitudinal ridges, glabrous. Leaves from the lower 20 cm only; sheaths straw-coloured to, particularly in the lowermost coriaceous ones, bright reddish brown, glabrous; blades to 30 cm long and 2−4 mm wide, flat or rounded-canaliculate, stiff and coriaceous, minutely scabrid on margin and midrib. Inflorescence a 4−6 cm wide anthela of 1 sessile spherical group of spikelets (capitulum) and 2−4 stalked globose groups of spikelets; each capitulum 9−16 mm in diam. and consisting of 20−50 spikelets; involucral bracts 5−7, erect or spreading, leaf-like, the largest 10−20 cm long and 1−2 mm wide; peduncles 2−4 cm long, somewhat flattened, glabrous; rhachis of capitula with prominent round cushion-like outgrowths from where the spikelets begins. Spikelets 5−8 × 1−1.5 mm, almost cylindrical, narrowly lanceolate with acute apex, 3−4-flowered, falling off entire when mature. Fertile glumes 3.5−4 mm long, ovate, with obtuse or subacute apex, light reddish brown to somewhat golden with indistinct lateral nerves; midrib 3-nerved and in the lowermost glume excurrent into a very short mucro. Stamens 2−3. Style 4−5 mm long, with 3 branches. Nutlet c. 1.5 × 0.5 mm, elliptic, triangular, grey to dark reddish brown, minutely papillose, apiculate.

On gently rolling orange sandhills with *Commiphora* and *Indigofera ruspolii*; 170 m. C2; not known elsewhere.

52. C. rubicundus Vahl (1805).

C. teneriffae Poiret (1806).

Tussocky annual with a small root-system, or forms with much swollen base perhaps perennial; stems 2−20 cm long and 0.5−1.5 mm thick, triangular, slightly scabrid below the inflorescence, sometimes woody at the very base. Leaf-blades 2−8 cm long and 1−4 mm wide, flat, scabrid on margin and midrib; sheaths pale or grey to purple, up to 10 mm wide, and frequently not enclosing the base all around. Inflorescence a congested anthela with sessile spikelets only; major involucral bract leafy, 1−5 cm long and 0.5−2.5 mm wide, usually longer than the inflorescence; anthela 0.7−2.5 cm in diam., consisting of 3−25 large and sometimes much crowded spikelets. Spikelets 4−15 × 4−6 mm, light to dark reddish brown, linear-lanceolate, squarrose, 12−40-flowered. Glumes 2.5−4 mm long, often of rather different length in each spikelet, ovate, reddish brown to chestnut with 5−8 very narrow paler nerves on each side of the midrib; midrib prominent at apex only, and here paler and excurrent into a short

recurved scabrid mucro. Stamens 3. Style unbranched, but often twisted at the apex, or with 2−3 short or long branches. Nutlet 0.9−1.1 × 0.6−0.75 mm, obovate, sharply triangular, greyish, minutely papillose.

In seasonally wet habitats, often on sand or on shallow soil over rocks and near temporary pools and swamps; near sea-level. S3; Old World tropics. Paoli 311; Hemming 422, 423.

53. C. pseudosomaliensis Kük. (1936); *Mariscus somaliensis* C.B. Cl. (1895); type: N1, without precise locality, Lort Phillips s.n. (K holo.).

Tussocky perennial with succulent cylindrical to narrow bottle-shaped stem-bases covered by membranous pale or light brown sheaths without keeled midrib; stem 3−25 cm long and 0.3−1.5 mm thick, triangular, glabrous or minutely scabrid. Leaves crowded near the plant base; blades 2−10 cm long and 0.5−1.5 mm wide, flat but folded when dry to appear filiform, densely scabrid on midrib and the pale narrow marginal border scabrid dentate. Inflorescence of 4−15 crowded sessile spikelets forming a head 1−4 cm in diam.; involucral bracts leaf-like, 2−15 cm long, spreading or reflexed. Spikelets 5−20 × 2−3 mm, flat, 10−25-flowered. Glumes 3−4 mm long, acute, triangular, light to dark reddish brown without a prominent pale marginal border, with many prominent lateral nerves and with the upper part of the midrib green or straw-coloured, prominently keeled and scabrid above, excurrent into a 0.6−0.8 mm long recurved mucro. Style with 3 short branches.

Open woodland or scrubland, probably in upland areas. N1; not known elsewhere.

Only known from the type-collection; very similar to *C. cunduduensis*.

54. C. cunduduensis Chiov. (1940). Fig. 68 A−C.

Densely tussocky perennial with crowded swollen stem bases covered by conspicuous pale leaf-sheaths; stems 3−15 cm long and 0.3−0.5 mm thick, triangular, smooth. Leaves from the basal 2−3 cm only, usually 5−8 per stem; blades 3−5 cm long and 0.5−1 mm wide, with a very conspicuous white marginal border reaching the tip of the blade. Inflorescence a terminal cluster of 4−10 sessile spikelets; the 2−4 leafy involucral bracts spreading or reflexed, usually 1−3 cm long only. Spikelets 5−18 × 2−4 mm, linear, occasionally somewhat curved, 10−20-flowered. Glumes 3−4 mm long, triangular, acute, pale to dark reddish brown, with 5−8 narrow pale nerves on each side of the prominent green midrib that is excurrent into a short straight or recurved mucro; midrib slightly keeled near apex, glabrous or with a few teeth. Stamens 3. Style unbranched or with 2−3 stigmas. Nutlet broadly obovate, triangular, 1−1.5 × 0.8−1.2 mm, greyish or dark reddish brown, glossy with minute tubercles.

Limestone escarpment, sand dunes, stony ground and gravelly plains; from sea level to 1900 m. N1−3; S1−3; also in eastern Ethiopia. Gillett 4053; Godding 109; Bally 7327.

This is very similar to *C. wissmannii* Schwartz from the Arabian peninsula, but differs in its smaller and paler glumes.

55. C. obbiadensis Chiov. (1928); type: C1, between "Scermarca Hassan" and "Tobungab", Puccioni & Stefanini 577 (FT holo.).

Tussocky perennial with c. 10 mm thick and 3−4 cm long swollen stem bases surrounded by a dense coat of old brownish leaf sheaths; stems to c. 20 cm long and 0.5−1 mm thick, triangular at least above but becoming more terete with age, glabrous. Leaves from the lower 5 cm only; upper sheaths often pale, the lower brownish; blades filiform, to 10 cm long and less than 0.5 mm wide, flat or with margin inrolled, prominently scabrid along margins. Inflorescence a terminal cluster, 15−25 mm in diam., consisting of c. 10 crowded spikelets; involucral bracts 2−4, erect or spreading, the largest 3−6 cm long. Spikelets 1−2 cm long and c. 2 mm wide, linear with acute or subacute apex, flattened, 12−20-flowered. Glumes 3−4 mm long, ovate to ligulate, reddish brown with many slender but distinct lateral nerves and a green midrib excurrent into a prominent 0.5−0.8 mm long straight or slightly recurved mucro. Style with 3 branches.

In dry grassland; probably below 200 m. C1; not known elsewhere.

This is a taxon of uncertain status only known from the immature type collection. Possibly Naylor 1078 from "Budbud" (C1) also belongs here.

56. C. baobab Lye (1996); type: S2, 10 km from "Uarsciek" towards "Itala", 2°18'N, 45°52'E, Moggi & Bavazzano 344 (FT holo.). Fig. 67 G−I.

Very tussocky perennial with numerous closely set c. 2 cm long and 5 mm wide narrow ovate pseudobulbs formed by crowded succulent leaf-sheaths; stems to c. 10 cm long and 0.5−1.5 mm thick, distinctly triangular with prominently and irregularly scabrid angles or almost terete. Leaves from the basal 2−2.5 cm only; blades to c. 5 cm long and 0.5−1.5 mm wide, very thick, flat or canaliculate, with distinct uncoloured denticulate marginal borders; upper part of sheaths whitish, the lower parts often darker and succulent. Inflorescence of 4−10 crowded sessile spikelets forming a head 1.5−2.5 cm in diam.; involucral bracts usually to c. 1 cm long only, spreading or reflexed. Spikelets 7−12 × 3−4.5 mm, lanceolate, prominently compressed, 10−20-flowered. Glumes 3−3.5 mm long, ovate, acute, medium to dark reddish brown with c. 6 slender pale nerves on each side of the green or straw-coloured midrib that is excurrent in a very short (0.2 mm) straight or slightly recurved mucro. Stamens 3. Style 3-

branched. Nutlet c. 0.7 × 0.5 mm, obovate, prominently triangular, dark reddish brown to almost black, minutely papillose.

In calcareous sand-dunes; below 150 m. S2; not known elsewhere. Moggi & Bavazzano 53.

This is somewhat similar to *C. cunduduensis* and *C. obbiadensis*, but has more densely set pseudobulbs and a very much smaller nutlet.

57. C. ossicaulis Lye (1996); type: C1, 20 km W of Xarardheere, 4°37'N, 47°41'E, Beckett 202 (K holo.). Fig. 68 D−F.

Perennial without an apparent rhizome, but with aggregated woody stem bases; stems 25−35 cm long and 0.6−1.5 mm thick, terete or slightly angular with many low longitudinal ridges, glabrous, yellowish below, green above. Leaves 10−20 cm long and c. 1 mm wide, folded or inrolled with a narrow pale scabrid marginal border, the numerous crowded old basal sheaths forming a bone-like cylinder; blades filiform, 0.3−0.5 mm wide. Inflorescence a simple anthela 4−8 × 2−7 cm, usually with 8−15 spreading or reflexed spikelets from a c. 1 cm long axis, rarely depauperate anthelas contain 2−5 spikelets only; involucral bract solitary, but may appear as 2−3 since the 1−2 lowermost spikelets have subtending bracts ending in a filiform, reflexed, 5−15 cm long leafblade. Spikelets 15−40 mm long (the largest in each anthela always at least 25 mm long) and 3−4 mm wide, linear with acute apex, mostly 20−35-flowered. Glumes 5−6 mm long, ovate-elliptic, medium to dark reddish brown with a prominent pale marginal border and 5−7 slender but prominent nerves on each side of the midrib that is green in the upper half and excurrent into a c. 1 mm long straight or slightly recurved green awn. Style with 3 branches. Nutlet c. 1 × 0.5 mm, obovate, rounded-triangular, medium reddish brown to greyish, minutely papillose.

In regenerating field on stabilised dune; 340−350 m. C1; not known elsewhere.

58. C. micromedusaeus Lye (1996); type: N3, gorge of "Wadi Nogal", 5 km from "Eil", Bally & Melville 15547 (K holo.). Fig. 68 G−I.

Small perennial or perhaps sometimes annual with a minute root system with densely pubescent roots; stems 2−8 cm long and 0.5−0.8 mm thick, obtusely triangular or with 1 flat and 1 concave side, glabrous with shallow longitudinal ridges, not swollen at the base. Leaves all basal; sheaths light brown to purplish with wide translucent margins, glabrous, densely crowded and forming a 1−1.5 cm long and 3−5 mm thick narrowly ovate cylinder or pseudobulb; blades probably 1−2 cm long but strongly and repeatedly spirally twisted, 0.2−0.5 mm wide, as dry reddish brown, very thick and glabrous except for the thin uncoloured and irregularly scabrid marginal border. Inflorescence a solitary terminal group of 2−6 sessile

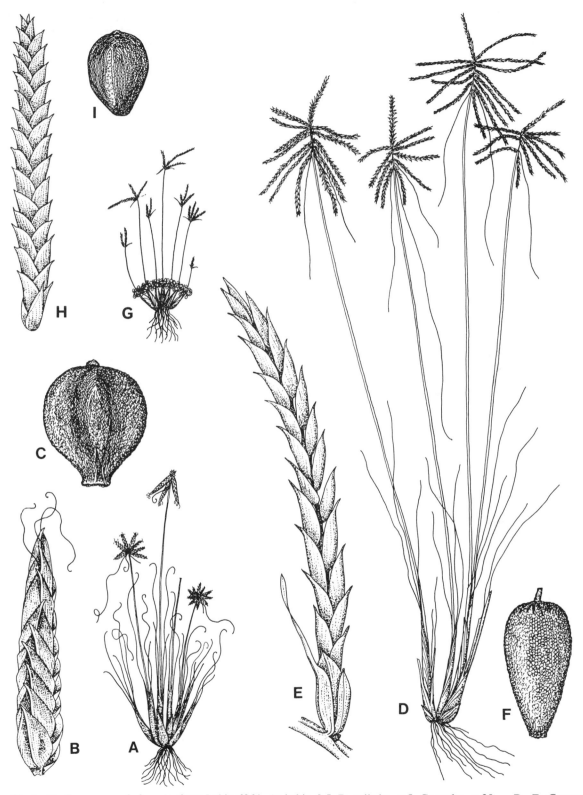

Fig. 68. A−C: *Cyperus cunduduensis*, from Lobin 6864. A: habit, 0.5. B: spikelet, × 5. C: nutlet, × 30. − D−F: *Cyperus ossicaulis*, from Beckett 202. D: habit, × 0.5. E: spikelet, × 5. F: nutlet, × 30. − G−I: *C. micromedusaeus*, from Bally & Melville 15547. G: habit, × 0.5. H: spikelet, × 5. I: nutlet, × 30. − Drawn by G. M. Lye.

erect or spreading spikelets; involucral bracts 2—3, up to 1 cm long with a glume-like base and a spirally twisted blade. Spikelets 6—20 × 2—2.5 mm, linear, 10—50-flowered; rhachis reddish brown, flattened but not winged, with flowers at c. 0.5 mm intervals. Glumes 2—2.5 mm long, triangular-ovate, light to medium reddish brown with an indistinct paler margin, with 5—7 prominent narrow nerves on each side of the midrib that is thickened above and excurrent into a short (to 0.5 mm long) slightly recurved mucro, glabrous except for the often scabrid midrib. Stamens 3. Style 2.5—3 mm long, with 3 branches. Nutlet 0.7—0.9 × 0.4—0.5 mm, obovate, triangular with 3 flat sides, greyish and smooth with a prominent cuticular layer, darker to almost blackish and minutely papillose when this layer is worn off.

In small holes in limestone pavement in *Acacia-Commiphora-Adenia ballyi* association; 50—150 m. N3; not known elsewhere.

59. C. gypsophilus Lye (1996); type: N2, 3 km E of "Las Anod", 8°25'N, 47°20'E, Hansen & Heemstra 6323 (K holo., C EA WAG iso.). Fig. 69 A—C.

Small perennial with a poorly developed short rhizome, but with woody stem-bases; roots with prominent tomentum; stems 10—20 cm long and 1—1.5 mm thick, glabrous, obscurely triangular above, almost terete below. Leaves c. 10 per stem, all basal or in the lower 3 cm of stem; sheaths pale to straw-coloured with a thick 7—15-nerved central part and scarious pale margin; blades 2—10 cm long and 1—1.5 mm wide, thick and almost semiterete with flat or canaliculate upper side, smooth or minutely scabrid on margin, often curved towards the apex. Inflorescence a dense terminal irregular cluster of spikelets 1—2 cm in diam. consisting of 8—15 crowded sessile spikelets; involucral bracts 1—3, erect, spreading or reflexed, leaf-like and to 5 cm long, often slightly curved at apex. Spikelets 8—10 × 2—3 mm, linear-lanceolate, only slightly compressed with regularly distichous glumes, 10—15-flowered. Glumes 3—4 mm long, ovate, pale to medium reddish brown; pale marginal border indistinct; lateral nerves c. 5 on each side of the rather obscure midrib that ends in the obtuse or subacute tip. Style with 3 very long branches. Nutlet c. 1 × 0.9 mm, greyish brown, obovate and prominently triangular with 3 slightly concave faces; the style-base persisting as a small apiculus; surface minutely papillose.

On gypsum hill; c. 800 m. N2; not known elsewhere.

60. C. somalidunensis Lye (1996); type: S2, 10 km from "Uarsciek" towards "Itala", Moggi & Bavazzano 377 (FT holo.). Fig. 69 D—F.

Small tussocky perennial with swollen stem-bases forming a regular row (indistinct rhizome) with only the 1—2 youngest carrying a stem; roots with prominent tomentum; stems 5—15 cm long and 0.7—1 mm thick, obscurely angular to somewhat compressed, usually scabrid at least on some ridges. Leaves many, mostly basal and 2.5 cm up the stems only; sheath reddish brown to almost black but with pale membranous margins and c. 7 prominent nerves that gradually split up into fibres; blades 3—10 cm long and 1—1.5 mm wide, flat but very thick and with prominent teeth along margins, often curved at apex; upper surface without nerves but with prominent rectangular cells; lower surface with indistinct nerves but without a prominent midrib. In- florescence a dense terminal irregular cluster of spikelets to c. 1 cm in diam., consisting of 5—15 crowded sessile spikelets; involucral bracts 2—3, spreading or reflexed, leaf-like and to c. 2 cm long. Spikelets 4—6 × c. 2 mm, ovate-lanceolate with acute apex, hardly compressed with irregularly distichous glumes, 5—9-flowered. Glumes 3—4 mm long, elliptic-ovate, medium reddish brown with an indistinct often incurved pale margin and 5—6 slender nerves on each side of the obscure only slightly keeled midrib ending in the obtuse apex or slightly excurrent. Style with 3 long branches. Nutlet 1.4—1.5 × 1.1—1.2 mm, greyish brown to black, elliptic and prominently triangular with 1 concave and 2 flat sides; surface smooth or minutely papillose.

Sandy plain, between consolidated and mobile dunes; below 150 m. S2; not known elsewhere.

61. C. recurvispicatus Lye (1996); C1, 28 km S of "Jeriban", 6°58'N, 48°52'E, Gillett, Hemming & Watson 22100 (K holo.). Fig. 69 G—I.

Tussocky perennial with a short woody rhizome and crowded stems; roots with tomentum; stems 10—15 cm long and 1—2.5 mm thick, somewhat compressed to terete or obscurely angular, glabrous, with leaves from the lower 1—3 cm only. Leaf-sheaths medium reddish brown, thick and coriaceous with many nerves and very prominent wide white membranous margins; leaf-blades to c. 20 cm long and 1.5—2 mm wide, flat but very thick, minutely scabrid along margin, often curled or coiled at apex. Inflorescence a lax anthela to 8 × 6 cm consisting of 1 sessile or subsessile group of spikelets and 1—4 stalked digitate groups of spikelets, each group consisting of 3—10 spreading and reflexed spikelets; involucral bracts short or to 20 cm long, spreading, leaf-like; peduncles to 4 cm long and 0.7 mm thick, terete. Spikelets 8—15 × 1—1.5 mm, linear, terete with acute apex, 12—16-flowered, at maturity disarticulating as 1 unit. Glumes c. 3.5 mm long, ovate, variegated golden and reddish brown with pale margin and patches, with 3—4 distinct lateral nerves on each side of the green midrib that is excurrent into a c. 0.2 mm long straight mucro. Style c. 4 mm long with 3 branches. Stamens 3. Nutlet 1.1—1.3 mm long, obtusely triangular, light reddish brown.

Very open *Acacia-Commiphora* bushland; 140 m. C1; not known elsewhere.

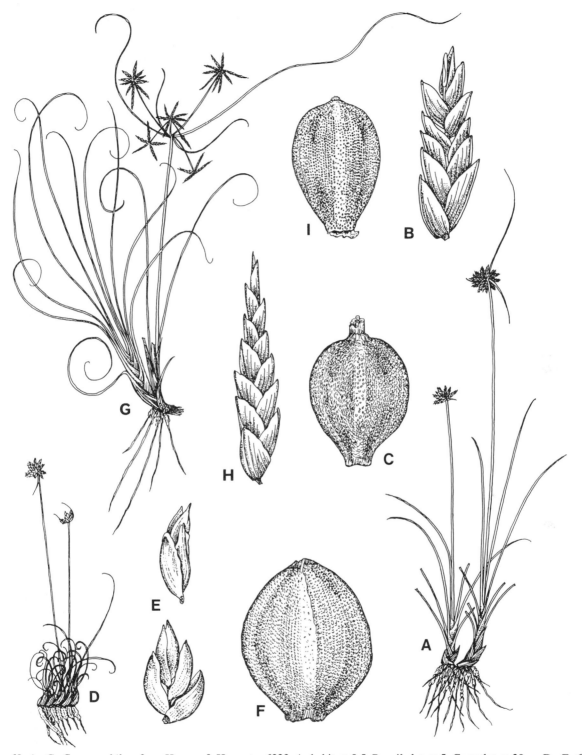

Fig. 69. A−C: *C. gypsophilus*, from Hansen & Heemstra 6323. A: habit, × 0.5. B: spikelet, × 5. C: nutlet, × 30. − D−F: *C. somalidunensis*, from Moggi & Bavazzano 377. D: habit, × 0.5. E: spikelet, front and side view, × 5. F: nutlet, × 30. − G−I: *C. recurvispicatus*, from Hemming & Watson 22100. G: habit, × 0.5. H: spikelet, × 5. I: nutlet, × 30. − Drawn by G. M. Lye.

62. C. phillipsiae (C.B. Cl.) Kük. (1936); *Mariscus phillipsiae* C.B. Cl. (1901); type: N1, without precise locality, Lort Phillips s.n. (K holo.).

Medium-sized perennial with a fleshy stem-base, but without a rhizome; stem 20−50 cm long, triangular, glabrous, in the lower part covered by wide greyish white leaf-sheaths. Leaves numerous, to 40 cm long and 2−6 mm wide, flat, strongly scabrid along the margin. Inflorescence of 4−10 sessile or stalked spikes; largest peduncles 0.5−3 cm long. Spikes 10−20 x 5−8 mm, with numerous densely set glumes; rhachis strongly nodular, glabrous. Spikelets 3−5 x 1−1.3 mm, somewhat angular, falling off entire when mature. Glumes 2.5−3 mm long, greyish white with or without green midrib and sometimes with an orange or reddish brown tinge; apex usually shortly mucronate. Style with 3 branches. Nutlet 1.2−1.3 x 0.7−0.8 mm, obovate, triangular, reddish brown, almost smooth to minutely papillose, when mature held tightly pressed between the stiff margins of the glumes.

In sandy soil in bushland and savanna; sea-level to c. 600 m. N1; C2; S3; Kenya. Moggi & Bavazzano 1632; Godfrey-Fausset 27; Glover & Gilliland 437.

63. C. vestitus Krauss (1845); *Mariscus vestitus* (Krauss) C.B. Cl. (1895).

Fairly robust succulent perennial with a thick ovate pseudobulb covered by pale to dark brown leaf-sheaths often splitting up into fibres, sometimes emitting slender stolons; stems 10−50 cm long and 0.7−2 mm thick, triangular, glabrous. Leaves up to 25 cm long and 2−6 mm wide, flat, strongly scabrid on margin and midrib. Inflorescence a 2−7 cm wide anthela consisting of 1−2 sessile and 3−8 stalked spikes; involucral bracts 2−5, leafy, erect or spreading, the largest 5−15 cm long and 2−7 mm wide; largest peduncle 1−7 cm long and 0.3−0.6 mm thick. Spikes 0.5−2.5 x 1−2 cm, with 6−15 spreading spikelets. Spikelets 5−14 x 1.5−2.5 mm, linear-lanceolate, reddish brown, 8−12-flowered. Glumes 3−3.5 mm long, ovate-oblong, reddish brown with 3−5 lateral nerves and a green midrib excurrent in a prominent recurved mucro c. 0.5 mm long. Style with 3 branches. Nutlet 1.5−1.8 x 0.9−1 mm, triangular, obovate, dark brown to blackish; surface papillose.

In grassland or open woodland; 600−1500 m. N1; eastern and southern Africa. Godfrey-Fausset 82; Gillett 4897.

64. C. cruentus Rottb. (1773).

Mariscus viridus Schweinf. (1867).

Tussocky perennial with few−many clustered oblong stem-bases on an obscure rhizome; stems 20−40 cm long and 1−2 mm thick, triangular, glabrous, with many withered loose leaf sheaths at the base and appearing swollen. Leaves with 5−30 cm long and 1−3 mm wide flat blades; sheaths pale above, brown below. Inflorescence a dense triangular to hemi- spherical contracted anthela 1−2 cm in diam.; involucral bracts 4−7, leafy, spreading or reflexed, the largest 10−30 cm long. Spikes obscure, c. 10 mm long and 7−9 mm wide, consisting of 10−25 crowded spikelets. Spikelets 4−5 x 1.5−2 mm, ovate, only slightly compressed, 3−4-flowered. Glumes 3−4 mm long, pale but light to medium reddish

brown at least in patches, with 5−8 prominent nerves on each side of the narrow green midrib that ends in the acute apex or is slightly excurrent. Style with 3 long branches. Nutlet 1.6−1.8 x c. 1 mm, oval, sharply triangular, minutely papillose.

Probably in bushland or open forest at intermediate altitude. N1; Ethiopia and North Africa. Cole s.n.

65. C. amauropus Steud. (1855).

Fairly robust succulent perennial with a short rhizome, and sometimes with 1−5 cm long stolons; stems 15−60 cm long and 0.5−2 mm thick (but often thicker near the base when fresh), triangular, glabrous. Leaves up to 25 cm long and 1−4 mm wide, flat, scabrid at least on margin; sheaths numerous at the base and producing a cylindrical pseudobulb, pallid or pale brown, but sometimes partly purplish. Inflorescence a 1−8 cm wide anthela consisting of several usually stalked short spikes, but the spikes may appear like digitate spikelet-clusters when consisting of 3−5 spikelets only; involucral bracts 2−5, leafy, erect or spreading, the largest 3−15 cm long and 1−3 mm wide; anthela of (1−)3−6 spikes on 0.2−4 cm long peduncles. Spikes 1−2 x 1−3 cm, with 5−10 spreading or reflexed spikelets. Spikelets 5−17 x 2−4 mm, linear-lanceolate, 8−18-flowered. Glumes 3−4 mm long, ovate-oblong, pale or dark reddish brown with 4−8 slender ribs on each side of the green or reddish brown midrib that is usually ending in the apex. Style with 3 branches. Nutlet 1.6−1.8 x 0.5−0.7 mm, triangular, oblong-elliptic, brown and densely papillose.

In dry bushland, often in bare soil, probably below 500 m. N1, 3; C1; northeastern and eastern Africa. Cole s.n.; Gillett & al. 22075A.

66. C. chaetophyllus (Chiov.) Kük. (1936); *Mariscus chaetophyllus* Chiov. (1916); type: S1, "Bur Meldac", Paoli 731 (FT holo.).

Densely tussocky perennial with a short woody rhizome producing many swollen stem-bases, some fertile but many with leaves only; stem very slender, often 5−20(−50) cm long and 0.3−0.7 mm thick, obtusely triangular to compressed, glabrous. Leaf-sheaths light to medium reddish brown or blackish with age, glabrous, sometimes splitting up into fibres, the inner succulent; blades to 30 cm long and 0.2−0.6 mm wide, filiform, flat or with margins inrolled, glabrous or minutely scabrid on margin. Inflorescence a single spike with 3−5 spreading spikelets from a slightly winged rhachis; lower 1−3 spikes each subtended by a filiform leaf-like spreading bract. Spikelets to c. 8 x 1−1.5 mm, lanceolate, slightly compressed, 8−15-flowered. Glumes 3−4 mm long, ovate-oblong, light to medium reddish brown with irregular pale margin and 5−8 slender nerves on each side of the green midrib that is excurrent into a c. 0.3 mm long mucro. Flower and nutlet not known.

Granitic rocks with shallow soils; 200−500 m. S1; not known elsewhere. Tardelli 37.

This is a taxon of uncertain status as the type is almost without an inflorescence and the other collection cited is very immature.

67. C. scleropodus Chiov. (1928); *Mariscus scleropodus* (Chiov.) Cuf. (1970); types: N3, "Giah", Puccioni & Stefanini 818 (FT lecto.).

Densely tussocky perennial; stems 5−25 cm long with prominently swollen bases covered by numerous thick dark brown to black leaf-sheaths not splitting up into fibres. Leaf-blades 5−30 cm long and 0.5−1 mm wide, twisted and convolute when dry, minutely scabrid on margin and midrib. Inflorescence a solitary dirty white head 10−20 mm in diam.; involucral bracts leafy, 3−4, spreading or reflexed, not dilated at the base, the largest 3−15 cm long. Spikelets 4−6 × 1.5−2 mm, lanceolate, flattened, 3−5-flowered. Glumes 3−3.5 mm long, lanceolate, light brown with wide pale margins and c. 10 lateral nerves and a narrow green midrib ending in or below the acute apex. Anthers 1.5−2 mm long. Style with 3 branches. Nutlet not seen.

In dry mountain region; c. 500 m. N3; not known elsewhere. Puccioni & Stefanini 795.

This is a taxon of uncertain status as the collections known are both immature.

68. C. paolii Chiov. (1915); *Mariscus paolii* (Chiov.) Chiov. (1916); *M. amomodorus* (K. Schum.) Cuf. var. *paolii* (Chiov.) Cuf. Enum.: 1449 (1970); type: S3, "Biejra", Paoli 206 (FT holo.).

Fairly robust tussocky perennial with few or many stems with swollen stem-bases; stems 15−45 cm long and 1−2 mm thick, triangular, glabrous, the base covered with leaf-sheaths that break up into numerous tough brown or blackish fibres. Leaves 10−30 cm long and 1.5−3 mm wide, flat, scabrid at least on margins; young sheaths pale brown and sometimes transparent. Inflorescence a white (but usually somewhat brownish as dry) head hemispherical or irregular in outline, 1−1.8 cm in diam. consisting of 30−50 crowded sessile spikelets; involucral bracts leafy, 3−5, reflexed, the largest 3−20 cm long and 1−3 mm wide, sometimes conspicuously dilated at the base. Spikelets 5−8 × 2−3 mm, ovate-elliptic, 5−8-flowered. Glumes 3.5−4.5 mm long, concave, white but usually pinkish brown when dry (except young inflorescences which often remain white), with 6−8 nerves on each side of the obscure midrib that ends in or below the obtuse apex. Style with 3 branches.

In seasonally wet grassland or wooded grassland. S2, 3; eastern Africa. Kazmi 5075.

The identity of this plant is uncertain as the mature nutlet is unknown. It is probably either the same as *C. submacropus* Kük. (in which case *C. paolii* is the correct name) or *C. firmipes* (C.B. Cl.) Kük. (which then will be the correct name).

69. C. dubius Rottb. (1773).

Tussocky perennial with many crowded stems and leaves; stems 8−40 cm long and 0.5−2 mm thick (the lower part is thicker when fresh), triangular, glabrous; base thick and bulbous. Leaves many, 5−30 cm long and 1.5−3.5 mm wide, flat and rather thick, scabrid at least on margin and midrib; upper sheaths grey, thin and membranous, the lower somewhat thicker, brown and occasionally splitting up into fibres. Inflorescence a solitary greenish white or dirty white head hemispherical or irregular in outline, 5−15 mm in diam., usually composed of 3−6 sessile congested spikes; involucral bracts 3−6, usually erect or spreading, the largest 4−15 cm long and 0.5−3.5 mm wide. Spikelets 2−5 × 1−2 mm, lanceolate, 3−6-flowered, usually densely clustered; rhachilla winged. Glumes 2−2.5 mm long, ovate, strongly concave, greenish with an uncoloured margin and 5−8 slender nerves on each side of the narrow midrib. Stamens 2−3. Style with 3 branches. Nutlet 1.2−1.4 mm long including a 0.1−0.2 mm long apiculus and 0.8−0.9 mm wide, obovate, triangular, brown with dark brown angles, strongly papillose.

In seasonally wet soil, often on rock-outcrops; probably 1000−1500 m. N1; widespread in tropical Africa and Asia.

No material seen by the author, but Drake-Brockman 81 from Golis Range was cited as this by Kükenthal in Pflanzenr. IV.20: 563 (1936). Also var. *coloratus* (Vahl) Kük. was said to occur in the same locality (Drake-Brockman 146).

70. C. eximius (C.B. Cl.) Mattf. & Kük. (1936); *Kyllinga eximia* C.B. Cl. (1902). Fig. 70 K.

Robust tussocky perennial with a short horizontal rhizome covered by the fibrous remains of old basal leaf-sheaths; stolons absent; stems 20−40 cm long and 1.5−2 mm thick, triangular, glabrous, the base slightly swollen and covered by numerous leaf-sheaths. Leaves 10−20 cm long and 3−8 mm wide, strongly scabrid at least on margin and midrib. Inflorescence a single white globose head 17−20 mm in diam.; involucral bracts 3−4, similar to the leaves, reflexed or spreading. Spikelets 5−8 × c. 2 mm, ovate, 2−5-flowered; 2 basal glumes and flowers separated on a c. 0.8 mm long spikelet-axis. Glumes 4.5−7 mm long, whitish, but with a light brownish tinge at least when dry, and with c. 5 nerves on each side of the scabrid unwinged midrib. Style with 2 branches. Nutlet 2−2.2 × c. 1.2 mm, obovate-flattened, dark reddish brown to blackish, minutely papillose.

On sandy soil in grassland or bushland or on rocks; sea-level to c. 550 m. S1−3; Ethiopia, Kenya. Bavazzano 856; Kazmi 655; Lobin 7003.

71. C. comosipes Mattf. & Kük. (1936); *Kyllinga leucocephala* Boeck. (1875).

Kyllinga eximia C.B. Cl. var. *kelleri* C.B. Cl. in Fl. Trop. Afr. 8: 288 (1902).

Tufted perennial with the stem-bases covered by the fibrous remains of old torn leaf-sheaths.; stems 15—50 cm long and 0.5—1.5 mm thick, sharply triangular with 1—3 longitudinal ridges on each of the 3 sides, densely scabrid at least on ridges above. Leaves with 10—25 cm long and 2—5 mm wide blades, strongly scabrid on midrib and margins. Inflorescence a globose or somewhat irregular whitish head 0.7—1.5 cm in diam., with a solitary spike or more commonly with 1—3 small inconspicuous lateral spikes surrounding the central spike; involucral bracts 2—4, leafy and reflexed or spreading, the largest 2—15 cm long. Spikelets 4—6 mm long, 2—4-flowered. Glumes 3—4 mm long, whitish with many prominent lateral nerves; keel unwinged, glabrous or with a few prostrate hairs. Style with 2 long branches. Nutlet 1.5—1.7 × 0.7—0.8 mm wide, flattened, minutely papillose.

In seasonally damp places in bushland and grassland. N3; S1; East Africa. Paoli 1028; Tardelli 218; Gillett & Hemming 24356.

72. C. brunneofibrosus Lye (1996); type: C2, 17 km E of Waajid (Uegit) on road to Oddur, 3°51'N & 43°23'E, Gillett & Hemming 24356 (K holo.). Fig. 70 A—C.

Densely tussocky perennial with a short woody rhizome with many stem bases densely covered by numerous old leaf-sheaths splitting up into dense "socks" of brownish fibres; roots many, greyish, up to 0.5 mm in diam.; stems 5—15 cm long and 0.5—0.8 mm thick, angular to terete, glabrous. Leaves 5—8 from within the fibrous "socks", and usually 1 from the lower part of the stem with light reddish brown, membranous sheath; blades to c. 10 cm long and 1—2 mm wide, flat, densely short-hairy on margin and midrib; lower surface with a slightly keeled midrib, the upper with a slightly sunken central part and several distinct lateral nerves. Inflorescence a terminal hemispheric head 5—10 mm in diam. consisting of numerous crowded sessile spikes; involucral bracts to 5 cm long and 2 mm wide, spreading or reflexed, leaf-like and densely short-hairy on margin and midrib. Spikelets 3—4 × 1—2 mm, ovate, brownish (at least as dry), 1—2(—3)-flowered. Glumes 3—4 mm long, ovate, as dry cinnamon-tinged but perhaps white when fresh; midrib distinct near the apex only, slightly excurrent into a 0.1—0.2 mm long straight or recurved mucro of the same colour as the glume or slightly paler; lateral nerves very indistinct. Stamens usually 3. Style with 2 branches. Nutlet c. 1 × 0.5 mm, elliptic, lenticular, reddish-brown, minutely papillose.

On shallow soil over limestone rocks and in sand or loam in bushland; 150—415 m. C1, 2; not known elsewhere. Beckett 136, 226; Elmi & Hansen 4087.

73. C. purpureoglandulosus Mattf. & Kük. (1936); *Kyllinga sphaerocephala* Boeck. (1875).

Perennial with stems solitary from tough, brown, persistent 1—2 mm thick scaly stolons producing new stems successively; stems 5—40 cm long and c. 1 mm wide, triangular, deeply ridged, glabrous or with a few spine-like hairs below the inflorescence, base thickened and covered with old darkened leaf-sheaths. Largest leaves 6—30 cm long and 2—5 mm broad, flat or inrolled, scabrid on margins and midrib, near apex also on upper surface. Inflorescence a single spike of 1—2-flowered spikelets or rarely a head of several spikes; inflorescence-bracts 2—4, leafy, the largest 2—8 cm long. Spikes 5—12 × 5—10 mm, spherical to ovate, white but fading to dirty greyish brown. Spikelets 3—4.5 mm long, 1—2-flowered; upper flower usually male. Glumes 2.5—3 mm long, white, but often with brownish dots, fading greyish brown, with 2—5 nerves on each side of the midrib; apex acute. Style with 2 branches. Nutlet elliptic to obovate, c. 1.4 × 0.7 mm, dark brown.

On maritime sand. S3; East Africa. Senni 103.

Senni 103 from "Ali Javio" was cited as this species by Chiovenda, Fl. Somala 2: 433 (1932). The collection has not been seen and the identification is most probably an error, as this is a species of high rainfall areas.

74. C. sesquiflorus (Torr.) Mattf. & Kük. subsp. **appendiculatus** (K. Schum.) Lye in Sedges and rushes of E. Afr., Appendix 3: 2 (1983). *Kyllinga odorata* Vahl var. *major* (C.B. Cl.) Chiov., Fl. Somala 2: 432 (1932).

Tufted perennial with mostly crowded stems from a usually short but sometimes up to 8 cm long creeping rhizome; stems 10—80 cm long and 0.7—2.8 mm thick, triangular, ridged, glabrous, the base usually swollen and covered by hardened scales that split but very rarely form fibres. Leaves flat or inrolled and with spine-like hairs on margin and midrib, the largest leaves 7—40 cm long and 2—5 mm wide; lower leaf-sheaths blade-less or with very short blade. Inflorescence a compound head, 7—15 × 5—17 mm, of a larger ovoid to ovoid-cylindrical central spike and usually much smaller lateral spikes, very rarely consisting of a single spike only; involucral bracts 3—6, leafy, usually spreading or reflexed, the largest 3—15 cm long. Spikes whitish but fading pale brownish. Spikelets 3—4.5 mm long, 2-flowered. Glumes 3—4 mm long, whitish with reddish dots and frequently with a greenish midrib, with 2—5 nerves on each side of the midrib; keel of midrib glabrous; apex usually acuminate with an excurrent midrib. Style with 2 branches. Nutlet usually 1.2—1.5(—1.9) × 0.7—0.9 mm, obovate to elliptic, blackish and minutely tuberculate.

Probably bushland at low altitude. S3; tropical Africa. Senni 422.

Senni 422 from "Ghersei" was cited as *K. odorata* var. *major* by Chiovenda, Fl. Somala 2: 433 (1932).

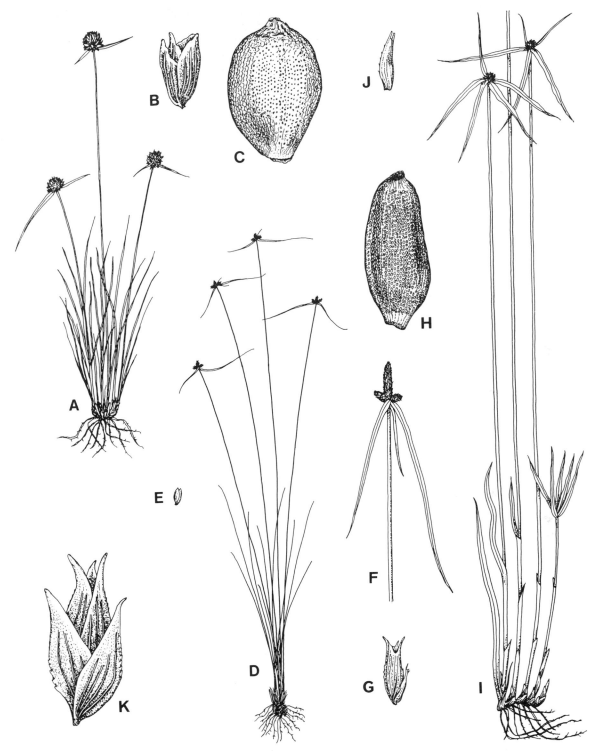

Fig. 70. A−C: *Cyperus brunneofibrosus*, from Gillett & Hemming 24356. A: habit, x 0.5. B: spikelet, x 10. C: nutlet, x 30. − D, E: *C. microstylis*, from Thulin & Mohamed 6803. D: habit, x 0.5. E: spikelet, x 5. − F−H: *C. oblongus* subsp. *jubensis*, from Thulin & Mohamed 6793. F: habit, x 0.5. G: spikelet, x 10. H: nutlet, x 30. − I, J: *C. erectus* subsp. *jubensis*, from Kazmi 5248. I: habit, x 0.5. J: spikelet, x 10. − K: *C. eximius*, spikelet, x 5, from Glover & Gilliland 462. − Drawn by G. M. Lye.

This collection has not been seen and the identification is most probably an error, as this is a taxon of high rainfall areas.

75. C. welwitschii (Ridley) Lye (1983).

Tufted perennial with the base of the stems swollen and densely covered by old leaf-sheaths, the oldest fibrous and thread-like; stems 5–25 cm long and 0.4–0.7 mm thick, triangular, ridged, glabrous. Leaves 5–10 cm long and 0.5–2 mm broad, flat, canaliculate or inrolled, with short prickles on margin and midrib. Inflorescence an irregular head, but often triangular in outline, 5–8 × 5–9 mm, consisting of several (often 3) greyish white spikes; central spike ovate to cylindrical, 3.5–4 mm wide. Spikelets 1.5–2 mm long, very asymmetric, 1-flowered. Glumes 1–1.8 mm long, whitish or pale yellow with or without golden brownish midrib, with 1–2 nerves on each side of the slightly or not winged but strongly ciliate midrib; apex with midrib slightly excurrent or midrib ending in apex. Style with 2 branches. Nutlet c. 1 mm long, elliptic.

On moist sandy soil near rivers or in shrubland; 190–300 m. S1; tropical Africa. O'Brien 22; Thulin & al. 6780, 7608.

76. C. microstylis (C.B. Cl.) Mattf. & Kük. (1936); *Kyllinga microstyla* C.B. Cl. (1895); type: N1, without precise locality, Lort Phillips s.n. (K holo.). Fig. 70 D, E.

Slender tufted perennial with swollen stem-bases often covered by the fibrous remains of old leaf-sheaths; stems usually 5–20 cm long and 0.3–0.7 mm thick, triangular, glabrous. Leaves to 10 cm long and 0.5–2 mm wide, flat, scabrid on margin and midrib. Inflorescence a small whitish head, 4–6 mm in diam., consisting of 1 central oval to rounded spike and 1–3 smaller lateral spikes; involucral bracts usually 3, spreading to reflexed, the largest 2–6 cm long, similar to the leaves. Spikelets 1–1.3 mm long, ovate, 1-flowered. Glumes 0.8–1 mm long, whitish with 1–2 lateral nerves on each side of the glabrous midrib that ends in the acute apex. Unbranched part of style 0.1–0.2 mm long with 2 0.2–0.4 mm long stigmas. Nutlet c. 0.6 × 0.3 mm, elliptic, compressed, brown, minutely papillose.

In silty or sandy soil; 200–400 m. N1; S1; Kenya. O'Brien 197; Friis & al. 4880; Thulin & Mohamed 6803.

77. C. oblongus (C.B. Cl.) Kük. (1936). subsp. jubensis (Mtoto.) Lye in Lidia 3: 132 (1994); *Kyllinga flava* C.B. Cl. subsp. *jubensis* Mtoto. in Ethiopian J. Sci. 13: 39 (1990); type S3, between "Bula Haji" and "Halinoadoi", 75–110 km from "Kismaiyo" towards "Badhadhe", Kazmi, Elmi & Rodol 731 (K holo.). Fig. 70 F–H.

Tufted perennial with a short rhizome; stems 25–50 cm long and 1–2 mm thick, triangular,

glabrous, only slightly swollen at the base. Leaves 5–6 per stem; sheaths brown, pale or straw-coloured, the 3–4 uppermost with up to 30 cm long and 2–4 mm wide flat blades. Inflorescence with a 1–2 cm long cylindrical spike subtended by 1–3 smaller lateral spikes, each with numerous crowded spikelets. Spikelets c. 2.5 mm long, 2-flowered. Glumes 1.8–2.5 mm long, pale to golden with a prominent green hairy midrib (hairs to 0.3 mm long) excurrent into a short straight or recurved mucro, and with 2 prominent nerves on each side of the midrib. Style with 2 c. 1 mm long stigmas. Nutlet c. 1.3 × 0.5 mm, obovate-elliptic, compressed, brown, minutely papillose.

On rocky outcrops and in degraded sandy bushland; c. 50–250 m. S1–3; not known elsewhere, but other subspecies in East Africa. Moggi & Tardelli 39; Thulin & Mohamed 6793; Lobin 6969.

78. C. erectus (Schumach.) Mattf. & Kük. (1936). subsp. jubensis (Chiov.) Lye (1995); *Kyllinga jubensis* Chiov. (1932); type: S3, "Uamo Ido", Tozzi 250 (FT holo.). Fig. 70 I, J.

Perennial with a creeping rhizome and stems densely set in a usually single row, the swollen basal parts of the stems persistent on the old dead rhizome; rhizome 4–10 cm long, and 3–4 mm wide, covered by scales and the densely set stems; stems 12–40 cm long and 0.8–1.7 mm thick, sharply triangular, ridged, glabrous, the base enclosed in several brownish bladeless sheaths. Leaf-blades flat, canaliculate or inrolled, usually 4–6 per stem, the largest to 5–20 cm long and 2–4 mm wide; margin and midrib densely set with short spine-like teeth especially above. Inflorescence a solitary ovate or hemispherical spike; involucral bracts 3–4, leafy, usually spreading or reflexed, the largest 5–15 cm long and 2–3.5 mm wide. Spikelets 2.5–3 mm long, 1–2-flowered. Glumes 2–3.5 mm long, very unequal, golden yellow with green midrib and 3–5 nerves on each side of the glabrous unwinged midrib; apex acuminate or with midrib excurrent and recurved. Style with 2 branches. Nutlet c. 1.2 × 0.6 mm, elliptic (but mature nutlet not seen).

Margin of pools; c. 20–100 m. S3; not known elsewhere, but subsp. *erectus* widespread in tropical Africa. Kazmi 5248.

79. C. aromaticus (Ridley) Mattf. & Kük. (1936). *Kyllinga polyphylla* Kunth (1837).

Robust perennial with a creeping rhizome and densely set stems; rhizome including scales c. 5 mm thick, the scales fairly thick, pale brown to dark purple or blackish, less than 10 mm long; stems 25–90 cm long and 1–3 mm thick (but wider across the leaf-sheaths), the basal part usually covered by purplish sheaths without leaf-blades. Upper leaf-sheaths with blades 3–15 cm long and 2–6 mm wide. Involucral bracts 5–8, usually long and spreading, the longest 6–15 cm; inflorescence an irregular hemispherical to globose head with a central spike and usually several smaller lateral spikes. Spikelets 3–4 mm long, 1–2-flowered, but only one producing nutlet. Glumes yellowish or straw-coloured with greenish midrib and frequently with dark brown dots or streaks especially near the midrib; 3–5 ribs on each side of the midrib. Style with 2 branches. Nutlet 1.2–1.5 mm long.

In somewhat shady habitats in upland areas; probably 1000–1500 m. N1; tropical Africa. Robecchi-Bricchetti 114.

80. **C. alatus** (Nees) F. Muell. (1874); *Kyllinga alata* Nees (1836).

C. cristatus (Kunth) Mattf. & Kük. (1936); *Kyllinga alba* Nees (1835).

Robust densely tufted perennial with a short rhizome producing usually numerous stems and leaves; stems 10–70 cm long and 0.4–1.7 mm thick, sharply triangular and deeply or obscurely ridged, glabrous (or with a few very short spine-like hairs), the base swollen and bulb-like and densely covered by old darkened leaf-sheaths often split into black fibres. Leaves 5–40 cm long and 2–7 mm broad, flat or inrolled; margin and midrib with short spine-like hairs especially above. Inflorescence a single globose or shortly ovate white head fading to pale dirty brown, 6–14 x 6–15 mm; involucral bracts 2–5, leafy, up to 15 cm long (but in smaller plants sometimes only 1 cm long), usually bent downwards. Spikelets 4–6.5 mm long, 1–2-flowered. Outer glumes equal, 3.5–6.5 mm long, whitish but often brown-spotted, with 2–3 mostly rather weak nerves on each side of the excurrent, strongly winged midrib; wings coarsely toothed, each tooth with a wide base and 1–2 narrow ciliate spines. Style with 2 branches. Nutlet 1.6–1.8 x 0.7–0.8 mm, blackish, minutely papillose.

In grassland or bushland, often in seasonally wet sand; near sea-level to c. 200 m. S2, 3; tropical and southern Africa. Paoli 257, 557; Moggi & Bavazzano 1708.

81. **C. aureoalatus** Lye (1995).

Densely tufted perennial with the base surrounded by strong fibres from old leaf-bases; stems 15–40 cm long and c. 1 mm thick, deeply ridged, glabrous or with a few scattered hairs below the inflorescence. Leaves 10–25 cm long and 1–3 mm broad, flat or inrolled, sometimes with minute dark dots; margin and midrib with short spine-like teeth especially above. Inflorescence a single 10–20 cm long and 6–10 mm wide golden brown spike of numerous sessile 2-flowered spikelets; involucral bracts 3–4, leafy, up to 12 cm long. Spikelets 3.5–4 mm long, with 2 bisexual flowers or the upper male. Glumes golden brown, but the long-acuminate midrib frequently greenish above; keel conspicuously

winged with coarse teeth each bearing 1–2 cilia. Nutlet c. 1 x 0.6 mm; style including its 2 branches 2 mm long.

Stony ground or sandy soil at low altitude. S3; tropical and southern Africa. Senni s.n.

82. **C. hyalinus** Vahl (1805); *Queenslandiella hyalina* (Vahl) Ballard (1933). Fig. 71 A–C.

Tufted annual with a shallow root system and 5–30 cm long stems. Leaves 5–15 cm long and 2–5 mm wide, flat, scabrid on margin and midrib; leaf-sheaths grey to reddish brown. Inflorescence of a few crowded spikes or with 1–few sessile spikes and 1–8 stalked spikes on 0.5–10 cm long peduncles; involucral bracts 3–6, leafy, erect or spreading, the largest 6–25 cm long and 1.5–6 mm wide, flat. Spikes 6–22 mm long and 6–16 mm wide, with (3–)5–10(–25) spreading spikelets. Spikelets 3–8 x 1.5–3.5 mm, ovate-lanceolate, appearing spiny with the spreading tips of the glumes. Glumes 2–3 mm long, golden or greyish with a green midrib excurrent in a recurved mucro and with 2–4 very prominent lateral nerves on each side of the midrib. Style with 2 branches. Nutlet 1.3–1.5 x 1–1.2 mm, flattened, somewhat cordate, brown to dark grey, minutely papillose.

Dunes, sandy grassland and as a weed of sandy soils; near sea-level. S2; scattered in eastern Africa, Madagascar, tropical Asia and Australia. Lobin 6949 & Kilian 2097.

83. **C. macrostachyos** Lam. (1791); *Pycreus macrostachyos* (Lam.) Raynal (1969).

Robust annual with solitary stem or tufted; stems 30–50 cm long and 1–4 mm thick, but to 8 mm thick across the sheaths, triangular, glabrous. Leaf-blades 10–40 cm long and 3–8 mm wide, flat, rather thick and soft below, scabrid on margin and midrib near the tip; sheaths rather wide, straw-coloured to brown with transparent scarious margins. Inflorescence a compound anthela of large sessile and stalked spikes; major branches to 10 cm long. Spikelets 5–10 x 1–2 mm, linear-lanceolate. Glumes 2–3 mm long, obovate, light brown or reddish brown with a very prominent wide pale marginal border. Style with 2 branches. Nutlet 1.5–2 x 0.7–1.2 mm, obovate to elliptic, grey, brown or blackish, minutely papillose.

Wet depressions in woodland; below 50 m. S3; pantropical. Moggi & Bavazzano 1617.

84. **C. pumilus** L. (1756).
var. **patens** (Vahl) Kük. in Pflanzenr. IV.20: 378 (1936); *Pycreus patens* (Vahl) Cherm. (1931).

Small tufted annual with short and very thin roots; stems 1–16 cm long and 0.3–0.7 mm thick, triangular, glabrous, not swollen or hardened at the base. Largest leaves 2–8 cm long and 1–1.5 mm wide, scabrid on margin and midrib near the tip, 1–3 per stem; leaf-sheaths grey to purplish, the lowest

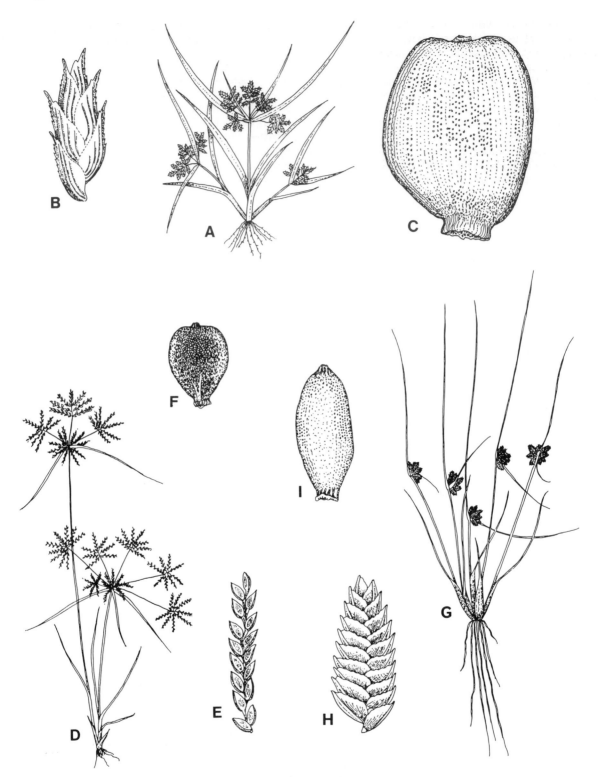

Fig. 71. A−C: *Cyperus hyalinus*, from Lobin 6949. A: habit, × 0.5. B: spikelet, × 10. C: nutlet, × 60. − D−F: *C. micropelophilus*, from Gillett & Hemming 24892. D: habit, × 0.5. E: spikelet, × 5. F: nutlet, × 60. − G−I: *C. dwarkensis*, from Kazmi & al. 844. G: habit, × 0.5. H: spikelet, × 10. I: nutlet, × 60. − Drawn by G. M. Lye.

without leaf-blade. Inflorescence of 1 sessile and 1−6 stalked spikelet-clusters, 1−10 cm wide, each with 3−20 rather laxly arranged spikelets; largest peduncles 0.5−5 cm long; major involucral bracts leafy, the largest 2−10 cm long. Spikelets 2−12 mm long and 1−2.5 mm wide, linear-lanceolate, grey to reddish brown, new spikelets sometimes arising from the prophylls of the old. Glumes 1−1.2 mm long, ovate, grey to reddish brown with green 3−5-nerved keel, conspicuously mucronate, imbricate when young, but later spreading and their wings infolding so as to lose contact with each other and expose the nutlets. Stamen usually 1, placed laterally. Style with 2 branches. Nutlet 0.5−0.6 × 0.3−0.4 mm, obovate, almost whitish when young, dark grey with a metallic shine when mature; surface with minute tubercles in longitudinal rows; apex apiculate.

In seasonally wet hollows, often in sandy soils; below 50 m. S3; tropical and southern Africa. Senni s.n.

85. C. micropelophilus Lye (1996); type: S1, "Bur Akaba", 2°48'N, 44°05'E, Gillett & Hemming 24892 (K holo.). Fig. 71 D−F.

Tufted slender annual with few−numerous stems; stems 5−20 cm long and c. 1 mm thick, triangular, glabrous, with 2−3 leaves in lower half. Leaf-blades 5−15 cm long and 1−3 mm wide, flat, scabrid at least on margin; sheaths rather loose, grey to purple. Inflorescence lax, 2−5 cm wide, consisting of 1 central sessile spikelet-cluster and 2−8 stalked spikelet-clusters; largest peduncle 2−5 cm long. Spikelets 2−10 × 1−2 mm wide, linear, reddish brown with 10−25 closely overlapping glumes. Glumes 1.1−1.3 mm long, ovate, reddish brown with green midrib ending below the obtuse apex. Style with 2 branches. Nutlet c. 0.6 mm × 0.5 mm, obovate to squarish, flattened, dark reddish brown, minutely papillose.

In damp spots where water trickles over the rock after rain; 200−350 m. S1; not known elsewhere.

This is somewhat intermediate between *C. pelophilus* Ridley and *C. pseudohildebrandtii* Kük. It has a much wider nutlet than that found in typical *C. pseudohildebrandtii* from Kenya and Tanzania.

86. C. dwarkensis Sahni & Naithani (1976); *Pycreus dwarkensis* (Sahni & Naithani) Hooper (1985). Fig. 71 G−I.

Small tufted annual with short and very thin roots; stems 1−8 cm long and up to 0.5 mm thick, angular to almost terete, glabrous. Leaves 3−4 per stem; sheaths reddish brown and usually the 2 upper producing 2−5 cm long up to 0.5 mm wide strongly inrolled glabrous blades. Inflorescence a dense head of 2−15 crowded spikelets and 1−2 erect and spreading leaf-like involucral bracts to c. 8 cm long. Spikelets 5−12 × 2.5−3 mm, linear-lanceolate, strongly compressed, variegated grey and reddish brown, 15−25-flowered. Glumes 1.5−2 mm long, greyish white with or without reddish brown patches near the partly

green 3-nerved midrib that is excurrent into a short slightly recurved awn; surface cells large and prominent. Style with 2 long stigmas. Stamens usually 2. Nutlet c. 1.2 × 0.7 mm, obovate in outline but prominently compressed and slightly twisted, reddish brown, minutely papillose, held tightly inside the glume and dispersed with it.

At the edges of small lakes or pools; 200−300 m. C2; India. Pignatti s.n.; Kazmi & al. 844.

7. SCHOENUS L. (1753).
Kükenthal in Feddes Repert. 44: 1−32, 65−101, 161−195 (1938).

Perennial herbs usually forming dense tussocks; stems usually terete and ending in a dense bracteate cluster of spikelets. Leaves all basal. Spikelets 1−5(−9)-flowered with subdistichous glumes, the lowermost ones sterile. Flowers usually hermaphrodite; perianth of 1−6 small bristles or absent. Stamens 1−6, but usually 3. Style with 3 stigmas, very rarely only 2. Nutlet small, usually 3-angular and without a persistent style-base.

About 75 species, mainly in Australia and New Zealand, only one species in Africa.

S. nigricans L. (1753). Fig. 72 A−C.

Densely tussocky plant with few−many stems and numerous leafy shoots; stems 15−70 cm long and 0.5−2 mm thick, almost terete with many rounded longitudinal ridges, glabrous. Leaves from the basal 12 cm only; basal sheaths chestnut-brown to reddish brown, often glossy; blades mostly 10−40 cm long and 0.4−1 mm wide, flat or somewhat canaliculate, hard and wiry, minutely scabrid at least on margin near the tip. Inflorescence a 5−20 mm wide and 10−15 mm long head of 5−20 crowded spikelets; largest involucral bract with a glume-like base and a 0.5−5 cm long green leafy awn. Spikelets 5−15 × 2−4 mm, somewhat flattened with distichous glumes, usually 2−3-flowered; lowermost glumes much shorter than the uppermost. Largest glumes 7−8 mm long, reddish brown to almost black, minutely scabrid on midrib, otherwise glabrous. Perianth-bristles very small or absent. Stamens 3. Style with 3 long strongly hairy stigmas. Nutlet c. 1.5 mm long, obtusely trigonous, whitish, shiny.

Seasonally wet places; probably between 1000 and 2000 m. N1; Eritrea, Ethiopia, South Africa, North Africa, Europe, Middle East, America. Cole s.n.

8. CAREX L. (1753).
Kükenthal in Pflanzenr. IV.20 (1909).

Perennial herbs with short or long creeping rhizomes, often forming compact and dense tussocks; stems triangular, rarely rounded. Leaves with prominent

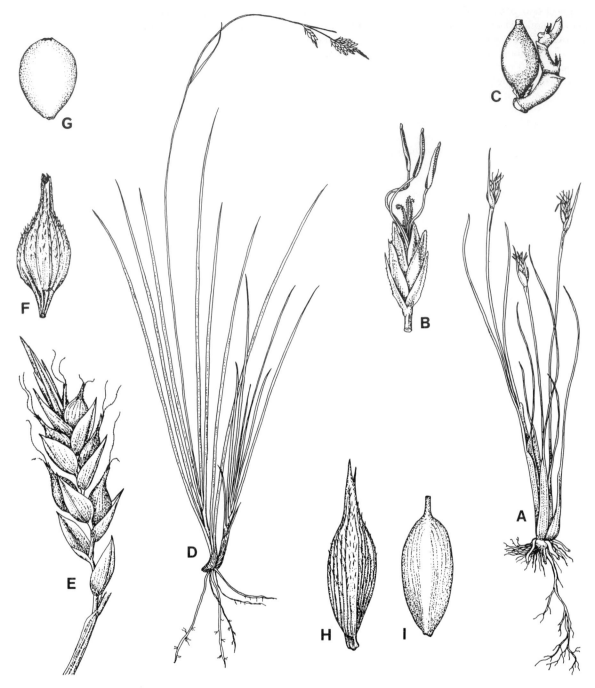

Fig. 72. A−C: *Schoenus nigricans*, from Polunin 5101. A: habit, × 0.5. B: spikelet, × 3. C: nutlet with rhachilla, × 10. − D−G: *Carex brunnea* subsp. *occidentalis*, from Thulin & al. 8982. D: habit, × 0.5. E: spike, × 5. F: utricle, × 10. G: nutlet, × 10. − H, I: *C. negrii*, from Coll. Agric. G 33. H: utricle, × 10. I: nutlet, × 10. − Drawn by G. M. Lye.

closed, glabrous or hairy sheaths, and with a small rim-like ligule at the junction of the sheath and the blade; blades usually well developed, rarely reduced to short triangular lobes, glabrous or hairy. Inflorescence either a single unisexual or bisexual spike or a variously composed panicle of spikes.

Individual spikes either unisexual or with male flowers below and female above or vice versa, more rarely with male flowers at both ends. Flowers unisexual, set in few- or many-flowered spikes and subtended by a bract usually called a glume. Perianth segments absent. Male flower of 3 stamens only.

Female flower consisting of an ovary with 2 or 3 style-branches, the ovary enclosed in a bottle-shaped prophyll (utricle), and only the stigmas projecting. Utricle with or without a short or long beak, glabrous or short-hairy, often with 2−3 prominent longitudinal ridges; surface smooth, micro- or macro-papillose. Nutlet triangular or biconvex, entirely enclosed in the utricle.

About 1500 species throughout most of the world, but rare at low altitudes in the tropics and most numerous in cold and temperate parts of the northern hemisphere.

1. Style with 3 branches; utricle narrowly lanceolate
    ....1. *C. negrii*
−   Style with 2 branches; utricle oval and lenticular
    .... 2. *C. brunnea*

1. C. negrii Chiov. (1912). Fig. 72 H, I.

Tall perennial with a horizontal rhizome often producing stems at up to 1 cm intervals; stems (10−)40−90 cm long and 1−2 mm thick, triangular, scabrid on angles, the larger part covered by leaf-sheaths. Leaves numerous, high up the stem; lower sheaths brownish, the upper green, mostly glabrous; blades usually 10−30 cm long and 2−3 mm wide, flat or inrolled, prominently scabrid at least near the apex. Inflorescence of 1−2 stalked or subsessile spikes from each of the uppermost sheaths; subtending bracts leafy and mostly longer than the spikes. Spikes 1−2.5 cm long and 3−4 mm wide, consisting of 2−10 female flowers at the base and 10−25 male flowers at the tip; male part of the spike longer than the female. Glumes c. 4 mm long, ovate, glabrous, yellowish brown or light reddish brown with a prominent pale marginal border and a green 1−3-nerved midrib ending in the acute apex; male glumes similar to the female. Style with 3 long stigmas. Utricle narrowly lanceolate, 3.5−4 × c. 1.5 mm including a 1−1.5 mm long distinct beak, green to light reddish brown with fairly prominent nerves on all sides, usually densely short-hairy at least in its upper half. Nutlet c. 2.5 × 1 mm, narrowly elliptic-triangular, almost smooth, dark reddish brown but with a pale base.

Dry upland forest; c. 1500−2000 m. N2; Ethiopia and Yemen. Bally & Melville 15993; Glover & Gilliland 1142.

2. C. brunnea Thunb. (1784).

subsp. **occidentalis** Lye (1996); type: N2, Karin Xaggarood, Thulin, Dahir & Hassan 8982 (UPS holo., K iso.). Fig. 72 D−G.

Fairly slender tussocky perennial with a compact horizontal woody rhizome and numerous crowded stems; stems 30−60 cm long and 1−1.5 mm thick, triangular, scabrid to subglabrous on angles. Leaves many; lower sheaths dark reddish brown with almost black nerves, sometimes splitting up into fibres; blades to 40 cm long and 3−4 mm wide, flat, scabrid on margin and ribs particularly towards the apex. Inflorescence of 1−3 slender stalked or subsessile spikes from each of the 5−8 uppermost sheaths; subtending bracts leafy and mostly much longer than the spikes (but the uppermost shorter). Spikes 1−3 cm long and c. 3 mm wide, consisting of 5−15 distantly set female flowers at the base and 2−6 male flowers at the tip; male part of the spike much shorter than the female. Glumes 3−4 mm long, ovate-lanceolate, light reddish brown with a pale 1−3-nerved somewhat scabrid midrib ending in the acute apex; male and female glumes similar, but male and upper female glumes often not scabrid. Style with 2 c. 3 mm long slender stigmas. Utricle oval and lenticular, c. 3.5 × 1.2−1.3 mm including a prominent 0.5−0.8 mm long cuneate base and a c. 1 mm long distinct beak, densely short-hairy except near the base; nerves many and prominent on both sides. Nutlet up to 2.5 × 1.5 mm, ovate-lenticular, almost smooth, dark brown or pale.

Rocky gully in deep shade in evergreen bushland with *Buxus, Juniperus, Olea, Pistacia* and *Acokanthera* on limestone; 1400 m. N2; Ethiopia.

There are other subspecies in Yemen, Madagascar, Mauritius and from India to Japan.

166. FLAGELLARIACEAE

by M. Thulin

Fl. Trop. E. Afr. (1971); Cuf. Enum.: 1506 (1971).

Glabrous lianes. Leaves alternate, with sheathing base and short petiole; blade ending in a long involute tip acting as a tendril. Inflorescences paniculate, terminal. Flowers small, bisexual, regular; tepals 6, whitish. Stamens 6, some sometimes reduced to staminodes; anthers 2-thecous, sagittate-basifixed, longitudinally dehiscent. Ovary superior, 3-celled, with 3 stylar branches. Fruit a small drupe, 1(−2)-seeded. Seeds globose with a small embryo, with endosperm.

Family with a single genus only.

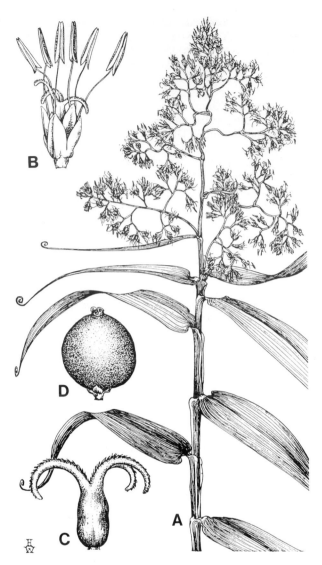

FLAGELLARIA L. (1753)

Description as for the family.
Genus of four species in the Old World tropics.

F. guineensis Schumach. (1827). Fig. 73.
Buuti-buuti, ova-mtir, tak-takeey (Som.).
Scandent climber to 5 m long or more. Leaves 12—18 cm long excluding the tendril, (1.4—)1.8—2.4(—2.8) cm wide, entire; leaf-sheaths slit to the base or almost so. Panicle 8—12 cm long; bracts ovate, c. 1 mm long. Tepals 2—3 mm long, subequal. Stamens exserted. Drupe globose, 5—9 mm in diam., red, with persistent style-base and perianth.
Bushland; up to c. 50 m. S3; E Kenya, E Tanzania, Mozambique, Zaire, West and South Africa, Madagascar. SMP 126; Moggi & Bavazzano 1705; Paoli 152.
Crushed and mixed with oil the plant is used against rheumatism in Somalia.

Fig. 73. *Flagellaria guineensis*. A: part of flowering shoot, × 0.5. B: flower, × 4. C: pistil, × 8. D: fruit, × 3. — Modified from Fl. Trop. E. Afr. (1971). Drawn by H. Wood.

167. POACEAE (GRAMINEAE)

by T. A. Cope

Cuf. Enum.: 1206—1414 (1968—1970); Fl. Trop. E. Afr. (1970, 1971, 1982); Cope, Key to Somali grasses (1985); Clayton & Renvoize, Genera Graminum (1986).

Annual or perennial herbs (rarely shrubs or trees, but not in Somalia), sometimes with rhizomes or stolons; stems cylindrical, jointed, usually hollow in the internodes, closed at the nodes; branches subtended by a leaf and with a 2-keeled hyaline leaflet (prophyll) at the base. Leaves solitary at the nodes, sometimes crowded at the base of the stem, alternate and 2-rowed, comprising sheath, ligule and blade; sheaths encircling the stem, with the margins free and overlapping or ± connate, frequently swollen at the base, the shoulder sometimes extended upwards into triangular auricles; ligule adaxial, at the junction of sheath and blade, membranous or reduced to a fringe of hairs, rarely absent; blades usually long and narrow, flat or sometimes inrolled or terete, usually passing gradually into the sheath, sometimes amplexicaul, rarely narrowed into a false petiole or articulated with the sheath. Inflorescence composed of spikelets arranged in a panicle or in racemes, the latter solitary, digitate or disposed along a central axis, usually terminal, sometimes (especially in Tribe 16. *Andropogoneae*) numerous, each being subtended by a bladeless sheath (spatheole) and the whole flowering branch system condensed into a leafy false panicle. Spikelets consisting of bracts distichously arranged along a slender axis (rhachilla); the 2 lowermost bracts (glumes) empty; the succeeding 1 to many bracts (lemmas) each enclosing a flower and opposed by a hyaline scale (palea), the whole (lemma, palea and flower) termed a floret; base of spikelet or floret sometimes with a horny downward prolongation (callus); glumes and lemmas often bearing 1 or more stiff bristles (awns). Flowers usually bisexual, sometimes unisexual, small and inconspicuous; perianth represented by 2, rarely 3, minute hyaline or fleshy scales (lodicules); stamens hypogynous, 1–6, rarely more, usually 3, with delicate filaments and 2-thecous anthers opening by a longitudinal slit or rarely a terminal pore; ovary 1-locular, with 1 anatropous ovule often adnate to the adaxial side of the carpel; styles usually 2, rarely 1 or 3, generally with plumose stigmas. Fruit mostly a caryopsis with thin pericarp adnate to the seed, rarely with free seed (e.g. 33. *Sporobolus*); caryopsis commonly combined with various parts of the spikelet, or less often the inflorescence, to form a false fruit; seed with starchy endosperm, an embryo at the base of the abaxial face, and a point or line (hilum) on the base or adaxial face marking the connection between pericarp and seed.

About 660 genera and 10000 species in 40 tribes; throughout the world.

The grasses form a natural and homogeneous family, remarkable both for the consistency of its basic theme, and for the number of variations that have been derived from it. The basic pattern of spikelet structure is consistent throughout the family, but much modified by reduction, suppression or elaboration of parts. For a comprehensive account of the family, see Clayton & Renvoize, Genera Graminum (1986).

The specimen citation for the grasses is done in a way that the number of specimens cited is a multiple of the number of flora regions where the species occurs. In a species that occurs, for example, in six regions and has six specimens cited, there is one specimen for each region and the specimens are cited in sequence. A species that has, say, four specimens cited for two regions will have two specimens for each region.

KEY TO TRIBES

1. Spikelets 1- to many-flowered, breaking up at maturity above the ± persistent glumes (not breaking up in cultivated cereals), or if falling entire then not 2-flowered with the upper floret bisexual and the lower male or barren; spikelets usually laterally compressed or terete 2
— Spikelets 2-flowered, falling entire at maturity, with the upper floret bisexual and the lower male or barren and in the latter case often much reduced; spikelets usually dorsally compressed, sometimes terete 19
2. Ovary with a fleshy, hairy apical appendage, the styles arising from beneath it 3
— Ovary without a fleshy, hairy apical appendage; styles terminal 4
3. Inflorescence a panicle ....6. *Bromeae*
— Inflorescence a bilateral raceme ....7. *Triticeae*
4. Lemma deeply cleft into 5–9 awns or lobes ....10. *Pappophoreae*
— Lemma entire or bilobed, awnless or 1- to 3-awned 5
5. Spikelets containing 1 fertile floret (except 36. *Lintonia* and 37. *Tetrapogon*), with or without 2 barren florets below it or 1 or more above 6
— Spikelets containing 2 or more fertile florets (only 1 in *Leptochloa rupestris*) 17
6. Glumes both suppressed; palea 1-keeled ....1. *Oryzeae*
— Glumes, or at least one of them, well developed; palea 2-keeled 7
7. Spikelets falling entire at maturity, either singly or in clusters, from the persistent axis of spike-like panicles or racemes, or the axis breaking up transversely at the nodes; lemma delicately 1- to 3-nerved (if spikelets falling entire with the pedicel or part of it from a contracted lobed panicle, see 7. *Polypogon*) 8
— Spikelets breaking up at maturity above the persistent glumes 9
8. Lower glume suppressed; inflorescence composed of a single terminal fragile raceme, the spikelets sunk in the axis and covered by the upper glume ....12. *Leptureae*
— Lower glume present (though often of extraordinary shape); inflorescence spike-like, composed of numerous short deciduous racemelets (sometimes these reduced to a single spikelet), these not sunk in the axis ....13. *Cynodonteae*
9. Inflorescence composed of racemes, these solitary, digitate or disposed along an axis 10
— Inflorescence a panicle, either open or contracted and spike-like 11
10. Spikelets arranged in 1-sided racemes (if racemes cuneate and disposed along an axis, see 23. *Dinebra*) ....13. *Cynodonteae*
— Spikelets embedded in alternate sides of a fragile or tough raceme ....11. *Eragrostideae*
11. Spikelets 1-flowered 12
— Spikelets 2- to 3-flowered 15

12. Lemma 3-awned .... 9. *Aristideae*
− Lemma 1-awned or awnless 13
13. Lemma hardened at maturity, awned from the tip .... 3. *Stipeae*
− Lemma hyaline or membranous at maturity 14
14. Lemma usually awned; glumes longer and firmer than the hyaline lemma; grain with adherent pericarp .... 5. *Aveneae*
− Lemma awnless; glumes and lemma similar in texture, the former often the shorter; grain with separable pericarp (on wetting) .... 11. *Eragrostideae*
15. Florets 2 per spikelet, the lower male or barren, the upper bisexual .... 15. *Arundinelleae*
− Florets 3 per spikelet, the 2 lowermost represented by barren lemmas 16
16. Barren lemmas coriaceous, at least the upper exceeding the fertile and transversely wrinkled .2. *Ehrharteae*
− Barren lemmas subulate, both shorter than the fertile .... 5. *Aveneae*
17. Tall reed-like grasses with large plumose panicles .... 8. *Arundineae*
− Slender grasses without large plumose panicles 18
18. Inflorescence composed of racemes, these solitary, digitate or disposed along an axis, sometimes in dense ovoid heads; if inflorescence an open panicle then lemma 3-nerved .... 11. *Eragrostideae*
− Inflorescence an open panicle, rarely a raceme (this sometimes with racemose branches below) but then lemmas tuberculate; lemma 5- to 7-nerved .... 4. *Poeae*
19. Spikelets solitary, rarely paired but then those of a pair alike, never in fragile racemes, nor geniculately awned; glumes usually membranous, the lower mostly small or rarely suppressed; upper lemma papery to polished and stony, awnless or at most mucronate .... 14. *Paniceae*
− Spikelets typically paired with 1 sessile and the other pedicelled, those of a pair usually dissimilar in shape and sex, with the pedicelled much reduced (rarely quite absent), occasionally the spikelets all alike, usually in fragile racemes and often the upper lemma of the sessile spikelet geniculately awned; glumes as long as the spikelet and enclosing the florets, ± rigid and firmer than the hyaline or membranous lemmas .... 16. *Andropogoneae*

Tribe 1. **ORYZEAE**

Ligule membranous. Inflorescence a panicle, occasionally with simple raceme-like primary branches, the spikelets all alike or the sexes separate. Spikelets 1-flowered, or 3-flowered with the 2 lower florets reduced to sterile lemmas, without rhachilla-extension, mostly laterally compressed, disarticulating above the glumes; glumes absent or just discernible as obscure lips at the tip of the pedicel; lemma membranous to coriaceous, 5- to 10-nerved, entire, with or without a straight awn; palea resembling the lemma, 3- to 7-nerved; stamens usually 6. Caryopsis linear to ovoid.

12 genera and about 70 species in tropical and warm temperate regions.

1. **ORYZA** L. (1753).

Annual or perennial. Panicle often with simple raceme-like primary branches, the spikelets all alike, bisexual. Spikelets with 1 fertile floret and 2 sterile lemmas, strongly laterally compressed; sterile lemmas up to half the length of the spikelet, subulate to narrowly ovate, coriaceous; fertile lemma coriaceous, strongly keeled, clasping the lateral nerves of the palea, awned or awnless; stamens 6.

About 20 species in the tropics and subtropics.

O. sativa L., rice, is cultivated along the Juba River in S1 and S3. It is an annual with persistent, usually awnless spikelets. The many varieties are discussed by Portères in J. Agric. Trop. Bot. Appl. 3: 341, 541, 627, and 821 (1956).

1. Annual; spikelets persistent (cultivated).......... .... *O. sativa* (see above)
− Perennial; spikelets deciduous (wild).............. .... *O. longistaminata*

O. longistaminata A. Chev. & Roehr. (1914). Fig. 74.
Perennial with extensive creeping rhizomes; stems up to 120 cm high, soft and spongy, up to 1 cm in diam. at the base, often decumbent and rooting from the lower nodes. Leaves up to 45 × 1.5 cm; ligule 15−45 mm long, acute. Panicle 20−30 cm long. Spikelets narrowly oblong, 7−9 mm long, scabrid to hispid, deciduous, obliquely articulated with the pedicel; sterile lemmas lanceolate, 2−3 mm long; awn 4−8 cm long, rigid.

Swamps, flooded grassland and riverbanks. S3; tropical Africa to Namibia and South Africa; also in Madagascar. Gillett & al. 5029; Rose Innes 720.

Tribe 2. **EHRHARTEAE**

Ligule usually a membrane, sometimes a line of hairs. Inflorescence a panicle or sometimes a unilateral raceme. Spikelets 3-flowered, the 2 lower florets reduced to sterile lemmas, rarely with a rhachilla-extension, laterally compressed, disarticulating above the persistent glumes; glumes membranous; sterile lemmas, or at least the upper, as long as the fertile floret, coriaceous, awned or awnless; fertile lemma firmly cartilaginous to coriaceous, 5- to 7-nerved, entire, awnless; palea nerveless or up to 2-nerved, rarely 3- to 5-nerved; anthers 1, 2, 3, 4 or 6. Caryopsis ellipsoid.

One genus with about 35 species in south temperate regions of the Old World.

2. **EHRHARTA** Thunb. (1779), nom. cons.

Annual or perennial. Glumes minute to large; sterile lemmas often transversely wrinkled, the upper often narrowed to a hook at the base, the lower subequal or shorter.

About 35 species, 25 mostly in South Africa, the rest Indonesia to New Zealand.

E. erecta Lam. (1786).
var. **abyssinica** (Hochst.) Pilg. in Notizbl. Bot. Gart. Berlin-Dahlem 9: 508 (1926); *E. abyssinica* Hochst. (1855). Fig. 75.

Loosely tufted or rambling perennial up to 100 cm high. Panicle open or contracted and narrow, rarely racemose. Spikelets 5−6(−6.5) mm long; glumes ovate, 5-nerved, acute, unequal, the upper slightly more than half the length of the spikelet, the lower shorter; sterile lemmas lanceolate in profile, glabrous, awnless, the lower narrowed to a blunt tip, rarely transversely wrinkled, the upper with appendages and lateral hair-tufts at the base, often rugose, rounded at the tip.

Forest shade; 1500−2050 m. N2; tropical Africa, southern Arabia and sporadically in southern India; replaced by var. *erecta* in southern Africa. Glover & Gilliland 1134; Newbould 980; Gillett & Watson 23817.

Tribe 3. **STIPEAE**

Ligule membranous. Inflorescence an open or contracted panicle, the spikelets all alike. Spikelets 1-flowered without rhachilla-extension, terete to laterally or dorsally compressed, disarticulating above the persistent glumes; glumes longer than the floret, hyaline to membranous, 1- to 7-nerved, acute to long-acuminate; lemma rounded on the back, 3- to 9-nerved, membranous to crustaceous, terete to lenticular and often enclosing the palea, awned from the entire or 2-toothed tip; palea usually as long as the lemma, without keels. Caryopsis fusiform.

Nine genera and about 400 species in temperate and warm temperate regions.

3. **STIPA** L. (1753).
Perennial, rarely annual. Floret fusiform, terete or rarely slightly laterally compressed, the callus usually pungent, rarely obtuse, bearded; lemma firmly membranous to coriaceous, the margins usually overlapping, entire to shortly bilobed; awn persistent or deciduous, 1- to 2-geniculate with twisted column, sometimes plumose.

About 300 species in temperate and warm temperate regions.

S. keniensis (Pilger) Freitag (1989); *Oryzopsis keniensis* Pilger (1926).
subsp. **somaliensis** Freitag in Kit Tan (ed.), Davis & Hedge Festschrift: 124 (1989); type: N2, "Daloh Forest near Erigavo", Glover & Gilliland 1095 (BM holo., EA K iso.).

S. dregeana Steud. var. *elongata* (Nees) Stapf in Dyer, Fl. Cap. 7: 573 (1899).

Densely tufted perennial up to 120 cm high; leaves 3−4 mm wide. Panicle effuse, 30−40 × 4−8 cm, the branches in whorls. Spikelets 6−7.5 mm long; glumes lanceolate-acuminate, subequal, the lower c. 0.5 mm longer than the upper; lemma 3.5−4.5(−5) mm long, with short obtuse callus and 2 minute apical teeth, loosely hairy throughout; awn 7−10 mm long, slightly twisted below, scabrid; anthers 2.5−3.5 mm long.

Juniperus forest; c. 2000 m. N1, 2; not known outside Somalia, but likely to occur in adjacent parts of Ethiopia; subsp. *keniensis* in East Africa. Drake-Brockman 479; Hansen & Heemstra 6254.

Fig. 74 (left). *Oryza longistaminata*. – From Mem. Bot. Surv. S. Afr. 58 (1991).

Fig. 75 (above). *Ehrharta erecta* var. *abyssinica*. A: habit, × 0.45. B: ligule, × 2.7. C: part of inflorescence showing 1 complete spikelet and 1 with the glumes remaining after the rest of the spikelet has fallen. – Modified from Fl. Zamb. 10:1 (1971). Drawn by D. Erasmus.

Tribe 4. **POEAE**

Ligule membranous. Inflorescence a panicle, the spikelets all alike. Spikelets of (1—)2—many fertile florets, the uppermost reduced, mostly laterally compressed, disarticulating below each floret; glumes persistent, not or scarcely exceeding the lowest lemma; lemmas membranous, 5- to 7-nerved, with or without a straight or curved awn from the tip; palea hyaline, as long as the lemma; stamens 3. Caryopsis usually ellipsoid.

49 genera and about 1200 species in temperate and cold regions of both hemispheres.

1. Lemma tuberculate; inflorescence a single raceme sometimes with racemose branches below ... 4. *Castellia*
— Lemma smooth; inflorescence a panicle .... 5. *Poa*

4. **CASTELLIA** Tineo (1846)

Annual. Inflorescence a raceme or with a few simple branches below, the spikelets sessile or nearly so, alternate in opposite rows. Spikelets 6- to 15-flowered, laterally compressed, disarticulating below each floret; glumes unequal, ± keeled, acute to obtuse, the lower 3-nerved, the upper 5-nerved; lemmas membranous, thinly 5-nerved, rounded on the back, awnless, densely tuberculate except towards the tip, obtuse to subacute.

A single species only.

C. tuberculosa (Moris) Bor (1948). Fig. 76.
Stems up to 75 cm high. Inflorescence 4—26 cm long. Spikelets ovate, 9—15 mm long; glumes glabrous, the lower 2.8—3.5 mm long, the upper 3.5—5 mm long; lemmas 4.2—5.7 mm long.

In evergreen bushland in rock crevices; c. 1450 m. N2; from the Canary Islands eastwards through the Mediterranean region to Sudan, Eritrea, Djibouti and Pakistan. Thulin, Dahir & Hassan 8939.

5. **POA** L. (1753).

Annual or perennial; leaves usually flat, often with a blunt or hooded tip. Panicle open or contracted.

Spikelets 2- to several-flowered; glumes slightly unequal, 1- to 3-nerved, keeled, glabrous; lemmas deeply concave, keeled on the back, 5- to 7-nerved, membranous with hyaline margins and tip, often hairy on the keel and nerves, rarely also on the back, awnless; callus short, often with a web of fine cottony hairs; rhachilla glabrous; palea-keels scaberulous to stiffly ciliolate; ovary glabrous.

Some 500 species in cool temperate regions throughout the world, extending into the tropics on mountain tops.

P. leptoclada Hochst. ex A. Rich. (1850). Fig. 77.
Slender tufted perennial up to 80 cm high, erect or straggling; leaves narrowly linear, acute; ligule 1—3.5 mm long. Panicle linear and spike-like, 4—30 cm long, the branches erect and ± appressed to the main axis. Spikelets 2.1—4.5 mm long; lemma 1.7—3.2 mm long, glabrous or hairy on the keel and nerves below, sometimes hairy on the back, acute; callus glabrous or with a tuft of cottony hairs; anthers 0.4—0.7 mm long.

In forest shade at high altitude. N2; Sudan and Ethiopia southwards on mountains to Zimbabwe, also in Cameroun and Arabia. Bally & Melville 15985; Glover & Gilliland 1135.

Tribe 5. **AVENEAE**

Ligule a membrane. Inflorescence an open or contracted panicle. Spikelets with 1—several fertile florets (only 1 in Somali representatives), laterally compressed, disarticulating below each floret or rarely falling entire with the pedicel or part of it; glumes persistent, usually longer than the adjacent lemma and often as long as the spikelet; lemmas hyaline to coriaceous, (3—)5- to 11-nerved, typically with a dorsal awn, this often geniculate with twisted column, but often awnless or weakly awned.

57 genera and some 1050 species in temperate and cold regions of both hemispheres.

A specimen of *Agrostis capillaris* L. associated with Somalia is present at K. The plant is said to originate from Mogadishu in 1958 and was grown in the USDA Plant Introduction Garden in Ames, Iowa. It is unlikely that this temperate species occurs in Somalia, and the most probable explanation is that the plant distributed from Ames was an accidental contamination of the garden. The record is omitted here.

1. Spikelets breaking up above the persistent glumes .... 6. *Calamagrostis*
— Spikelets falling entire with the pedicel or part of it .... 7. *Polypogon*

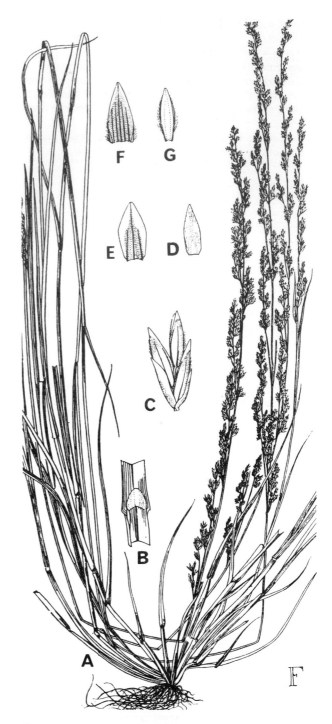

Fig. 76 (above). *Castellia tuberculosa.* A: habit, × 1. B: spikelet, × 6. – From Thulin & al. 8939. Drawn by L. Petrusson.

Fig. 77 (right). *Poa leptoclada.* A: habit, × 0.5. B: ligule, × 1.5. C: spikelet, × 6. D: lower glume, × 6. E: upper glume, × 6. F: lemma, × 6. G: palea, × 6. – Modified from Fl. Zamb. 10:1 (1971).

6. CALAMAGROSTIS Adans. (1763).

Perennial. Panicle usually contracted to spike-like, rarely open. Spikelets 1-flowered, with or without rhachilla-extension; glumes equal or unequal, as long as the spikelet, membranous, 1(–3)-nerved, acute to acuminate; lemma membranous to coriaceous, sometimes hyaline, (3–)5-nerved, awnless, mucronate or with a straight or geniculate dorsal awn; callus bearded with hairs shorter than or much exceeding the lemma; palea 1/3 as long to as long as the lemma.

Some 270 species in temperate regions throughout the world, and on mountains in the tropics.

C. canescens (Weber) Roth (1789).

C. lanceolata Roth var. *somalensis* Chiov., Fl. Somala 2: 451 (1932); type: S2, "Afgoi", Guidotti s.n. (FT holo.).

Loosely tufted with rhizomes, up to 150 cm high; leaves linear, up to 4 mm wide, scaberulous. Panicle 16 cm long, narrow and loose. Spikelets 4–4.5 mm long, tinged with purple; glumes lanceolate, scaberulous; lemma hyaline, 2.3 mm long, faintly

5-nerved, denticulate and shortly awned at the tip; callus-hairs longer than the lemma; palea 3/5 the length of the lemma.

S2; Europe. Guidotti s.n.

The only known specimen from Somalia was found in a sample of fodder. The species is unlikely to be native; more probably, it was introduced in fodder from Italy and was self-sown.

7. POLYPOGON Desf. (1798)

Annual or perennial. Panicle contracted to spike-like, sometimes open. Spikelets 1-flowered without rhachilla-extension, falling entire together with the pedicel or part of it; glumes equal, as long as the spikelet, chartaceous, 1-nerved, scabrid, entire to bilobed, often awned; lemma hyaline, 5-nerved, the nerves sometimes excurrent from the truncate tip, awnless or with a subapical awnlet or geniculate dorsal awn; palea 1/2 as long to as long as the lemma.

18 species in warm temperate regions, and on mountains in the tropics.

1. Perennials ....3. *P. viridis*
 − Annuals 2
2. Awns of the glumes 4−7 mm long, at least twice the length of the body of the glume...............
.... 1. *P. monspeliensis*
 − Awns of the glumes 0.6−3 mm long, seldom more than the length of the body of the glume, occasionally 1/2 as long again ....2. *P. fugax*

1. **P. monspeliensis** (L.) Desf. (1798). Fig. 78.

Annual up to 80 cm high. Panicle narrowly ovate to narrowly oblong, cylindrical or lobed, 1.5−16 cm long, very dense and bristly. Spikelets 2−3 mm long; glumes slightly notched at the tip, scabrid below, minutely hairy on the margins, with a fine straight awn 4−7 mm long; lemma about 1/2 the length of the glumes, smooth, awnless or with an awn up to 2 mm long.

N1; warm temperate Old World. Drake-Brockman 512, 513.

2. **P. fugax** Nees ex Steud. (1854).

Annual up to 60 cm high, often decumbent and rooting from the nodes. Panicle narrowly ovate, oblong or cylindrical, usually lobed, 3−15 cm long, dense but scarcely bristly. Spikelets 1.8−2.4 mm long; glumes slightly notched at the tip, scabrid below, minutely hairy on the margins, with a fine straight awn 0.6−3 mm long; lemma about 1/2 the length of the glumes, smooth, awnless or with an awn up to 2 mm long.

Banks of streams and irrigation ditches, often floating out on the water; 1500−1800 m. N2; Southwest Asia from Iraq eastwards to Burma, but

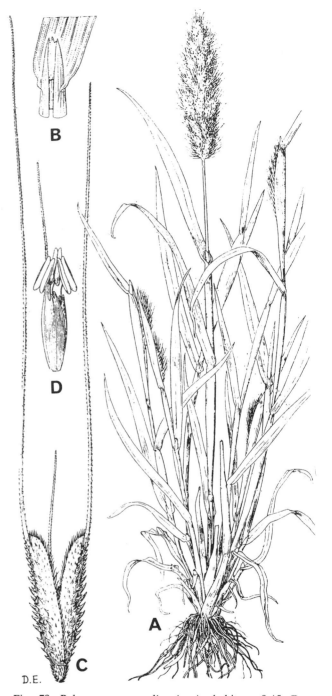

Fig. 78. *Polypogon monspeliensis*. A: habit, × 0.45. B: ligule, × 4.5. C: spikelet, × 18. D: floret, × 18. − Modified from Fl. Zamb. 10:1 (1971). Drawn by D. Erasmus.

mainly in the Himalayas. McKinnon 240; Glover & Gilliland 597, 919.

The specimens cited above − the only three known from Somalia − all come from "Medishe" in the vicinity of "Erigavo". The species may well have been introduced for its fodder value.

Fig. 79. *Bromus leptocladus*. A: habit and inflorescence, × 0.45. B: ligule, × 1. C: spikelet, × 1. D: lower glume, × 2.7. E: upper glume, × 2.7. F: lemma, × 2.7. G: palea, × 2.7. — Modified from Fl. Zamb. 10:1 (1971). Drawn by D. Erasmus.

3. **P. viridis** (Gouan) Breistr. (1966).

Agrostis semiverticillata (Forssk.) C. Chr. (1922).

Stoloniferous perennial up to 100 cm high. Panicle pyramidal, lobed, 2−15 cm long, dense, the branches subverticillate. Spikelets 1.5−2.5 mm long; glumes obtuse, scabrid on the back, awnless; lemma c. 1 mm long, denticulate, awnless.

Damp areas and streambanks; c. 1750 m. N2; Mediterranean region eastwards to Northwest India. McKinnon 271.

The species was probably introduced into Somalia, in the same general area as *P. fugax*, for its fodder value.

Tribe 6. **BROMEAE**

Ligule a membrane. Inflorescence a panicle, the spikelets all alike. Spikelets of several to many fertile florets with imperfect florets above, laterally compressed, disarticulating below each floret, rarely the florets falling in a cluster; glumes persistent, shorter than the lowest lemma, entire; lemmas herbaceous to coriaceous, 5- to 13-nerved, 2-toothed at the tip, with a straight or recurved subapical awn, occasionally several-awned, rarely awnless; ovary capped by a fleshy hairy appendage.

Three genera and some 160 species in temperate regions.

8. **BROMUS** L. (1753)

Annual or perennial; sheaths with the margins connate for most of their length, usually hairy. Panicle open or contracted. Spikelets cuneate to ovate, breaking up below each floret, with a single awn (in Somalia).

Some 150 species in temperate regions of both hemispheres, but mainly in the north, and on tropical mountains.

B. leptoclados Nees (1841). Fig. 79.
Hirmi (Som.).

Loosely tufted perennial up to 2 m high. Panicle oblong, 15−30 cm long, loose and nodding, sometimes somewhat contracted. Spikelets narrowly oblong to cuneate, 13−30 mm long; glumes narrow, the lower 3−6 mm long, 1-nerved, the upper 7−14 mm long, 3-nerved, both with an awn-point up to 5 mm long; lemmas narrowly elliptic, 7−14 mm long, herbaceous with hyaline margins, 3- to 5(−7)-nerved, the intermediate nerves often faint; awn 2−12 mm long, straight; palea finely ciliolate on the keels.

In forest shade. N2; throughout the African highlands and on mountains in Arabia. Glover & Gilliland 1096.

Tribe 7. **TRITICEAE**

Ligule a membrane. Inflorescence a single bilateral raceme or quasi-raceme, the spikelets alternate in 2 opposite rows, single or in groups of 2−3 at each node, broadside to the rhachis and all alike; rhachis tough, rarely fragile. Spikelets 1- to many-flowered with the uppermost florets reduced, laterally compressed, disarticulating below each floret; glumes persistent, shorter or narrower than the adjacent lemma, usually coriaceous, sometimes awn-like; lemma coriaceous to membranous, 5- to 11-nerved, with or without a straight or recurved awn from the tip; ovary capped by a small fleshy hairy appendage.

18 genera and some 330 species in temperate and warm temperate regions, mostly in the northern hemisphere.

9. **TRITICUM** L. (1753)

Ligule a membrane. Inflorescence a linear raceme bearing solitary spikelets on a fragile or tough rhachis. Spikelets 3−9-flowered; glumes oblong to ovate, coriaceous, 5−11-nerved, 1−2-keeled (but sometimes becoming rounded below as the grain expands), obtuse, truncate or 2-toothed, the lateral nerves diverging into the teeth, mucronate or awned; lemmas rounded on the back or keeled near the tip, the tip similar to that of the glumes.

Between 10 and 20 species in the Mediterranean region and western Asia, more than half of them cultivated.

T. aestivum L. (1753).

Annual with non-shattering spikelets on a tough rhachis; glumes with a compressed keel in the upper half only, rounded below (often the midnerve prominent to the base, but the glume otherwise without a ridge below).

Bread-wheat is − or has been − cultivated at about 1500 m on the plains east of Borama (N1). Bally 9999.

Tribe 8. **ARUNDINEAE**

Ligule a line of hairs (a membrane in *Arundo*); mostly tussock-forming perennials with basal leaves (except *Arundo* and *Phragmites*). Inflorescence a panicle, the spikelets all alike (in Somali species). Spikelets with several fertile florets and imperfect florets above, disarticulating between the florets and above the persistent membranous glumes; lemmas rounded on the back, (1−)3- to 11-nerved, hyaline to coriaceous, entire or bilobed, with or without a straight or geniculate awn from the tip or sinus; palea 2-nerved. Caryopsis ellipsoid, sometimes with a free or separable pericarp; hilum narrowly oblong to linear.

40 genera with some 300 species worldwide, but best developed in the southern hemisphere.

1. Ligule a membrane; lemma hairy; rhachilla glabrous .... 10. *Arundo*
   − Ligule a line of hairs; lemma glabrous; rhachilla hairy .... 11. *Phragmites*

10. ARUNDO L. (1753)

Tall rhizomatous perennial reed; leaves cauline; ligule membranous with minutely ciliolate margin. Panicle large, plumose. Glumes as long as the spikelet, 3- to 5-nerved; floret-callus short, glabrous; lemmas membranous, 3- to 7-nerved, plumose below the middle, entire or 2-toothed, with a short straight awn from between the teeth.

Three species from the Mediterranean region to China, widely introduced elsewhere.

A. donax L. (1753).
Stems up to 5 m high; leaves conspicuously distichous, rounded or cordate at the base, up to 60 × 5 cm. Panicle 30−60 cm long. Spikelets 10−15 mm long; lemmas (6−)8.5−13 mm long, with an awn c. 1.5 mm long.

S2. A Mediterranean species, introduced in Somalia as it is to many countries around the world. Bavazzano 213.

11. PHRAGMITES Adans. (1763)

Tall rhizomatous perennial reed; leaves cauline, the blades deciduous; ligule a very short membrane with a long-ciliate margin. Panicle large, plumose. Spikelets with lowest floret male or empty; glumes shorter than the lowest lemma, 3- to 5-nerved; floret-callus linear, plumose; fertile lemmas hyaline, 1- to 3-nerved, glabrous, long-caudate (though very fragile and the long narrow tip often breaking off), entire.

Three or four species worldwide.

1. Leaves usually smooth on the lower surface; upper glume 5.5−9 mm long; rhachilla-hairs copious, silky, 8.5−12 mm long ....1. *P. australis*
   − Leaves scabrid on the lower surface, at least towards the tip; upper glume 4−6 mm long; rhachilla-hairs sparse, 4−7(−8) mm long..........
   .... 2. *P. karka*

1. **P. australis** (Cav.) Trin. ex Steud. (1841).
   subsp. **altissimus** (Benth.) Clayton in Taxon 17: 169 (1968); *Arundo altissima* Benth (1826).
   ? *A. maxima* Forssk. (1775); *P. communis* Trin. var. *isiacus* (Del.) Coss. in Coss. & Durieu, Expl. Sci. Algérie 2: 175 (1855); *P. vulgaris* Crép. var. *maximus* (Forssk.) Chiov., Fl. Somala 1: 338 (1929).

   Stems up to 6 m high; leaves 20−60 cm × 8−32 mm, flat, smooth beneath, the tip filiform and flexuous. Panicle 30−45 cm long, the lowest node usually few-branched, some of the branches bearing spikelets nearly to the base. Spikelets with rhachilla-hairs 8−12.5 mm long, copious, silky; lower glume ovate, 2.7−5 mm long; upper glume narrowly elliptic-oblong, 5.5−9 mm long, obtuse to tridenticulate; first lemma 9.5−17 mm long; fertile lemma 10−14 mm long.

   Along irrigation channels and at the edges of cultivated fields. N2, 3; S2, 3; Ethiopia, Kenya and the southern edge of the Sahara northwards to the shores of the Mediterranean and the Arabian Peninsula. Puccioni & Stefanini 852; Kazmi & Mohamed 5620; Rose Innes 884; Paoli 408.

2. **P. karka** (Retz.) Trin. ex Steud. (1841).
   Gul bilanwe (Som.).

   Stems 2−4(−10) m high; leaves 30−80 cm × 12−40 mm, flat, scabrid on the lower surface at least towards the tip, the tip attenuate and stiff (occasionally smooth beneath or with filiform tip). Panicle (20−)30−50 cm long, the lowest node often many-branched in a whorl, the branches bare of spikelets in the lower part. Spikelets with rhachilla-hairs 4−7(−8) mm long, rather sparse; lower glume ovate, 2−4 mm long; upper glume narrowly to very narrowly elliptic, 4−6 mm long; first lemma (7.5−)10−12 mm long; fertile lemma 8.5−11 mm long.

   Tug banks and river flood plains; up to 550 m. N1; S2, 3; tropical Africa, tropical Asia and northern Australia. Gillett 4562; Rose Innes 648; Vatova 2.

Tribe 9. **ARISTIDEAE**

Ligule a line of hairs. Inflorescence an open to contracted panicle, the spikelets all alike. Spikelets 1-flowered without rhachilla-extension, laterally compressed or terete, disarticulating above the persistent glumes; glumes longer than the body of the lemma (rarely shorter in *Aristida*), membranous to scarious, mostly acute to acuminate; lemma terete (rarely laterally compressed), 1- to 3-nerved, coriaceous, wrapped around and concealing the palea, 3-awned, the awns ± connate at the base and often raised upon a twisted column, the laterals sometimes reduced; palea less than half the length of the lemma, sometimes little longer than the lodicules; stamens 3, rarely 1. Caryopsis usually fusiform, the hilum linear.

Three genera and c. 300 species in the tropics and subtropics.

1. Central awn, and sometimes also the laterals, plumose .... 12. *Stipagrostis*
− All three awns quite glabrous .... 13. *Aristida*

12. **STIPAGROSTIS** Nees (1832)

Perennial, rarely annual, sometimes with knotty rhizomatous base or suffruticose; leaves mostly inrolled, sometimes pungent. Glumes 1- to 11-nerved; callus of floret long and pungent; lemma convolute; awns with or without a column, the central always, the laterals sometimes, plumose.

Some 50 species in Africa, south-western and central Asia, and Pakistan.

A record of *Aristida plumosa* L. var. *brachypoda* (Tausch) Trin. & Rupr. (= *Stipagrostis plumosa* (L.) Munro ex T. Anders.) from the "Ahl Mountains" by Stapf in Kew Bull. 1907: 217 (1907) was based on a vegetation description by Engler in Sitzungsber. Kgl. Preuss. Akad. Wiss. 10: 404 (1904), where *Aristida brachypoda* Tausch was mentioned as seen by Hildebrandt. As no material is available to substantiate this record it is excluded here.

1. Column hairy 2
− Column glabrous 3
2. Callus with 2 collars of hair, a lower of very short hairs just behind the tip of the callus, and an upper of much longer hairs at the base of the body of the lemma, the callus glabrous between the collars ....1. *S. hirtigluma*
− Callus with 1 continuous collar of hair, the hairs increasing in length upwards ...... 2. *S. uniplumis*
3. Central awn plumose to the base, 6−7 cm long .... 3. *S. paradisea*
− Central awn glabrous in the lower half, 2−4.5 cm long .... 4. *S. xylosa*

1. S. hirtigluma (Steud. ex Trin. & Rupr.) De Winter (1963); *Aristida hirtigluma* Steud. ex Trin. & Rupr. (1842).

Harfo, sarem (Som.).

Annual up to 45 cm high; internodes and sheaths glabrous or scaberulous. Panicle contracted although the spikelets are on rather long filiform pedicels. Glumes unequal, narrowly lanceolate, 3-nerved, obtuse, the midnerve shortly excurrent, hairy on the back, the lower 7.5−9.5 mm long, the upper 10−11.5 mm long; lemma, including the callus, 3−4.5 mm long, papillose-scabrid; callus with 2 collars of hair, the lower, just behind the naked tip, with very short hairs all about the same length, the upper, at the summit of the callus, about half as long as the body of the lemma, the callus glabrous between the collars; column pilose, especially towards the junction of the awns, rarely almost glabrous, 7−13 mm long; central awn 4−7 cm long, plumose throughout except for the excurrent naked tip, sometimes only thinly hairy or almost glabrous below; lateral awns 1−1.6 cm long, glabrous.

Acacia-Commiphora bushland on calcareous or gypseous hills, sometimes on sandy alluvium or as a weed of plantations; 175−1280 m. N1−3; C1, 2; S1; tropical and southern Africa to India. Hemming 2122; Thulin & Warfa 5419; Bally & Melville 15603; Thulin 5653; Beckett 393; Fries & al. 4776.

2. S. uniplumis (Licht.) De Winter (1963). Fig. 80.
Aristida papposa Trin. & Rupr. (1842).

Tufted perennial, often short-lived, up to 75 cm high; internodes glabrous. Panicle contracted although the spikelets are on long filiform branches and pedicels. Glumes unequal, narrowly lanceolate, 3-nerved, glabrous or thinly hairy, the lower 7−8 mm long, shortly bifid with the central nerve excurrent, the upper 9−10.5 mm long, narrowed above into a short awn; lemma, including the callus, 2.5−3 mm long, papillose-scabrid; column pilose, especially (and sometimes only) at the junction of the awns, 6.5−10 mm long; central awn 3−4 cm long, plumose throughout, sometimes thinly so below, with or without an excurrent naked tip; lateral awns 0.8−1.2 cm long, glabrous.

Loose *Acacia-Commiphora* bushland over limestone; 300−1070 m. N1, 2; C1, 2; tropical and southern Africa to India. Hansen & Heemstra 6142; Thomson 92a; Gillett & al. 21913; Martin s.n.

3. S. paradisea (Edgew.) De Winter (1963).

Tufted perennial up to 40 cm high, much-branched at the base; internodes glabrous. Panicle contracted. Glumes unequal, lanceolate-subulate, 3-nerved,

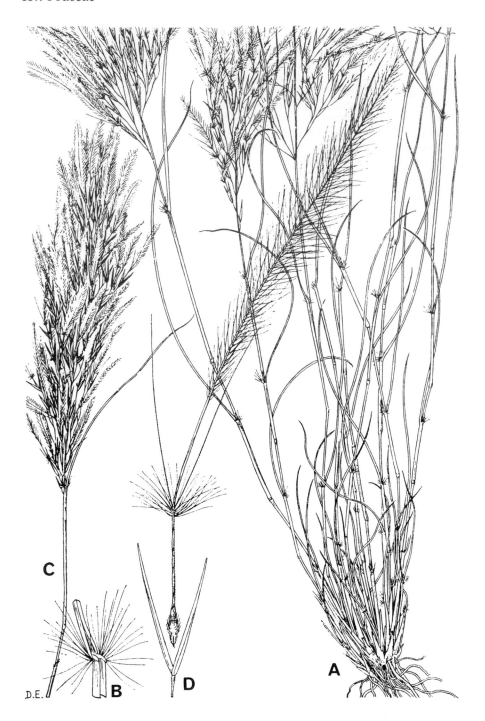

Fig. 80. *Stipagrostis uniplumis*. A: habit, × 2/3. B: ligule, × 4. C: inflorescence, × 2/3. D: spikelet, with glumes detached. — Modified from Fl. Trop. E. Afr. (1970). Drawn by D. Erasmus.

glabrous, the lower 14−19 mm long, the upper 12−14 mm long; lemma, including the callus, 4−4.5 mm long, smooth; column glabrous, 11−14 mm long; central awn 6−7 cm long, plumose throughout except for the excurrent naked tip; lateral awns 1.6−2.5 cm long, glabrous.

Limestone and sandstone hills. N3; Egypt and Arabia, possibly also Sudan. Bally & Melville 15869; Glover & Gilliland 904.

4. **S. xylosa** Cope (1992); type: N2, "Erigavo district, Hubera", McKinnon 110 (K holo.).

Hadaf, laah (Som.).

Loosely tufted perennial up to 25 cm high, branched and woody below, the leaves mostly confined to the base but not forming a tight cushion; internodes glabrous. Panicle contracted. Glumes unequal, 3-nerved, narrowly lanceolate-acuminate, glabrous, the lower 12−15 mm long, the upper 9−12 mm long; lemma, including the callus, c. 3.5 mm long, smooth; column glabrous, 4–8 mm long; central awn 3.5–4.5 cm long, plumose in the upper half but with an excurrent naked tip, the feather nevertheless very obtuse in outline; lateral awns 1–1.5 cm long, glabrous.

Acacia bushland on stony plains; 0–1500 m. N1–3; Arabia (Yemen). Gillett 4750; McKinnon 110; Thulin & Warfa 6135.

13. ARISTIDA L. (1753)

Annual or perennial; leaves flat or rolled. Glumes 1(−5)-nerved; callus of floret obtuse to pungent; lemma convolute or involute; awns with or without a column, persistent or deciduous, glabrous, flat or terete, sometimes the laterals reduced or ± suppressed.

Some 250 species in the tropics and subtropics.

1. Lemma or column not articulated at its summit (sect. *Aristida*) 2
 − Lemma or column articulated at its summit 5
2. Lateral awns absent or much reduced, if up to 13 mm long then much finer than the central; tip of lemma with a semicircular bend below the column ....1. *A. abnormis*
 − Lateral awns well developed, similar to the central; lemma straight 3
3. Annual or short-lived perennial..................
 ....4. *A. adscensionis*
 − Densely tufted perennial 4
4. Lemma, including the callus, 6–9.5 mm long; column 6–8 mm long .... 2. *A. somalensis*
 − Lemma, including the callus, 10–13 mm long; column absent or up to 4 mm long..............
 .... 3. *A. stenostachya*
5. Articulation at the top of the column, just below the awns (sect. *Pseudarthratherum*)..............
 ....18. *A. mutabilis*
 − Articulation at the base of the column (sect. *Arthratherum*) 6
6. Annual 7
 − Tufted perennial 8
7. Glumes unequal, linear-lanceolate, long-acuminate, the lower 15–25 mm long, the upper 13–20 mm long; lemma 4–7.5 mm long.........
 ....5. *A. funiculata*
 − Glumes subequal, linear, tapering to a fine awn, c. 35 mm long including the awn; lemma c. 10 mm long .... 6. *A. leptura*
8. Awns very unequal, the laterals not much more than half as long as the central 9
 − Awns about equal, the laterals only slightly shorter than the central 11
9. Panicle delicate, open .... 7. *A. anisochaeta*
 − Panicle dense, spike-like 10

10. Lemma-body contracted in the upper half, smooth, 2.5–2.8 mm long; column mostly less than 20 mm long .... 8. *A. kelleri*
 − Lemma-body not contracted, ± cylindrical, papillose, 3.3–4.7 mm long; column mostly 20–25 mm long .... 9. *A. triticoides*
11. Panicle tightly congested, very dense 12
 − Panicle at most loosely contracted 13
12. Panicle cuneate below but embraced by the uppermost sheath; upper glume entire; column dilated at the distal end .... 10. *A. protensa*
 − Panicle tapered below, not cuneate, fully exserted from the uppermost sheath; upper glume 2-toothed at the tip; column not dilated at the distal end ....11. *A. pycnostachya*
13. Low-growing densely tufted plant up to 20(−25) cm high, with much-branched woody base; awns 6–7 cm long, stiff and divergent, the whole inflorescence comprising most of the height of the plant ....12. *A. migiurtina*
 − Loosely or densely tufted plants usually over 30 cm high and often up to 100 cm; inflorescence much shorter than the stem 14
14. Lemma-body contracted in the upper half, the narrowed portion scabrid .... 13. *A. paoliana*
 − Lemma-body not contracted, ± cylindrical 15
15. Suffruticose, with woody glaucous stems rising to 100 cm and branched from the upper nodes.......
 ....14. *A. sieberiana*
 − Not suffruticose; stems wiry, rising to about 50 cm, mostly branched only at the base 16
16. Awns very slender, flexuous, 9.5–10.5 cm long..
 ....15. *A. tenuiseta*
 − Awns 4.5–6.5 cm long 17
17. Upper glume, including the awn, 25–33 mm long, the awn 1/3–1/2 as long as the body.....
 ....16. *A. schebehliensis*
 − Upper glume, including the awn (if present), 13–22 mm long, the awn less than 1/4 as long as the body, usually much shorter and frequently quite absent .... 17. *A. stenophylla*

1. **A. abnormis** Chiov. (1903).
 A. redacta auct. non Stapf.
 Tiif (Som.).

Annual up to 40 cm high, but usually much less. Panicle loose and open, sometimes contracted but not spike-like. Glumes subequal (the upper fractionally the longer), narrowly lanceolate, 5–7.5 mm long, scaberulous, the lower usually deciduous before the floret is mature; lemma, including the callus and the 4–10 mm long column, 12–22 mm long, scabrid, with a semicircular bend below the column; column without an articulation; central awn 0.7–2.3 cm long; lateral awns up to 1.3 cm long, usually much shorter and sometimes absent altogether.

Sparsely vegetated limestone hillsides; 100–350 m. N1–3; Djibouti, Ethiopia and Arabia. Cole s.n.; Barbier 1183; Thulin & Warfa 5923.

2. **A. somalensis** Stapf (1907); type: N1, "Golis Range", Drake-Brockman 127 (K holo.).

Sodaheleh (Som.).

Tufted perennial up to 80 cm high. Panicle ovate, effuse, the branches bare of spikelets in the lower half. Glumes unequal, finely acuminate, the lower 8–12 mm long, the upper (11–)13–18 mm long; lemma, including the callus but not the column, 6–9.5 mm long, scaberulous above, the tip gradually extended into a column 6–8 mm long; column without an articulation; central awn 3–4 cm long; lateral awns a little shorter.

Granitic slopes; 1370–1580 m. N1; Ethiopia and Kenya. Farquharson 35; Wood 72/107.

3. **A. stenostachya** Clayton (1968).

Densely tufted perennial up to 120 cm high. Panicle contracted or spike-like, sometimes interrupted below. Glumes subequal (the lower marginally the longer), 10–19 mm long, enclosing the lemma; lemma, including the callus but not the 0–4 mm long column, 10–13 mm long, smooth or faintly scaberulous on the midnerve, with or without a slightly twisted column; lemma or column without an articulation; central awn 2–3 cm long; lateral awns a little shorter.

Clearings in bushland on sandy soil. S3; Kenya, Tanzania and Zambia. Rose Innes 756.

4. **A. adscensionis** L. (1753).

A. caerulescens Desf. (1798).

A. adscensionis var. *festucoides* (Poir.) Henrard in Meded. Rijks-Herb. 54: 177 (1926).

A. adscensionis var. *aethiopica* (Trin. & Rupr.) T. Durand & Schinz, Consp. Fl. Afric. 5: 799 (1895); *A. aethiopica* (Trin. & Rupr.) Chiov. (1907).

Bille, birreh, ebateetee, harfo, madweed, tinleh (Som.).

Annual or short-lived perennial up to 75 cm high (usually much less). Panicle usually contracted about the primary branches, these either spreading or appressed to the main axis (but the panicle never spike-like). Glumes unequal, linear-lanceolate, the lower 4–8.5 mm long, acute, the upper 7–11.5 mm long, obtuse or emarginate to apiculate; lemma, including the callus, (5–)11.5–17 mm long, laterally compressed, scabrid on the keel and sometimes also on the flanks, passing into the awns without either column or articulation; awns terete, the central (0.7–)1.5–2.5 cm long, the laterals similar or a little shorter.

Poor sandy and stony soils with open shrubland or mixed grassland, cultivated areas, flood-plains and waste ground; 0–1650 m. N1–3; C1, 2; S1–3; throughout the tropics. Hansen & Heemstra 6146; Kazmi & al. 45; Bally & Melville 15652; Elmi 514; Beckett 270; Gillett & Hemming 24312A; Alstrup & Michelsen 60; Hemming & Deshmukh 215.

An extremely variable species ranging from annual to short-lived perennial; perennial plants were at one time separated as *A. caerulescens*, but it is now apparent that the distinction is induced by the environment and is not therefore of any taxonomic value.

5. **A. funiculata** Trin. & Rupr. (1842).

Bille, birreh (Som.).

Annual up to 30 cm high. Panicle sparse, contracted, scarcely exserted from the uppermost sheath. Glumes unequal, linear-lanceolate, long-acuminate, the lower 15–25 mm long, the upper 13–20 mm long; lemma, including the callus, 4–7.5 mm long, terete, smooth, articulated at the summit and bearing a column 2–4.5 cm long; awns subequal, filiform, 4.5–8.5 cm long.

Dry sandy or stony soils, often in *Acacia-Commiphora* bushland; 165–200 m. N2; C1, 2; S1; tropical Africa westwards to Senegambia, then eastwards through Arabia to Pakistan and India. Hemming 1353; Wieland 4626; Beckett 392; Fries & al. 4775.

Readily distinguished from other members of sect. *Arthratherum* by its annual habit combined with inverse position of the glumes with the lower longer than the upper. *A. leptura* is the only other annual in the section, but this has subequal, very long glumes.

6. **A. leptura** Cope (1992); type: S3, Kismayu to Kolbio road, Rose Innes 762 (K holo.).

Annual up to 55 cm high. Panicle sparse, loosely contracted, fully exserted from the uppermost sheath. Glumes subequal, linear, gradually tapering to a long fine awn, c. 35 mm long; lemma, including the callus, c. 10 mm long, terete, scaberulous above, articulated at the summit and bearing a column c. 3 cm long; awns subequal, filiform, 6–6.5 cm long.

Disturbed sandy soil at the edge of a marsh. S3; Tanzania.

Of all Somali *Aristida* species, the closest relative of this one would appear to be *A. funiculata* since it is the only other annual in sect. *Arthratherum*.

7. **A. anisochaeta** Clayton (1969).

Xalfo (Som.).

Tufted, short-lived perennial with much-branched wiry stems up to 60 cm high. Panicle delicate, open, the spikelets loosely gathered towards the tips of the slender, capillary branches. Glumes very unequal, sharply acute to minutely emarginate-mucronate, the lower linear-lanceolate, 5–7 mm long, the upper linear-caudate, 10–12.5 mm long; lemma, including the callus, 2.5–3 mm long, fusiform, blackish at maturity, scaberulous on the nerves above, articulated at the summit and bearing a column 5–6.5 mm long; central awn 5–6 cm long, stout; lateral awns 8–12 mm long, very fine.

Acacia-Commiphora bushland on red sandy soils often over limestone; 150–380 m. C1, 2; Ethiopia.

Elmi & Hansen 4078; Gillett & Hemming 24601; Elmi 493; Thulin & Dahir 6427.

8. **A. kelleri** Hack. (1900). Fig. 81.
    Bajeh, machew (Som.).
Densely tufted perennial with branching rootstock; stems wiry, up to 45 cm high. Panicle spike-like with densely crowded spikelets, not fully exserted from the uppermost sheath. Glumes unequal, linear, the lower 5.2−8 mm long, 2-toothed, mucronate from between the teeth, the upper 10−12 mm long, bifid, with an awn 3−5 mm long from between the acuminate lobes; lemma, including the callus, 2.5−3 mm long, terete, smooth and glabrous, narrowed above the middle, articulated at the summit and bearing a weakly twisted column mostly less than 2 cm long; central awn 4−5.5 cm long, curving outwards; lateral awns about half as long, thinner in texture, erect.

Open *Acacia-Commiphora* bushland on orange sand over limestone, often dominant; 200−1000 m. N1; C1, 2; Ethiopia and north-eastern Kenya. Hansen & Heemstra 6186; Gillett & al. 21922; Kazmi & al. 5340.

Often confused with *A. triticoides*, but distinguished with relative ease by the shorter lemma with smooth narrowed tip, panicle not fully exserted from the uppermost sheath, and shorter, less tightly twisted column.

9. **A. triticoides** Henrard (1933); type: N1, "Upper Sheikh", Appleton 104 (K holo.).
    Mahjen (Som.).
Densely tufted perennial with branching rootstock; stems up to 65 cm high. Panicle dense, spike-like. Glumes very unequal, the lower linear-lanceolate, 5.5−8 mm long, finely acute or awn-pointed, the upper linear, 12−16 mm long, emarginate to bifid with an awn-point up to 2 mm long from between the lobes; lemma, including the callus, 3.3−4.7 mm long, terete, broad and papillose above, articulated at the summit and bearing a tightly twisted column mostly more than 2 cm long; central awn 4−6 cm long, recurved at the base; lateral awns 3−4 cm long, thinner in texture, erect.

Open *Acacia-Commiphora* bushland on red sand over limestone or gypsum, sometimes in rocky gullies; 60−1100 m. N1−3; C1, 2; S1; Ethiopia, Arabia (Yemen) and Pakistan. Gillett 4783; Hemming 1341; Lavranos 7234; Gillett & al. 22107; Rose Innes 842; Gillett & Hemming 24255.

10. **A. protensa** Henrard (1928); type: C1, near "Obbia", Drake-Brockman 957 (K holo.).
Densely tufted perennial up to 40 cm high. Panicle dense, spike-like, cuneate below, not fully exserted from the uppermost sheath. Glumes very unequal, the lower linear-lanceolate, 12−14 mm long, awn-pointed, the upper linear-caudate, expanded below, up to 25 mm long, 2-toothed at the tip and with an awn 5−10 mm long from between the teeth; lemma, including the callus, 5−6 mm long, terete, smooth or minutely scaberulous above, articulated at the summit and bearing a column 6−8.5 mm long, this slightly dilated at the distal end; awns subequal, 8−9 cm long.

C1; not known elsewhere. Drake-Brockman 950, 952.

Known only from "Harajab" near "Obbia", and represented by just three collections.

A very distinctive species recognized by the short dense panicle cuneate below, the linear upper glume with expanded base and a slightly dilated distal end of the column. *A. pycnostachya* is superficially similar, but differs in several ways: it is taller and more tussocky; the panicle is elongate, fully exserted and without the cuneate base; the upper glume is not 2-toothed, but entire; and the column is not dilated at the distal end.

11. **A. pycnostachya** Cope (1992); type: C1, Ceel Dheer, Beckett 390 (K holo.).
    Meyro (Som.).
Densely tufted perennial with thick woody base, up to 90 cm high. Panicle dense, spike-like, tapered below, fully exserted from the uppermost sheath. Glumes unequal, linear-lanceolate, long-acuminate-aristate, the lower 23.5−27 mm long, the upper 29−38 mm long; lemma, including the callus, c. 7.5 mm long, terete, smooth, articulated at the summit and bearing a column c. 16 mm long, this not dilated at the distal end; awns subequal, 7−8.5 cm long.

Sandy soils in *Acacia-Dichrostachys* bushland, and as a weed of manioc; up to c. 200 m. C1; S2; known only from near Hobyo, Ceel Dheer and Mogadishu. Wieland 4315; Raimondo 10/82.

Rather similar to *A. protensa*, but differing in a number of important characters (see under that species for details). Two of the three sheets cited bear immature plants, and it is rather hard to be sure of their relationship to the type specimen. Superficially the three look identical, and clearly none is *A. protensa*.

12. **A. migiurtina** Chiov. (1928); type: N3, "Hafun", Puccioni & Stefanini 8 (FT holo.).
    Gud lebah (Som.).
Tufted perennial up to 20(−25) cm high (including the awns). Panicle contracted, at least as long as the supporting stem and, with the awns included, comprising most of the height of the plant. Glumes very unequal, long-acuminate, awn-tipped, the lower 8−14 mm long, entire, the upper 14−23 mm long, minutely bifid; lemma, including the callus, 4−5.5 mm long, terete, smooth, articulated at the summit and bearing a column 1−2 cm long; awns subequal, the central 6.5−8 cm long, the laterals 5−6 cm long.

Open sandy or stony soils, sometimes where seasonally flooded; 190−1520 m. N1−3; southern Arabia.

Fig. 81. *Aristida kelleri*. A: habit, × 0.5. B: spikelet, × 2. C: lemma, × 6. – From Fl. Trop. E. Afr. (1970). Drawn by D. Erasmus.

A

B

C

D.E.

Gillett 4029; Hansen & Heemstra 6214; Hemming 1847.

The growth habit is unique in Somalia. The leaves are short and curved and mostly confined to a loose basal cushion, and over-topped by the relatively large obconical panicle.

13. **A. paoliana** (Chiov.) Henrard (1927); *A. stipiformis* Poir. var. *paoliana* Chiov. in Ann. Bot. 13:

371 (1915); type: S2, Mogadishu, Paoli 115 (FT holo.).

A. hemmingii Clayton (1969); type: C1, "Galkayo", Hemming 1422 (K holo.).

Ula dheere (Som.).

Loosely tufted perennial with knotty base and profusely branched woody stems, up to 90 cm high. Panicle loose and open, sometimes contracted. Glumes unequal, the lower linear-lanceolate, 4–5.3 mm long, emarginate to obtusely bilobed, with a mucro up to 1.5 mm long from between the lobes, the upper linear, 10−13 mm long, emarginate to 2-toothed, shortly awned in the sinus; lemma, including the callus, 4−6 mm long, markedly narrowed above the middle, the narrowed portion scabrid, articulated at the summit and bearing a column 1−1.5 cm long; awns subequal, the central 3.5−5 cm long, the laterals a little shorter.

Acacia-Commiphora bushland on orange sand overlying limestone; 100−420 m. C1, 2; S2; Ethiopia and Kenya. Wieland 4578; Kuchar 17657; Moggi & Bavazzano 1031.

Differs from *A. sieberiana* by the narrowed upper portion of the lemma, and from *A. stenophylla* and *A. schebehliensis* by the woody stems freely branched from all nodes, especially the upper.

14. **A. sieberiana** Trin. (1821).

A. sieberiana var. *nubica* Trin. & Rupr., Sp. Gram. Stipac.: 161 (1842).

A. pallida Steud. (1854).

Birreh, marchain (Som.).

Loosely tufted robust perennial with woody stems branched above, up to 100 cm high. Panicle loosely contracted. Glumes unequal, the lower lanceolate, 9.5−15.5 mm long, bifid at the tip, with an awn (1−)2−6 mm long in the sinus, the upper linear, 15−20 mm long, usually hairy in the middle part, deeply bifid at the tip, the lobes 3−8 mm long, with an awn (1−)3−10 mm long in the sinus; lemma, including the callus, 8−12 mm long, terete, scaberulous to coarsely scabrid above, sometimes smooth, articulated at the summit and bearing a column (1.2−)1.8−3 cm long; awns subequal, the central 4.5−8.5 cm long, the laterals a little shorter.

Sandy soils in deciduous bushland. N1; C1, 2; S2; Senegal and Cameroun eastwards to Tunisia and Palestine. Glover & Gilliland 212; Hemming 3385; Kuchar 17747; Bettini s.n.

In the past this species has been confused with *A. paoliana* and *A. schebehliensis*. It differs from the former by the cylindrical lemma not narrowed above, and from the latter by its woody, suffruticose habit.

15. **A. tenuiseta** Cope (1992); type: C1, "Obbio", Wieland 4439 (K holo.).

Densely tufted perennial up to 30 cm high. Panicle loosely contracted. Glumes very unequal, narrowly lanceolate, minutely 2-toothed, awned in the sinus, the lower 8−9.5 mm long, with an awn 2−3 mm long, the upper 21−23 mm long, with an awn 7−10 mm long; lemma, including the callus, 6−6.5 mm long, terete, not narrowed above the middle, articulated at the summit and bearing a column 1.2−2.5 cm long; awns subequal, very slender, flexuous, 9.5−10.5 cm long.

Sandy coastal plains and dunes; 0−30 m. C1; not known elsewhere. Wieland 4404; Beckett 360.

Rather similar to *A. schebehliensis* on account of

the relatively long-awned glumes, but the awns of the lemma are much longer and much finer. All known material was collected near Hobyo.

16. **A. schebehliensis** Henrard (1928); type: S2, between "Giabadgeh" and "Muccoidere", Paoli 1327 (FT holo.).

Tussocky perennial with branching wiry stems up to 50 cm high. Panicle loosely contracted. Glumes very unequal, narrowly lanceolate, long-acuminate, the lower (10.5−)12−13 mm long, with an awn (5−)8−9 mm long, the upper 18−20 mm long, with an awn (6.5−)13−14 mm long; lemma, including the callus, 10−11 mm long, terete, not narrowed above the middle, papillose towards the tip, articulated at the summit and bearing a column 1−3 cm long; awns subequal, 5−8 cm long.

S2, 3; not known elsewhere. Paoli 1327; Senni 278.

The upper glume is significantly longer than that of *A. stenophylla* and includes a proportionately much longer awn. In this respect it is similar to *A. tenuiseta*, but this has longer, much finer lemma-awns.

17. **A. stenophylla** Henrard (1928); type: C1, "Obbia" to "Wuarande", Robecchi-Bricchetti s.n. (RO holo.).

Birreh, maad, marchain, xalfo (Som.).

Tussocky perennial with branching wiry stems up to 45 cm high. Panicle loosely contracted. Glumes very unequal, broadly linear, obtuse, abruptly acute, emarginate-mucronate or bifid and shortly awned in the sinus, the lower 5.5−9 mm long, the upper 13−20 mm long, the awns up to 4 mm long; lemma, including the callus, 4.5−8 mm long, terete, not narrowed above the middle, smooth or minutely papillose towards the tip, articulated at the summit and bearing a column 1−2.3 cm long; awns subequal, the central 4−6.5 cm long, the laterals a little shorter.

Open shrubland or perennial grassland in orange sand over limestone; 30−900 m. N1, 3; C1, 2; S2; Ethiopia. Hansen & Heemstra 6185; Hemming 1742; Elmi & Hansen 4053; Kuchar 17794; Kuchar 16317.

Although the species is usually recognized by the blunt awnless glumes, some specimens show a distinct tendency to form a short awn mostly less than 3 mm long, but occasionally up to 4 mm. Variation between the two extremes is quite continuous so it seems better at this stage to enlarge the circumscription of *A. stenophylla* than to attempt to accommodate the extremes in separate taxa.

18. **A. mutabilis** Trin. & Rupr. (1842).

A. meccana Hochst. ex Trin. & Rupr. (1842).

Dub derigan, half, maruet (Som.).

Annual up to 35 cm high. Panicle open or contracted, the spikelets clustered at the tips of the branches. Glumes unequal, lanceolate, acute, acuminate or shortly 2-toothed, tipped with a mucro or short awn up to 1 mm long, the lower 3.5−5.5 mm long, the upper 5.5−7 mm long; lemma, including the callus, 5−8 mm long, terete, smooth, narrowed above the middle into a column a little less than half the total length of the lemma, the column articulated at the summit; awns slender, 1.5−2.5 cm long.

Subdesert bushland and short grassland on stony and sandy soils overlying limestone, gypsum or volcanic rocks; 30−900 m. N1; C1; S1−3; tropical Africa to India. Bally & Melville 16129; Beckett 219; Hemming & Deshmukh 324; Munro 80; Rose Innes 740.

Tribe 10. **PAPPOPHOREAE**

Ligule a line of hairs. Inflorescence a dense, often narrow panicle, the spikelets all alike. Spikelets several-flowered, slightly laterally compressed, the lower 1 or 2 florets bisexual, the upper progressively reduced, disarticulating above the glumes but not between the florets; glumes persistent, thinly membranous, distinctly 3- to 9-nerved, at least as long as the body of the lowermost lemma, entire; lemma 9- to 11-nerved, broad, rounded on the back, the nerves produced into 5−9 awns or hyaline lobes; palea-keels ciliate.

Five genera and 41 species in the tropics.

1. Lemma 9-awned .... 14. *Enneapogon*
− Lemma 5-awned, the awns alternating with 6 hyaline lobes ....15. *Schmidtia*

14. ENNEAPOGON P. Beauv. (1812)

Perennials, sometimes annuals; leaves usually narrow, often inrolled. Spikelets with only 1 fertile floret; lemmas chartaceous to coriaceous, hairy below the middle, with 9 ciliate (rarely scaberulous) awns; uppermost florets reduced to a brush-like appendage.

28 species in the tropics and subtropics, especially Australia and Africa.

1. Awns of fertile lemma scaberulous throughout
   .... 1. *E. scaber*
− Awns of fertile lemma ciliate for most of their length, scaberulous only towards the tip 2
2. Third lemma vestigial, 0.3−2(−3.5) mm long including the awns 3
− Third lemma well developed (and often accompanied by a vestigial fourth), 2.5−9(−10.5) mm long including the awns 5
3. Perennial; stems wiry, arising from a sub-bulbous base and characteristically branched a few cm above the base .... 3. *E. scoparius*
− Annual or perennial; stems herbaceous, densely or loosely tufted, neither wiry nor branched above a sub-bulbous base 4
4. Anthers 0.4−0.8(−1.2) mm long; densely tufted annual or perennial, the stems often invested at the base in old leaf-sheaths, these usually disintegrating into fibres .... 2. *E. desvauxii*
− Anthers 1−1.7 mm long; loosely tufted annual, the base not invested in old sheaths or fibres
   .... 4. *E. cenchroides*
5. Fertile lemma sparsely to densely hairy on the back, the hairs evenly distributed .. 5. *E. persicus*
− Fertile lemma with 3 dense patches of hair on the back, one along the midnerve and one along each margin ....6. *E. lophotrichus*

1. **E. scaber** Lehm. (1831).
   Tufted perennial up to 35 cm high; basal sheaths remaining intact. Panicle loosely contracted. Lower glume 7-nerved, 3.7−5.2 mm long; upper glume 5-nerved, 4.5−6.5 mm long; fertile lemma 5−6.5 mm long (including awns), hairy on the back all over; awns scaberulous; anthers 1.3−2.7 mm long; third lemma vestigial, 0.3−0.6 mm long.
   Sandstone ravines; c. 300 m. N1; North Africa, southern Arabia and southern Africa. Gillett 4823.

2. **E. desvauxii** P. Beauv. (1812).
   Pappophorum fasciculatum Chiov. (1928); type: C1, "Uarandi", Puccioni & Stefanini 506 (FT holo.).
   Caws leeye, harfo (Som.).
   Tufted perennial or sometimes annual, up to 27 cm high; basal sheaths persistent, forming a pseudo-bulbous base to the stem and usually ultimately disintegrating into a tuft of fibres. Panicle densely contracted, spike-like. Lower glume 7-nerved, 1.8−3.5 mm long; upper glume 5-nerved, 2−4.5 mm long; fertile lemma 3.4−6 mm long (including awns), hairy on the back all over; awns ciliate; anthers 0.4−0.8(−1.2) mm long; third lemma vestigial, 0.3−0.8 mm long.
   Mostly on limestone hills, occasionally on granite or stabilized coastal sand; 0−1500 m. N1−3; C1; Africa to India and China, Central and South America. Hemming 2129; Glover & Gilliland 1083; Bally & Melville 15758; Wieland 4293.

3. **E. scoparius** Stapf (1900).
   Caws mulaax (Som.).
   Tufted perennial up to 80 cm high, with sub-bulbous base; stems wiry, characteristically branched a few cm above the base; basal sheaths remaining intact. Panicle densely contracted, spike-like. Lower glume usually 5-nerved, 2.4−3.1 mm long; upper

glume usually 3-nerved, 3−4.6 mm long; fertile lemma 4.4−5.6 mm long (including awns), hairy on the back all over; awns ciliate; anthers 0.8−1.3 mm long; third lemma vestigial, 0.8−1.3 mm long.

Commiphora bushland on sand; c. 170 m. C2; tropical Africa and southern Arabia (Yemen). Kuchar 17647.

4. **E. cenchroides** (Roem. & Schult.) C.E. Hubb. (1934). Fig. 82.

E. mollis Lehm. (1831).

Annual up to 75 cm high; basal sheaths remaining intact. Panicle loosely contracted, often lobed at the base. Lower glume 5- to 7-nerved, 2.6−3.8 mm long; upper glume 3-nerved, 3.5−5 mm long; fertile lemma 4−7 mm long (including awns), hairy on the back all over; awns ciliate; anthers 1−1.7 mm long; third lemma vestigial, occasionally well developed (but barren), 0.3−2(−3.5) mm long.

Open *Acacia* woodland and *Acacia-Commiphora* bushland; 670−1430 m. N1, 2; C1; Africa, Arabia and India. Gillett 4870; Gillett & Beckett 23548; Thulin & Dahir 6523.

5. **E. persicus** Boiss. (1844).

E. schimperianus (Hochst. ex A. Rich.) Renvoize (1968).

E. elegans (Nees ex Steud.) Stapf (1907).

Pappophorum glumosum Hochst. (1855).

Aiya makarreh, aus, dikil, jabioki (Som.).

Tufted perennial up to 50 cm high; basal sheaths remaining intact. Panicle loosely or densely contracted, spike-like or lobed. Glumes often deeply suffused with purple, the lower 7-nerved, 3.5−7.5 (−10) mm long, the upper 7-nerved, 4.5−10.3 mm long; fertile lemma 6−13 mm long (including awns), hairy on the back all over; awns ciliate; anthers 0.6−1 mm long; third lemma often well developed (but barren), 2.5−9(−10.5) mm long, sometimes accompanied by a vestigial fourth lemma.

Perennial grassland, shrubland or open woodland on sandy, gravelly or alluvial soils overlying limestone; 160−1500 m. N1−3; C1, 2; S1, 2; tropical Africa, south-western Asia to southern India. Hansen & Heemstra 6183, 6250; Lavranos & Bavazzano s.n.; Elmi & Hansen 4070; Gillett & Hemming 24615; Hemming & Deshmukh 378; Moggi & Bavazzano 407.

6. **E. lophotrichus** Chiov. ex H. Scholz & P. König (1983).

Annual up to 20 cm high; basal sheaths remaining intact. Panicle loosely contracted, oblong. Glumes often flushed with pink, the lower 7- to 9-nerved, 3.5−5 mm long, the upper 5- to 7-nerved, 5−7 mm long; fertile lemma 6−7.5 mm long (including awns), hairy on the back in 3 dense patches, one along the midnerve and one along each margin; awns ciliate; anthers 0.3−0.5 mm long; third lemma well de-

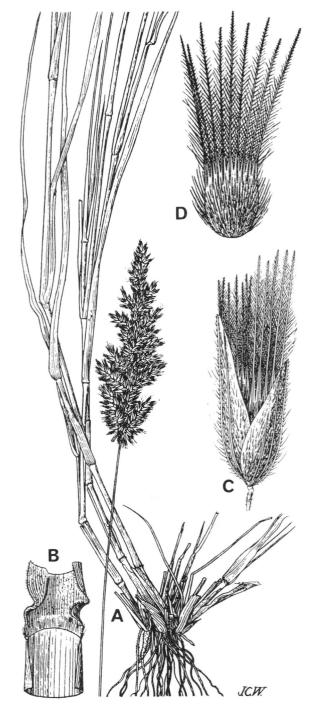

Fig. 82. *Enneapogon cenchroides*. A: habit, × 2/3. B: ligule, × 3. C: spikelet, × 12. D: lemma, × 12. − Modified from Fl. Zamb. 10:1 (1971). Drawn by J. C. Webb.

veloped (but barren), 3.2−5.3 mm long, usually accompanied by a vestigial fourth lemma.

Limestone hillsides and talus-slopes; 350−900 m. N1, 3; C1; Djibouti, Ethiopia, Egypt and Arabia. Glover & Gilliland 844; Thulin & Warfa 6001; Scortecci s.n.

15. SCHMIDTIA Steud. (1852), nom. cons.

Annual or perennial; leaves and sheaths bearing gland-tipped hairs and often viscid. Spikelets 4—6(—9)-flowered; glumes strongly (7—)9—11-nerved; lemmas subcoriaceous, hairy below the middle, 6-lobed, the hyaline lobes (which are themselves sometimes awn-tipped) alternating with 5 scaberulous awns; uppermost 1 or 2 florets sterile and reduced.

Two species in Africa, one of them also in Socotra and Pakistan.

S. pappophoroides Steud. (1852). Fig. 83.

Shortly rhizomatous perennial, often also with long surface stolons, up to 90 cm high, swollen at the base and often suffrutescent. Panicle loose or slightly contracted. Lower glume 4.6—7.5 mm long; upper glume 6—9 mm long; lowest lemma 8.5—14 mm long (including awns); awns 4.5—8 mm long.

Open bushland on sandy or stony soil; c. 580 m. N2; S3; Socotra and Ethiopia to southern and western Africa, also Cape Verde Islands and Pakistan. Fausset 45; Tardelli & Bavazzano 513.

Tribe 11. **ERAGROSTIDEAE**

Ligule membranous or a line of hairs. Inflorescence a panicle or of tough unilateral racemes (a 'bottle-brush' in 27. *Harpachne*; spikelets embedded in 19. *Oropetium*), these digitate or scattered along an axis, rarely solitary, the spikelets all alike. Spikelets sometimes 1-flowered, typically several- to many-flowered with the lower florets fertile and the uppermost reduced, usually laterally compressed, commonly disarticulating below each floret but with a wide variety of other abscission-modes; glumes mostly persistent (except 26. *Eragrostis* and 33. *Sporobolus*), usually membranous, 0- to 1-nerved and shorter than the lowest lemma (but some exceptions), entire; lemmas membranous to coriaceous, 1- to 3-nerved (except 16. *Aeluropus*), entire or 2- to 3-lobed and then occasionally with small subsidiary teeth between the lobes, with or without 1—3 straight or flexuous terminal awns. Fruit sometimes with a free pericarp.

77 genera and about 1000 species in the tropics and subtropics.

1. Spikelets 2- to several-flowered, if 1-flowered the inflorescence either a solitary raceme or composed of several racemes arranged along an axis 2
 — Spikelets strictly 1-flowered; inflorescence an open or spike-like panicle 19
2. Lemmas 9- to 11-nerved 3
 — Lemmas 3-nerved (sometimes 1 or more subsidiary nerves present on either side of the keel in 29. *Eleusine*). 4
3. Spikelets disarticulating between the florets, the racemes persistent .... 16. *Aeluropus*
 — Spikelets not disarticulating, the whole raceme deciduous .... 25. *Drake-Brockmania*
4. Tip of lemma emarginate to 2- to 3-lobed, or the flanks hairy between the lateral nerves and the margin (with clavate hairs on the back below in 28. *Coelachyrum*), or florets conspicuously bearded from the callus 5
 — Tip of lemma entire, the nerves and flanks glabrous (rarely minutely ciliate on the margin itself); florets not bearded from the callus 14
5. Leaves rigid and pungent; inflorescence a short dense ovoid head .... 20. *Odyssea*
 — Leaves not conspicuously rigid and pungent although sometimes rather firm 6
6. Grain strongly flattened, concavo-convex, with a free pericarp .... 28. *Coelachyrum*
 — Grain seldom flattened and then not with a free pericarp 7
7. Inflorescence a solitary raceme, this not deciduous; spikelets glabrous 8
 — Inflorescence of 2 or more racemes 9
8. Spikelets 1-flowered, embedded in the axis ....19. *Oropetium*
 — Spikelets 2- or more-flowered, not embedded in the axis ....18. *Tripogon*
9. Racemes persistent 10
 — Racemes deciduous 13
10. Lower glume much longer than the lowest lemma .... 22. *Trichoneura*
 — Lower glume not exceeding the lowest lemma 11
11. Spikelets disarticulating above the glumes but not between the florets see tribe ....13. *Cynodonteae*
 — Spikelets disarticulating between the florets, or the rhachilla persistent, or the spikelets 1-flowered in racemes that are not digitate 12
12. Florets bearded from the callus ....21. *Halopyrum*
 — Florets not bearded from the callus ....17. *Leptochloa*
13. Glumes as long as the spikelet, enclosing the florets; racemes arranged along a central axis ..... 23. *Dinebra*
 — Glumes not or scarcely exceeding the adjacent lemmas; racemes digitate .... 24. *Ochthochloa*

14. Inflorescence a panicle, or if a solitary raceme then not a 'bottle-brush' with reflexed spikelets.............
....26. *Eragrostis*
– Inflorescence of 1 or more racemes 15
15. Raceme solitary, a 'bottle-brush' with reflexed spikelets ....27. *Harpachne*
– Racemes 2 or more 16
16. Racemes terminating in a rigid naked point .... 31. *Dactyloctenium*
– Racemes terminating in a fertile or abortive spikelet 17
17. Racemes arranged on a long central axis ....33. *Desmostachya*
– Racemes digitate or whorled 18
18. Lemma-tip awnless and the lateral nerves not excurrent; rhachilla fragile, the spikelets disarticulating between the florets .... 29. *Eleusine*
– Lemma-tip awn-pointed and the lateral nerves slightly excurrent; rhachilla tough, the lemmas disarticulating leaving the persistent paleas .... 30. *Acrachne*
19. Spikelets fusiform, glabrous; glumes and lemmas rounded on the back; fruit without a beak 33. *Sporobolus*
– Spikelets strongly laterally compressed, shortly hairy; glumes and lemmas keeled; fruit with a conspicuous beak ....34. *Urochondra*

16. AELUROPUS Trin. (1822)

Stoloniferous perennials; ligule a very short membrane fringed with hairs. Inflorescence capitate to spike-like, comprising short densely spiculate racemes appressed to a central axis (often a single ovoid raceme in *A. lagopoides*). Spikelets several-flowered, disarticulating below each floret; glumes shorter than the lemmas, the lower 1- to 3-nerved, the upper 5- to 7-nerved; lemmas rounded on the back, chartaceous, strongly 9- to 11-nerved, glabrous or hairy on the margins, entire or emarginate, apiculate. Caryopsis ellipsoid.

Three or four species from Mauritania and the Mediterranean region eastwards to northern China, and southwards to Ethiopia and Sri Lanka.

A. lagopoides (L.) Trin. ex Thwaites (1874). Fig. 84.
   A. repens (Desf.) Parl. (1848).
   A. massauensis (Fresen.) Mattei (1910).
   Garo (Som.).
Subshrubby or sward-forming, with long creeping stolons, up to 15(–30) cm high; leaves narrow, spreading, subdistichous, soft and tapering or, more usually, distichous, rigid and pungent, hairy or glabrous. Inflorescence a globose, elliptic or oblong head, sometimes reduced to a solitary raceme of closely crowded spikelets, (0.5–)1–2(–2.5) × (0.5–)1–1.5 cm. Spikelets elliptic-oblong, 4- to 8-flowered, 2.5–4.5 mm long; glumes elliptic, unequal, villous, the lower 1.4–1.7 mm long, the upper 1.8–2.2 mm long; lemmas broadly elliptic, apiculate, 2.4–2.8 mm long, villous on the margins.
Sandy seashores, salt-flats and estuaries; sea-level. N1, 2; Mediterranean region to India. Gillett 4741; Hemming 2057.

Fig. 83. *Schmidtia pappophoroides*. A: habit, × 0.6. B: ligule, × 3.5. C: inflorescence, × 0.6. D: spikelet, × 3.5. E: lower glume, × 7.5. F: lemma, × 7.5. – Modified from Fl. Zamb. 10:1 (1971). Drawn by J. C. Webb.

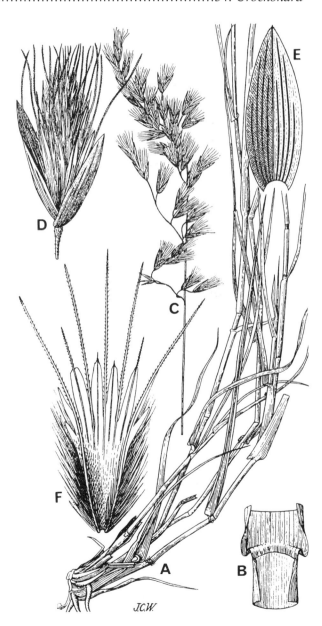

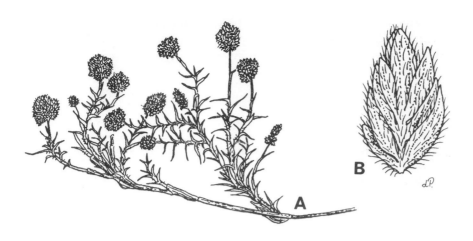

Fig. 84. *Aeluropus lago-poides*. A: habit, × 1. B: spikelet, × 12. — From Thulin & al. 8241 (Yemen). Drawn by L. Petrusson.

17. LEPTOCHLOA P. Beauv. (1812)

Annual or perennial; ligule membranous, sometimes with a ciliate fringe. Inflorescence open, comprising several slender racemes arranged along a central axis. Spikelets laterally compressed or subterete, 2- to several-flowered, or 1-flowered without rhachilla-extension; lemmas keeled or rounded, glabrous or appressed-hairy on the nerves, obtuse or 2-toothed, rarely acute, sometimes mucronate, rarely with a short awn; stamens 2 or 3. Caryopsis laterally or dorsally compressed.

40 species throughout the tropics and in warm temperate parts of America and Australia.

1. Spikelets 6−15 mm long, subterete with rounded lemma, loosely arranged in the indistinctly secund racemes ....1. *L. fusca*
− Spikelets 1.6−5.5 mm long, laterally compressed with keeled lemma, overlapping in the clearly secund racemes 2
2. Spikelets 1-flowered; leaves narrowly lanceolate, often deflexed ....4. *L. rupestris*
− Spikelets 2- to several-flowered; leaves linear ...3
3. Perennial; spikelets 4−5.5 mm long..........
......................................2. *L. obtusiflora*
− Annual; spikelets 2−2.5 mm long .. 3. *L. panicea*

1. **L. fusca** (L.) Kunth (1829); *Diplachne fusca* (L.) P. Beauv. ex Roem. & Schult. (1817).
Diplachne fusca var. *alba* (Steud.) Chiov., Fl. Somala 1: 337 (1929).
Densely tufted aquatic or semi-aquatic perennial up to 1.5 m high, rooting and branching from the lower nodes; leaves linear, finely tapered. Inflorescence 20−35 cm long, composed of 10−30 racemes each 7−15 cm long and bearing loosely arranged, indistinctly secund spikelets. Spikelets 6−15 mm long, 7- to 11-flowered, subterete; lower glume 1.4−2.2 mm long; upper glume 2−3.4 mm long; lemma 2.2−4 mm long, rounded on the back,pilose on the nerves below, tipped with a short mucro

0.1−0.3 mm long. Caryopsis dorsally compressed.
Margins of pools and streams; c. 600 m. C2; Ethiopia, tropical and subtropical Old World. Puccioni & Stefanini 274 (not seen).
The single collection from Somalia cited by Chiovenda (1929) has not been seen and the record needs confirmation.

2. **L. obtusiflora** Hochst. (1855). Fig. 85.
Anadug, aus urun, buldorle, hubnali, luguli, raroh (Som.).
Tufted perennial up to 2 m high; leaves linear, finely tapered. Inflorescence 10−20 cm long, composed of up to 20 slender flexuous racemes each 5−16 cm long, clustered towards the tip of the main axis and bearing clearly secund spikelets. Spikelets 4−5.5 mm long, 6- to 9-flowered, laterally compressed; lower glume 1.3−1.8 mm long; upper glume 1.7−2.4 mm long; lemma 1.8−2.4 mm long, keeled, shortly hairy on the nerves, emarginate to bluntly 2-lobed. Grain concavo-convex with free pericarp.
Acacia bushland; 320−1520 m. N1, 2; C2; S2, 3; tropical Africa. Hansen & Heemstra 6105; Gillett 4080; Kuchar 17755; Bigi 38; Warfa & Warsame 1104.

3. **L. panicea** (Retz.) Ohwi (1941).
L. filiformis auct. non (Lam.) P. Beauv.
Annual up to 110 cm high; leaves linear, long-attenuate. Inflorescence 20−30 cm long, composed of numerous straight, slender ascending racemes each 4−11 cm long, scattered along the main axis and bearing clearly secund spikelets. Spikelets 1.9−2.5 mm long, 2- to 5-flowered (usually 3-flowered), laterally compressed; lower glume 0.7−1.5 mm long; upper glume 0.9−1.6 mm long; lemma 0.8−1.2 mm long, keeled, minutely hairy on the back, bluntly 2-toothed. Caryopsis broadly elliptic, trigonous in cross-section.
S3; tropical Africa and tropical Asia. Scassellati & Mazzocchi 28 (not seen).
The single collection from Somalia, cited as *L. filiformis* by Chiovenda, Res. Sci. Miss. Stefanini-

Paoli: 227 (1916), has not been seen and the record needs confirmation. *L. filiformis* is a South American species, but *L. panicea* was previously often mis-identified as this.

4. **L. rupestris** C. E. Hubb. (1941); type: N1, "Wobleh", Gillett 4981 (K holo.).

Slender perennial with long straggling wiry rhizomes bearing tufts of leafy branches at the nodes, up to 60 cm high; leaves narrowly lanceolate, acute, widely diverging or deflexed. Inflorescence 6–25 cm long, composed of numerous very slender, straight or slightly flexuous racemes, each 2–5 cm long, scattered along the main axis and bearing clearly secund spikelets. Spikelets 1.6–2.4 mm long, 1-flowered, laterally compressed; glumes subequal, as long as the spikelet, only slightly exceeding the floret, if at all; lemma keeled, minutely hairy on the nerves, entire, acute. Caryopsis not known.

Evergreen scrub on mountain slopes, and among rocks by running water; 1300–1700 m. N1; southern Ethiopia, Uganda, Kenya and Arabia (Yemen). Gillett 4838, 4981; Wood 72/95.

18. TRIPOGON Roem. & Schult. (1817).

Perennials; ligule a narrow membrane with a ciliate fringe. Inflorescence a single unilateral raceme. Spikelets 2- to several-flowered, laterally compressed, broadside to the rhachis, linear to elliptic, both glumes well developed, the upper sometimes 3-nerved; lemmas slightly keeled or rounded, glabrous, 2-toothed, mucronate or awned, the awn straight or rarely flexuous, often the teeth also awned; palea sometimes winged; stamens 1–3. Caryopsis subterete.

Some 30 species in the Old World tropics and one in tropical America.

T. subtilissimus Chiov. (1906). Fig. 86.
Harfo, mahan suq, sehansoho (Som.).

Tufted perennial up to 23 cm high. Raceme 3.5–9.5(–11) cm long, flexuous or curved, slender and sometimes feathery. Spikelets 4.2–11 mm long, 5- to 19-flowered; lower glume 0.8–2.2 mm long; upper glume 1.6–3.8 mm long; lemma 2-toothed and awned, often with additional lobes between the teeth, 1.1–2.6 mm long, the midnerve excurrent into an awn 0.8–1.4(–2.5) mm long, the lateral nerves excurrent for 0.1–1.6 mm; anthers 2, 0.2–0.5 mm long.

Open *Acacia-Commiphora* shrubland or *Sporobolus*-dominated grassland on sandy and silty soils overlying calcareous or gypseous rocks; 170–1740 m. N1, 2; C2; S1; Ethiopia, Kenya and Arabia (Yemen and Oman). Gillett 4800; Bally & Melville 16022; Beckett 180; Fries & al. 4819.

A variable species in which the lemma-tip ranges from shallowly lobed with barely excurrent lateral

Fig. 85. *Leptochloa obtusiflora*. A: habit, × 0.9. B: spikelet, × 11. — From Drummond & Hemsley 4192 (Kenya). Drawn by L. Petrusson.

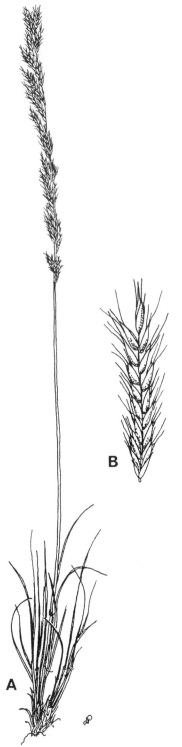

Fig. 86. *Tripogon subtilissimus*. A: habit, × 0.9. B: spikelet, × 5.5. — From Thulin & Mohamed 6973. Drawn by L. Petrusson.

nerves, to deeply divided with acuminate lobes and the nerves drawn out into conspicuous awns. The raceme similarly ranges from rather slender (when

lemma shallowly lobed) to rather dense and feathery (when lemma deeply lobed). It is hard to believe that these two extremes represent the same taxon, but there are too many intermediates for a satisfactory taxonomic division to be made.

19. OROPETIUM Trin. (1842)

Small tufted perennials, rarely annual; ligule membranous with a ciliate margin. Inflorescence a single straight or coiled raceme, the spikelets in two opposite, subopposite or adjacent ranks, sunk in the rhachis, this tough or fracturing into segments of 1−4(−8) spikelets. Spikelets 1-flowered, with or without a minute rhachilla-extension concealed in callus-hairs, dorsally compressed, edgeways to the rhachis; lower glume obscure or absent; upper glume exceeding and concealing the floret, coriaceous, 1- to 3-nerved, acute to awned; lemma lightly keeled, hyaline, glabrous to pilose, emarginate to 2-toothed, mucronate. Grain with pericarp reluctantly separable.

Six species in Africa, Arabia and India.

1. Rhachis tough, tightly wavy; lemma 0.7−1 mm long .... 1. *O. thomaeum*
- Rhachis fragile, straight or loosely flexuous; lemma 1.5−3 mm long 2
2. Raceme straight or curved; upper glume acute, 2−3 mm long; lemma 1.5−2.5 mm long.......... ....2. *O. capense*
- Raceme usually flexuous or coiled; upper glume acuminate to finely awned, 5−20 mm long including the awn; lemma 2.4−3 mm long......... .... 3. *O. minimum*

1. O. thomaeum (L.f.) Trin. (1820).

Dwarf, densely tufted perennial up to 7 cm high. Raceme straight or curved, 1−4 cm long; rhachis tough, always tightly wavy and somewhat spongy. Spikelets in opposite ranks, deeply embedded; upper glume acute, 1.8−2.5 mm long; lemma 0.7−1 mm long, obscurely mucronulate.

Granitic slopes; c. 200 m. S1; Ethiopia, Kenya and Tanzania to Pakistan, India, Burma and Sri Lanka. Alstrup & Michelsen 157; Rose Innes 683.

2. O. capense Stapf (1900).

Nilo-kois (Som.).

Dwarf, densely tufted perennial up to 9 cm high. Raceme straight or curved, 2−3 cm long; rhachis loosely sinuous, fracturing into segments of 1−4(−8) spikelets. Spikelets in opposite, subopposite or adjacent ranks, deeply to superficially embedded; upper glume acute, 2−3 mm long; lemma 1.5−2.5 mm long, obscurely mucronulate.

Rock crevices on limestone or igneous slopes; 1200−1600 m. N1−3; S1; tropical and southern Africa. Gillett 4964; Thulin, Dahir & Hassan 8959;

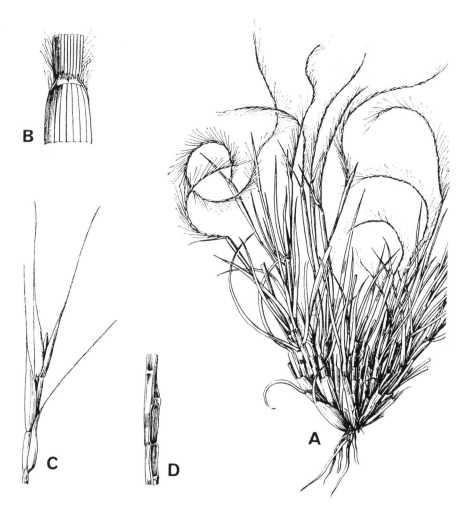

Fig. 87. *Oropetium minimum*.
A: habit, × 1. B: ligule, × 4.
C: upper part of spike, × 6.
D: part of rhachis of spike. —
Modified from Hook. Ic. Pl.
34: t. 3341 (1937). Drawn by
S. Ross-Craig.

Bally & Melville 15735; Thulin, Hedrén & Abdi Dahir 7614.

3. **O. minimum** (Hochst.) Pilg. (1945); *Lepturus minimus* Hochst. (1855); *Chaetostichium mimimum* (Hochst.) C. E. Hubb. (1937). Fig. 87.

Harfo, mahansugen, mahasougar (Som.).

Dwarf, densely tufted perennial up to 9 cm high. Raceme curved, flexuous or coiled, rarely straight, 3—5 cm long; rhachis almost straight, fracturing into segments of 1 or 2 spikelets. Spiklelets in subopposite or adjacent ranks, superficially embedded; upper glume acuminate or drawn out into a fine flexuous awn, 5—20 mm long including the awn; lemma 2.4—3 mm long, with a mucro 0.3—0.5 mm long.

Acacia-Commiphora bushland overlying calcareous or gypseous rocks; 400—1000 m. N1—3; tropical Africa eastwards to Chad, also Arabia. Hansen & Heemstra 6169; McKinnon 90A; Beckett 1035.

20. ODYSSEA Stapf (1922)

Rhizomatous perennials with imbricate sheaths and distichous stiff, pungent leaves; ligule a line of hairs. Inflorescence an elliptic to globose head of short crowded racemes. Spikelets several-flowered, slightly laterally compressed; lemma scarious, rounded on the back, silky-villous on the nerves, 2-toothed, mucronate. Grain ellipsoid with free pericarp.

Two species in Africa and Arabia.

1. Stems branched throughout, the plant forming rounded bushes up to 1.3 m high; lemma 3.8—4.3 mm long .... 1. *O. mucronata*
- Stems branched only at the base, the plant forming dense mats up to 30 cm high; lemma 2.3—3.3 mm long .... 2. *O. paucinervis*

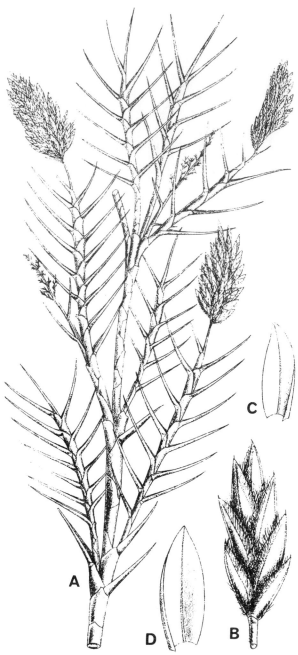

Fig. 88. *Odyssea mucronata*. A: habit, × 1. B: spikelet with glumes removed, × 6. C: lower glume, × 9. D: upper glume, × 9. — Modified from Hook. Ic. Pl. 31, t. 3100 (1922).

1. **O. mucronata** (Forssk.) Stapf (1922). Fig. 88.
 Afrug, gubangub (Som.).

Rounded spiny bush up to 1.3 m high or capable of accumulating sand to form mounds up to 5 m high and 2 m across; leaves inrolled, very stiff and pungent, up to 3 cm long. Panicle 1.5−3 × 0.8−2.5 cm; spikelets 5- to 7-flowered, 7−9 mm long; lower glume 2.7−3 mm long; upper glume 3.2−3.6 mm long; lemma 3.8−4.3 mm long.

Coastal sand-dunes. N1, 2; Red Sea coasts of Eritrea and Arabia (Yemen). Glover & Gilliland 852; Hemming 2046.

2. **O. paucinervis** (Nees) Stapf (1922).
Mat-forming grass with long stout deeply penetrating rhizomes bearing dense tufts of spiny glaucous shoots at the nodes; stems up to 30 cm high, branched only at the base; leaves loosely inrolled, up to 6 cm × 5 mm, somewhat stiff and pungent. Panicle narrowly elliptic to elliptic-oblong, 1.5−7 cm long; spikelets 4- to 9-flowered, 5−9 mm long; lower glume 1.8−2.4 mm long; upper glume 2.7−3.3 mm long; lemma 2.3−3.3 mm long.

Coastal dunes. S2; Tanzania and tropical Africa south of the Zaire River. Maunder 4.

The one specimen representing this species in Somalia is in poor condition and its determination, based almost entirely on habit, is tentative.

21. HALOPYRUM Stapf (1896)

Perennial with robust stolons; ligule a line of hairs; leaves stiff, inrolled, filiform at the tip. Inflorescence comprising short racemes appressed to an elongated axis. Spikelets several-flowered, the callus of each floret conspicuously bearded; glumes 3- to 7-nerved; lemmas rounded on the back, coriaceous, asperulous, entire or minutely 2-toothed, mucronate. Caryopsis elliptic, concavo-convex.

One species only.

H. mucronatum (L.) Stapf (1896). Fig. 89.
 Sido (Som.).

Tough stoloniferous perennial forming dense tussocks; stems up to 1.5(−2) m high, rigid, woody, branching to produce clusters of shoots at the nodes. Leaves narrowly linear to setaceous, stiff, glaucous. Panicle narrow, up to 26 cm long. Spikelets 8- to 16-flowered, 1.5−2 cm long; glumes 6.5−9.5 mm long, coriaceous, acute; lemmas 7.5−8.5 mm long, the callus and rhachilla-tip bearded with white hairs 3−5 mm long.

Coastal sand. N1−3; C1; S2, 3; on the shores of the Indian Ocean from Mozambique to Sri Lanka. Simmons B4; Gillett & Watson 23867; Hemming 1630; Gillett & al. 22189; Alstrup & Michelsen 34; Bally 9510.

22. TRICHONEURA Anderss. (1855)

Annual or perennial; ligule membranous. Inflorescence composed of stiff spreading or short appressed racemes on a central axis. Spikelets cuneate, several-flowered, disarticulating between the florets; glumes narrow, subequal, 1-nerved, longer than the adjacent lemmas and often as long as the spikelet, usually

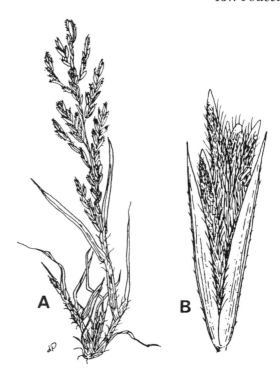

Fig. 90. *Trichoneura mollis*. A: habit, × 1. B: spikelet, × 12. − From Gilbert 2084 (Ethiopia). Drawn by L. Petrusson.

T. mollis (Kunth) Ekman (1912). Fig. 90.

Slender annual up to 50 cm high. Inflorescence 5−25 cm long, fairly compact with stiff ascending racemes 1.5−5 cm long. Spikelets 5- to 9-flowered, 6.2−8 mm long, green or tinged with red; glumes as long as or exceeding the florets, caudate at the tip; lemmas oblong, 2.5−3.5 mm long, sparsely hairy on the back; awn 6−20 mm long; palea capitate-pilose.

C2; Senegal and Mauritania eastwards to the Red Sea coasts of Africa and Arabia, also in northern Oman. Fausset s.n.; Moggi & al. 226A.

There are also unconfirmed reports of *T. mollis* from northern Somalia.

23. DINEBRA Jacq. (1809)

Annuals; ligule membranous. Inflorescence composed of elongated or cuneate racemes on a central axis, these deciduous or with deciduous secondary branchlets. Spikelets 1- to several- flowered, laterally compressed, cuneate, eventually disarticulating between the florets; glumes subequal, exceeding and enclosing the florets, often coriaceous, sometimes 3-nerved, acute to aristate; lemmas keeled, thinly membranous, pubescent on the nerves or sometimes glabrous, acute to emarginate, with or without a mucro. Caryopsis elliptic, trigonous in section.

Three species in Africa, Madagascar and India.

Fig. 89. *Halopyrum mucronatum*. A: habit, × 2/3. B: ligule, × 2. C: spikelet, × 4. D: floret, × 3.5. − Modified from a drawing for Fl. Zamb. by J. C. Webb.

tapering to a mucro or awn; lemmas rounded or lightly keeled, membranous, ciliate on the lateral nerves, 2-toothed, mucronate or shortly awned; palea often capitate-pilose between the keels. Caryopsis narrow, dorsally flattened, concavo-convex.

Seven species in tropical Africa and Arabia, Texas and the Galapagos Islands.

Fig. 91. *Dinebra retroflexa* var. *condensata*. A: habit, ×
2/3. B: spikelet, × 12. C: lemma, × 16. D: palea, × 16. —
Modified from Fl. Trop. E. Afr. (1974). Drawn by M. E.
Church.

D. retroflexa (Vahl) Panzer (1814).
var. **condensata** S.M. Phillips in Kew Bull. 28: 412
(1973). Fig. 91.

Loosely tufted, up to 50 cm high, slender, usually
straggling and ascending from a decumbent base,
much-branched and rooting from the lower nodes.

Inflorescence 8.5—17 cm long, linear with short
oblong to cuneate densely crowded racemes 1.5—6
mm apart; racemes 0.6—1.6 cm long, at first
ascending, eventually reflexing and finally deciduous.
Spikelets 1- to 2-flowered, narrowly cuneate, 5—6.5
mm long; glumes coriaceous, asymmetric, overlapping abaxially and obscuring the florets; lemmas
narrowly ovate, 1.8—2.4 mm long, appressed-
pubescent along the lateral nerves and on the back
below.

Roadsides and cultivated fields; 170—250 m. S1, 3;
East tropical Africa southwards to South Africa.
Thulin, Hedrén & Abdi Dahir 7621; Wieland 1228;
Fiori 66; Terry 3393.

Differs from var. *retroflexa* by the shorter, more
densely crowded racemes and the generally 1- or
2-flowered spikelets. The typical variety occurs over
much of tropical Africa northwards to Egypt and
eastwards into Iraq and India.

24. OCHTHOCHLOA Edgew. (1842)

Stoloniferous perennial; ligule a short membrane
with a long ciliate fringe. Inflorescence composed of
(2—)3—5 short secund digitate racemes, these
deciduous at maturity. Spikelets several-flowered,
strongly laterally compressed, disarticulating above
the glumes but not between the florets; glumes
unequal, shorter than the adjacent lemmas, the upper
with a thickened 3-nerved keel; lemmas keeled,
villous on the margins and keel below, acute or with a
short awn-point. Grain ellipsoid, smooth, with free
pericarp.

One species only.

O. compressa (Forssk.) Hilu (1981); *Eleusine
compressa* (Forssk.) Asch. & Schweinf. (1922). Fig.
92.

Hari hari (Som.).

Sprawling perennial with stolons up to 1 m long and
prostrate or ascending stems up to 40 cm high; leaves
glaucous, flat or inrolled, short, stiff and pungent.
Racemes 2—4 cm long. Spikelets 3- to 8-flowered,
6—8 mm long; lower glume 2.5—3 mm long; upper
glume 3.6—5.6 mm long; lemmas 4—5.6 mm long.

Sandy beaches and hillsides; 0—430 m. N1; S2;
Djibouti, Ethiopia, Sudan, Egypt and through Arabia
to Pakistan and north-west India. Gillett 4512;
Hemming 2357.

25. DRAKE-BROCKMANIA Stapf (1912)

Decumbent annual or perennial; ligule a membrane.
Inflorescence composed of (2—)3—10 short compact
deciduous racemes on a central axis, this sometimes
short and subcapitate. Spikelets strongly laterally
compressed, several- to many-flowered, densely

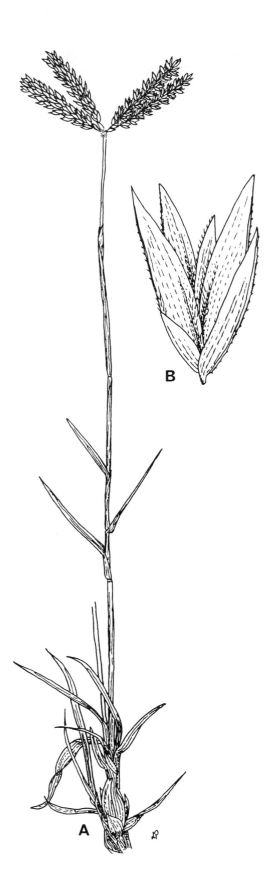

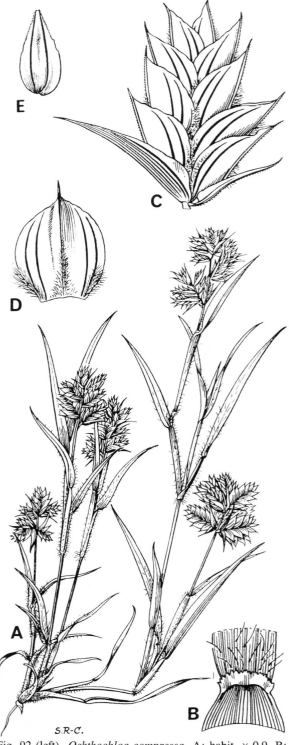

Fig. 92 (left). *Ochthochloa compressa*. A: habit, × 0.9. B: spikelet, × 11. − From Hedberg 92015 (Saudi Arabia). Drawn by L. Petrusson.

Fig. 93 (above). *Drake-Brockmania somalensis*. A: habit, × 1. B: ligule, × 8. C: spikelet, × 6. D: lemma, × 6. E: palea, × 6. − Modified from Hook. Ic. Pl. 35: t. 3455 (1947). Drawn by S. Ross-Craig.

177

imbricate, eventually disarticulating between the florets but usually the whole raceme falling before many florets have been shed; glumes unequal, chartaceous, the lower 1- to 3-nerved, the upper conspicuously 6- to 9-nerved, usually exceeding the adjacent lemma and often as long as the spikelet, the acuminate tip often reflexed; lemmas 3- to 7-nerved, keeled, villous on the margins and keel below, entire, cuspidate or mucronate; palea-keels winged or not. Caryopsis ellipsoid.

Two species in East tropical Africa and Arabia.

D. somalensis Stapf (1912); type: N1, "Bulhar", Drake-Brockman 616, 617, 646 & 647 (all K syn.). Fig. 93.

Eleusine somalensis Hack. (1900).

Prostrate annual, sometimes rooting from the nodes, the flowering stems ascending, up to 15 cm high. Inflorescence subcapitate, comprising 2−6 racemes on an axis 0.3−3 cm long; racemes 0.7−1.7 cm long, spreading at first, becoming reflexed. Spikelets 5- to 9-flowered, broadly oblong to cuneate, 6−11 mm long, pale yellow-green with dark green upper glume- and lemma-nerves; lower glume narrowly lanceolate, 2.9−4 mm long, acute, mucronate; upper glume oblong-lanceolate, 5−8 mm long; lemmas obliquely ovate, 4−7 mm long, cuspidate or mucronate; palea-keels gibbous, broadly winged.

Bare sandy soil. N1; S1−3; Ethiopia, Sudan, Kenya, Tanzania and Arabia (Farasan Island). Drake-Brockman 924; Thulin & Bashir Mohamed 6963; Friis & al. 5025; Kilian & Lobin 7015.

Eleusine somalensis is not the basionym of *Drake-Brockmania somalensis*. It is based on a type from Ethiopia, and the choice of the same epithet is coincidental.

26. ERAGROSTIS Wolf (1776)

Annuals or perennials; ligule a line of hairs, rarely a membrane. Inflorescence an open or contracted panicle, very rarely of racemes along an axis. Spikelets 2- to many-flowered, orbicular to vermiform, variously disarticulating; glumes often deciduous, 1- or rarely 3-nerved; lemmas 3-nerved, keeled or rounded on the back, membranous to coriaceous, glabrous to asperulous or rarely hairy, entire, truncate to acuminate, sometimes mucronate; palea-keels sometimes winged or ciliate; anthers 2−3. Fruit mostly globose to ellipsoid, usually a caryopsis but sometimes the pericarp free.

About 350 species in the tropics and subtropics.

1. Glaucous perennial forming spiny cushions or small bushes; leaves distichous, pungent; stems clothed in imbricate sheaths ...... 11. *E. mahrana*
   − Annual or perennial, but not forming spiny cushions or small bushes 2

2. Densely tufted perennial with creeping woody rhizomes and stolons .... 12. *E. incrassata*
   − Annual or perennial, but without woody rhizomes or stolons 3
3. Florets persistent and the grain retained on the mature panicle ....14. *E. tef*
   − Florets variously deciduous at maturity 4
4. Spikelets disarticulating below the glumes and falling entire 5
   − Spikelets variously breaking up above the glumes 6
5. Tussocky perennial 50−120 cm high; spikelets ovate to suborbicular .... 1. *E. superba*
   − Annual up to 30 cm high; spikelets oblong with serrate outline .... 2. *E. sennii*
6. Spikelets shedding their florets from the tip downwards, the rhachilla-segments falling with the florets 7
   − Spikelets shedding their florets from below upwards, the rhachilla tough, remaining upon the pedicel or eventually breaking up 14
7. Palea ciliate on the keels, the long hairs clearly visible beyond the margins of the lemma 8
   − Palea glabrous or scabrid on the keels 11
8. Perennial .... 3. *E. caespitosa*
   − Annual 9
9. Panicle loose and open .... 6. *E. lepida*
   − Panicle spike-like, lobed or capitate, the spikelets densely crowded 10
10. Panicle spike-like or lobed and interrupted, not overtopped by the uppermost leaves; lemmas broadly obtuse, mucronate .... 4. *E. ciliaris*
    − Panicle capitate, subtended and long-overtopped by the (2−)3 uppermost leaves; lemmas 3-toothed or at least with all 3 nerves excurrent..............
    .... 5. *E. tridentata*
11. Ligule a membrane .... 7. *E. japonica*
    − Ligule a line of hairs 12
12. Lemma 2.3−2.6 mm long ......10. *E. kuchariana*
    − Lemma 1.2−2 mm long 13
13. Panicle delicate, effuse, the spikelets on slender pedicels; anthers 0.2 mm long .... 8. *E. aspera*
    − Panicle stiff, moderately dense, the spikelets shortly pedicelled and often loosely grouped about the primary branches; anthers 0.5−0.7 mm long .... 9. *E. ambleia*
14. Palea falling at about the same time as the lemma leaving the naked, zig-zag rhachilla...............
    .... 13. *E. aethiopica*
    − Palea persisting on the rhachilla long after the lemma has fallen (sometimes the rhachilla disarticulating above the glumes at an early stage with most of the florets still attached, or the upper part becoming fragile once the lower lemmas have begun to fall) 15
15. Tufted perennials 16
    − Coarse or delicate annual, or if a short-lived perennial then lower glume 0.3−1(−1.4) mm long 17

16. Leaves flat, stiff and pungent, scabrid on both surfaces; lemmas obtuse; anthers 3, c. 0.2 mm long .... 21. *E. lutensis*
– Leaves flat or inrolled, soft, not pungent, smooth on both surfaces; lemmas sharply acute to apiculate; anthers 2, 0.6–0.7 mm long.......... .... 24. *E. perbella*
17. Short-lived perennial with stiff glaucous leaves often forming a basal cushion .... 23. *E. papposa*
– Coarse or delicate annual; leaves mostly soft and green, not forming a basal cushion 18
18. Plant sticky from glandular hairs, especially on the sheaths; lemmas oblong, truncate.......... ....16. *E. gloeophylla*
– Plant without glandular hairs, not sticky, but crateriform glands often present on the leaf-margins 19
19. Lemma gibbous on the margin; spikelet with serrate outline ....15. *E. vatovae*
– Lemma not gibbous on the margin; spikelet with smooth outline 20
20. Lemma truncate at the tip; grain glossy.......... ....19. *E. psammophila*
– Lemma subacute to obtuse; grain reticulate ... 21
21. Grain subrotund, 0.4–0.6 mm in diam.; leaf-margins with or without crateriform glands..... ....17. *E. cilianensis*
– Grain broadly oblong to elliptic-oblong, 0.5–1 mm long 22
22. Leaves with crateriform glands along the margins ....18. *E. minor*
– Leaves without crateriform glands along the margins 23
23. Lemmas thin, chartaceous, the lateral nerves not thickened, scaberulous only near the tip, 1.7–2.3 mm long ....20. *E. barrelieri*
– Lemmas thinly coriaceous, with thickened scabrid lateral nerves, scaberulous on the flanks throughout, 2.3–2.6 mm long...................... ....22. *E. trachyantha*

1. **E. superba** Peyr. (1860).

Tough, tussocky perennial up to 120 cm high; leaves glaucous, flat or inrolled. Panicle lanceolate to narrowly oblong, 9–30 cm long, sparsely branched to subracemose, the spikelets on short pedicels. Spikelets ovate to suborbicular, 6–16 mm long, 8- to 29-flowered, falling entire; glumes subequal, lanceolate, 2.3–5.8 mm long; lemmas boat-shaped, cartilaginous, 3–5.8 mm long, narrowly obtuse; palea-keels thickened, each with a membranous ciliolate wing; anthers 3, c. 2 mm long.

Coastal bushland. S3; tropical and southern Africa. Senni 381 (not seen).

Recorded from Somalia by Chiovenda, Fl. Somala 2: 458 (1932). The single collection cited is from "Baddada".

2. **E. sennii** Chiov. (1932); type: S3, "Mangab", Senni 239 (FT holo.).

E. abrumpens Kabuye (1973).

Annual up to 30 cm high; leaves flat or inrolled. Panicle 2–10 cm long, sparse and often racemose, the spikelets clustered towards the tip of the stem on short pedicels, or with few-spiculate branches below. Spikelets oblong, 8–18 mm long, the outline conspicuously serrate, 9- to 32-flowered, falling entire; glumes subequal, narrowly lanceolate, 2.5–4 mm long, acuminate; lemmas narrowly ovate in profile, cartilaginous, 3–3.8 mm long, shortly acuminate with recurved tip; palea-keels thickened, each with a narrow wing; anthers 2, 0.75–0.9 mm long.

Evergreen bushland and heavily used grassland, and as a weed of disturbed and cultivated soils; 25–300 m. S1–3; Ethiopia and Kenya. Eagleton 91; Rose Innes & Trump 1013; Gillett & al. 25316.

3. **E. caespitosa** Chiov. (1915); type: S1, "Mallable", Paoli 756 (FT holo.).

Tufted wiry perennial up to 50 cm high, the base invested with hard yellow glabrous to silky-tomentose scales; leaves flat. Panicle narrowly oblong, 3–13 cm long, rather dense. Spikelets oblong, 2.5–5.5 mm long, 4- to 15-flowered, disarticulating between the florets, the rhachilla fragile; glumes subequal, lanceolate, 1–2 mm long, subacute; lemmas elliptic-oblong in profile, thinly membranous, 1.3–1.7 mm long, smooth or scaberulous, with or without a few bristles on the keel, broadly obtuse and shortly mucronate; palea-keels pectinate-ciliate, the hairs 0.4–0.7 mm long; anthers 3, c. 0.8 mm long.

Dry overgrazed soils; c. 200 m. S1; Kenya, Uganda and Tanzania.

Although widespread in East Africa between 300 and 1600 m, it is known from Somalia only from the type specimen.

4. **E. ciliaris** (L.) R. Br. (1818).

E. ciliaris var. *arabica* (Jaub. & Spach) Asch. & Schweinf., Ill. Fl. Égypte 2: 172 (1887).

Annual up to 50 cm high; leaves flat. Panicle 1–12.5 cm long, contracted, spike-like or lobed and interrupted. Spikelets oblong-elliptic to ovate, 2–3.5 mm long, 6- to 12-flowered, the florets loose and divergent, covered by the spreading palea-hairs, disarticulating between the florets, the rhachilla fragile; glumes equal, lanceolate, 0.9–1 mm long; lemmas oblong in profile, thinly membranous, 0.9–1.5 mm long, with a few pectinate hairs on the lower part of the keel at least in the upper lemmas, broadly obtuse, mucronate; palea-keels pectinate-ciliate with tubercle-based hairs 0.4–0.8 mm long; anthers 2, 0.1–0.2 mm long.

Semi-desert shrubland, especially of *Acacia-Commiphora*, on red sand over limestone or gypsum, coastal dunes, saline plains and areas of cultivation;

0−1200 m. N1−3; C1; S1−3; throughout the tropics. McKinnon 277B; Thulin 4298; Thulin & Warfa 5624; Wieland 4529; Wieland 1001; Gillett & Beckett 23211B; Kazmi 5245.

5. E. tridentata Cope (1992); type: C1, near Hobyo, Thulin & Abdi Dahir 6655 (K holo., UPS iso.).

Annual up to 18 cm high; leaves flat. Panicle 0.6−1.3 cm long, densely contracted, capitate, subtended and long-overtopped by the (2−)3 uppermost leaves. Spikelets elliptic-oblong, 2.2−4.4 mm long, 4- to 8-flowered, the florets loose and divergent, disarticulating between the florets, the rhachilla fragile; glumes subequal, lanceolate, 1.1−1.7 mm long; lemmas oblong in profile, thinly membranous, 1.6−1.8 mm long, broadly obtuse, the nerves excurrent as very short mucros, often the apex of the lemma concave between the nerves; palea-keels pectinate-ciliate with tubercle-based hairs c. 0.5 mm long; anthers 2, c. 0.2 mm long.

Coastal plain on flat open limestone rocks; 50 m. C1; known only from the type.

This remarkable species, discovered only recently (1989) has a most unusual aspect for a species of *Eragrostis* with the capitate inflorescence sunk between the elongated uppermost leaves, rather in the manner of some annual species of *Juncus*.

6. E. lepida (A. Rich.) Hochst. ex Steud. (1854).

Annual up to 40 cm high; leaves flat. Panicle lanceolate to narrowly elliptic, 3−10(−15) cm long, open, delicate. Spikelets elliptic to oblong, 2.2−3 mm long, 4- to 10-flowered, the florets loose and divergent, covered by the spreading palea-hairs, disarticulating between the florets, the rhachilla fragile; glumes unequal, lanceolate, the lower 0.6−0.9 mm long, acute, the upper 0.9−1.1 mm long, often mucronate; lemmas oblong in profile, thinly membranous, (0.8−)1.1−1.2 mm long, scaberulous, truncate, mucronate; palea-keels pectinate-ciliate with tubercle-based hairs 0.3−0.4 mm long; anthers 2, 0.1−0.15 mm long.

Disturbed ground, often near water; 900−1000 m. N2, 3; S3; Djibouti, Sudan, Ethiopia, northern Kenya and Saudi Arabia. Hildebrandt 877; Bally & Melville 15829; Rose Innes 666.

7. E. japonica (Thunb.) Trin. (1830).

Annual or short-lived perennial up to 150 cm high; leaves flat with membranous ligule. Panicle very variable in shape, ranging from linear and contracted to loose and open with ascending branches, 5−55 cm long, the branches solitary or whorled. Spikelets elliptic to narrowly oblong, 1.3−3 mm long, 4- to 14-flowered, disarticulating between the florets, the rhachilla fragile; glumes subequal, narrowly ovate, 0.5−0.8 mm long, subacute to obtuse; lemmas broadly oblong-elliptic in profile, thinly membranous, 0.7−1 mm long, broadly obtuse; palea-keels smooth or scabrid; anthers 2, c. 0.2 mm long.

S1, 2; tropical Africa to SE Asia. Pollacci & Maffei 34, 118.

8. E. aspera (Jacq.) Nees (1841).

Coarse annual up to 110 cm high; leaves flat. Panicle elliptic to ovate, 10−50 cm long, open and very diffuse, the spikelets on long capillary pedicels, the branches solitary or whorled. Spikelets oblong to linear, 3−8 mm long, 5- to 20-flowered, disarticulating between the florets, the rhachilla fragile, but sometimes a few lemmas falling before their palea; glumes subequal, oblong-lanceolate, 0.8−1.4 mm long, obtuse; lemmas broadly elliptic in profile, membranous, 1.1−1.5 mm long, broadly obtuse to truncate; palea-keels scabrid; anthers 3, c. 0.2 mm long.

On limestone rocks or in *Acacia* woodland on red sand over limestone; c. 450−1200 m. N2; S1; tropical and southern Africa to India. Newbould 931; Thulin & Bashir Mohamed 6847.

9. E. ambleia Clayton (1972).

Annual up to 90 cm high; leaves flat. Panicle elliptic, 15−25 cm long, the branches stiff and horizontally spreading, hairy in the axils, the spikelets shortly pedicelled and often loosely grouped about the primary branches. Spikelets linear, 4−10 mm long, 7- to 15-flowered, disarticulating between the florets, the rhachilla fragile; glumes subequal, lanceolate to narrowly ovate, 1.5−2 mm long, subobtuse to obtuse; lemmas broadly oblong-elliptic in profile, 1.6−2 mm long, broadly truncate; palea-keels scabrid; anthers 3, 0.5−0.7 mm long.

Open bushland on sandy soil. S3; Kenya. Rose Innes 757.

A little-known species only recently discovered in Somalia close to the Kenya border not far from Kolbio.

10. E. kuchariana S.M. Phillips (1991); type: C2, "Jalalaksi", Kuchar 17656 (K holo.).

Caws dameer (Som.).

Annual up to 60 cm high; leaves flat or inrolled, the sheaths covered with clavate glandular hairs. Panicle 25−30 cm long, diffuse or rarely contracted, the spikelets on pedicels 0.5−2 mm long. Spikelets linear, 12−24 mm long, 14- to 30-flowered, disarticulating between the florets, the rhachilla fragile, often several florets falling together; glumes unequal, narrowly lanceolate-oblong, the lower 1.7−1.9 mm long, the upper 2.2−2.3 mm long; lemmas lanceolate-oblong, firmly chartaceous, 2.3−2.6 mm long, obtuse or narrowly truncate; palea-keels scabrid; anthers 3, c. 0.8 mm long.

Orange sand with *Commiphora*. C1, 2; not known elsewhere. Thulin, Hedrén & Abdi Dahir 7343, 7418.

11. E. mahrana Schweinf. (1894), as 'mabrana'.
E. hararensis Chiov. (1896).
Gubangub (Som.).

Tough glaucous perennial forming spiny cushions or small bushes up to 30 cm high, the branches clothed in persistent imbricate sheaths; leaves distichous, pungent, the blade often deciduous from the sheath. Panicle ovate, 3−7 cm long, the spikelets on short pedicels less than 1 mm long. Spikelets linear to narrowly oblong, 4−14 mm long, 5- to 30-flowered, the longer spikelets often falcately curved, rhachilla fragile, the florets disarticulating in groups from above, some lemmas also falling, their palea persistent for a short time; glumes subequal, lanceolate, 1.3−1.5 mm long, acute; lemmas oblong-elliptic in profile, membranous, 1.9−2.6 mm long, scaberulous, broadly obtuse; palea equalling or exceeding the lemma, scabrid on the keels; anthers 3, 1−1.2 mm long.

Limestone pavement, alluvium in water-courses and dry sandy or gravelly soils; 0−560 m. N1−3; Ethiopia and Arabia (Yemen and Oman). Hemming 2378; Hemming & Watson 3176; Bally & Melville 15650.

12. E. incrassata Cope (1992); type: N2, "Erigavo", McKinnon 232 (K holo., BM iso.).

Ris (Som.).

Tufted, glaucous perennial up to 40 cm high; base of plant much-branched, woody, spreading by long woody rhizomes and stolons; leaves distichous, inrolled, sometimes flat. Panicle oblong or narrowly lanceolate, 3−9 cm long. Spikelets linear or narrowly oblong, 6−17 mm long, up to 16-flowered; glumes subequal, ovate or lanceolate, 2.2−2.6 mm long, acute; lemmas ovate-elliptic or ovate in profile, membranous, 2.3−2.8(−3.4) mm long, subacute or rarely obtuse; palea-keels scabrid; anthers 3, 1.3−1.4 mm long.

Damp ground in water-courses; c. 1750 m. N2; known only from the area around "Erigavo". McKinnon 269; Glover & Gilliland 761, 1124.

A very distinctive species forming extensive mats, or tufts connected by woody rhizomes and stolons. The mode of disarticulation of the spikelet is not yet known.

13. E. aethiopica Chiov. (1899); types: S Somalia, "Uebi", Bricchetti 193, 247 & 254 (FT syn.).

Caws cad, gelmis (Som.).

Annual up to 65 cm high; leaves flat. Panicle elliptic, 6−25 cm long, very diffuse with filiform pedicels, the lowermost branches usually whorled, glabrous or shortly hairy in the axils. Spikelets linear to narrowly oblong, 3−6.8 mm long, 4- to 19-flowered, shedding their florets from below upwards, the palea falling with or soon after the lemma, the rhachilla tough; glumes very unequal, the lower a minute nerveless scale 0.3−0.4 mm long, the upper lanceolate, 0.6−0.7 mm long; lemmas narrowly elliptic in profile, thinly membranous, 0.7−1.1 mm long, obtuse; palea-keels smooth; anthers 3, 0.1−0.2

mm long. Grain ellipsoid, 0.4−0.5 mm long.

Red sand, silt and clay; 120−760 m. N1, 2; C1, 2; S1−3; Djibouti and Ethiopia southwards to South Africa, also in Arabia (Yemen). Hansen & Heemstra 6180; Glover & Gilliland 229; Herlocker 452; Moggi & al. 228; Hemming & Deshmukh 329; Kazmi 5065; Tardelli & Bavazzano 837.

14. E. tef (Zucc.) Trotter (1918).

Annual up to 100 cm high; leaves flat. Panicle contracted and narrowly elliptic to diffuse and ovate, 10−60 cm long, the branches ascending, flexuous, the lowermost often whorled. Spikelets narrowly oblong, 4−9 mm long, 4- to 12-flowered, the florets remaining intact on the tough rhachilla and retaining the plump grain, occasionally some lemmas and paleas eventually falling; glumes unequal, lanceolate, acuminate, the lower 1−2.8 mm long, the upper 1.5−3 mm long; lemmas oblong-lanceolate in profile, membranous, 2−3 mm long, scaberulous on the keel and on the flanks above, acute; palea-keels scaberulous; anthers 3, (0.2−)0.3−0.5 mm long. Grain turgid, 1−2 mm long.

Probably a casual introduction. ?N1; Ethiopia, introduced or adventive elsewhere. Farquharson s.n.

As a cereal, tef is apparently not cultivated outside Ethiopia. The single known occurrence of the species in northern Somalia is from 1931. There are no modern records.

15. E. vatovae (Chiov.) S.M. Phillips (1982); Acrachne vatovae Chiov. (1940); type: S1, "Baidoa", Vatova 61 (FT holo.).

Annual up to 18 cm high; leaves flat. Panicle c. 8 cm long, contracted, branched below with the spikelets clustered on short pedicels, racemose above. Spikelets ovate to oblong, c. 11 mm long, c. 16-flowered, with markedly serrate outline, shedding their lemmas from below upwards, the paleas persistent on the tough rhachilla; glumes subequal, lanceolate, 3−3.5 mm long, acuminate; lemmas gibbously ovate in profile with straight keel, membranous, c. 3 mm long, acute; palea-keels scabrid.

Probably bushland at c. 400 m altitude. S1; known only from the type.

A little-known species only relatively recently recognized as an Eragrostis.

16. E. gloeophylla S.M. Phillips (1987); type: C2, "Bulo Burti", Roffy 60037/3 (K holo.).

Baldoleh (Som.).

Annual up to 45 cm high; leaves flat or inrolled, the sheaths, underside of blades and sometimes the stems coated in glandular hairs, the sticky exudate produced attracting sand grains and other debris. Panicle 2−10 cm long, contracted, the spikelets subsessile on short lateral branches. Spikelets narrowly oblong, 5−16 mm long, 6- to 19-flowered, shedding their lemmas from below upwards, but the rhachilla fragile and

Fig. 94. *Eragrostis cilianensis*. A: habit, × 2/3. B: part of leaf showing marginal glands, × 4. C: spikelet, × 6. D: caryopsis, × 24. — Modified from Fl. Trop. E. Afr. (1974). Drawn by M. E. Church.

often breaking away entire or in segments soon after the lemmas have begun to fall; glumes unequal, narrowly lanceolate, the lower 1.5−2.5 mm long, the

upper 2−3 mm long; lemmas narrowly oblong in profile, membranous, 2−3.2 mm long, truncate; palea-keels scabrid; anthers 3, 0.3−0.5 mm long. Grain plumply ellipsoid, 0.5 mm long.

Red sand over limestone; 160−350 m. C1, 2; Ethiopia (Ogaden). Beckett 10A; Elmi & Hansen 4024.

A close relative of *E. cilianensis*, but differing by the oblong, truncate lemmas and the covering of viscid glands, particularly on the sheaths and underside of the leaves.

17. **E. cilianensis** (All.) Vignolo ex Janch. (1907). Fig. 94.
E. megastachya (Koeler) Link (1827).
Agar, domar, doye, ramag (Som.).
Annual up to 75 cm high; leaves flat or inrolled, sometimes with crateriform glands along the margins. Panicle lanceolate to narrowly elliptic, 5−15(−20) cm long, usually ± contracted with short pedicels, occasionally very contracted to a dense ovoid or cylindrical head, the branches and pedicels often bearing crateriform glands. Spikelets oblong-ovate (linear when many-flowered), 4.5−15(−20) mm long, 8- to 30(−45)-flowered, shedding their lemmas from below upwards, the paleas persistent on the tough rhachilla, often the rhachilla fragile in the upper part, rarely breaking away entire soon after the lemmas have begun to fall; glumes subequal, lanceolate, 1.5−2.7 mm long, acute, often with crateriform glands along the nerve; lemmas elliptic-ovate in profile, papery, 2−2.8 mm long, scaberulous, subacute to obtuse; palea-keels scabrid; anthers 3, 0.3−0.5 mm long. Grain usually subrotund, 0.4−0.6 mm in diam.

Mixed shrubland in deep red sand over gypsum, limestone gullies and coastal plains, and in areas of disturbance and cultivation; 25−1200 m. N1−3; C1, 2; S1−3; tropical and warm temperate Old World. Hansen & Heemstra 6141; Bally & Melville 16029, 15616; Beckett 216; Gillett & Beckett 23318; Hemming & Deshmukh 289; Rose Innes 809; Gillett & al. 25317.

18. **E. minor** Host (1809).
E. pappiana (Chiov.) Chiov. (1908).
Tigat had (Som.).
Annual up to 60 cm high; leaves flat or inrolled, with crateriform glands along the margins. Panicle elliptic to ovate, 3−20 cm long, rather open but with short pedicels, the branchlets and pedicels bearing crateriform glands. Spikelets linear to narrowly oblong, 4.8−9 mm long, 8- to 16-flowered, shedding their lemmas from below upwards, the paleas persistent on the tough rhachilla; glumes subequal, lanceolate, 1.3−1.8 mm long, acute, sometimes with crateriform glands along the nerve; lemmas ovate in profile, papery, 1.4−1.8 mm long, scaberulous, obtuse; palea-keels scabrid; anthers 3, c. 0.3 mm

long. Grain usually broadly oblong, (0.5—)0.6—0.8 mm long.

Sandy or stony soils along water-courses; c. 50 m. N1; subtropical and warm temperate Old World. Drake-Brockman 658; Glover & Gilliland 840.

Differs from *E. cilianensis* mainly by the narrower spikelets, shorter lemmas and oblong grain, and from *E. barrelieri* mainly by the presence of crateriform glands along the leaf-margins.

19. **E. psammophila** S.M. Phillips (1991); type: C1, "Dusa Mareb" to "Galkayo" road, Thulin & Warfa 5363 (K holo., UPS iso.). Fig. 95.

Annual up to 17 cm high; leaves slender, the sheaths tuberculate-pilose. Panicle ovate, 5—9 cm long, the branches solitary or paired, the pedicels 0.5—4 mm long. Spikelets narrowly oblong, 4—6 mm long, 6- to 10-flowered, shedding their lemmas initially from below upwards with the paleas persistent on the rhachilla, but the upper part often breaking away early; glumes unequal, the lower oblong, 0.9—1.1 mm long, the upper lanceolate, 1.2—1.5 mm long; lemmas oblong in profile, chartaceous, 1.5—1.8 mm long, truncate-emarginate; palea-keels scabrid; anthers 3, 0.25—0.3 mm long. Grain ellipsoid, c. 0.5 mm long.

Acacia-Commiphora bushland on red sand; c. 250 m. C1; not known elsewhere. Thulin 5651.

20. **E. barrelieri** Daveau (1894).

Agar (Som.).

Annual up to 60 cm high; leaves flat or inrolled, always without crateriform glands on the margins. Panicle lanceolate to elliptic, 3—15 cm long, the spikelets evenly spaced or gathered into fascicles, the branchlets and pedicels often bearing crateriform glands. Spikelets linear, 6—15 mm long, 8- to 25-flowered, shedding their florets from below upwards, the paleas persistent on the tough rhachilla; glumes unequal, lanceolate, acute, without crateriform glands on the nerve, the lower 1—1.5 mm long, the upper 1.3—2.1 mm long; lemmas oblong-lanceolate in profile, papery, 1.7—2.3 mm long, scaberulous above, obtuse; palea-keels scabrid; anthers 3, c. 0.2 mm long. Grain elliptic-oblong, 0.6—1 mm long.

Flood-plains and escarpments, and on disturbed soils; 750—1200 m. N1—3; Mediterranean region, tropical Africa and south-western Asia. McKinnon 277A, 250A; Bally & Melville 15826.

21. **E. lutensis** Cope (1995); type: N1, "Ber", Glover & Gilliland 32 (K holo.). Fig. 96 C.

Tufted perennial up to 40 cm high; leaves flat, stiff and pungent, scabrid on both surfaces, without crateriform glands on the margins. Panicle elliptic, 8—10 cm long, the spikelets shortly pedicelled on the racemose primary branches, the branches and pedicels without crateriform glands. Spikelets oblong to linear, 4—7.5 mm long, 5- to 9-flowered, shedding their florets from

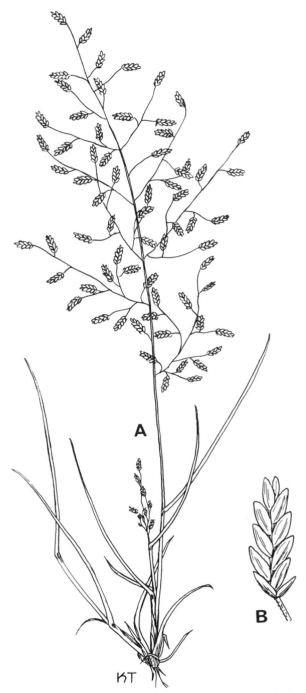

Fig. 95. *Eragrostis psammophila*. A: habit, × 1. B: spikelet, × 6. — From Thulin & Warfa 5363. Drawn by K. Thunberg.

below upwards, the paleas persistent on the rhachilla; glumes unequal, lanceolate, acute, without glands on the nerve, the lower 1.6—2.1 mm long, the upper 2.1—2.3 mm long; lemmas oblong-lanceolate in profile, chartaceous, 2.2—2.4 mm long, scaberulous above, obtuse; palea-keels scabrid; anthers 3, c. 0.2 mm long. Grain unknown.

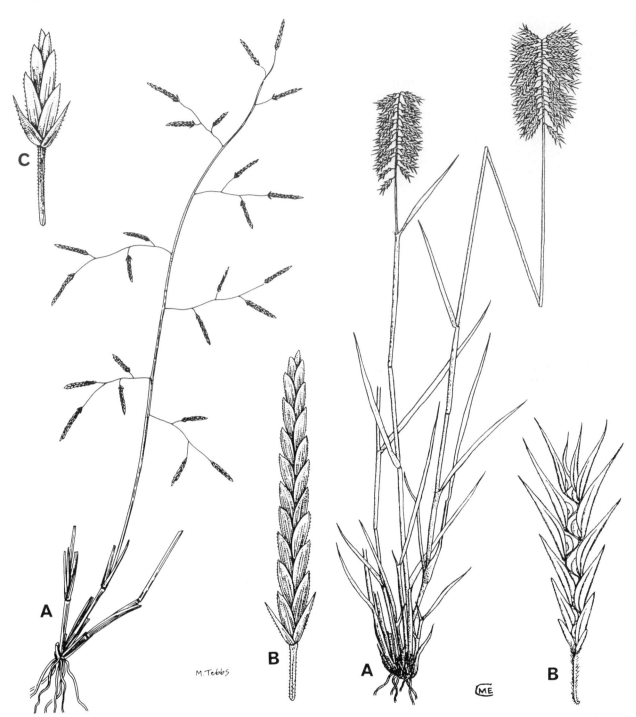

Fig. 96. A, B: *Eragrostis trachyantha*. A: habit, × 1. B: spikelet, × 9. C: *E. lutensis*, spikelet, × 9. – Modified from Kew Bull. 50 (1995). Drawn by M. Tebbs.

Fig. 97. *Harpachne schimperi*. A: habit, × 2/3. B: spikelet with attached pedicel, × 4. – Modified from Fl. Trop. E. Afr. (1974). Drawn by M. E. Church.

In shallow water in flood plain; c. 1000 m. N1; known only from the type.

22. **E. trachyantha** Cope (1995); type: S1, 1°34'N, 42°57'E, Beckett & White 1536 (K holo.). Fig. 96 A, B.

Annual up to 40 cm high, erect or geniculately ascending; leaves inrolled, without crateriform glands on the margins. Panicle lanceolate to narrowly ovate, up to 30 cm long (far exceeding the length of the supporting stem), the spikelets remote and shortly (0.5–9 mm) pedicelled on the distant racemose primary branches, the branches and pedicels without crateriform glands. Spikelets linear, 8−20 mm long, 10- to 26-flowered, shedding their florets from below upwards, the paleas persistent on the rhachilla, though the latter becoming fragile almost as soon as the lemmas start to fall; glumes unequal, lanceolate, very acute, without glands on the nerve, the lower 1.5−2.2 mm long, the upper 2.3−2.7 mm long; lemmas oblong-lanceolate in profile, subcoriaceous with thickened nerves, 2.3−2.6 mm long, scabrid on the keel, scaberulous on the lateral nerves and flanks, obtuse to subacute; palea-keels scabrid; anthers 3, c. 0.3 mm long. Grain broadly oblong, 0.7 mm long.

Deciduous and evergreen shrubland; c. 95 m. S1; known only from the type.

23. **E. papposa** (Roem. & Schult.) Steud. (1840).

E. aulacosperma (Fresen.) Steud. (1840).

Dihi, harfo, manoun, ramass, tingleh, warram bilu (Som.).

Compactly tufted, short-lived perennial up to 60 cm high; leaves glaucous, stiff, flat or inrolled, in the latter state usually ± confined to the base of the plant, older leaves often disarticulating from the sheath. Panicle ovate, 5−22 cm long, either open with filiform branches and capillary pedicels or ± compact with shorter, stouter pedicels. Spikelets linear, 3.2−8.5 mm long, 5- to 18(−22)-flowered, shedding their florets from below upwards, the paleas persistent on the tough rhachilla; glumes unequal, linear-lanceolate, the lower 0.3−1(−1.4) mm long, the upper 1−1.4(−1.8) mm long; lemmas narrowly ovate in profile, membranous, 1.2−1.9 mm long, obtuse; palea-keels scaberulous; anthers 3, 0.1−0.2 mm long.

Juniperus or *Acacia* shrubland over limestone, gypsum or gneiss, sandy and gravelly soils in watercourses, and as a weed of cultivation; 900−2100 m. N1−3; Mediterranean region south to Tanzania and east to Pakistan. Hansen & Heemstra 6112; Hemming 2317; Bavazzano s.n.

24. **E. perbella** K. Schum. (1895).

E. araiostachya Chiov. (1932); type: S3, "Baddada", Senni 384 (FT holo.).

Loosely tufted perennial up to 100 cm high; leaves flat or inrolled. Panicle oblong to ovate, 9−35 cm long, moderately dense, the spikelets shortly pedicelled. Spikelets narrowly oblong, 7−18 mm long, 12- to 30-flowered, shedding their lemmas from below upwards, but the rhachilla usually fracturing above the glumes before all the lemmas have fallen; glumes subequal, lanceolate, 1.7−2.7 mm long, sharply acute; lemmas broadly ovate in profile, membranous, 2−3 mm long, sharply acute to apiculate; palea-keels scabrid; anthers 2, 0.6−0.7 mm long.

Coastal bushland. S3; Kenya and Tanzania.

In Somalia known only from the type of *E. araiostachya*.

27. **HARPACHNE** Hochst. ex A. Rich. (1850)

Perennials; ligule a line of hairs. Inflorescence a single 'bottle-brush' raceme with reflexed spikelets hanging on slender pedicels. Spikelets several-flowered, falling entire together with the pungent or hooked pedicel, the florets gradually increasing in size upwards; lemmas keeled, glabrous except for the ciliolate margin, entire, acute to setaceously acuminate; palea gibbous, the keels winged. Caryopsis laterally compressed, obliquely elliptic.

Two species in North and North-east Africa and Arabia.

H. schimperi Hochst. ex A. Rich. (1850). Fig. 97.

Densely tufted, up to 40 cm high. Panicle oblong, 4−7.5 cm long, the spikelets densely crowded; pedicels 2−6.5 mm long, villous. Spikelets 6- to 13-flowered, 11−14 mm long, wedge-shaped; glumes unequal, oblong-linear, the lower 1.6−2.1 mm long, the upper 3.3−4.5 mm long; lowest lemma lanceolate, 3.8−4.9 mm long, acute, the upper lemmas progressively narrower, becoming setaceously acuminate and up to 6.8 mm long.

N1; Ethiopia and Sudan southwards to Zambia, also in Arabia (Asir region). Drake-Brockman 181.

28. **COELACHYRUM** Hochst. & Nees (1842)

Annual or perennial, often stoloniferous; ligule membranous, sometimes with a ciliate fringe. Inflorescence rarely a panicle, more often of open or dense racemes, these disposed along an axis or subdigitate. Spikelets several-flowered; glumes 1- to 3-nerved; lemma lightly keeled at first, but becoming broadly rounded as the grain expands, thinly chartaceous (thinly coriaceous in *C. yemenicum*), puberulous to villous on the flanks, lateral nerves and sometimes also on the keel (clavate-hairy on the back in *C. yemenicum*), obtuse, sometimes mucronate; palea sometimes villous on the keels. Grain broadly elliptic to subrotund, strongly flattened, concavo-convex, the pericarp free.

Seven species in Africa, Arabia and Pakistan.

1. Spikelets subsessile or pedicelled and rather distant along the raceme-like branches of an open panicle 2
− Spikelets subsessile and closely overlapping in subdigitate or scattered racemes 3
2. Lemma glabrous or shortly ciliate on the margins .... 1. *C. piercei*
− Lemma densely asperulous all over............... .... 2. *C. longiglume*
3. Racemes scattered along an axis which is several times longer than the longest raceme; lemma with short clavate hairs on the back below............ .... 5. *C. yemenicum*

— Racemes subdigitate or on an axis which is only a little longer than the longest raceme; lemmas sparsely to densely clothed with simple, not clavate, hairs on the lateral nerves and flanks ... 4

4. Tufted annual without stolons; hairs on lemma rather short and inconspicuous..3. *C. brevifolium*
— Tufted perennial with woody stolons; hairs on lemma long and shaggy ....4. *C. poiflorum*

1. **C. piercei** (Benth.) Bor (1952).

C. stoloniferum C.E. Hubb. (1941); type: N1, "Eil Demet", Gillett 4324 (K holo.).

Dooyo (Som.).

Tufted perennial, sometimes stoloniferous, up to 45 cm high. Inflorescence an open panicle, 5.5−11 cm long, the branches often racemose, the spikelets borne on pedicels 1−4 mm long. Spikelets 7- to 16-flowered, ovate, 4.5−7.5 mm long; glumes unequal, ovate-elliptic, the lower (1.7−)2.2−3 mm long, the upper (2.1−)3−3.5 mm long, mucronate; lemmas broadly elliptic, 2.4−3 mm long, smooth and glabrous or sometimes shortly hairy on the margins, broadly rounded to truncate, mucronate.

Grassland or evergreen bushland on sand overlying limestone, or fixed coastal dunes; 0−1000 m. N1; C1; S2. Arabian Peninsula and Pakistan (Baluchistan coast). Gillett & al. 22214A; Gillett & Beckett 23222; Hansen 6026.

2. **C. longiglume** Napper (1963). Fig. 98.

Slender annual up to 70 cm high. Inflorescence an ovate open panicle, 6−12 cm long, the branches sometimes racemose, the spikelets borne on pedicels 1−6 mm long. Spikelets 5- to 10-flowered, broadly elliptic, 5.4−6.8 mm long; glumes unequal, the lower lanceolate to elliptic-oblong, 2.8−5 mm long, the upper elliptic-oblong, 3.6−5.7 mm long; lemmas broadly elliptic, 3−4.3 mm long, densely asperulous all over, broadly rounded to truncate and often mucronulate.

Gypseous, saline plain with *Limonium*; c. 200 m. S1; northern Kenya. Thulin & Bashir Mohamed 6998.

3. **C. brevifolium** Hochst. & Nees (1842).

Annual, without stolons, up to 35 cm high. Inflorescence 2−4 cm long, comprising 3−6 subdigitate racemes on an axis scarcely longer than the longest raceme, the spikelets subsessile. Spikelets 5- to 8-flowered, broadly ovate, 3−3.4 mm long; glumes unequal, broadly ovate to broadly elliptic, mucronate, the lower 1.2−1.7 mm long, the upper 1.6−2.5 mm long; lemmas broadly elliptic to subrotund, 1.8−2 mm long, shortly and inconspicuously hairy, broadly rounded to truncate, mucronate.

Coastal sand amongst *Jatropha crinita*; c. 40 m. S2; Sudan, Eritrea and the Arabian Peninsula. Raimondo 2; Senni 623.

4. **C. poiflorum** Chiov. (1897).

Beriween, domar, saddaho (Som.).

Tufted perennial with woody stolons, up to 35 cm high. Inflorescence 2−4 cm long, comprising 3−5 dense racemes scattered along an axis at most a little longer than the longest raceme, the spikelets subsessile. Spikelets 6- to 16-flowered, ovate, 2.6−5.6 mm long, often blackish; glumes unequal, the lower ovate, 1.5−2 mm long, the upper broadly ovate, 2−2.3 mm long, mucronate or shortly (c. 0.3 mm) awned; lemmas broadly ovate-elliptic, 2.1−2.3 mm long, with long shaggy hairs, obtuse to truncate, mucronate.

Grassland and open *Acacia* shrubland overlying limestone, gypsum or gneiss; 750−1900 m. N1−3; C1; Djibouti, Ethiopia and the Arabian Peninsula. Hemming 2132; Hansen & Heemstra 6234; Nugent 41; Wieland 4483.

5. **C. yemenicum** (Schweinf.) S.M. Phillips (1982). Fig. 99.

Cypholepis yemenica (Schweinf.) Chiov. (1908).

Leptochloa appletonii Stapf (1907); type: northern Somalia, Drake-Brockman 99, 101, 147 and Appleton s.n. (all K syn.).

Doma-aru (Som.).

Slender, densely tufted perennial without stolons, up to 100 cm high. Inflorescence narrow, 3.5−19 cm long, comprising 2−8 racemes widely disposed on an axis several times longer than the longest raceme, the spikelets subsessile. Spikelets 7- to 12-flowered, elliptic to narrowly lanceolate-elliptic, 5−10 mm long; glumes subequal, 2.1−4 mm long, obtuse to acute, the lower narrowly oblong-lanceolate, the upper oblong-elliptic; lemmas ovate, 2.5−4.7 mm long, appressed hairy on the back below with short clavate hairs, obtuse to subacute.

Grassland and open shrubland on rocky slopes of limestone, gypsum, sandstone or gneiss; 1200−1900 m. N1−3; southern Africa northwards to Ethiopia, Djibouti and the Arabian Peninsula. Gillett 4872; Bally & Melville 16033, 15783.

29. **ELEUSINE** Gaertn. (1788).

Annual or perennial; ligule membranous, usually with a ciliate fringe; leaves folded and sheaths strongly keeled. Inflorescence comprising digitate or subdigitate racemes, the axis shorter than the longest raceme, the racemes with imbricate spikelets and terminating in a fertile spikelet. Spikelets several-flowered, disarticulating between the florets (except in *E. coracana*); lemmas strongly keeled, sometimes the keel thickened and containing 1−3 closely spaced additional nerves, membranous, glabrous, obtuse or acute. Grain ellipsoid to subglobose, trigonous in outline, flat or concave on the hilar side, with free pericarp.

Nine species, mostly in Africa, but one in South America and one cosmopolitan weed.

Fig. 98. *Coelachyrum longiglume*. A: habit, 2/3. B: spikelet, × 8. C: lemma, × 10. – Modified from Fl. Trop. E. Afr. (1974). Drawn by M. E. Church.

E. coracana (L.) Gaertn. (1788) (syn. *E. tocussa* Fresen.) is a cultivated cereal. It has very broad racemes of closely overlapping non-shattering spikelets in which the large plump grain is exposed between the gaping lemma and palea; it is probably a derivative of *E. indica* subsp. *africana*. It is, or has been, grown in southern Somalia under the name

Fig. 99. *Coelachyrum yemenicum*. A: habit, × 2/3. B: spikelet, × 6. C: lemma, × 12. – Modified from Fl. Trop. E. Afr. (1974). Drawn by M. E. Church.

Fig. 100. *Eleusine indica* subsp. *africana*, habit and spikelet. − From Mem. Bot. Surv. S. Afr. 58 (1991).

'wemba' or 'uembe', and its grain is used for making porridge.

1. Perennial; midnerve of lemma simple; leaves with hair-tufts scattered along the margins........
    ....1. *E. floccifolia*
−   Annuals; midnerve of lemma with 1−3 subsidiary nerves close to each side of it forming a thickened keel; leaves without hair-tufts on the margins, at most loosely pilose 2
2. Racemes 4−7 mm wide; spikelets elliptic, disarticulating between the florets; grain oblong to broadly oblong, blackish, never exposed when ripe ....2. *E. indica*
−   Racemes 9−15 mm wide; spikelets ovate, non-shattering, very closely overlapping; grain plump, almost globose, usually brown, often exposed between the gaping lemma and palea.............
    .... *E. coracana* (see above)

1. **E. floccifolia** (Forssk.) Spreng. (1824).
    Gurgor (Som.).
    Densely tufted rhizomatous perennial up to 70 cm high; leaves with scattered tufts of short white hair along the margins. Inflorescence comprising 2−5 (−10) slender subdigitate racemes 2.5−8.5(−12.5) cm long (usually 1 or 2 of them set below the rest).

Spikelets 4- to 7-flowered, elliptic, 3.3−6.8 mm long; glumes unequal, dark olive-grey, acute, the lower 1−1.2 mm long, the upper 2.6−4.2 mm long; lemmas narrowly elliptic in profile, pale with a tinge of grey, 2.8−4.6 mm long, acute, the midnerve simple. Grain minutely rugulose.

Damp pastures and grassy glades amongst junipers, especially on limestone; 1350−2250 m. N1, 2; Ethiopia, Kenya (introduced) and Arabia (Yemen). Hemming 3229; Wood 71/23; Hansen & Heemstra 6243; Newbould 776.

2. **E. indica** (L.) Gaertn. (1788).
subsp. **africana** (Kenn.-O'Byrne) S.M. Phillips in Kew Bull. 27: 259 (1972). Fig. 100.
    Baldole (Som.).
    Robust annual up to 85 cm high. Inflorescence comprising 1−10(−17) slender digitate racemes 3.5−15.5 cm long, 4−7 mm wide (a few often set below the rest). Spikelets 3- to 9-flowered, elliptic, 4.6−7.8 mm long, pale green or with a brownish tinge; glumes unequal, acute, the lower 2−3.2(−3.9) mm long, broadly to very narrowly winged on the keel, the upper 3−4.7 mm long; lemmas lanceolate in profile, 3.7−4.9 mm long, acute to subacute, with 1 or 2 additional nerves on either side of the keel. Grain obliquely ridged, uniformly granular.

Eroded limestone slopes, disturbed ground; up to c. 1300 m. N1; S3; eastern and southern Africa and southern Arabia. Hemming 2218 (N1).

30. ACRACHNE Wight & Arn. ex Chiov. (1907)

Annual; ligule membranous with a ciliate fringe. Inflorescence comprising digitate or whorled racemes, these with imbricate spikelets and terminating in an abortive spikelet. Spikelets several-flowered, shedding the lemmas but retaining the persistent rhachilla and paleas; lemmas strongly keeled, firmly membranous, glabrous, entire or 2-toothed, tipped with a stout awn-point. Grain ellipsoid with free pericarp.

Three species in the Old World tropics.

A. racemosa (Heyne ex Roem. & Schult.) Ohwi (1947). Fig. 101.
    Dhalad (Som.).
    Stems up to 75 cm high. Racemes digitate, often with additional whorls below, 1.5−10 cm long. Spikelets 6- to 25-flowered, 5.5−13 mm long; glumes unequal, the lower 1.2−3 mm long, acute and mucronate, the upper 1.5−3 mm long, acuminate and awned, the awn 1/3−2/3 the length of the body; lemmas narrowly ovate in profile, 2−2.8 mm long, the awn 0.3−0.9 mm long.

Acacia shrubland and as a weed of cultivation. N1; S1, 2; Old World tropics. Glover & Gilliland 1000; Hemming & Deshmukh 294; Dahir & Rose Innes ODA4.

31. DACTYLOCTENIUM Willd. (1809)

Annual or perennial; ligule membranous. Inflorescence composed of paired or digitate racemes, these bearing imbricate spikelets and terminating in a bare pointed rhachis-extension, eventually disarticulating from the stem though sometimes very tardily. Spikelets several-flowered, disarticulating above the glumes, but usually not between the florets; upper glume with an oblique awn from just below the tip; lemmas strongly keeled, membranous, glabrous, acute to shortly awned, the tip often recurved. Grain with free pericarp.

13 species around the Indian Ocean from Natal to northern India, one species in Australia and one cosmopolitan weed.

1. Annual, sometimes stoloniferous 2
− Stoloniferous perennial 3
2. Inflorescence open, the racemes 1.2−6.5(−7.5) cm long, linear to narrowly oblong, ascending or radiating; lemmas acute and with a stout cusp or mucro; grain transversely rugose 1. *D. aegyptium*
− Inflorescence compact, the racemes 0.8−1.8 (−2.6) cm long, oblong to broadly oblong, clustered in a dense head; lemmas conspicuously acuminate-mucronate; grain granular or granular-striate .... 2. *D. aristatum*
3. Leaves short, rigid, distichous and pungent.......
.... 4. *D. robecchii*
− Leaves soft or rigid, but if rigid then neither distichous nor pungent 4
4. Inflorescence a compact head; racemes oblong, 0.8−2 cm long, slightly falcate ...3. *D. scindicum*
− Inflorescence open; racemes linear to narrowly oblong, 2−7 cm long, straight or slightly falcate .... 5. *D. geminatum*

1. **D. aegyptium** (L.) Willd. (1809).

Cynosurus ciliaris Hook. f. (1896), nom. prov.; *Dactyloctenium ciliare* Chiov. (1929, 1932), nom. nud., based on the preceding.

Hurbunle, sadeho (Som.).

Annual, often stoloniferous and mat-forming, up to 70(−100) cm high; leaves papillose-hispid, especially on the margins. Inflorescence comprising (1−)3−9 linear to narrowly oblong racemes 1.2−6.5(−7.5) cm long, these ascending or radiating. Spikelets broadly ovate, 3.5−4.5 mm long; glumes subequal, 1.5−2.2 mm long, the upper with an awn from half as long to as long as the body; lemmas narrowly ovate in profile, 2.6−4 mm long, acute and with a stout cusp or mucro up to 1 mm long; anthers 0.25−0.8 mm long. Grain transversely rugose.

Sandy and alluvial soils in flood-plains and areas of cultivation; 0−1800 m. N1, 3; S1−3; tropical and warm temperate Old World. Glover & Gilliland 22; Hemming 1600; Eagleton 32; Dahir ODA12; Gillett & al. 25314.

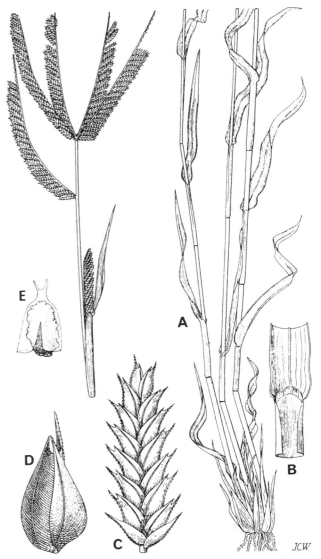

Fig. 101. *Acrachne racemosa*. A: habit, × 0.55. B: ligule, × 1.6. C: spikelet, × 7. D: lemma, × 11. E: grain showing free pericarp, × 16. − Modified from a drawing for Fl. Zamb. by J. C. Webb.

2. **D. aristatum** Link (1827).

D. radulans auct. non (R. Br.) P. Beauv.

Caws farqoole, sadeho (Som.).

Sprawling, geniculately ascending annual, often rooting from the lower nodes, up to 40 cm high; leaves conspicuously papillose-hispid. Inflorescence comprising (2−)4−7(−11) oblong or broadly oblong racemes 0.8−1.8(−2.6) cm long, these clustered in a dense, often ovoid head. Spikelets broadly ovate, 4.1−5.2 mm long; glumes subequal, 1.7−2.3 mm long, the upper with an awn as long as or longer than the body; lemmas lanceolate to narrowly ovate in profile, (3−)3.3−4.3 mm long, acuminate and often with a stout mucro up to 1 mm long; anthers 0.3−0.5 mm long. Grain granular or granular-striate.

Fig. 102. *Dactyloctenium geminatum*. A: habit, × 0.6. B: spikelet, × 9.5. C: grain, × 16. — Modified from Fl. Trop. E. Afr. (1974). Drawn by M. E. Church.

Coastal dunes, sandy soils, limestone outcrops and disturbed ground; 0—550 m. N1, 3; C1, 2; S1—3; eastern Africa to north-western India. Hemming 2358; Puccioni & Stefanini 2; Thulin, Hedrén & Abdi Dahir 7274; Kuchar 17594; Moggi & Bavazzano 90; Friis & al. 5030; Rose Innes 784.

3. **D. scindicum** Boiss. (1859).
D. glaucophyllum Courbai (1862).
Dooyo, saddexo (Som.).
Stoloniferous mat-forming perennial up to 45 cm high; leaves tough and rather glaucous, flat or folded, scattered-papillose-hispid. Inflorescence comprising 3—4(−5) slightly falcate, oblong racemes 0.8—2 cm long, these forming a compact head. Spikelets broadly lanceolate to ovate, 4—8 mm long; glumes sub-

equal, 1.5—2.5 mm long, the upper with an awn half as long to as long as the body; lemmas lanceolate in profile, 3—3.8 mm long, acute and with a mucro up to 0.8 mm long; anthers 1.1—2 mm long. Grain transversely rugose.

Grassland and *Acacia-Commiphora* shrubland on shallow sandy or alluvial soils overlying limestone or gypsum; 50—1400 m. N1—3; C1, 2; S1—3; north-eastern Africa through Arabia to north-western India. Gillett 3903; Bally & Melville 15900; Lavranos 7250; Herlocker 450; Rose Innes 828; Wieland 1136; Elmi 463; Hemming 413.

4. **D. robecchii** (Chiov.) Chiov. (1929); *Eleusine robecchii* Chiov. (1896); types: N3, Robecchi-Bricchetti s.n. (FT syn.).
Gubangub (Som.).
Tough, mat-forming or subshrubby perennial with scaly stolons, up to 30 cm high; leaves short, distichous, rigid and pungent. Inflorescence comprising (2−)3—4(−5) oblong racemes 0.8—2 cm long, these forming a compact head. Spikelets broadly ovate, 2.8—4.2 mm long; glumes subequal, 1—2.5 mm long, the upper with an awn half as long to as long as the body; lemmas elliptic in profile, 2.4—4 mm long, acute, apiculate or mucronate; anthers 1.2—1.4 mm long. Grain granular.

Alluvial plains and rocky slopes, especially on limestone; 300—1400 m. N1—3; Yemen, Socotra and Oman. Bettini s.n.; McKinnon 103; Hansen & Heemstra 6269.

5. **D. geminatum** Hack. (1899). Fig. 102.
D. glaucophyllum var. *somalicum* Chiov. in Ann. Bot. 13: 371 (1915); types: S2, Scassellati 142, Paoli 379, 1334 (all FT syn.).
D. bogdanii S.M. Phillips (1974).
Stoloniferous mat-forming perennial up to 110 cm high; leaves glaucous, flat, tough and glabrous (except for the papillose-hispid margins) or softer and pilose. Inflorescence comprising (1−)2—4(−6) linear to narrowly oblong, straight or slightly falcate racemes 2—7 cm long, these ascending or diverging. Spikelets elliptic to ovate, 3—6.7 mm long; glumes subequal, 1.3—2.2 mm long, the upper with an awn 2/3−1 1/4 times the length of the body; lemmas lanceolate in profile, 2.8—4 mm long, subacute to cuspidate, often with a mucro up to 0.75 mm long; anthers 1.1—2.3 mm long. Grain transversely rugose.

Sandy and alluvial soils, along water-courses, in cultivated fields, on alkaline flats and limestone escarpments; 0—450 m. C2; S1—3; eastern Africa. Kuchar 17150; Beckett 1804; Rose Innes 700; Hemming & Deshmukh 197.

In Kenya and Tanzania this species shows a marked tendency towards differentiation into two taxa. *D. geminatum* s. str. typically has 2 racemes and a subacute lemma with glabrous or scaberulous keel; the segregate taxon, called *D. bogdanii*, has up to 6

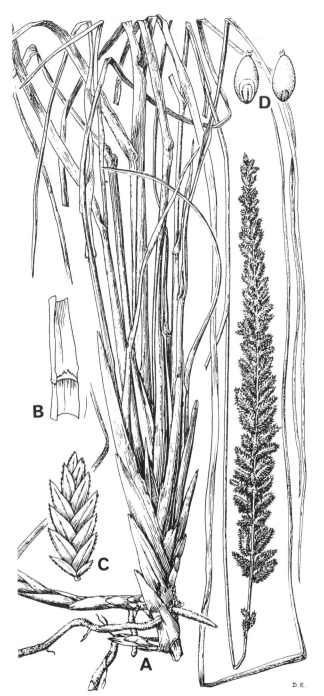

Fig. 103. *Desmostachya bipinnata*. A: habit, × 2/3. B: ligule, × 3. C: spikelet, × 6. D: grains, × 12. − Modified from a drawing by D. Erasmus.

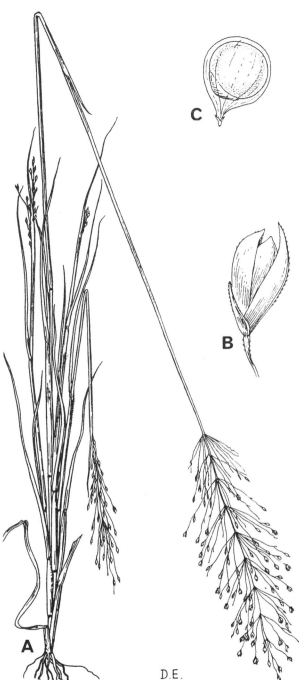

Fig. 104. *Sporobolus panicoides*. A: habit, × 2/3. B: spikelet, × 16. C: grain, moistened to show mucilaginous pericarp, × 16. − Modified from Fl. Trop. E. Afr. (1974). Drawn by D. Erasmus.

racemes and an acute or cuspidate lemma with the scabrid keel extended into a mucro up to 0.75 mm long. However, in Somalia the distinguishing features break down and the recognition of more than one taxon is untenable. Clearly *Dactyloctenium* is actively speciating in eastern Africa and further taxonomic complications will be encountered further south.

32. DESMOSTACHYA (Hook. f.) Stapf (1898)

Perennial; ligule a line of hairs. Inflorescence comprising numerous racemes on a long central axis, the spikelets densely imbricate. Spikelets several-flowered, sessile, falling entire; glumes acute; lemmas keeled, glabrous, acute. Caryopsis ovoid.

One species only.

191

D. bipinnata (L.) Stapf (1900). Fig. 103.

Harsh, tussock-forming rhizomatous perennial up to 150 cm high; lower sheaths leathery, often densely flabellate. Inflorescence up to 60 cm long, the racemes 1—4 cm long, erect or curving outwards from the main axis. Spikelets narrowly ovate to linear-oblong, 3—10 mm long; glumes unequal, the lower 0.7—1.5 mm long, the upper 1.1—2 mm long; lemmas straw-coloured or suffused with purple, 1.8—2.7 mm long.

Coastal saltmarsh; c. 5 m. N2; Old World tropics. Gillett & Watson 23865.

33. SPOROBOLUS R. Br. (1810)

Annual or perennial; ligule a line of hairs. Panicle open or contracted, rarely spike-like, exserted from the uppermost sheath. Spikelets 1-flowered, usually without rhachilla-extension, fusiform, disarticulating below the floret; glumes deciduous, awnless, the upper usually the longer; lemma thinly membranous and often shiny, 1-nerved, entire, awnless; anthers 2—3. Grain globose to ellipsoid, rounded or truncate, not beaked; pericarp free, commonly swelling when wet and expelling the seed to the tip of the spikelet.

About 160 species in the tropics and subtropics.

1. Panicle-branches, or at least the lowermost, in whorls; panicle open or sometimes loosely contracted 2
— Panicle-branches not whorled, not even the lowermost 6
2. Perennial .... 5. *S. ioclados*
— Annual 3
3. Grain spherical or subglobose 4
— Grain obovate 5
4. Grain 0.85—1(-1.5) mm in diam.; spikelets 2—2.7 mm long .... 1. *S. panicoides*
— Grain 0.3 mm in diam.; spikelets 0.7 mm long .... 2. *S. minimus*
5. Leaves 0.5—0.8(−1.3) mm wide .... 3. *S. minutus*
— Leaves 3—7 mm wide .....4. *S. coromandelianus*
6. Panicle dense, spike-like or the primary branches bearing spikelets or branchlets from the extreme base 7
— Panicle loose and diffuse, or if with the spikelets clustered then the primary branches bare of spikelets or branchlets for at least 1/4 their length ... 13
7. Leaves pungent, often tightly rolled; panicle spike-like (if plant forming dense cushions up to 10 cm deep and with short broad flat leaves, see 16. *S. tourneuxii*) 8
— Leaves not pungent 9
8. Panicle smoothly cylindrical; leaves not conspicuously distichous .... 6. *S. spicatus*
— Panicle untidily cylindrical; leaves conspicuously distichous .... 7. *S. virginicus*
9. Lower glume as long as the spikelet or almost so, 1-nerved; the upper similar 10

— Lower glume much shorter than the spikelet, up to 1/2 as long, hyaline and nerveless; the upper up to 3/4 as long 11
10. Stems robust, 3.5—8 mm in diam. at the base; glumes acute .... 8. *S. consimilis*
— Stems wiry, c. 1 mm in diam. at the base; glumes acuminate .... 9. *S. helvolus*
11. Basal sheaths becoming fibrous with age.......... ....10. *S. pellucidus*
— Basal sheaths broad and papery, not becoming fibrous with age 12
12. Panicle dense, the primary branches appressed ....11. *S. brockmanii*
— Panicle open, the primary branches spreading or ascending ....12. *S. fimbriatus*
13. Spikelets clustered about the branches or branchlets 14
— Spikelets evenly distributed, not clustered, the panicle diffuse 21
14. Spikelets appressed or secund along much of the length of the branches or branchlets 15
— Spikelets gathered into clumps at the tips of the branches or branchlets 18
15. Basal sheaths becoming fibrous with age........ ....13. *S. angustifolius*
— Basal sheaths not becoming fibrous with age ...16
16. Stems 1—2 m high, 3—6 mm in diam. at the base ....14. *S. macranthelus*
— Stems up to 1 m high, 1—3 mm in diam. at the base 17
17. Leaves firm, 2—5 mm wide 12
— Leaves soft, 6—9 mm wide .....15. *S. agrostoides*
18. Panicle ± contracted, the primary branches short and stiff 19
— Panicle loose and open, the branches and branchlets slender, flexuous 20
19. Plants densely tufted, forming cushions up to 10 cm deep; leaves short, broad and densely imbricate ....16. *S. tourneuxii*
— Plants loosely tufted with extensively creeping stolons; leaves slender, not at all imbricate...... ....17. *S. ruspolianus*
20. Basal sheaths becoming indurated and often invested with fibres; plants densely tufted from a short oblique rhizome ....18. *S. nervosus*
— Basal sheaths neither indurated nor fibrous; plants with slender creeping rhizomes and weak, often geniculate, stems .... 19. *S. confinis*
21. Leaves pungent 22
— Leaves not pungent 23
22. Lower glume broadly ovate, embracing the base of the spikelet and not more than 1/4 its length; spikelets 1.2—2 mm long .... 23. *S. compactus*
— Lower glume oblong, narrower than the spikelet and at least 1/2 its length; spikelets 1.7—2.3(−2.8) mm long ....24. *S. somalensis*
23. Delicate annual ....20. *S. tenuissimus*
— Tufted perennial 24

24. Basal sheaths at most shortly ciliate on the margins, eventually disintegrating into thin brown fibres, these not woolly-tomentose within .... 21. *S. festivus*
— Basal sheaths woolly-tomentose with curly hairs on the margins, eventually disintegrating into coarse yellow fibres, the woolly hairs retained within .... 22. *S. airiformis*

1. **S. panicoides** A. Rich. (1850). Fig. 104.
Agar (Som.).
Annual up to 75 cm high; leaves flat. Panicle narrowly elliptic, 4−19 cm long; primary branches in whorls, those of the lowermost whorl bearing only 1 or 2 spikelets, sometimes barren. Spikelets 2−2.7 mm long; lower glume narrowly ovate, (0.8−)1.3−1.7 mm long; upper glume ovate, as long as the spikelet; lemma elliptic-ovate, 1.6−2 mm long; anthers 3, 0.9−1.4 mm long. Grain spherical, 0.85−1(−1.5) mm in diam.
Stony hillsides; up to 1600 m. N1; S3; East tropical Africa and Arabia. Hemming 2269; Tardelli & Bavazzano 797.

2. **S. minimus** Cope (1992).
Dwarf annual up to 14 cm high; leaves flat, rigid, 1.5−4 mm wide. Panicle oblong, 1.5−4 cm long; primary branches in whorls, the lowest not barren. Spikelets 0.7 mm long; lower glume elliptic, as long as the spikelet; upper glume broadly elliptic, fractionally shorter than the lower; lemma similar to the upper glume; anthers 3, 0.1 mm long. Grain subglobose, 0.3 mm in diam.
Sparse *Acacia* bushland overlying gypsum, and in dry water-courses of limestone silt and sand; 0−10 m. N1, 3; S1; Arabia. McKinnon 83; Beckett 562; Hemming & Deshmukh 336.

3. **S. minutus** Link (1827).
Annual up to 16 cm high. Leaves flat, 0.5−0.8(−1.3) mm wide, glabrous. Panicle narrowly elliptic, 2−4 cm long; primary branches in whorls, the lowest never barren. Spikelets 0.95−1.4 mm long; lower glume elliptic-ovate, 0.3−0.5 mm long; upper glume narrowly ovate, as long as the spikelet; lemma ovate, 0.8−1.2 mm long; anthers 2, c. 0.2 mm long. Grain obovate, 0.5−0.6 mm long.
Sandy and gravelly, often saline, soils; c. 180 m. N2; C1; S1, 3; eastern Africa and Arabia. Glover & Gilliland 194; Wieland 4577; Thulin & Bashir Mohamed 6964; Ciferri 142.

4. **S. coromandelianus** (Retz.) Kunth (1829).
Annual up to 20 cm high; leaves flat, 2−4 mm wide. Panicle ovate, 2−8 cm long; primary branches in whorls, the lowest never barren. Spikelets 1−1.4 mm long; lower glume oblong, 0.1−0.5 mm long; upper glume oblong-elliptic, as long as the spikelet; lemma similar to the upper glume but a little shorter;

anthers 3, 0.2−0.3 mm long. Grain obovate, 0.7−0.8 mm long.
Coastal dunes; 0−15 m. S2; Old World tropics, mainly India and southern Africa. Paoli 1341.

5. **S. ioclados** (Nees ex Trin.) Nees (1841).
S. marginatus Hochst. ex A. Rich. (1850).
S. arabicus Boiss. (1853).
S. genalensis Chiov. (1940); type: S2, Bigi 22 (FT syn.).
Iyon-jabi, timo-hwelli, timo-nagodleh, xilfo (Som.).
Tussocky perennial, often with creeping stolons, up to 80 cm high; leaves flat or rolled, harsh or soft, often pungent; basal sheaths persistent, chartaceous, often keeled and flabellate. Panicle narrowly ovate to pyramidal, 4−15 cm long; primary branches in whorls (at least the lowermost), bare of spikelets in the lower 1/4−1/3. Spikelets 1.5−3(−3.3) mm long; lower glume narrowly ovate to narrowly oblong, 0.5−1 mm long, obtuse; upper glume narrowly ovate to lanceolate, (1/2−)2/3 as long to as long as the spikelet; lemma lanceolate-elliptic, as long as the spikelet or almost so; anthers 3, 0.9−1.4 mm long. Grain ellipsoid, 0.8−1.2 mm long.
Grassland and low shrubland on silty, sandy and stony soils, gypsum plains and limestone pavements; 15−1600 m. N1, 2; C1, 2; S1−3; tropical Africa to India and Sri Lanka. Wood 71/11; Collenette 107; Herlocker 331; Rose Innes 799; Gillett & Hemming 24329; Kuchar 17689; Hemming & Deshmukh 65.

6. **S. spicatus** (Vahl) Kunth (1829).
Af-rukh (Som.).
Tufted, sometimes mat-forming perennial; stems wiry, up to 70 cm high, often with fascicles of shoots at the nodes, arching over and rooting to form long looping stolons; leaves stiff and pungent, usually inrolled but sometimes flat, often painfully spiny. Panicle spike-like, smoothly and compactly cylindrical, 15−20 cm long, the branches closely appressed. Spikelets (1.7−)2.2−2.8 mm long; lower glume broadly ovate to lanceolate, 0.5−0.9 mm long; upper glume narrowly ovate to narrowly lanceolate, 2/3 as long to as long as the spikelet; lemma oblong-elliptic, as long as the spikelet; anthers 3, 1−1.5 mm long. Grain ellipsoid, 0.7−1.1 mm long.
Sandy deserts, gypsum plains and coastal dunes, and in areas of brackish, saline or alkaline water; 0−950 m. N1−3; C1, 2; S2, 3; Africa to Arabia and India. Bally & Melville 16126; Beckett 869; Collenette 186; Thulin & Abdi Dahir 6507; Rose Innes 803; Gillett & Hemming 24497; Rose Innes 750.

7. **S. virginicus** (L.) Kunth (1829).
Rhizomatous perennial up to 30 cm high; leaves conspicuously distichous (especially on sterile shoots), inrolled, stiff and pungent. Panicle spike-

like, untidily cylindrical, 2—10 cm long, the branches closely appressed. Spikelets 1.7—2.6 mm long; lower glume lanceolate, acute, 1.2—2 mm long, but variable even within the panicle; upper glume narrowly ovate-elliptic, acute, as long as the spikelet; lemma similar to the upper glume and equal to it; anthers 3, 1—1.3 mm long. Grain subglobose, c. 0.7 mm in diam.

Saline soils on the coast; sea-level. S2, 3; tropics and subtropics throughout the world. Pignatti s.n.; Rose Innes 749.

Doubtfully distinct from the Mediterranean *S. arenarius* (Gouan) Duval-Jouve (syn. *S. pungens* (Schreb.) Kunth) which has an ovate rather than cylindrical panicle embraced below by the uppermost sheath.

8. S. consimilis Fresen. (1837).

S. nogalensis Chiov. (1928); type: N3, Puccioni & Stefanini 994 (FT holo.).

S. robustus auct. non Kunth.

Aus dur, bieyis, darif bal (Som.).

Robust perennial tussock-grass with short rhizomes, sometimes forming looping stolons, 1.3—3 m high, the stems 3.5—8 mm in diam. at the base; leaves flat, harshly scabrid along the margins. Panicle linear-lanceolate, 25—35 cm long, the primary branches bearing spikelets or branchlets from the base. Spikelets 1.7—2.3 mm long; lower glume 1-nerved, lanceolate, 1.3—2.3 mm long, acute; upper glume narrowly ovate, almost as long as the spikelet; lemma similar to the upper glume, as long as the spikelet; anthers 3, 1—1.3 mm long. Grain narrowly oblong, 0.7—1 mm long, tetragonal in section.

Seasonally flooded areas, often on alkaline or saline soils; 0—950 m. N1—3; Arabia and tropical Africa southwards to South Africa and Namibia. Gillett 4766; Glover & Gilliland 133; Hemming 1633.

Often confused with *S. robustus* Kunth, a species from western African seashores somewhat intermediate between *S. consimilis* and *S. virginicus*.

9. S. helvolus (Trin.) T. Durand & Schinz (1895).

S. senegalensis Chiov. var. *glaucifolius* (Hochst. ex Steud.) Chiov., Fl. Somala 1: 334 (1929).

S. podotrichus Chiov. (1896); type: S Somalia, "Webi", Robecchi-Bricchetti 241 (FT holo.); *S. senegalensis* var. *podotrichus* (Chiov.) Chiov., loc. cit. (1929).

S. trichophorus Gand. (1922); type: S2, Fiori 44 (LY holo.).

S. senegalensis var. *microstachys* Chiov., Pl. Nov. Aethiop.: 251 (1928); type: C2, Puccioni & Stefanini 206 (FT holo.).

Aggagar, dabro, domar, dub derigan, jarbo, saydho (Som.).

Tufted perennial up to 60 cm high, with slender stolons, the stems wiry, about 1 mm in diam. at the base; leaves flat, glaucous. Panicle linear to narrowly lanceolate, 4—12 cm long, the primary branches

bearing spikelets or densely spiculate branchlets from the base, mostly appressed to the axis but sometimes somewhat spreading. Spikelets 1.4—2 mm long; glumes narrowly lanceolate to lanceolate, acuminate, the lower 1-nerved, sometimes a little shorter than the upper; lemma narrowly ovate, acute, as long as the upper glume; anthers 3, 0.6—0.8 mm long. Grain ellipsoid, c. 0.5 mm long.

Low shrubland on sandy or silty loam overlying limestone or gypsum; 10—760 m. N1—3; C1, 2; S1—3; tropical Africa to India. Hansen & Heemstra 6190; Glover & Gilliland 1040; Hansen & Heemstra 6310; Bally & Melville 15352; Rose Innes 841; Gillett & Hemming 24268; Rose Innes & Trump 935; Rose Innes 667.

10. S. pellucidus Hochst. (1855).

Garogaro, ramass (Som.).

Densely tufted perennial up to 75 cm high; basal sheaths disintegrating into a cushion of fibres. Panicle narrowly lanceolate, 6—20 cm long, the primary branches 1—2 cm long, spreading or ascending, subsecund, branched from the base, the branchlets densely spiculate from the base. Spikelets 1.7—2 (—2.2) mm long; lower glume oblong and obtuse to ovate and acute, 0.5—0.7 mm long; upper glume narrowly ovate, acute, 1/2—3/4 the length of the spikelet; lemma ovate-elliptic, acute, as long as the spikelet; anthers 3, 0.8—0.9 mm long. Grain obovoid to ellipsoid, 0.6—0.8 mm long.

Acacia-Commiphora bushland in red sand overlying limestone; 50—1800 m. N1, 2; C2; S1—3; tropical Africa and Arabia. Hemming 2143; Glover & Gilliland 978A; Beckett 179; Friis & al. 4876; Thulin & Dahir 6738; Popov 1064.

11. S. brockmanii Stapf (1907); type: N1, "Golis Range", Drake-Brockman 11 (K holo.).

S. capensis (P. Beauv.) Kunth var. *altissimus* Chiov. in Ann. R. Ist. Bot. Roma 6: 168, t. 11 (1896); type: N Somalia, Robecchi-Bricchetti 572 (FT holo.).

S. indicus var. *microspiculus* Chiov., Fl. Somala 1: 335 (1929); type: C1, Puccioni & Stefanini 576 (FT holo.).

S. indicus auct. non (L.) R. Br.

Gagarariedu, garragarro, saddeho (Som.).

Tufted perennial up to 85 cm high, arising from a short oblique rhizome, the stems 1—1.75 mm in diam. at the base; basal sheaths rather broad and papery; leaves mostly basal, the midrib scarcely more prominent than the secondary nerves. Panicle linear, spike-like, (6—)10—21 cm long, the primary branches 1—4 cm long, the spikelets appressed along the whole length of the secondary branchlets, these clustered about the whole length of the primary or the latter naked in the lower 1/3. Spikelets 1.7—2.1(—2.4) mm long; lower glume lanceolate or lanceolate-oblong, 0.8—1.3 mm long; upper glume similar in shape, about 2/3—3/4 the length of the spikelet; lemma narrowly ovate, as long as the spikelet; anthers 3, 1—1.4 mm long. Grain not known.

Grassland and juniper forest on the deeper soils at 1300—1650 m, and coastal plains at 0—50 m. N1, 2; C1; not known elsewhere. Glover & Gilliland 612; Lavranos 6876; Thulin & Abdi Dahir 6651.

Two populations exist in Somalia: one at altitudes of between 1300 and 1650 m in the northern mountains, and another on coastal plains between Hobyo and Ceel Dheere. Apart from the obvious difference in habitat, the coastal population differs in having the narrower leaves with a long tapering point characteristic of *S. africanus* (Poir.) A. Robyns & Tournay, but it does have the longer upper glume found in the northern population. It is possible that the two populations represent different taxa, but the whole complex, including *S. fimbriatus* and *S. agrostoides*, is very poorly understood at present.

12. S. fimbriatus (Trin.) Nees (1841).

Debo welodle (Som.).

Tufted perennial up to 100 cm high, arising from a short oblique rhizome, the stems 2—3 mm in diam. at the base; basal sheaths broad and papery; leaves mostly basal, usually flat, the midrib often prominent as a white streak above, the secondary nerves obscure. Panicle linear to lanceolate, 15—50 cm long, the primary branches 2—9(—15) cm long, the spikelets appressed along the whole length of the secondary branchlets, these clustered about the whole length of the primary or the latter naked in the lower 1/3. Spikelets 1.4—2.2 mm long; lower glume narrowly oblong to lanceolate, nerveless, 0.6—1.4 mm long; upper glume narrowly ovate, 2/3 as long to as long as the spikelet; lemma narrowly ovate, slightly shorter than to as long as the spikelet; anthers 3, 0.9—1.2 mm long. Grain obovoid, truncate at the top, c. 0.6 mm long, tetragonal in section.

Commiphora scrub at c. 800 m. N1; Sudan southwards to South Africa. Gillett 4537.

Two further collections, Drake-Brockman 396 and 397, can probably be referred to this species, although the upper glume is scarcely more than half as long as the lemma. Such a glume-length is characteristic of *S. africanus*, but the habit of this plant — with its very compact panicle — is quite different.

13. S. angustifolius A. Rich. (1850).

Densely tufted perennial up to 60 cm high; basal sheaths becoming fibrous with age. Panicle linear to narrowly ovate, 5—16 cm long, the spikelets ± secund and loosely contracted about the branchlets, the primary branches bare of spikelets or branchlets in the lower 1/4—1/3. Spikelets 2—3 mm long; lower glume oblong to narrowly ovate, 0.8—1.4 mm long; upper glume narrowly ovate, 1/2—3/4 the length of the spikelet; lemma narrowly ovate, as long as the spikelet; anthers 3, 0.7—1.2 mm long. Grain ellipsoid, truncate, 1—1.2 mm long.

Shallow soil. S1; tropical Africa and Arabia. Wieland 1173.

14. S. macranthelus Chiov. (1932); type: S3, Senni 276 (FT holo.).

Robust, tufted perennial 1—2 m high, the stems 3—6 mm in diam. at the base; basal sheaths broad and papery; leaves narrow, stiff and somewhat glaucous, the midrib commonly forming a prominent white streak 1—2 mm wide on the upper surface, the margins harshly scabrid. Panicle linear to narrowly ovate, 35—75 cm long, the spikelets appressed or somewhat spreading and secund along the branches and branchlets, the former bare for the lower 1—2 cm. Spikelets 1.6—2.4 mm long; lower glume narrowly ovate, 0.8—1.2 mm long; upper glume narrowly ovate, 3/4—4/5 as long as the spikelet; lemma ovate, as long as the spikelet; anthers 3, c. 1 mm long. Grain ellipsoid, c. 0.6 mm long, tetragonal in section.

Deciduous bushland; c. 100 m. S3; eastern and southern tropical Africa. In Somalia known only from the type from Kolbio.

Similar to *S. fimbriatus* and distinguished by the robust stems, stiff narrow leaves and much-branched panicle.

15. S. agrostoides Chiov. (1897); type: S1, "Sidlei", Riva 1207 (FT holo.).

Tufted perennial up to 140 cm high, arising from a short oblique rhizome; basal sheaths broad and papery; leaves 6—9 mm wide, soft and shining, usually smooth. Panicle lanceolate to narrowly ovate, 15—40 cm long, graceful, rather diffuse, the primary branches filiform and bare for much of their length, the spikelets ± appressed along the length of the secondary and tertiary branchlets. Spikelets 1.6—2 mm long; lower glume lanceolate, 0.7—1.3 mm long; upper glume narrowly ovate, 3/4—4/5 the length of the spikelet; lemma narrowly ovate, as long as the spikelet; anthers 3, c. 0.8 mm long. Grain ellipsoid, c. 0.7 mm long.

Alluvial soils and as a weed of vines and sugar-cane. S1—3; East Africa from Somalia to the eastern border of Zaire. Terry 3472; Rose Innes 660; Paoli 314.

16. S. tourneuxii Coss. (1889). Fig. 105.

Gorof, gubangub, tuur-cadde (Som.).

Tufted perennial, often forming dense cushions up to 10 cm deep; stems subwoody and much-branched, the fertile rising to 20 cm; basal sheaths persistent, densely imbricate; leaves flat, 1—2(—3) cm × 2—3 mm, glaucous, pungent. Panicle ovate or narrowly ovate, 1—4 cm long, open or contracted, the primary branches short and stiff, ascending or spreading, bare of spikelets in the lower 1/3—1/2 (not obvious when panicle contracted), the spikelets gathered into clumps at the tips. Spikelets 1.8—2.1 mm long; glumes subequal, 1/2—2/3 the length of the spikelet, the

Fig. 105. *Sporobolus tourneuxii*. A: habit, × 2/3. B: ligule, × 6. C: spikelet, × 24. — Modified from Fl. Iranica 70, t. 66 (1970).

lower ovate to lanceolate, the upper oblong-ovate to narrowly elliptic; lemma oblong-lanceolate to elliptic, as long as the spikelet; anthers (2−)3, 1−1.5 mm long. Grain obovoid, 0.5−0.6 mm long.

Shallow soils overlying limestone or gypsum; 580−1680 m. N1−3; North Africa through Arabia to north-western India. Beckett 1159; Lavranos 9125A; Kazmi & al. 5718.

17. **S. ruspolianus** Chiov. (1906).
S. fruticulosus Stapf (1907); type: N1, Appleton s.n. (K holo.).
Caws-digimaale, sifar (Som.).
Subshrubby, loosely tufted extensively stoloniferous perennial, the slender woody stems arising from a knotty base and usually branched from the lower nodes, up to 40 cm high; lower sheaths brown, papery, persistent, not at all imbricate, the very base of the stem clothed in whitish cataphylls; leaves flat, narrow, glaucous, not pungent. Panicle ovate or narrowly ovate, 2.5−6.5 cm long, loose or contracted, the primary branches short and stiff, ascending or spreading, bare in the lower 1/3−1/2, the spikelets gathered into clumps at the tips. Spikelets 1.4−2.3 mm long; glumes narrowly ovate, acute or obtuse, the lower 0.9−1.3 mm long, the upper 2/3−3/4 the length of the spikelet; lemma narrowly ovate, as long as the spikelet; anthers 3, 0.9−1.1 mm long.

Seasonally flooded calcareous or gypseous soils; 160−2150 m. N1−3; C1; S1, 3; Ethiopia, Socotra and Oman. Gillett 4148; McKinnon 91; Kazmi & al. 5694; Herlocker 443; Wieland 1120; Kazmi 5128.

18. **S. nervosus** Hochst. (1855). Fig. 106.
S. longibrachiatus Stapf (1907); type: N1, Appleton s.n. (K holo.).
Hamasut, madwedle, maheen suga, ramass (Som.).
Densely tufted perennial arising from a short oblique rhizome, up to 45 cm high; basal sheaths yellowish, indurated, often covered by the fibrous remains of decaying sheaths; leaves 1−3 mm wide. Panicle ovate or narrowly ovate, 7−20 cm long, the primary branches slender, flexuous, bare below, the spikelets secund and clustered at the tips of the branchlets. Spikelets (1.5−)1.7−2.1(−2.4) mm long; lower glume lanceolate, acute, 0.8−1 mm long; upper glume narrowly ovate, acute, 2/3−4/5 the length of the spikelet; lemma narrowly ovate, as long as the spikelet; anthers 3, 0.8−1 mm long. Grain ellipsoid, c. 0.6 mm long.

Acacia-Commiphora scrub on shallow soils overlying gypsum or limestone; alluvial plains; 170−1830 m. N1−3; C1, 2; S1; tropical Africa, Namibia and Arabia. Gillett 4820; Keogh 129; Thulin & Warfa 5854; Beckett 221; Martin 9; Thulin & Bashir Mohamed 6845A.

Similar to *S. angustifolius* but with generally smaller spikelets and relatively longer upper glume, and with the spikelets clustered at the tips of the branchlets rather than secund along their whole length. Spikelets over 2.1 mm are unusual, but the specimens concerned are clearly not *S. angustifolius*; it is doubtful whether it is worth recognising them as a distinct taxon because of so many other uncertainties in the systematics of this part of *Sporobolus*.

19. **S. confinis** (Steud.) Chiov. (1908).

Perennial with long slender creeping rhizomes; stems erect or geniculately ascending, up to 60 cm high; basal sheaths neither fibrous nor indurated; leaves 2−6 mm wide. Panicle narrowly ovate, 8−15 cm long, the primary branches slender, flexuous, bare below, the spikelets secund and clustered at the tips of the branchlets. Spikelets 1.6−2 mm long; lower glume lanceolate to narrowly oblong, 0.7−1.2 mm long; upper glume narrowly ovate, 3/4−4/5 the length of the spikelet; lemma narrowly ovate, as long as the spikelet; anthers 3, 0.9−1.1 mm long. Grain ellipsoid, 0.7−0.8 mm long.

Grassland with dwarf shrubs; 50 m. S3; tropical East Africa and Arabia. Deshmukh 400.

Known from Somalia from a single incomplete specimen.

20. **S. tenuissimus** (Schrank) Kuntze (1893).

Weak annual up to 80 cm high. Panicle narrowly oblong, 10−40 cm long, very delicate and effuse, the branches and pedicels capillary. Spikelets 0.8−1.1 mm long; lower glume oblong, raggedly truncate, 0.1−0.3 mm long; upper glume ovate-oblong, subacute, about 1/2 the length of the spikelet; lemma ovate, as long as the spikelet; anthers 3, 0.1−0.2 mm long. Grain obovoid, 0.4−0.5 mm long, truncate.

S2; tropical Africa, tropical Asia eastwards to Burma and tropical America. Pignatti s.n.

21. **S. festivus** Hochst. ex A. Rich. (1850).

Geedho (Som.).

Densely tufted perennial up to 60 cm high; basal sheaths becoming fibrous with age, glabrous within, their margins glabrous or shortly ciliate. Panicle narrowly ovate, 3−22 cm long, very delicate and diffuse, the branches capillary. Spikelets 1−1.5 mm long; lower glume narrowly ovate to oblong, 0.4−0.6 mm long; upper glume narrowly ovate, 1/2−2/3 the length of the spikelet; lemma narrowly ovate, acute, as long as the spikelet; anthers 3, 0.6−0.8 mm long. Grain ellipsoid to obovoid, 0.4−0.7 mm long.

N1, 3; tropical Africa and Arabia southwards to South Africa and Namibia. Appleton s.n.; Barbier 1092.

22. **S. airiformis** Chiov. (1939); type: N1, "Buramo" to "Warieto", Gillett 4898 (K syn.).

Densely tufted perennial up to 35 cm high; basal sheaths woolly-tomentose with curly hairs on the margins, horny and yellow, eventually forming a

Fig. 106. *Sporobolus nervosus*. A: habit, × 0.6. B: spikelet, × 12.5. − Modified from Fl. Trop. E. Afr. (1974). Drawn by D. Erasmus.

cushion of coarse fibres, the woolly hairs retained within. Panicle ovate, 5−8 cm long, diffuse, the branches capillary. Spikelets 1.8−2 mm long; lower glume narrowly ovate to oblong-ovate, 0.5−0.8 mm long; upper glume narrowly ovate, 1/3−1/2 the length of the spikelet; lemma narrowly ovate, obtuse or subacute, as long as the spikelet; anthers 3, 1−1.2 mm long. Grain obovoid, c. 1 mm long.

Lightly wooded rocky slopes at c. 1350−1500 m. N1, 2; Ethiopia and Arabia. Gillett 4898; Bally & Melville 15907.

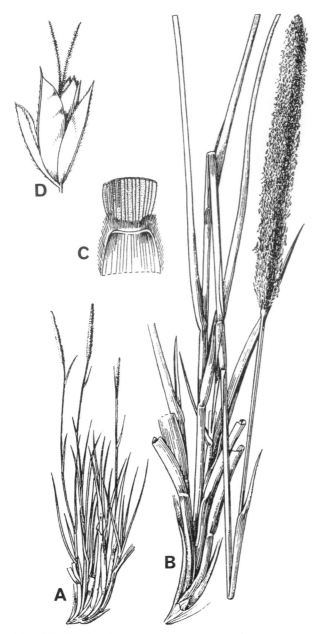

Fig. 107. *Urochondra setulosa*. A: habit, × 1/4. B: part of plant, × 1. C: ligule. D: spikelet, × 16. − Modified from Hook. Ic. Pl. 35: t. 3457 (1947).

23. S. compactus Clayton (1971); type: N2, "Erigavo", McKinnon 89 (K holo.).

Dih (Som.).

Creeping perennial with short scaly stolons, forming compact cushions; stems up to 30 cm high; leaves ± inrolled, rigid, glaucous and pungent. Panicle ovate, 4−10 cm long, diffuse, the branches capillary. Spikelets 1.2−1.7 mm long; lower glume broadly ovate, 0.3−0.5 mm long, embracing the base of the spikelet; upper glume oblong-elliptic, as long as the spikelet, obtuse; lemma oblong-ovate, as long as the spikelet, obtuse; anthers 3, 1−1.3 mm long. Grain ellipsoid, 0.7−1 mm long.

Gypsum plains; 140−1740 m. N2; C1; not known elsewhere. McKinnon 267; Beckett 319.

24. S. somalensis Chiov. (1896).

S. variegatus Stapf (1907); type: N1, "Bohotle" to "Upper Sheikh", Appleton s.n. (K holo.).

Dihi (Som.).

Creeping perennial with short scaly stolons, forming compact cushions; stems up to 30 cm high; leaves usually flat, stiff, glaucous and pungent. Panicle ovate, 2−9 cm long, diffuse, the branches capillary. Spikelets 1.3−2.2 mm long; lower glume oblong, 0.5−1.2 mm long, narrower than the spikelet; upper glume oblong, 1/2−2/3 the length of the spikelet, obtuse; lemma ovate-oblong, as long as the spikelet, truncate and erose at the tip; anthers 3, 1−1.5 mm long. Grain narrowly ellipsoid, c. 1.4 mm long.

Grassland with scattered *Acacia-Commiphora* shrubland on alluvial soils, often overlying limestone; 600−1370 m. N1−3; C1; Ethiopia (Ogaden). Gillett 4082; Hansen & Heemstra 6213; Nugent 49; Hemming 1413.

34. UROCHONDRA C.E. Hubb. (1947)

Perennial; ligule a line of hairs; leaves rigid, inrolled and pungent. Panicle spike-like, cylindrical, exserted from the uppermost sheath. Spikelets 1-flowered without rhachilla-extension, strongly laterally compressed and keeled; glumes narrow, almost as long as the floret, awnless; lemma membranous, 1-nerved, mucronate. Grain ellipsoid, surmounted by a pallid beak, formed from the thickened style-bases, about a third its length; pericarp free, swelling when wet.

A single species only.

U. setulosa (Trin.) C.E. Hubb. (1947); *Sporobolus setulosus* (Trin.) A. Terracc. (1893). Fig. 107.

Afrug, afrus, ferawein (Som.).

Shortly rhizomatous plant up to 90 cm high. Panicle 4.5−16 cm long. Spikelets oblong or obovate-oblong, 2−3 mm long; glumes linear or lanceolate-oblong, obtuse or acute, ciliate on the keel, the lower 1.5−2.5 mm long, the upper 2.6−3 mm long; lemma as long as the spikelet, ciliate on the keel.

Open sandy plains, with saline or alkaline soil; 0−580 m. N1−3; C1, 2; S1−3; coastal regions of the Indian Ocean and the Red Sea from north-eastern Africa to north-western India. Drake-Brockman 445; Glover & Gilliland 118; Puccioni & Stefanini 7; Gillett & al. 22113; Kuchar 17347; Paoli 1016; Bally & Melville 15284; Rose Innes & Trump 904.

Tribe 12. **LEPTUREAE**

Ligule membranous with a ciliate margin. Inflorescence a single cylindrical bilateral raceme, the spikelets all alike, alternate, borne edgeways on and embedded in hollows in the fragile rhachis. Spikelets 1-flowered with rhachilla-extension, dorsally compressed, falling entire; lower glume minute or suppressed (but well developed in the terminal spikelet); upper glume appressed to the axis, exceeding and covering the sunken floret, coriaceous, 5- to 12-nerved, obtuse to caudately awned; lemma 3-nerved, membranous. Caryopsis ellipsoid with free pericarp.

One genus only.

35. LEPTURUS R. Br. (1810)

Perennial, often stoloniferous. Description otherwise as for tribe.

About 10 species in the Old World tropics.

L. repens (G. Forst.) R. Br. (1810). Fig. 108.

Perennial with branched creeping stolons, up to 60 cm high; leaves flat or inrolled, stiff, mostly glaucous, finely acute to pungent. Racemes 4–20 cm long, straight or slightly curved. Spikelets (8–)10–14 (–22)mm long (the terminal up to 28 mm); upper glume narrowly lanceolate to lanceolate or subulate, 8- to 12-nerved, smooth or scaberulous, finely acute to acuminate, tipped with a rigid awn; lemma ovate or ovate-oblong, 4–5.5 mm long, minutely hairy at the base and sometimes on the margins, obtuse; anthers 1.5–2.5 mm long.

Coastal sands. S3; coastal regions of eastern Africa, the Mascarene Islands, Sri Lanka, Malaysia, northern Australia and Polynesia. Moggi & Bavazzano 102; Moggi & al. 396; Tardelli & Bavazzano 564.

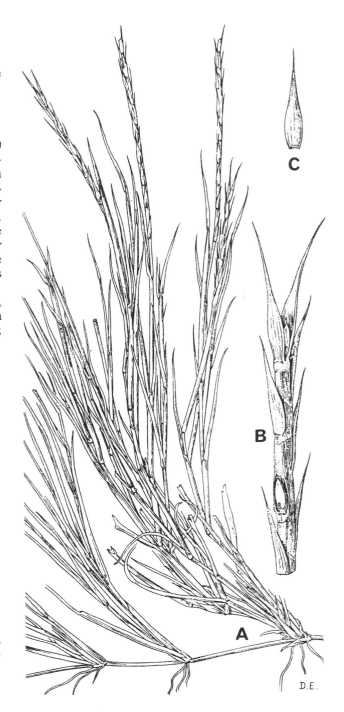

Fig. 108. *Lepturus repens*. A: habit, × 2/3. B: part of spike, × 2. C: upper glume, × 2. — Modified from Fl. Trop. E. Afr. (1974). Drawn by D. Erasmus.

Tribe 13. CYNODONTEAE

Ligule a short membrane with a ciliate margin. Inflorescence of tough unilateral racemes, these solitary, digitate or scattered along an axis, often deciduous, the spikelets all alike. Spikelets with 1 fertile floret (except 36. *Lintonia* & 37. *Tetrapogon*) with or without additional male or barren florets, laterally or dorsally compressed, disarticulating above the glumes but not between the florets, or falling entire; glumes herbaceous to membranous, 1- to 3(−5)-nerved, shorter than the floret or enclosing it, sometimes the lower absent; lemma membranous to coriaceous, 3-nerved (5- to 11-nerved in 36. *Lintonia*), often ciliate on the nerves, entire or 2-lobed, with or without 1−3 terminal or subapical awns, these usually straight. Fruit sometimes with a free pericarp.

59 genera and about 300 species in the tropics, extending into the North American prairies and temperate and subtropical coasts.

1. Spikelets containing 2−5 fertile florets 2
− Spikelets containing 1 fertile floret 3
2. Body of lemma 5- to 11-nerved .... 36. *Lintonia*
− Body of lemma 3-nerved .... 37. *Tetrapogon*
3. Racemes persistent, the spikelets breaking up at maturity 4
− Racemes deciduous or spikelets falling entire 10
4. Lemma concealed, both glumes exceeding and closed around the florets ...... 42. *Microchloa*
− Lemma exposed, one or both glumes shorter than the floret 5
5. Spikelets 1-flowered, awnless .... 43. *Cynodon*
− Spikelets 2- to several-flowered, or the lemma sinuously awned 6
6. Lemma laterally compressed; grain trigonous to subterete 7
− Lemma and grain dorsally compressed 9
7. Lemma produced into a long sinuous awn .... 44. *Schoenefeldia*
− Lemma awnless or with a straight awn 8
8. Upper glume acute to faintly 2-toothed, awnless or rarely with a very short awn; lemma awned, usually pallid .... 38. *Chloris*
− Upper glume obtuse to 2-lobed, distinctly awned; lemma awnless or almost so, dark brown to golden in colour .... 39. *Eustachys*
9. Fertile lemma 1-awned .... 40. *Enteropogon*
− Fertile lemma with deep hyaline awned lobes and also awned from between the lobes................ .... 41. *Afrotrichloris*
10. Spikelets 2-flowered; racemes cuneate .... 45. *Melanocenchris*
− Spikelets 1-flowered 11
11. Lower glume very small or suppressed; upper glume and lemma about equal, with raised nerves bearing hooked bristles .... 46. *Tragus*
− Lower glume well developed 12
12. Glumes bristle-like .... 51. *Tetrachaete*
− Glumes not bristle-like 13
13. Racemes comprising 2−3 spikelets separated by short internodes .... 47. *Dignathia*
− Racemes comprising 1 spikelet or a pair of spikelets side by side 14
14. Spikelets awned .... 50. *Perotis*
− Spikelets awnless 15
15. Racemes pedunculate .... 48. *Leptothrium*
− Racemes sessile .... 49. *Pseudozoysia*

36. LINTONIA Stapf (1911)

Perennial. Racemes 2 or more, digitate or on a short central axis. Spikelets ovoid, with 2−4 fertile florets followed by progressively smaller sterile florets; fertile lemmas membranous, pubescent at least on the flanks, with 5−11 indurated nerves, the laterals coalescent below, emarginate to 2-lobed, with a short subapical awn.

Two species in the dry savannas of eastern Africa.

L. nutans Stapf (1911). Fig. 109.

Tufted perennial with short rhizomes, up to 75 cm high. Racemes (1−)2−4(−6), 4−11.5 cm long. Spikelets 4- to 10-flowered, 5.3−11 mm long; glumes 1-nerved, unequal, the lower narrowly lanceolate, 1.9−4.5 mm long, acuminate, the upper oblong, 3.3−5.8 mm long, obtuse; lemmas broadly elliptic to obovate, 4.4−6.7(−9.2) mm long, tough, 7- to 11-nerved, appressed villous between the nerves below, strigillose above; awn 0.75−10.5 mm long.

Grain elliptic, 2–2.2mm long.

Abandoned fields. S1, 2; eastern and south-eastern Africa. Warfa & Yusuf 1126; Senni 697.

37. TETRAPOGON Desf. (1799)

Annual or perennial. Racemes 1–4, digitate, often hairy. Spikelets laterally compressed, cuneate, with 2–5 fertile florets, the rhachilla terminating in a clavate cluster of sterile lemmas; fertile lemmas rounded or keeled, coriaceous, ciliate on the nerves and keel, sometimes subglabrous, entire or 2-toothed, with a subapical awn. Grain with free pericarp.

Five species in Africa, south-western Asia and India.

1. Lowest lemma shortly ciliate or subglabrous
.... 1. *T. tenellus*
– Lowest lemma conspicuously hairy 2
2. Racemes embraced by the inflated uppermost sheath; glumes conspicuous, 5–12 mm long
.... 4. *T. cenchriformis*
– Racemes exserted; glumes inconspicuous, 2–5 mm long 3
3. Spikelet-hairs pallid .... 2. *T. villosus*
– Spikelet-hairs ferruginous .... 3. *T. ferrugineus*

1. **T. tenellus** (K.D. Koenig ex Roxb.) Chiov. (1908); *Chloris tenella* K.D. Koenig ex Roxb. (1820).
   T. macranthus (Jaub. & Spach) Benth. (1881).
   T. triangulatus (A. Rich.) Schweinf. var. *agowensis* Hochst. ex Chiov. in Ann. R. Ist. Bot. Roma 6: 171 (1896), nom. nud.
   T. triangulatus var. *sericatus* Hochst. ex Chiov., loc. cit., t. 15 (1896).
   Aiya makarreh, aus cherin, halfo, hol, jeebin (Som.).
   Annual up to 50 cm high; basal sheaths keeled but scarcely flabellate. Racemes solitary, occasionally paired, exserted or partially enclosed in the slightly inflated uppermost sheath, 2.5–7 cm long. Glumes inconspicuous, 3–5.5(–6.5) mm long; lowest lemma 3.7–5.5 mm long, shortly ciliate on the nerves and keel or subglabrous; awn 4.5–10.5 mm long.

   Open deciduous bushland on fertile soils, also gardens and abandoned fields; 50–1600 m. N1–3; C1, 2; S1–3; tropical Africa, through Arabia to India. Gillett 4901; Glover & Gilliland 157; Nugent 39; Beckett 203; Martin 12; Wieland 1234; Kazmi 5078; Rose Innes 672.

2. **T. villosus** Desf. (1799).
   Aus damer, aus khansa, aya mukarre, buri wena, hamk-hari, hraf wurub (Som.).
   Densely tufted perennial up to 60 cm high; basal sheaths keeled and conspicuously flabellate. Racemes usually paired (rarely solitary or up to 4) but often interlocked back to back and appearing solitary,

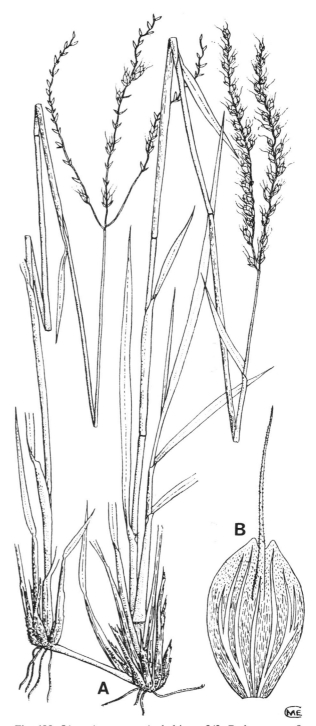

Fig. 109. *Lintonia nutans*. A: habit, × 2/3. B: lemma, × 8. – Modified from Fl. Trop. E. Afr. (1974). Drawn by M. E. Church.

exserted or partially enclosed in the slightly inflated uppermost sheath, 3–8.5 cm long. Glumes inconspicuous, 2–4 mm long; lowest lemma 2.5–3.7 mm long, ciliate on the nerves and keel with pallid hairs 3–8 mm long; awn 8.5–13 mm long.

Sparsely vegetated hillsides and screes of limestone or gneiss, and gypsum plains; 550–1950 m. N1–3. Macaronesia through North and North-east Africa to Arabia and India. McKinnon 38, 132; Bally & Melville 15620.

3. T. ferrugineus (Renvoize) S.M. Phillips (1987); *Chloris ferruginea* Renvoize (1973).

Densely tufted perennial up to 75 cm high, the stems arising from an oblique rhizome; basal sheaths lightly keeled but scarcely flabellate. Racemes up to 8, subdigitate, 4–10 cm long, well exserted from the uppermost sheath, Glumes inconspicuous, 4–5.5 mm long; lowest lemma c. 2.5 mm long, coriaceous, brown, covered with short ferruginous hairs and with a tuft of longer hairs at the tip; awn 3.5–4 mm long.

Limestone hills; 220 m. C2; Kenya and Ethiopia. Beckett 189.

4. T. cenchriformis (A. Rich.) Clayton (1962). Fig. 110.

T. spathaceus (Hochst. ex Steud.) Hack. ex T. Durand & Schinz (1895).

T. geminatus (Hochst.) Chiov. (1896); *T. macranthus* forma *geminatus* (Hochst.) Chiov. in Pirotta, Fl. Eritrea: 353 (1907).

Agar eriorod (Som.).

Loosely to densely tufted annual or short-lived perennial up to 60 cm high; basal sheaths keeled, loosely flabellate. Racemes solitary, rarely paired, partially enclosed in the very inflated uppermost sheath, 3–6 cm long. Glumes conspicuous, 5–12 mm long; lowest lemma 4–6 mm long, ciliate on the nerves and keel with hairs c. 4 mm long; awn 7–11 mm long.

Overgrazed *Acacia-Commiphora* bushland on limestone hillsides and red sand; 600–1680 m. N1, 2; C2; S1, 3; Macaronesia, tropical Africa and Arabia. Glover & Gilliland 103; Bally & Melville 16264; Rose Innes 819; Hemming & Deshmukh 332; Moggi & Bavazzano 1400.

38. CHLORIS Sw. (1788)

Annual or perennial. Racemes digitate or rarely crowded on an elongated axis, the spikelets pectinate or appressed. Spikelets laterally compressed, with 1 fertile floret, sometimes with a smaller male floret, the rhachilla terminating in 1 or more reduced lemmas; glumes acute; fertile lemma keeled, cartilaginous to coriaceous, mostly pallid, often decoratively ciliate on the margins, entire or bilobed, conspicuously awned from the tip or just below it. Grain ellipsoid and trigonous to lanceolate and subterete, the pericarp free (though sometimes reluctantly).

About 55 species in tropical and warm temperate regions of both hemispheres.

1. Racemes numerous, densely crowded on a long axis ....1. *C. roxburghiana*
- Racemes digitate 2
2. Leaves oblong, obtuse 3
- Leaves linear, tapering 5
3. Delicate annual .... 2. *C. pycnothrix*
- Robust, tufted or stoloniferous perennial 4
4. Fertile lemma ciliate on the margins................ .... 3. *C. mossambicensis*
- Fertile lemma quite glabrous ......4. *C. jubaensis*
5. Fertile lemma obliquely obovate in profile, the keel slightly gibbous, with a crown of spreading hairs 1.5–4 mm long just below the tip........... .... 5. *C. virgata*
- Fertile lemma lanceolate to obovate-elliptic in profile, not gibbous, without a crown of spreading hairs though often long- or short-ciliate on the margins 6
6. Spikelets 3- to 4-flowered, 2-awned; fertile lemma ciliate on the margins above; 2nd lemma similar to the fertile but smaller; 3rd and 4th (when present) reduced to clavate awnless scales .... 6. *C. gayana*
- Spikelets 2-flowered, 2-awned; fertile lemma thinly pilose on the flanks; 2nd lemma a minute awned rudiment (at most half the length of the fertile), sometimes reduced to an awned rhachilla-segment ....7. *C. mensensis*

1. C. roxburghiana Schult. (1824).

C. myriostachya Hochst. (1855).

Altuko, anadug, ane kuduq, budewene, iamakari, saddexa (Som.).

Densely tufted perennial up to 1.5 m high; leaves linear, tapering. Inflorescence a compact head of numerous racemes 3–8 cm long on an axis 6–18 cm long. Spikelets 3- to 4-flowered, 3- to 4-awned; fertile lemma elliptic in profile, 1.3–2 mm long, sparsely ciliate on the keel and margins, with an awn 8–17 mm long; 2nd to 4th lemmas progressively smaller, reduced to elliptic glabrous awned scales.

In shrubland on sandy or alluvial soils; 15–1000 m. N1; C1; S1–3; eastern and southern Africa, Arabia and southern India. Gillett 3197; Elmi s.n.; Paoli 381; Alstrup & Michelsen 24; Rose Innes 787.

2. C. pycnothrix Trin. (1824).

Domar, harfo (Som.).

Annual up to 25 cm high, erect or geniculately ascending, often rooting at the nodes; leaves oblong, very obtuse. Inflorescence a head of 4–6 digitate racemes, these 3–6 cm long. Spikelets 2-flowered, 1- to 2-awned; fertile lemma narrowly elliptic in profile, 2.3–2.6 mm long, scabrid on the margins, keel and flanks, with an awn 8–18 mm long; callus rounded, ciliate; 2nd lemma similar to the fertile but much smaller, awned or awnless.

Acacia woodland on limestone hillsides, and as a garden weed; 1300–1400 m. N1; tropical Africa,

Fig. 110. *Tetrapogon cenchriformis*. A: habit, × 2/3. B: lower glume, × 8. C: upper glume, × 8. D: spikelet, with glumes removed, × 8. – From Fl. Trop. E. Afr. (1974). Drawn by M. E. Church.

Arabia and South America. Gillett 3917.

3. C. mossambicensis K. Schum. (1895).

Perennial up to 85 cm high, erect or ascending, tufted, shortly rhizomatous or stoloniferous; leaves oblong, obtuse. Inflorescence a head of (2–)4–5(–7) digitate racemes, these 3–8 cm long. Spikelets 2-flowered, 2-awned; fertile lemma obovate-oblanceolate in profile, 2–3(–4) mm long, ciliate on the margins towards the tip and along the keel, glabrous on the flanks, with an awn (4–)7–11 mm long; callus pungent, pubescent; 2nd lemma smaller than the fertile, clavate, glabrous, with an awn 2.5–7.5 mm long.

Consolidated dunes. S3; Kenya, Tanzania and Mozambique. Moggi & Bavazzano 2716.

Fig. 111. *Chloris virgata.*
A: habit, × 2/3. B: lower
glume, × 14. C: upper
glume, × 14. D: spikelet,
with glumes removed. —
From Fl. Trop. E. Afr.
(1974). Drawn by M. E.
Church.

4. C. jubaensis Cope (1995); type: S3, N of Jana
Cabdalle, Rose Innes 793 (K holo.). Fig. 112 A, B.

Perennial up to 95 cm high, erect or ascending,
tufted or stoloniferous; leaves oblong, obtuse. Inflorescence a head of 5−8 digitate or subdigitate racemes,
these 4−8 cm long. Spikelets 2- to 3-flowered, 2- to
3-awned; fertile lemma obovate-elliptic in profile,
3.4−3.8 mm long, glabrous, with an awn 6−7 mm
long; callus pungent, pubescent; 2nd lemma an
inflated clavate glabrous scale with an awn 4.5−6

mm long; 3rd lemma, if present, similar to the 2nd
but smaller and with a very short awn.

Bushland on dark grey sandy loam with manganese
concretions. S3; not known elsewhere. Gorini 514;
Moggi & Bavazzano 1454.

5. C. virgata Sw. (1797). Fig. 111.

C. virgata var. *elegans* (Kunth) Stapf in Dyer, Fl.
Cap. 7: 642 (1900).

C. multiradiata Hochst. (1855).

Fig. 112. A, B: *Chloris jubaensis*. A: habit, × 1. B: spikelet, with glumes removed, × 9. – C: *C. gayana*, spikelet, with glumes removed, × 9. – From Kew Bull. 50: 112 (1995). Drawn by M. Tebbs.

M Tebbs

Agar, aus dug, rareh biyud (Som.).

Annual up to 70 cm high, erect or ascending, occasionally rooting at the lower nodes; leaves linear, tapering. Inflorescence a head of 4−8 digitate racemes, these 2.5−5 cm long. Spikelets 3-flowered, 2-awned; fertile lemma obliquely obovate in profile, the keel slightly gibbous, 2−3.6 mm long, shortly ciliate on the margins, flanks and keel, with a crown of spreading hairs 1.5−4 mm long just below the tip, and with an awn 2.5−8.5 mm long; 2nd lemma an oblong glabrous scale; 3rd lemma a clavate glabrous awnless scale.

Acacia woodland on sand overlying limestone or gypsum; 200−1500 m. N1, 2; C2; S1−3; throughout the tropics. Wood 73/169; McKinnon 249; Beckett 271; Hemming & Deshmukh 268; Dahir & Rose Innes ODA9; Paoli 316.

205

Fig. 113. *Eustachys paspaloides*. A: habit, × 2/3. B: glumes, × 24. C: florets, × 24. — From Fl. Trop. E. Afr. (1974). Drawn by D. Erasmus.

6. **C. gayana** Kunth (1830). Fig. 112 C.

Stoloniferous perennial up to 1 m high; leaves linear, tapering. Inflorescence a head of 9–12 digitate racemes, these 3–12 cm long. Spikelets 3- to 4-flowered, 2-awned; fertile lemma lanceolate in profile, 2.9–3.2 mm long, sparsely to densely ciliate on the margins and keel, with an awn 4–5 mm long; 2nd floret with a palea and often a male flower, the lemma lanceolate, ciliate on the margins; 3rd lemma a scabrid oblong or clavate awnless scale; 4th lemma, if present, a glabrous clavate awnless scale.

Alluvial soils or in open bushland; up to 1300 m. N2; S2, 3; tropical and southern Africa and Arabia. Thulin, Dahir & Hassan 9112; Warfa & Warsame 1092; Gorini s.n.

7. **C. mensensis** (Schweinf.) Cuf. (1968).

C. somalensis Rendle (1899); type: N1, Lort Phillips s.n. (BM holo.).

Baldostie godie, iya makarai (Som.).

Tufted perennial up to 1 m high; leaves linear, tapering. Inflorescence a head of 3–4 digitate racemes, these 5−9(−14) cm long. Spikelets olive-green, 2-flowered, 2-awned; fertile lemma lanceolate in profile, 3.5−5.2 mm long, glabrous below, thinly pilose on the flanks above, with an awn 7−12 mm long; 2nd lemma an elliptic awned rudiment or reduced to an awned rhachilla-segment.

Rocky hillsides with former *Juniperus* woodland; 1600−2000 m. N1; Ethiopia. Bally & Melville 16219; Drake-Brockman 480; Wood 72/109.

39. EUSTACHYS Desv. (1810)

Annual or perennial; leaves obtuse, the sheaths strongly keeled. Racemes 2 to many, digitate, the spikelets pectinate. Spikelets laterally compressed, with 1 fertile floret, rarely a smaller male floret, and a rhachilla terminating in a small clavate lemma; upper glume obtuse to 2-lobed, with a short apical awn; fertile lemma keeled, cartilaginous to coriaceous, dark brown to golden, acute to emarginate, awnless or with a very short apical awn-point. Grain ellipsoid, trigonous, the pericarp reluctantly free.

10 species in the tropics, mainly in the New World.

E. paspaloides (Vahl) Lanza & Mattei (1910). Fig. 113.

Perennial up to 70 cm high, erect or ascending; sheaths flabellate, often in bunches in the lower part of the stem. Racemes 2.5−8 cm long; fertile lemma ovate in profile, 1.5−2.1 mm long, glabrous or sparsely pubescent, ± ciliate on the keel and lateral nerves; 2nd lemma a little smaller, truncate.

Grassy fixed dunes overlying limestone; 50 m. C1; tropical and southern Africa and Arabia. Gillett & al. 22213.

40. ENTEROPOGON Nees (1836)

Annual or perennial. Racemes solitary or digitate. Spikelets dorsally compressed, with 1 fertile floret, sometimes also a smaller male floret, the rhachilla terminating in an awned rudiment or a cluster of rudiments; fertile lemma broadly rounded to almost flat, with prominently raised midnerve, sub-coriaceous, often 2-toothed, awned. Grain narrowly elliptic, dorsally compressed, concavo-convex in section, the pericarp free.

17 species in the tropics, distinguished from *Chloris* by the dorsally compressed spikelets.

1. Callus of fertile lemma bearded with hairs 3−4 mm long .... 1. *E. barbatus*
   − Callus of fertile lemma pilose or bearded with hairs less than 2 mm long 2
2. Sheaths loosely hirsute on the margins, strongly keeled 3
   − Sheaths glabrous, at most lightly keeled 4

3. Awn of fertile lemma 10−20 mm long..............
   .... 4. *E. sechellensis*
   − Awn of fertile lemma 2.5−5 mm long.............
   .... 5. *E. monostachyos*
4. Awn of fertile lemma 1−5 mm long..............
   .... 2. *E. rupestris*
   − Awn of fertile lemma 10−18 mm long..............
   .... 3. *E. macrostachyus*

1. **E. barbatus** C.E. Hubb. (1941); type: N1, Hargeisa, Gillett 4196 (K holo.).

Tufted perennial up to 80 cm high; sheaths keeled, loosely hairy along the margins. Raceme solitary, 7−18 cm long, finely silky-hairy. Spikelets 2- or sometimes 3-flowered; fertile lemma 6−7 mm long, copiously bearded from the callus with hairs 3−4 mm long, with an awn 4−5 mm long; 2nd lemma similar to the fertile but smaller, the 3rd rudimentary or absent.

Acacia woodland degraded by overgrazing; c. 1000 m. N1, 2; Kenya and Ethiopia. Bally & Melville 16188; Fausset 43.

2. **E. rupestris** (J.A. Schmidt) A. Chev. (1935).
   E. somalensis Chiov. (1896); type: N1, "Habr Awal", Robecchi-Bricchetti 563 (FT holo.).
   E. ruspolianus Chiov. (1897).

Aus gorof, awse ad, baldole, gauwadiri, gorror, gowdere (Som.).

Tufted perennial up to 90 cm high, the stems wiry, often bare below, much-branched, the basal buds clad in short pale cataphylls; sheaths glabrous, not keeled. Raceme solitary, 6−17 cm long. Spikelets 2- or sometimes 3-flowered; fertile lemma 4.5−8 mm long, shortly bearded on the callus with hairs less than 2 mm long, with an awn 1−5 mm long; 2nd lemma similar to the fertile but smaller and awnless; 3rd lemma rudimentary or absent.

Open *Acacia-Commiphora* woodland on deep red sand overlying limestone, and seasonally flooded grassy plains; 160−1400 m. N1−3; C1; S1−3; Cape Verde Is. and Mauritania, and Sudan southwards to Botswana and Namibia. Hansen & Heemstra 6191; Maconochie 3711; Hemming 1386; Bally & Melville 15361; Wieland 1273; Bettini s.n.; Balladelli s.n.

3. **E. macrostachyus** (Hochst. ex A. Rich.) Munro ex Benth. (1881).

Tufted perennial up to 1 m high; basal sheaths glabrous, at most lightly keeled. Racemes solitary (very rarely up to 3), 7−20 cm long. Spikelets (2−)3-flowered; fertile lemma 7−10 mm long (4−5 mm in 2-flowered forms), shortly bearded on the callus with hairs less than 2 mm long, scaberulous, with an awn 10−18 mm long; 2nd lemma similar to the fertile but smaller; 3rd lemma an awned rudiment.

Open bushland, sandy plains, clay-alluvium flood-plains, canal-banks, rice- and cane-fields;

Fig. 114. *Enteropogon sechellensis*. A: habit, × 2/3. B: spikelet, with glumes detached, × 8. C: lowest lemma, × 8. — From Fl. Trop. E. Afr. (1974). Drawn by M. E. Church.

20−1400 m. N1; C1, 2; S1−3; tropical Africa and tropical Arabia. Godding 61; Elmi & Hansen 4052A; Moggi & Bavazzano 767, 1291; Norris 32; Rose Innes 789.

4. **E. sechellensis** (Bak.) T. Durand & Schinz (1895). Fig. 114.

Tufted perennial up to 1 m high, the stems wiry; sheaths laterally compressed and keeled, loosely hairy mainly along the margins. Raceme solitary, 6−15 cm long. Spikelets 3-flowered; fertile lemma 5−8 mm long, shortly bearded on the callus with hairs less than 2 mm long, with an awn 10−20 mm long; 2nd and 3rd lemmas progressively reduced, otherwise similar to the fertile.

Stabilized dunes of red sand; 0−50 m. S2, 3; coastal regions of eastern Africa from Somalia to Mozambique, Seychelles and Madagascar. Kazmi 5100; Rose Innes 630.

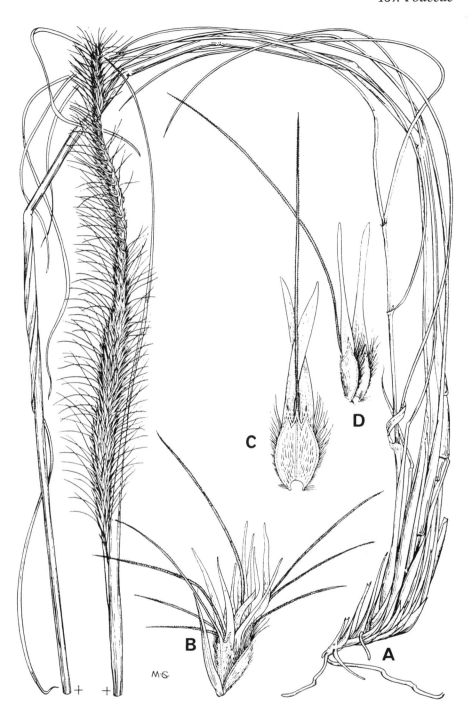

Fig. 115. *Afrotrichloris hyaloptera*. A: habit, × 2/3. B: spikelet, × 3. C: fertile lemma, flattened, × 4. D: fertile lemma, side view, × 4. — Modified from Kew Bull. 21: 106 (1967). Drawn by M. Grierson.

5. E. monostachyos (Vahl) K. Schum. (1895).

Tufted perennial up to 1 m high; sheaths laterally flattened and strongly keeled, loosely hairy along the margins with fine white hairs. Raceme solitary, 8—15 cm long. Spikelets 2- to rarely 3-flowered; fertile lemma 6—8 mm long, shortly bearded on the callus with hairs less than 2 mm long, with an awn 2.5—5 mm long; 2nd lemma similar to the fertile but smaller.

Cliff-face in a gorge. C1; tropical Africa and India. Wieland 4707.

Represented in Somalia by a single collection that corresponds with subsp. *africanus* Clayton in Kew Bull. 21: 109 (1967). This has a generally 2-flowered spikelet and the awn of the fertile lemma is 2.5—5 mm long. Subsp. *monostachyos*, from India, has a normally 3-flowered spikelet and an awn up to 8 mm long.

Fig. 116. A, B: *Microchloa indica*. A: habit, × 2/3. B: ligule, × 9. — C–F: *M. kunthii*. C: habit, × 2/3. D: part of inflorescence, × 4. E: spikelet, × 18. F: lemma, × 18. — Modified from a drawing made for Fl. Zamb. by J. C. Webb.

41. AFROTRICHLORIS Chiov. (1915)

Perennial. Racemes solitary, rarely paired. Spikelets with 1 fertile floret and a cluster of sterile lemmas; fertile lemma broadly rounded on the back, thinly coriaceous below, pilose, cleft to beyond the middle into 2 hyaline acute or aristulate lobes, awned from the sinus. Grain elliptic, dorsally compressed, the pericarp free.

Two species in Somalia, one also in Ethiopia (Ogaden).

1. Lobes of fertile lemma (14–)17–35 mm long, 1-nerved, attenuate into a fine awn; central awn (1.5–)2.5–4 cm long; lemma appressedly pubescent on the body; raceme 5–10(–14) cm long .... 1. *A. martinii*
– Lobes of fertile lemma 6–9 mm long, nerveless, acute; central awn 1.5–3 cm long; lemma ciliate along the lateral nerves, pubescent between; raceme 14–22 cm long .... 2. *A. hyaloptera*

1. **A. martinii** Chiov. (1915); type: S2, Mogadishu, Paoli 89 (FT holo.).

Ouse mulleh (Som.).

Tufted perennial with wiry stems up to 60 cm high. Raceme 5−10(−14) cm long. Fertile lemma appressedly pubescent on the body, this 3.5−5 mm long; lobes 1-nerved, (14−)17−35 mm long, attenuate into a fine awn; central awn (1.5−)2.5−4 cm long.

Fixed dunes with grass or low shrubs, mostly near the coast; 0−360 m. C1; S2; not known elsewhere. Gillett & Hemming 24560; Herlocker 330; Hansen 6027; Rose Innes 880.

2. **A. hyaloptera** Clayton (1967); type: C2, N of "Bulo Burti", Roffey 60041/5 (K holo., EA iso.). Fig. 115.

Aus guruf, caws mulaax (Som.).

Tufted perennial with wiry stems up to 70 cm high. Raceme 14−22 cm long. Fertile lemma ciliate along the lateral nerves, pubescent between, the body 3.5−4 mm long; lobes nerveless, 6−9 mm long, acute; central awn 1.5−3 cm long.

Mixed bushland on orange sand; 120−320 m. C1, 2; Ethiopia (Ogaden). Elmi & Hansen 4052; Gillett & al. 22298; Kuchar 17754; Thulin & Dahir 6457.

42. MICROCHLOA R. Br. (1810)

Annual or perennial. Racemes solitary, rarely 2, bearing inclined or pectinate spikelets. Spikelets subterete to subcylindrical, 1-flowered without rhachilla-extension, or with a second well developed male floret; glumes subequal, enclosing the florets; lemma shorter than the glumes, keeled, thinly membranous, acute to bilobed, sometimes mucronulate. Caryopsis ellipsoid.

Six species throughout the tropics.

1. Annual .... 1. *M. indica*
− Perennial, the base clothed in the fibrous remains of old sheaths ....2. *M. kunthii*

1. **M. indica** (L.f.) P. Beauv. (1812). Fig. 116 A, B.
Harfo (Som.).

Annual up to 50 cm high, usually growing in loose individual tufts, rarely forming mats. Raceme solitary, becoming curved with age, 1.5−15 cm long. Glumes 1.7−2.9 mm long; anthers 0.3−0.7 mm long. Caryopsis c. 0.9 mm long.

Shallow soil overlying gypsum; c. 1700 m. N2; throughout the tropics. McKinnon 903.

Only collected once in Somalia.

2. **M. kunthii** Desv. (1831). Fig. 116 C−F.
M. abyssinica Hochst. ex A. Rich. (1850).
Agar (Som.).

Tufted mat-forming perennial up to 30 cm high, the base usually clothed in the fibrous remains of old

sheaths. Raceme solitary, becoming curved with age, 3.5−11(−25) cm long. Glumes 2.4−3.8 mm long; anthers 0.5−1.2 mm long. Caryopsis c. 1.5 mm long.

Stony hillsides and disturbed ground; 1400−1600 m. N1; throughout the tropics. McKinnon 42A; Wood 71/36.

43. CYNODON Rich. (1805), nom. cons.

Perennial, mostly rhizomatous or stoloniferous, or both, and sward-forming. Racemes digitate, sometimes in 2 or more closely spaced whorls. Spikelets strongly laterally compressed, 1-flowered, with or without a rhachilla-extension (this very rarely bearing a vestigial floret); glumes very short to as long as the floret; lemma keeled, firmly cartilaginous, entire, awnless. Caryopsis ellipsoid, laterally compressed.

Eight species in the Old World tropics, one pantropical species extending into warm temperate regions.

C. ruspolianus Chiov. (1897) (= *C. plectostachyus* (K. Schum.) Pilg.) was described from 'Somalia', but the type locality, Mil-Mil is in Ethiopia (Ogaden) and the plant is otherwise unknown from the area covered by this Flora.

1. Plant with underground rhizomes as well as surface stolons ....1. *C. dactylon*
− Plant without underground rhizomes, but with surface stolons .... 2. *C. nlemfuensis*

1. **C. dactylon** (L.) Pers. (1805). Fig. 117.
Dactylon officinale Vill. (1786).
Gymnopogon digitatus Nees ex Steud. (1840).
Cynodon glabratus Steud. (1854).
Darris, domar, harfo, hrari, rys, sadeho, serdi (Som.).

Stoloniferous sward-forming perennial with slender rhizomes; stems slender, up to 40 cm high. Racemes (2−)3−6, digitate, 1.5−6 cm long. Glumes shorter than the floret; lemma 2−3 mm long, subglabrous to silky-pubescent on the keel.

Gravel, sand and alluvium, especially along water-courses, waste and trampled ground and as a weed of cultivation; 0−1750 m. N1−3; C1; S2; tropical and warm temperate regions, the most widespread of tropical lawn-grasses. Gillett 3930; McKinnon 265; Hemming 1769; Gillett & al. 22173; Paoli 125.

2. **C. nlemfuensis** Vanderyst (1922).

Stoloniferous perennial without rhizomes; stolons stout and woody, lying flat on the ground; stems robust to fairly slender, up to 60 cm high, soft, not woody. Racemes 4−13 in 1 or 2 whorls, 4−10 cm long. Glumes shorter than the floret, rarely as long; lemma 2−3 mm long, silky-pubescent to softly ciliate on the keel.

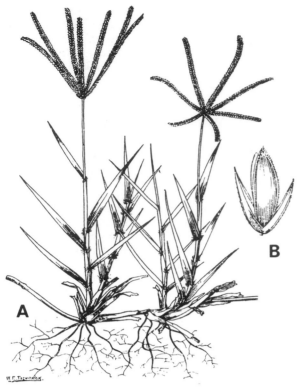

Fig. 117. *Cynodon dactylon*. A: habit, × 1. B: spikelet, × 8. – Modified from Fl. W. Trop. Afr. 3:2 (1972). Drawn by W. E. Trevithick.

Fig. 118. *Melanocenchris abyssinica*, × 0.7. – From Stud. Fl. Egypt, ed. 2 (1974).

var. **nlemfuensis.**

Stems moderately robust to somewhat wiry, 1–1.5 mm in diam.; leaves green or glaucous, 2–5 mm wide; racemes 4–9, each 4–7 cm long.

Disturbed and cultivated soils, becoming a serious weed of cereals and plantations; 0–85 m. N1; S1–3; tropical Africa, mostly in the east; introduced elsewhere. Bricchetti 558; Moggi & Bavazzano 1307; Paoli 1346; Rose Innes 642.

var. **robustus** Clayton & Harlan in Kew Bull. 24: 189 (1970).

Stems stout, 2–3 mm in diam. at the base; leaves thin, green, 5–6 mm wide; racemes 6–13, each 6–10 cm long.

S2; tropical East Africa, introduced elsewhere. Virgo 12.

44. SCHOENEFELDIA Kunth (1830)

Annual or perennial. Racemes solitary or digitate. Spikelets laterally compressed, strictly 1-flowered or with a rhachilla-extension terminating in a minute long-awned rudiment; glumes longer than the floret; lemma keeled, cartilaginous, 2-toothed, with a long sinuous awn. Grain ellipsoid, laterally compressed, the pericarp free.

Two species in Africa and Madagascar through Arabia to Pakistan and India.

S. transiens (Pilg.) Chiov. (1916). Fig. 119.

Xalfo (Som.).

Densely tufted perennial up to 1.2 m high. Racemes 2–4, 10–20 cm long. Spikelets 2-flowered; glumes 2–5 mm long; lemma 3–4 mm long, shortly ciliate on the nerves, with an awn 2.5–4 cm long, the awns at length characteristically delicately braided.

Deciduous or evergreen shrubland usually on clay soils, sometimes on orange sand; 60–180 m. C2; S1–3; East Africa southwards to Zimbabwe. Beckett 173; Moggi & Bavazzano 71A; Alstrup & Michelsen 6; Rose Innes 775.

45. MELANOCENCHRIS Nees (1841)

Annual or perennial. Racemes disposed along an axis, cuneate, deciduous, ending in a forked bristle, generally with 1–2 fertile and 2–3 progressively smaller sterile spikelets. Fertile spikelets dorsally compressed, 2-flowered, the upper floret male or sterile, with rhachilla-extension; glumes placed side by side, hairy, the lower awn-like, the upper similar but with a narrowly expanded base; lemma chartaceous, 3-awned; palea 2-awned. Caryopsis ellipsoid.

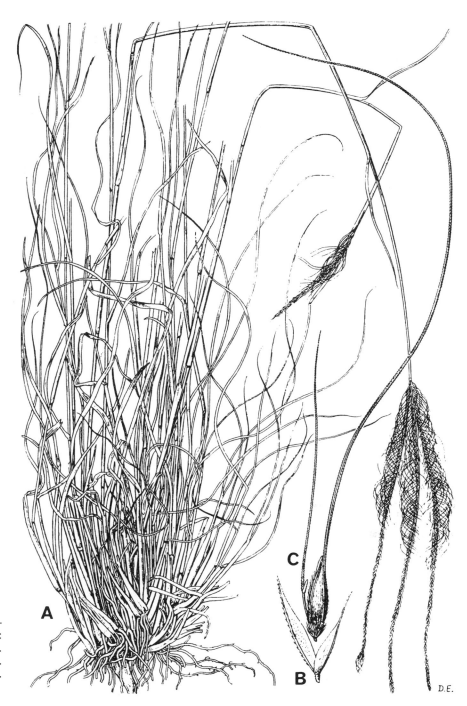

Fig. 119. *Schoenefeldia transiens*. A: habit, × 0.5. B: glumes, × 7. C: florets, × 7. — From Fl. Trop. E. Afr. (1974). Drawn by D. Erasmus.

Three species from tropical Africa to India and Sri Lanka.

M. abyssinica (R. Br. ex Fresen.) Hochst. (1855). Fig. 118.

Annual up to 20 cm high. Racemes 3−5(−6), 6−14 mm apart, 10−15 mm long including the awns. Glumes of lowermost spikelets as long as the raceme, densely ciliate below; fertile lemma (4−)6−7 mm long including the short awns.

Acacia-Commiphora shrubland on gravelly calcareous soil; 175 m. S1; north-eastern Africa through Arabia to north-western India. Friis & al. 4769.

46. TRAGUS Hall. f. (1768), nom. cons.

Annual or perennial. Inflorescence a cylindrical false raceme; racemelets deciduous, shortly pedunculate, of 2−5 spikelets, these contiguous or on a short

213

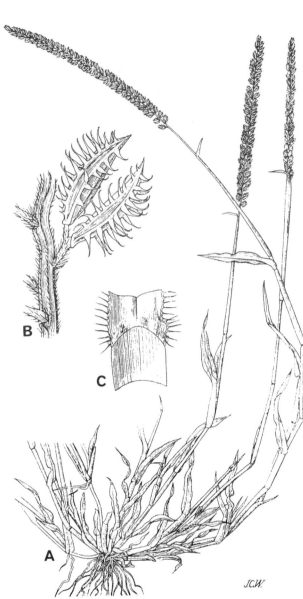

Fig. 120. *Tragus berteronianus*. A: habit, × 0.6. B: ligule, × 4. C: spikelet pair, × 8. — Modified from a drawing for Fl. Zamb. by J. C. Webb.

rhachis, sometimes the upper reduced. Spikelets 1-flowered; lower glume a minute scale or suppressed; upper glume scarcely exceeding the floret, rounded, its 5–7 nerves forming prominent ribs, some of them bearing stout hooked prickles, acute to acuminate; lemma acute.

Seven species throughout the tropics.

1. Spikelets in a racemelet of 2–4 fertile and 1–2 sterile, separated by distinct internodes, the rhachis prolonged; upper glume 7-nerved..........
....1. *T. racemosus*
– Spikelets in pairs, the rhachis not prolonged; upper glume 5- or 7-nerved 2

2. Spikelets subequal, arising from the same level without a rhachis internode; upper glume 7-nerved ....2. *T. heptaneuron*
– Spikelets unequal, separated by a distinct internode; upper glume 5-nerved ..3. *T. berteronianus*

1. **T. racemosus** (L.) All. (1785).
   T. paucispinus Hack. (1901).
   Harfo yer yer, nafir (Som.).
   Annual up to 25 cm high. Inflorescence (2–)5–7.5 cm long; racemelets comprising 2–4 fertile and 1–2 sterile spikelets separated by distinct internodes, the rhachis prolonged. Spikelets 4–5.5 mm long, acuminate; upper glume 7-nerved, the prickles hooked and with turgid or swollen base.
   Desert soils, drainage channels, disturbed ground and rich loam on field-margins; 1300–1550 m. N1; southern Europe, temperate Asia and northern tropical Africa. Gillett 3929; Hansen & Heemstra 6124; Maconochie 3743.

2. **T. heptaneuron** Clayton (1972).
   Nafir, rarmay (Som.).
   Annual up to 30 cm high. Inflorescence 3–9 cm long; racemelets comprising 2 subequal spikelets arising at the same level without a rhachis-internode or prolongation. Spikelets 3–3.5 mm long, acute; upper glume 7-nerved, the prickles hooked above, terete below.
   Open shrubland on sandy, sometimes saline, soil; up to 50 m. C1, 2; S2, 3; Kenya and Tanzania. Herlocker 433; Beckett 275A; Elmi & Hansen 4003; Hemming 424.

3. **T. berteronianus** Schult. (1824). Fig. 120.
   T. alienus auct. non (Spreng.) Schult.
   Harfo (Som.).
   Annual up to 20 cm high. Inflorescence 2–7.5 cm long; racemelets comprising 2 spikelets separated by a short but distinct internode, the upper shorter than the lower, the rhachis not prolonged. Spikelets 2–3 mm long, acute; upper glume 5-nerved, the prickles hooked and with turgid or swollen base.
   Open shrubland on sandy soil, silty clay and cultivated ground; 200–1550 m. N1–3; C1, 2; S2; Africa, south-western Asia, China and America. Bally & Melville 16154; Collenette 240; Puccioni & Stefanini 3; Herlocker 433A; Rose Innes 839; Moggi & Bavazzano 324.

47. DIGNATHIA Stapf (1911)

Annual or perennial. Inflorescence a cylindrical false raceme; racemelets deciduous, pedunculate, comprising 1–2 spikelets and a rudiment separated by the internodes of a short curvaceous rhachis. Spikelets 1-flowered; glumes longer than the floret, laterally compressed, thickly indurated, ± gibbous (rarely straight), scaberulous to lanose, the upper the longer and caudate to a stiff point or awn; lemma keeled, shortly awned.

Five species in tropical Africa, southern Arabia and India.

1. Glumes densely long-ciliate along the keel and sometimes also the margins 2
- Glumes at most shortly hairy along the margins and keel 3
2. Inflorescence (2−)3.5−7 cm long, pallid, usually exserted, sometimes partially embraced by the slightly inflated upper-most sheath .. 1. *D. ciliata*
- Inflorescence 1.5−2.5 cm long, tinged with red, embraced below by the very inflated often bladeless uppermost sheath .... 2. *D. villosa*
3. Spikelets 2−3.5 mm long; glumes scaberulous ....3. *D. gracilis*
- Spikelets 4−7 mm long; glumes appressed-pubescent ....4. *D. hirtella*

1. **D. ciliata** C.E. Hubb. (1934); type: N Somalia, "Haud", Fausset 46 (K holo.).

Aus cherin, fodarder (Som.).

Subshrubby perennial with branched woody stems naked below, up to 60 cm high. Inflorescence exserted or sometimes partially embraced below by the slightly inflated uppermost sheath, (2−)3.5−7 cm long; racemelets comprising 1 fertile spikelet and a barren rudiment of 2 glumes. Spikelets narrowly ovate, 4.5−7 mm long; glumes similar, narrowly lanceolate, densely long-ciliate with yellowish tubercle-based hairs on the keel, shortly ciliate on the margins, caudate-acuminate with the tips divergent.

Sand-dunes with low shrubs; 230−580 m. N1; C2; Ethiopia (mostly Ogaden). Glover & Gilliland 366; Thulin & Dahir 6458.

2. **D. villosa** C.E. Hubb. (1936); type: N Somalia, without precise locality, Fausset s.n. (K holo.).

Aus bured, yeris (Som.).

Subshrubby perennial with branched woody stems naked below, up to 20(−25) cm high. Inflorescence 1.5−2.5 cm long, embraced below by the very inflated often bladeless uppermost sheath, this often deeply suffused with red; racemelets comprising 1 fertile spikelet and a barren rudiment of 2 glumes. Spikelets ovate or narrowly ovate, 4−5.5(−7) mm long; glumes similar, linear-lanceolate, densely long-ciliate with shaggy yellowish tubercle-based hairs on the keel and margins, aristate-acuminate, the tips not divergent.

Acacia-Commiphora shrubland on red sand overlying limestone or sometimes gypsum; 100−1000 m. N1, 3; C1, 2; not known elsewhere. Gillett 4201; Hemming 1741; Beckett 213; Bally & Melville 15348.

3. **D. gracilis** Stapf (1911). Fig. 121.

Annual up to 50 cm high though usually much less.

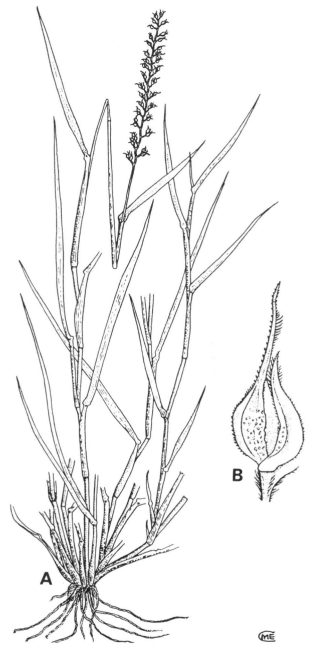

Fig. 121. *Dignathia gracilis*. A: habit, x 2/3. B: spikelet, x 16. − Modified from Fl. Trop. E. Afr. (1974). Drawn by M. E. Church.

Inflorescence 3−8 cm long; racemelets comprising (1−)2 fertile spikelets and a rudiment of 2 gaping glumes. Spikelets suborbicular with rostrate tip, 2−3.5 mm long; glumes scaberulous on the body, spinously ciliate on the keel and margins, the lower crescent-shaped, the upper with a strongly gibbous body 1.5−2 mm long drawn out into a stiff pointed beak.

Sand-dunes with low shrubs; 0−170 m. C2; S2, 3; Kenya and Mozambique. Kuchar 17648; Hansen 6042; Rose Innes 753.

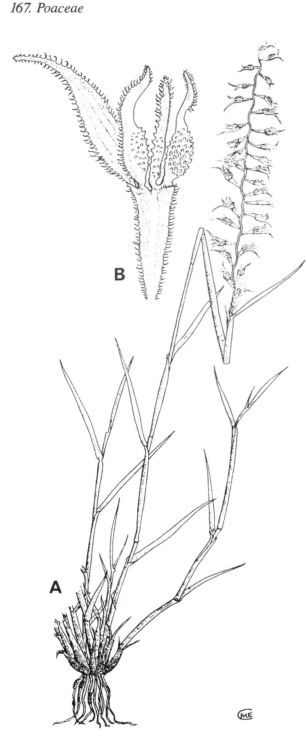

Fig. 122. *Leptothrium senegalense*. A: habit, × 2/3. B: spikelet pair, × 6. − Modified from Fl. Trop. E. Afr. (1974). Drawn by M. E. Church.

4. **D. hirtella** Stapf (1911).

Halfa, sakbir (Som.).

Annual up to 30 cm high though usually much less. Inflorescence 2−7 cm long; racemelets comprising (1−)2 fertile spikelets and a rudiment. Spikelets

suborbicular with rostrate tip, 4−7 mm long; glumes appressed-pubescent on the body, shortly ciliate on the margins, the lower crescent- or s-shaped, the upper with a gibbous body 1.5−2 mm long drawn out into a stiff awn-like beak.

Dense shrubland on sandy soils, and areas of cultivation and heavy disturbance; 50−230 m. C1, 2; S1−3; Ethiopia, southern Arabia and India. Herlocker 52; Beckett 272; Thulin & Bashir Mohamed 6791; Elmi & Hansen 4004; Tardelli & Bavazzano 791.

48. LEPTOTHRIUM Kunth (1829)

Perennial. Inflorescence an open false raceme; racemelets comprising (1−)2 spikelets side by side on the truncate tip of a cuneate peduncle. Spikelets 1-flowered; glumes longer than the floret, indurated, smooth or tuberculate-spinulose, the lower usually modified into a long flat recurved acuminate tail (less so in the second spikelet of a pair, rarely no longer than the upper), the upper laterally compressed and enfolding the floret (very rarely resembling the lower); lemma 1-nerved, acute.

One species from Senegal to Pakistan and one in the Caribbean.

L. senegalense (Kunth) Clayton (1972); *Latipes senegalensis* Kunth (1830). Fig. 122.

Latipes inermis Chiov. (1928); type: C1, "Obbia", Puccioni & Stefanini 354 (FT holo.).

Chabioke, chebiore, chediore, rarmay (Som.).

Short-lived perennial forming tough bunches, up to 75 cm high. Inflorescence 6−11 cm long, the racemelets ± distant on the wavy main axis, brightly coloured purple and light green; peduncle cuneate, flattened, 1−4 mm long, ciliate with hooked hairs on the margins. First spikelet: lower (outer) glume 3−6.5 mm long, the margins ciliate with hooked hairs above; upper glume 3−4 mm long, densely tuberculate and often spinulose, pectinate with hooked spines near one margin, rostrate at the tip. Second spikelet: lower (outer) glume 2.5−4 mm long, tuberculate-spinulose; upper glume 3−4.5 mm long, tuberculate-spinulose, rostrate above and with a crest of hooked prickles.

Acacia-Commiphora shrubland on red or orange sand overlying limestone; coastal dunes; 0−1400 m. N1−3; C1, 2; S1−3; tropical Africa and southwestern Asia eastwards to Pakistan. Bally & Melville 16153; Elmi & Hansen 4009; Bally & Melville 15649; Beckett 107A; Rose Innes 807; Wieland 1151; Thulin 4747; Rose Innes 631.

49. PSEUDOZOYSIA Chiov. (1928)

Perennial. Inflorescence a cylindrical false raceme embraced below by the inflated uppermost sheath;

racemelets comprising 2 sessile contiguous spikelets. Spikelets 1-flowered; glumes longer than the floret, thickly indurated, shallowly tuberculate, the lower ovate, the upper subglobose and enclosing the floret; lemma obtuse.

A single species only.

P. sessilis Chiov. (1928); type: C1, "Obbia", Puccioni & Stefanini 355 (FT holo.).

Tufted short-lived perennial up to 15 cm high. Inflorescence 2.5−3 cm long; glumes 3 mm long.

Coastal dunes; 0 m. C1; not known elsewhere.

A scarcely-known genus represented only by the type of its one species, itself a rather poor specimen that is best not dissected too much; hence the very brief description.

50. PEROTIS Ait. (1789)

Annual or sometimes perennial. Inflorescence a delicate cylindrical raceme, the spikelets stipitate or rounded at the base. Spikelets solitary, all alike; glumes subequal, lanceolate to narrowly lanceolate, membranous, longer than the floret, long-awned; lemma 1-nerved, acute.

12 species in the Old World tropics.

1. Base of spikelet abruptly rounded, without a definite callus ....1. *P. patens*
− Base of spikelet drawn out into a distinct stipitiform callus 2
2. Awns 35−45 mm long; glumes pilose on the flanks ....4. *P. somalensis*
− Awns 5−20 mm long; glumes minutely hispidulous to coarsely scabrid on the flanks 3
3. Glumes 2.5−3.5 mm long; callus 0.5−1 mm long .... 2. *P. hildebrandtii*
− Glumes 5.6−7.6 mm long; callus 2−2.3 mm long .... 3. *P. acanthoneuron*

1. **P. patens** Gand. (1920).

P. indica auct. non (L.) Kuntze.

P. latifolia auct. non Ait.

Loosely tufted annual or short-lived perennial up to 60 cm high, the stems tough and wiry; leaves lanceolate to narrowly ovate. Raceme 5−30 cm long, the spikelets dense, ascending or spreading horizontally. Spikelets 1.2−2.7 mm long, rounded at the base and without a definite callus; glumes scaberulous to minutely hispidulous, the midnerve of the upper inconspicuous; awns flexuous, 9−17 mm long. Caryopsis terete.

N1; tropical and southern Africa and Madagascar. Drake-Brockman 430; Lort Phillips s.n.

Known from Somalia only from the two specimens cited, neither of which gives any indication of locality, but almost certainly from N1.

Fig. 123. *Perotis hildebrandtii*. A: habit, × 2/3. B: spikelet, × 10. − Modified from Fl. Trop. E. Afr. (1974). Drawn by M. E. Church.

2. **P. hildebrandtii** Mez (1921). Fig. 123.

P. indica forma *glabra* Chiov., Result. Sci. Miss. Stefanini-Paoli: 182 (1916); type: S1, Paoli 1222 (FT holo.).

P. indica forma *hirta* Chiov., loc. cit. (1916); types: S1, Paoli 742, 1167 (FT syn.).

Annual up to 50 cm high, usually geniculately

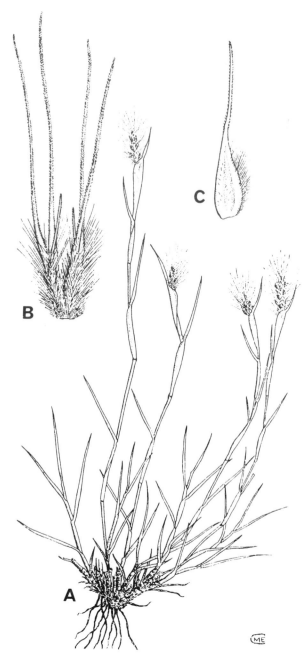

Fig. 124. *Tetrachaete elionuroides*. A: habit, × 1. B: spikelet pair, × 8. C: lemma, × 12. − Modified from Fl. Trop. E. Afr. (1974). Drawn by M. E. Church.

ascending, the stems rather wiry; leaves broadly linear to narrowly lanceolate. Raceme 2−20 cm long, the spikelets loose, spreading horizontally. Spikelets 2.5−3.5 mm long, with a conspicuous stipitiform callus 0.5−1 mm long; glumes irregularly puberulous to sparsely hispidulous on the flanks, the midnerve of the upper broad, green, slightly depressed, spinulose-scabrid; awns flexuous, 5−15 mm long. Caryopsis strongly flattened.

Acacia-Commiphora shrubland on red sand and as a weed of cultivation; up to 200 m. S1−3; tropical Africa and Seychelles. Thulin & Bashir Mohamed 7084; Bally & Melville 15273; Rose Innes 759.

3. **P. acanthoneuron** Cope (1995); type: S2, Bally & Melville 15273 (K holo.).

Tufted annual up to 70 cm high, erect or geniculately ascending; leaves lanceolate to narrowly ovate. Raceme 12−17 cm long, usually partially embraced below the uppermost sheath, the spikelets ascending. Spikelets 5.6−7.6 mm long, with a linear, stipitiform callus 2−2.3 mm long; glumes coarsely scabrid on the flanks, the scabridities ± in longitudinal lines, the midnerve conspicuously spinose with curved prickles, that of the upper glume deeply depressed; awns slender, flexuous, 10−20 mm long. Caryopsis strongly flattened.

Shrubland on pale pink sandy soil or as a weed of cultivation. S2; not known elsewhere. Bettini s.n.; Raimondo 8/82.

4. **P. somalensis** Chiov. (1916); type: S1, "El Magu", Paoli 624 (FT holo., K iso.).

Annual up to 45 cm high, erect, the stems rather stout; leaves broadly linear to narrowly lanceolate. Raceme 8−10 cm long, the spikelets dense, ascending. Spikelets 2.5−3.5 mm long, with a conspicuous stipitiform callus 0.5−0.7 mm long; glumes pilose on the flanks, the midnerve of the upper inconspicuous; awns flexuous, 35−45 mm long. Caryopsis strongly flattened.

S1, 3; not known elsewhere. Rose Innes 769; Senni 305.

Known from only three specimens, none of which gives any indication of habitat.

51. **TETRACHAETE** Chiov. (1902)

Annual. Inflorescence a spike-like false raceme embraced below by the inflated uppermost sheath; racemelets comprising 2 sessile contiguous spikelets. Spikelets 1-flowered, the glumes much longer than the floret, bristle-like, plumose; lemma ovate, strongly keeled, glabrous, coriaceous, 3-nerved, long-awned.

One species only.

T. elionuroides Chiov. (1903). Fig. 124.
Makoore (Som.).

Wiry annual up to 25 cm high. Inflorescence 1−3.5 cm long, white-plumose. Glumes 10−17 mm long, straight, rigid, villously plumose below; lemma 2−2.5 mm long, tipped with an awn 3−4.5 mm long.

Stony soils overlying limestone and gypsum, also on alluvium and heavy clay; up to 600 m. C2; S1−3; Ethiopia, Kenya, Tanzania, southern Arabia. Thulin & Warfa 5322; Eagleton 92; Rose Innes & Trump 1002; Hemming & Deshmukh 333.

Tribe 14. **PANICEAE**

Ligule a short membrane with ciliate fringe, rarely absent. Inflorescence an open to spike-like panicle or of unilateral racemes (these rarely compound), usually terminal, the spikelets all alike, sometimes paired but those of a pair similar, when racemose the lower glume usually turned away from the rhachis (abaxial). Spikelets 2-flowered without rhachilla-extension, usually dorsally compressed, falling entire, rarely awned; glumes membranous or herbaceous, rarely coriaceous, the upper often as long as the spikelet, the lower usually shorter and sometimes rudimentary; lower floret male or barren, its lemma usually membranous or herbaceous and as long as the spikelet, rarely indurated, with or without a palea; upper floret bisexual, the lemma and palea ± indurated; lodicules 2, fleshy; stamens 3, rarely fewer; stigmas 2. Caryopsis with large embryo, usually with round or oval hilum.

100 genera and c. 2000 species, mainly in the tropics.

1. Spikelets not subtended by deciduous bristles, spines or bracts, if subtended by bristles then these persistent on the branches 2
− Spikelets, singly or in clusters, subtended by a deciduous involucre of 1 or more bristles or spines, or several glumaceous bracts 16
2. Upper lemma coriaceous to bony at maturity with narrow inrolled margins 3
− Upper lemma cartilaginous to chartaceous or rarely hyaline, the margins flat and usually hyaline 13
3. Spikelets subtended by persistent bristles or the raceme-rhachis prolonged into a subulate point 4
− Spikelets not subtended by bristles nor the raceme-rhachis prolonged into a subulate point 6
4. Lower lemma sulcate; keels of lower palea becoming indurated .... 61. *Holcolemma*
− Lower lemma not sulcate; keels of lower palea not becoming indurated 5
5. Inflorescence a panicle, all or most of the spikelets subtended by bristles .... 59. *Setaria*
− Inflorescence of racemes, only the terminal spikelet subtended by a bristle .... 60. *Paspalidium*
6. Spikelets supported on a globular bead .... 57. *Eriochloa*
− Spikelets without a bead 7
7. Inflorescence a panicle ....52. *Panicum*
− Inflorescence of unilateral racemes, the spikelets usually single or paired but sometimes in irregular clusters or in short secondary racemelets 8
8. Glumes and lemmas awned 9
− Glumes and lemmas awnless or at most mucronate 10
9. Upper lemma awnless, coriaceous .... 53. *Echinochloa*
− Upper lemma awned, crisply chartaceous .... 54. *Alloteropsis*
10. Lower glume absent .... 58. *Paspalum*
− Lower glume present 11
11. Tip of upper palea reflexed or slightly protuberant .... 53. *Echinochloa*
− Tip of upper palea not reflexed 12
12. Upper lemma usually muticous, if shortly mucronate then spikelets plump (if some spikelets borne singly then lower glume adaxial, if lower glume a minute truncate cuff see 57. *Eriochloa*) ....55. *Brachiaria*
− Upper lemma mucronate; spikelets plano-convex, cuspidate (if some spikelets borne singly then lower glume abaxial) .... 56. *Urochloa*
13. Spikelets dorsally compressed; inflorescence of racemes 14
− Spikelets laterally compressed; inflorescence a panicle 15
14. Racemes persistent .... 64. *Digitaria*
− Racemes deciduous .... 65. *Taeniorhachis*
15. Upper lemma dorsally compressed, the stigmas emerging terminally .... 62. *Tricholaena*
− Upper lemma laterally compressed, the stigmas emerging laterally .... 63. *Melinis*
16. Involucre composed of several glumaceous bracts .... 68. *Anthephora*
− Involucre composed of bristles or spines 17
17. Involucral bristles free throughout, ± filiform .... 66. *Pennisetum*
− Involucral bristles flattened into spines and connate below, often forming a cup ....67. *Cenchrus*

52. **PANICUM** L. (1753)

Annual or perennial; leaves flat or inrolled, linear to ovate. Inflorescence a panicle, usually much-branched but occasionally contracted about the primary branches. Spikelets dorsally or weakly laterally compressed, ovate to oblong; glumes hyaline to membranous, usually the lower shorter than, and the upper as long as the spikelet; lower floret male or barren, its lemma usually resembling the upper glume, with or without a palea; upper lemma about as long as the spikelet, crustaceous, the margins inrolled and clasping only the edges of the palea. Caryopsis ellipsoid, dorsally compressed.

About 470 species throughout the tropics, extending into temperate North America.

P. repens L. was reported for Somalia by Stapf (1907), but the single specimen cited, Drake-Brockman 98, was misidentified. Stapf himself corrected the name to *P. coloratum* var. *minus* when he revised the genus for Fl. Trop. Afr. (1920).

P. bossii Tropea (1911), described from S3, has not yet been identified; the description is inadequate and the type specimen (Macaluso s.n. from "Giumbo") has not been seen.

1. Spikelets asymmetric in profile ..10. *P. laticomum*
 − Spikelets symmetric in profile 2
2. Panicle and upper part of stem bearing slender clavellate hairs .... 1. *P. deustum*
 − Plant glabrous or hairy, but the hairs not clavellate 3
3. Upper lemma transversely rugose 4
 − Upper lemma smooth or granulose 5
4. Spikelets rounded on the back .... 2. *P. maximum*
 − Spikelets with a groove on the back ..3. *P. infestum*
5. Leaves stiff, subulate and conspicuously distichous; plant tufted, shrubby .... 7. *P. pinifolium*
 − Leaves not stiff and subulate 6
6. Upper lemma glossy brown 6. *P. atrosanguineum*
 − Upper lemma pallid 7
7. Lower glume 1/2−3/4 the length of the spikelet ...8
 − Lower glume up to 1/3 the length of the spikelet ...9
8. Coarse robust annual; spikelets not gaping.........
.... 4. *P. hippothrix*
 − Shrubby perennial; spikelets turgid, gaping........
.... 5. *P. turgidum*
9. Spikelets 2−3 mm long, obtuse or acute..........
.... 8. *P. coloratum*
 − Spikelets 3−4 mm long, acuminate................
.... 9. *P. subalbidum*

1. P. deustum Thunb. (1794).

Tufted, shortly rhizomatous perennial up to 200 cm high, pilose with slender clavellate hairs in and below the panicle; leaves linear to narrowly lanceolate, 15−50 cm × 5−35(−40) mm, cordate or straight at the base. Panicle ovate to oblong, 18−40 cm long, the primary branches stiff and usually ascending, the secondary very short. Spikelets oblong, 3.5−5(−5.5) mm long, often tinged with purple, blunt; lower glume broadly ovate, 1/2−2/3 the length of the spikelet, 5- to 7-nerved, separated from the upper by a short internode; upper lemma pallid, dull or glossy.

Amongst boulders; up to c. 1800 m. N2; S3; Sudan and Ethiopia southwards to South Africa. Gorini 500; Glover & Gilliland 608.

2. P. maximum Jacq. (1781).

Arabsa, baldoli, caws lagoley, weineh (Som.).

Loosely to densely tufted perennial up to 150(−200) cm high, erect from a shortly rhizomatous base or ascending and rooting from the lower nodes; leaves linear to narrowly lanceolate, 12−40 cm × 4−35 mm. Panicle oblong or pyramidal, 12−45 cm long, usually much-branched, the branches ascending to spreading, the lowermost in a whorl. Spikelets oblong, 3−4.5 mm long, glabrous or pubescent, blunt or acute; lower glume broadly ovate, 1/3−1/2 the length of the spikelet, 3-nerved, obtuse or acute; upper lemma conspicuously transversely rugose.

Shady places in open bushland on seasonally flooded sandy and alluvial soils; 0−1700 m. N1; S1−3; tropical and southern Africa and Arabia. Glover & Gilliland 20; Thulin, Hedrén & Abdi Dahir 7550; Rose Innes 652; Hemming & Deshmukh 79.

3. P. infestum Anderss. (1863).

P. sennii Chiov. (1932); type: S3, "Jack-Omisso", Senni 386 (FT holo.).

P. pseudoinfestum Chiov. (1932); types: S3, "Afmadu", Gorini 61, 62, 65 (syn.).

Densely tufted perennial with shortly rhizomatous woody rootstock, up to 200 cm high, usually pilose or hispid with tubercle-based hairs, rarely glabrous; leaves linear, 15−50 cm × 2−10 mm. Panicle oblong, 10−30 cm long, the primary branches ascending, spicate, the secondary usually absent or very short. Spikelets oblong, 2.5−4.1 mm long, glabrous, with a groove on the back, apiculate; lower glume broadly ovate, 1/4−1/3 the length of the spikelet, 1- to 3-nerved;upper lemma conspicuously transversely rugose.

Open bushland and grassland, and in areas of cultivation. S1−3; tropical to South Africa. Tardelli 133; McLeish 888; Rose Innes 635.

4. P. hippothrix K. Schum. (1895).

Coarse robust annual up to 100 cm high; leaves linear, 15−50 cm × 2−8 mm, straight at the base. Panicle often incompletely exserted from the uppermost sheath, ovate to oblong, 25−40 cm long, the spikelets sparsely distributed on stiff wiry spreading or ascending branches. Spikelets narrowly ovate, 3.5−4.5 mm long, acuminate; lower glume ovate, 2/3−3/4 the length of the spikelet, 5-nerved, acuminate; upper lemma pallid, glossy.

Open spaces between thickets. S3; Kenya and Tanzania. Rose Innes 766.

5. P. turgidum Forssk. (1775). Fig. 125.

Dallan, darif, dungara (Som.).

Glaucous perennial forming rounded bushes up to 100(−200) cm high and often as much across; stems erect or ascending, woody, branched at the nodes, sometimes forming fastigiate tufts; leaves linear-lanceolate, (0.5−)2−15 cm × 1−6 mm, flat, folded or inrolled, glabrous and glaucous, stiff and pungent, often much shorter than the sheath, rarely filiform and up to 30 cm long. Panicle subpyramidal, 2.5–15(–30) cm long, lax, the branches distant and eventually spreading. Spikelets ovoid, 3.5–4.5(–5) mm long, glabrous, acute or acuminate, turgid and often gaping at anthesis; lower glume broadly ovate, 3/4 as long to almost as long as the spikelet, 5- to 9-nerved; upper lemma pallid, smooth and glossy.

Sandy plains and deserts, dunes and seashores, and in sandy pockets in rocky outcrops; 0–1370 m. N1–3; northern and north-eastern Africa eastwards to Pakistan. Glover & Gilliland 522; Thulin 4295; Bally & Melville 15653.

6. P. atrosanguineum Hochst. ex A. Rich. (1850).

Annual up to 40 cm high, ascending; leaves lanceolate, 6–14 cm × 5–14 mm, straight at the base, pilose or hispid with tubercle-based hairs, acute. Panicle ovate or oblong, 6–20 cm long, much-branched, the branches fine and spreading. Spikelets ovate-oblong, 1.8–2.2 mm long, glabrous, often tinged with purple, shortly acuminate; lower glume broadly ovate, 2/3 the length of the spikelet, 3- to 5-nerved, acuminate; upper lemma dark brown, smooth and glossy.

Open deciduous woodland in the south, a weed of cultivation in the north; 450–1550 m. N1; S1, 3; tropical Africa, eastwards through Arabia to north-western India. McKinnon 16; Thulin & Bashir Mohamed 6846; Rose Innes 622.

7. P. pinifolium Chiov. (1906). Fig. 126.

P. appletonii Stapf (1907); type: C1, near "Obbia", Appleton (K).

Tufted perennial bunch-grass; stems tough, stout and straight, branched above, often densely so, up to 70 cm high; leaves distant below, densely crowded above, conspicuously distichous, the lower inrolled and pungent, the upper subulate. Panicle ovate, 1–2 cm long, scarcely exserted from the uppermost sheath. Spikelets globose, 1.8–2 mm long; lower glume broadly ovate, 1/2 the length of the spikelet, 0- to 1-nerved; upper lemma pallid, smooth and glossy.

Coastal grassland and bare sand, coral cliffs and in shallow sand overlying limestone pavement; 0–300 m. N3; C1; S2, 3; Kenya and Zanzibar. Thulin & Warfa 5904; Thulin & Dahir 6583; Alstrup & Michelsen 33; Moggi & Bavazzano 32.

8. P. coloratum L. (1753).

Tufted, pubescent or subglabrous perennial, the base usually knotty or slightly swollen, often with persistent scales; stems erect, up to 200 cm high, sometimes more, rarely decumbent; leaves linear, 3–30 cm × 2–12 mm, straight or subamplexicaul at the base, flat, glabrous or hairy, acute or acuminate. Panicle ovate, 4–30 cm long, usually much-branched, contracted or spreading, the branches

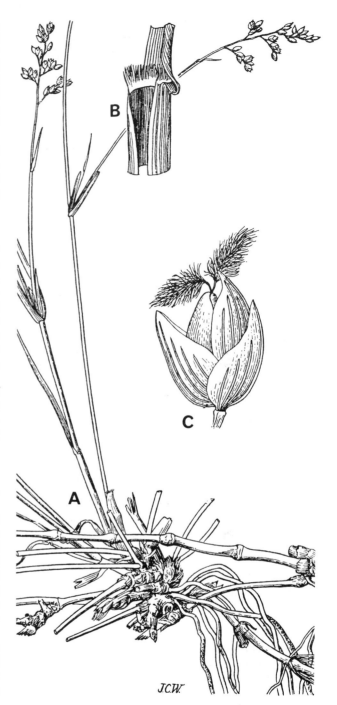

Fig. 125. *Panicum turgidum*. A: habit, × 2/3. B: ligule, × 6. C: spikelet, × 8. – Modified from a drawing by J. C. Webb.

ascending. Spikelets ovate-elliptic, 2–3 mm long, green, often tinged with purple, obtuse, acute or occasionally acuminate; lower glume ovate, 1/4–1/3 the length of the spikelet, 1(–3)-nerved, cuff-like, clasping, acute; upper lemma pallid, smooth and glossy.

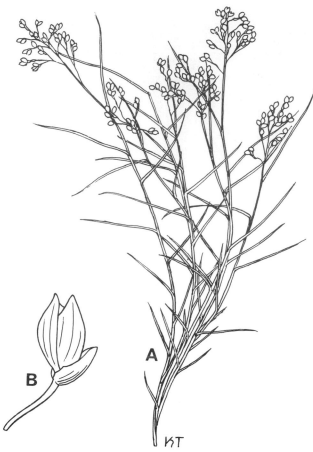

Fig. 126. *Panicum pinifolium*. A: habit, × 1. B: spikelet, × 12. – From Thulin & al. 7184. Drawn by K. Thunberg.

1. Coarsely hispid or rarely subglabrous plants 50–200 cm high or more, usually erect and unbranched; leaves 15–30 cm × 5–12 mm; panicle 15–30 cm long .... var. *coloratum*
– Subglabrous or pilose plants up to 40(–90) cm high, often branching above the base; leaves 3–10(–15) cm × 2–5 mm; panicle 4–7(–10) cm long .... var. *minus*

var. coloratum.
Seasonally inundated marshes on heavy clay soils. S2, 3; tropical and subtropical Africa. Munro 64; Rose Innes & Trump 931, 1014; Peveling 230.

var. minus Stapf ex Chiov., Result. Sci. Miss. Stefanini-Paoli 1: 183 (1916); type: S2, "Dafet", Paoli 1270 (FT holo.).
Agar, caws gubad, gurgurro, jalbo, medu (Som.).
Short grassland or open bushland on sand overlying limestone or gypsum; 50–1500 m. N1–3; C1, 2; S1, 2; tropical East Africa. Hemming 1432; Hansen & Heemstra 6208; Bally & Melville 15773; Thulin & al. 7266; Elmi & Hansen 4016; Wieland 1517; Rose Innes & Trump 1006.

A very variable species that can, in eastern Africa, be partitioned into two varieties. The distinction between them is not always clear-cut, but in Somalia it is quite marked. Var. *coloratum* is not all that common and is confined to S2 and S3; var. *minus*, on the other hand, is widespread and unrecorded only from S3.

9. **P. subalbidum** Kunth (1831).
P. proliferum Lam. var. *longijubatum* Stapf in Fl. Cap. 7: 406 (1899); *P. longijubatum* (Stapf) Stapf (1920).
Annual or short-lived perennial; stems robust, up to 200 cm high and 5–10 mm in diam. at the base, erect or decumbent, often rooting from the lower nodes; leaves linear, 20–50 cm × 7–15 mm, cordate to straight at the base. Panicle ovate or oblong, 20–50 cm long, sparsely to moderately branched, spreading, the spikelets evenly dispersed. Spikelets ovate, 3–4 mm long, acuminate; lower glume broadly ovate, 1/4–1/3 the length of the spikelet, 1- to 3-nerved, cuff-like and clasping, obtuse or acute; upper lemma pallid, smooth and glossy.
Riverside at low altitude. S3; throughout tropical Africa. Scassellati & Mazzocchi s.n.
Only recorded from the Juba valley at "Elvalda" by Chiovenda, Result. Sci. Miss. Stefanini-Paoli 1: 226 (1916), as *P. proliferum* var. *longijubatum*. The single collection cited has not been seen and the record needs confirmation.

10. **P. laticomum** Nees (1841).
Annual; stems up to 200 cm long, slender, scrambling, decumbent or geniculately ascending; leaves lanceolate, 6–10 cm × (5–)10–28 mm, abruptly or asymmetrically narrowed at the base, acuminate. Panicle ovate, obovate or oblong, (6–) 10–20(–24) cm long, finely and profusely branched. Spikelets narrowly ovate, 1.5–2 mm long, asymmetric in profile, sparsely to densely hairy, rarely glabrous, acute; lower glume broadly ovate, 1/2 the length of the spikelet, 3-nerved, acute, separated from the upper by a short internode; upper glume 2/3 the length of the spikelet; upper lemma granulose, pallid, exposed at the tip.
In deep shade in riverine forest. S3; tropical Africa southwards to Natal. Maunder 76.

53. ECHINOCHLOA P. Beauv. (1812), nom. cons.

Annual or perennial; ligule a line of hairs or absent. Inflorescence composed of racemes arranged along a central axis. Spikelets paired or in short secondary racemelets, typically densely packed in 4 rows, narrowly elliptic to subrotund, flat on one side, gibbous on the other, often hispidulous, cuspidate or awned at the tip; glumes acute to acuminate, the lower about 1/3 the length of the spikelet; lower floret male or barren, its lemma often stiffly awned, with or without a palea; upper lemma crustaceous, smooth and glossy, its margins inrolled and clasping only the edges of the palea, terminating in a short membranous laterally compressed incurved beak; upper palea acute, the tip shortly reflexed and slightly protuberant from the lemma. Caryopsis broadly elliptic, dorsally flattened.

30—40 species in tropical and warm temperate regions of the world.

All species of *Echinochloa* are difficult to name because there is so much variation within species and intergradation between them. *E. pyramidalis* and *E. stagnina* always have a ligule, *E. haploclada* may or may not. *E. colona* is always annual.

1. Ligule absent 2
− Ligule represented by a line of hairs, at least in the lower leaves; perennials 4
2. Racemes distinctly compound with short secondary branchlets, the inflorescence untidily ovate; spikelets 2−3(−3.5) mm long, often with a short curved awn .... 1. *E. cruspavonis*
− Racemes not or indistinctly compound 3
3. Annual; spikelets 1.5−3(−3.5) mm long..........
.... 2. *E. colona*
− Perennial; spikelets 1.5−2.5 mm long (if 3 mm or more see 4. *E. pyramidalis*) .... 3. *E. haploclada*
4. Spikelets seldom over 2.5 mm long, crowded, commonly rotund, often awned 3. *E. haploclada*
− Spikelets seldom under 3 mm long, elliptic 5
5. Spikelets awnless, rarely with a subulate point up to 3 mm long, plump, 2.5−4 mm long; stems robust, erect .... 4. *E. pyramidalis*
− Spikelets awned, tapering, narrowly ovate, 3.5−6 mm long; stems spongy, decumbent or floating .... 5. *E. stagnina*

1. **E. cruspavonis** (Kunth) Schult. (1824).

Robust perennial (rarely annual); stems stout, up to 200 cm high, often decumbent and rooting in mud from the lower nodes; ligule absent. Inflorescence large, loose, untidily ovate, 10−30 cm long, the racemes mostly compound with short branchlets; racemes 3−15 cm long. Spikelets elliptic, 2−3(−3.5) mm long, hispid; lower floret male or barren, acute to acuminate or with a short curved awn 1−3(−7) mm long; upper lemma 2−2.5(−3) mm long.

Weed of irrigated fields. S2, 3; tropical and South Africa, and tropical America. Hansen 6060; Gorini 497.

2. **E. colona** (L.) Link (1833). Fig. 127.

Agar, belbeteti, biyeis, domar, dorar, shalbir (Som.).

Annual; stems up to 100 cm high, erect or ascending; ligule absent. Inflorescence typically linear, 1−15 cm long, the racemes neatly 4-rowed; racemes seldom over 3 cm long, simple, commonly half their

Fig. 127. *Echinochloa colona*. A: habit, × 2/3. B: ligule, × 2. C: portion of raceme, × 4. − Modified from Fl. Trop. E. Afr. (1982). Drawn by D. Erasmus.

length apart and appressed to the axis, but sometimes subverticillate and spreading (rarely forming a lanceolate head, but then the spikelets purplish with the lower floret male). Spikelets ovate-elliptic to subglobose, 1.5−3(−3.5) mm long, pubescent (rarely shortly hispid); lower floret male or barren, the lemma acute to cuspidate (rarely with a subulate point up to 1 mm long); upper lemma 2−3 mm long.

Floodplains, marshes and riverbanks, damp or shady depressions, coastal dunes and as a weed of irrigated fields; 0−1520 m. N1−3; C2; S1−3; throughout the tropics and subtropics. Hemming 1283; McKinnon 256; Thulin & Warfa 5989; Kuchar 15657; Gillett & Hemming 24656A; Bavazzano 66; Rose Innes 645.

3. E. haploclada (Stapf) Stapf (1920).
E. haploclada var. *stenostachya* Chiov., Fl. Somala 2: 445 (1932); type: S3, Kismayu, Balladelli 269 (K iso.).

Dooraar, gar-garo, sabool (Som.).

Rhizomatous perennial; stems up to 300 cm high, often wiry; ligule absent or a line of hairs. Inflorescence lanceolate (rarely linear), 7−25 cm long; racemes 1−5 cm long, densely crowded with appressed spikelets. Spikelets elliptic to subglobose, small, 1.5−2.5(−3) mm long, ± hispid; lower floret male, acute or with a short curved awn up to 5 mm long; upper lemma 1.5−2.3 mm long.

Marshes, floodplains and paddy fields; 20−670 m. N1; C1; S1−3; Sudan and Ethiopia southwards to Zimbabwe. Hemming 2035; Kuchar 17829; Beckett 1482; Kuchar 17493; Kazmi 5250.

4. E. pyramidalis (Lam.) Hitchc. & Chase (1917); *Panicum pyramidale* Lam. (1791).
P. crusgalli L. var. *polystachium* Aschers. & Schweinf. forma *aristatum* Chiov. in Ann. R. Ist. Bot. Roma 7: 63 (1897) and f. *muticum* Chiov., loc. cit. (1897).

Reed-like rhizomatous perennial; stems robust, up to 400 cm high, firm, erect; ligule a line of hairs. Inflorescence ovate to narrowly lanceolate with the racemes overlapping or linear with the racemes ± distant, 8−40 cm long; racemes simple or compound, 3−20 cm long. Spikelets narrowly ovate to broadly elliptic, plump, 2.5−3.5(−4) mm long, glabrous to hispid; lower lemma acute to acuminate, rarely with a subulate awn-point up to 3 mm long; upper lemma 2−3 mm long.

Seasonally flooded alluvial clay; c. 100 m. S2; tropical and South Africa and Madagascar. Rose Innes 612.

5. E. stagnina (Retz.) P. Beauv. (1812).
Rhizomatous perennial; stems spongy, up to 200 cm high, decumbent and rooting from the nodes, often floating on water; ligule a line of hairs. Inflorescence ovate to narrowly lanceolate, 6−25 cm long, typically open with the racemes secund, flexuous and ± nodding, but displaying much variation; racemes simple, 2−8 cm long. Spikelets narrowly ovate,

3.5−6 mm long, hispid often from tubercles (rarely glabrescent); lower floret male or barren, the lemma tapering to an awn (1−)3−20(−50) mm long; upper lemma 3−5 mm long.

Riverbanks and irrigation ditches, often spreading across open water by means of floating stems. S2; tropical Africa eastwards to Assam and Indo-China. Moggi & Bavazzano 2757; Terry 3371.

54. ALLOTEROPSIS J. Presl (1830)

Annual or perennial; leaves inrolled to lanceolate; ligule membranous or a line of hairs. Inflorescence composed of irregular racemes, these digitate or in whorls on a short central axis. Spikelets ovate to elliptic, dorsally compressed; glumes unequal, acute to shortly awned, the lower shorter than the spikelet, the upper as long as the spikelet and ciliate on the margins; lower floret male, its lemma resembling the upper glume but eciliate, the palea short and deeply bifid; upper lemma crisply chartaceous, acuminate to a short awn, the margins inrolled and covering only the edges of the acute palea.

Five species in the Old World tropics.

A. cimicina (L.) Stapf (1919). Fig. 128.
Annual up to 120 cm high, erect or ascending; leaves narrowly lanceolate to narrowly ovate, ± amplexicaul. Inflorescence of 4−11 racemes, usually in 1 whorl; racemes 7−25 cm long, bare of spikelets below. Spikelets narrowly ovate-elliptic, 3.5−5.5 mm long; lower glume 1/2−3/4 the length of the spikelet; upper glume subulate-acuminate, cartilaginous, smooth and glossy; upper lemma with an awn 2−4.5 mm long, the palea papillose with globular hairs.

Dry open places. S3; throughout the Old World tropics. Kazmi 5246; Moggi & al. 351; Tardelli & Bavazzano 889.

55. BRACHIARIA (Trin.) Griseb. (1853)

Annual or perennial; leaves linear to lanceolate. Inflorescence of racemes along a central axis, the rhachis filiform to ribbon-like. Spikelets single or paired, rarely in fascicles or secondary racemelets, sessile or pedicelled, adaxial, plump, sometimes the lowest internode elongated and accrescent to the sheathing base of the lower glume and forming a short stipe; lower glume shorter than the spikelet; lower floret male or barren; upper lemma coriaceous to crustaceous, obtuse to acute, occasionally mucronate, its margins inrolled and covering only the edges of the palea; upper palea obtuse to subacute, its tip tucked within the lemma. Caryopsis elliptic, dorsally compressed.

Some 100 species in the tropics, mainly in the Old World.

1. Rhachis of racemes flat, ± ribbon-like, narrowly winged ....1. *B. mutica*
   - Rhachis of racemes solid, ± triquetrous or crescentic in section 2
2. Plants perennial 3
   - Plants annual 9
3. Spikelets in pairs or on short side branches, loosely or irregularly arranged on the raceme .. 4
   - Spikelets single (occasionally in pairs), neatly and densely packed on the raceme; leaves linear 6
4. Spikelets 7–8.5 mm long; leaves narrowly lanceolate ....6. *B. longiflora*
   - Spikelets 3–5 mm long 5
5. Leaves narrowly lanceolate, cordate at the base ....7. *B. chusqueoides*
   - Leaves linear, not cordate at the base ..8. *B. arida*
6. Racemes 1-rowed; spikelets glabrous to sparsely pubescent, 4–6 mm long, obtuse to subacute .... 2. *B. brizantha*
   - Racemes 2-rowed; spikelets pubescent to villous 7
7. Upper glume and lower lemma nearly always with a transverse fringe of hairs near the tip, cartilaginous, sharply acute to subulate.......... .... 3. *B. serrata*
   - Upper glume and lower lemma not transversely fringed, membranous, acute to acuminate 8
8. Leaves 5–40 cm long, green, soft, not conspicuously distichous .... 4. *B. lachnantha*
   - Leaves 2–7 cm long, glaucous, rigid with thickened margins, conspicuously distichous...... ....5. *B. stefaninii*
9. Upper lemma smooth, shining, obtuse; spikelets imbricate; lower glume up to 1/5 the length of the spikelet ....9. *B. eruciformis*
   - Upper lemma granulose to rugose, subacute to mucronate 10
10. Glumes separated by a distinct internode; upper lemma coarsely rugose; racemes slender, distant, secund ....10. *B. leersioides*
    - Glumes adjacent, or if slightly separated then upper lemma not coarsely rugose 11
11. Racemes compound, bearing the spikelets in dense fascicles or on short secondary branchlets .... 14. *B. deflexa*
    - Racemes simple, the spikelets borne singly or in pairs, exceptionally the longest racemes with a few secondary branchlets 12
12. Spikelets 1.5–2.5 mm long, without a stipe 13
    - Spikelets 2.5–5 mm long, sometimes with a stipe 14
13. Spikelets silvery-villous, borne singly; upper lemma granulose .... 11. *B. breviglumis*
    - Spikelets glabrous, borne in pairs; upper lemma rugose, mucronulate ....12. *B. reptans*
14. Spikelets mostly single, neatly 1- to 2-rowed, hispidulous, 4–5 mm long ... 13. *B. leucacrantha*
    - Spikelets mostly paired, ± loosely or irregularly arranged, glabrous to pubescent 15

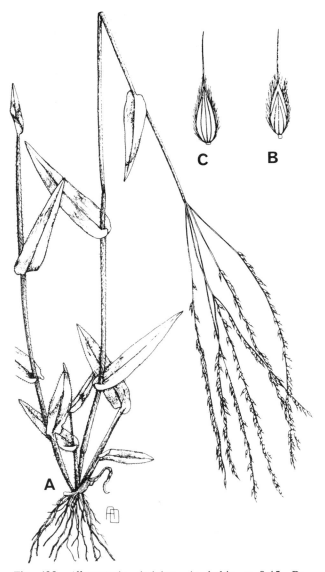

Fig. 128. *Alloteropsis cimicina.* A: habit, × 0.45. B: spikelet, showing lower glume, × 3.6. C: spikelet, showing upper glume, × 3.6. – Modified from Fl. Trop. E. Afr. (1982). Drawn by A. Davies.

15. Pedicels, or some of them, longer than the spikelets; spikelets distant, spreading, the racemes sometimes compound .... 14. *B. deflexa*
    - Pedicels shorter than the spikelets; spikelets contiguous and appressed to the racemes, these usually simple, but the longest may have branchlets at the base 16
16. Upper glume and lower lemma thinly cartilaginous, usually smooth and glossy, but sometimes papillose-pubescent; spikelets 2.5–4.5 mm long ....15. *B. ovalis*
    - Upper glume and lower lemma membranous, glabrous or pubescent; spikelets 2.5–3.5 mm long ....16. *B. ramosa*

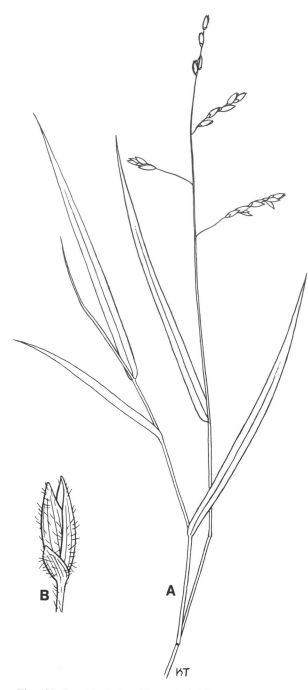

Fig. 129. *Brachiaria longiflora.* A: habit, × 0.7. B: spikelet, × 4. — From Thulin & al. 7145. Drawn by K. Thunberg.

1. **B. mutica** (Forssk.) Stapf (1919).

Sprawling perennial up to 125 cm high, prostrate and rooting from the lower nodes; leaves broadly linear. Inflorescence of 5−20 racemes on an axis 7−20 cm long; racemes 2−10 cm long, bearing paired spikelets in several untidy rows on a narrowly (0.5−1 mm) winged rhachis (sometimes the spikelets on short secondary branchlets below or borne singly above). Spikelets elliptic, 2.5−3.5 mm long, glabrous, acute; lower glume 1/4−1/3 the length of the spikelet; upper glume adjacent to the lower; upper lemma rugulose, obtuse, with an obscure mucro.

Probably wet grassland at low altitude. S2; throughout the tropics, obviously introduced in Somalia. Guidotti s.n.

Reported for Somalia by Chiovenda, Fl. Somala 2: 443 (1932), but the only collection cited, Guidotti s.n., has not been seen.

2. **B. brizantha** (Hochst. ex A. Rich.) Stapf (1919).

Tufted perennial up to 200 cm high, erect or sometimes geniculately ascending; leaves linear to broadly linear. Inflorescence of (1−)2−16 racemes on an axis 3−20 cm long; racemes 4−20 cm long, bearing the spikelets singly in a single row on a rhachis crescentic in section and c. 1 mm wide. Spikelets plumply elliptic, 4−6 mm long, glabrous or sometimes sparsely pubescent, obtuse to subacute, with a slight stipe at the base; lower glume c. 1/3 the length of the spikelet, clasping, acute or obtuse; upper glume separated from the lower by a very short internode; upper lemma granulose, acute.

?N1; tropical and southern Africa, introduced elsewhere in the tropics.

Reported from northern Somalia in Cuf. Enum.: 1312 (1969), but no collection has been seen and the record needs confirmation.

3. **B. serrata** (Thunb.) Stapf (1919).

Panicum serratum Thunb. var. *gossypium* (A. Rich.) T. Durand & Schinz, Consp. Fl. Afric. 5: 765 (1895).

Densely tufted perennial up to 100 cm high, the basal sheaths silky-tomentose; leaves narrowly lanceolate, glabrous to softly pilose, the margins ± cartilaginous with scattered hooked teeth. Inflorescence of 5−15 racemes on an axis 3−15 cm long; racemes short and compact, 0.5−2 cm long, bearing the spikelets singly on a triquetrous rhachis. Spikelets ovate-elliptic, 2.3−4.5 mm long, with a transverse fringe of pink or silvery hairs 1−2 mm long near the tip (rarely the fringe missing), otherwise pubescent, cuspidate or shortly awned, without a stipe; lower glume 1/2 the length of the spikelet, subacute; upper glume thinly cartilaginous; lower lemma similar to the upper glume but membranous about the midnerve; upper lemma striate, acute.

Open bushland on stony hillsides; 160−1650 m. N1; tropical and South Africa. Hansen & al. 6447; Hemstra 3012.

4. **B. lachnantha** (Hochst.) Stapf (1919); *Panicum lachnanthum* Hochst. (1855).

Garagarad (Som.).

Densely tufted perennial up to 100 cm high, the basal sheaths silky-tomentose; leaves broadly linear to narrowly lanceolate, 5−40 cm × 3−10 mm, green, soft, not conspicuously distichous. Inflorescence of

2−15 racemes on an axis 2−15 cm long; racemes 2−5 cm long, bearing the spikelets singly (or the longer racemes with paired spikelets at the base) on a pubescent to villous triquetrous rhachis. Spikelets elliptic, 3−4.5 mm long, pubescent to villous, acute, with a very short stipe; lower glume 1/4−1/3 the length of the spikelet, clasping; upper glume and lower lemma membranous; upper lemma faintly striate, subacute.

Black soils in *Juniperus* forest and open grassland; 1400−1650 m. N1; S2, 3; Ethiopia, Kenya, Uganda and Tanzania. Hansen & al. 6416; Cufodontis 504; Kazmi 5172.

5. B. stefaninii Chiov. (1928); type: N3, "Dafurieroi" to "Gheideli", Puccioni & Stefanini 909 (FT holo.).

Densely tufted perennial up to 40 cm high, profusely branched above a thick woody base, erect, the basal sheaths velvety-pubescent; leaves broadly linear to narrowly lanceolate, 2−7 cm × 2−4 mm, glaucous, rigid with thickened margins, conspicuously distichous. Inflorescence of 2−4 racemes on an axis 2−5 cm long; racemes 1−1.5 cm long, bearing the spikelets singly on a glabrous or puberulous triquetrous rhachis. Spikelets narrowly ovate to ovate, 3−4 mm long, villous, acuminate, with a very short stipe; lower glume 1/3 the length of the spikelet, clasping; upper glume and lower lemma membranous; upper lemma striate, acute and faintly mucronulate.

Limestone pavement and sandy plains; c. 300 m. N3; not known elsewhere. Bally & Melville 15608; Hemming 1859.

6. B. longiflora Clayton (1980). Fig. 129.

Tufted perennial from a woody rootstock, up to 120 cm high, the stems stiff and wiry; leaves narrowly lanceolate. Inflorescence of 2−4 racemes on an axis 3−7 cm long; racemes 2−5 cm long, bearing paired spikelets loosely spaced along a thinly villous triquetrous rhachis. Spikelets narrowly elliptic, 7−8.5 mm long, sparsely pubescent, acute, with a stipe 0.5−1 mm long; lower glume 1/4(−1/3) the length of the spikelet, clasping, separated from the upper by a short internode; upper glume and lower lemma chartaceous; upper lemma faintly striate, obtuse.

Open bushland on fixed dunes; c. 120 m. S2, 3; Ethiopia and Kenya. Rose Innes 872; Thulin, Hedrén & Abdi Dahir 7145.

7. B. chusqueoides (Hack.) Clayton (1980).

Scandent or creeping perennial up to 75 cm high, the stems slender and wiry; leaves narrowly lanceolate, cordate at the base. Inflorescence of 2−7 racemes widely spaced on an axis 2−13 cm long; racemes 1.5−7 cm long, bearing paired spikelets (at least in the middle of the raceme) loosely contiguous on a triquetrous rhachis. Spikelets elliptic, 3−5 mm long, glabrous, acute, with a short stipe 0.2−0.5 mm

long; lower glume 1/3−1/2 the length of the spikelet, clasping; upper glume separated from the lower by a very short internode; upper lemma rugose, acute.

In wet ground. S2; tropical Africa southwards to South Africa, and in Arabia (Yemen). Bavazzano 264, 1238.

8. B. arida (Mez) Stapf (1919); type: N Somalia, without precice locality, Hildebrandt 1483 (?B syn.).

Tufted, shortly rhizomatous perennial up to 50 cm high, erect or sprawling; leaves linear. Inflorescence of 4−5 racemes widely spaced on an axis 8−14 cm long; racemes 2.5−4 cm long, bearing paired spikelets (at least in the middle of the raceme) loosely contiguous on a triquetrous rhachis. Spikelets elliptic, 4−4.5 mm long, glabrous, acute or subacute, with a stipe 0.2−0.3 mm long; lower glume c. 1/3 the length of the spikelet, clasping; upper glume separated from the lower by a very short internode; upper lemma rugose, acute.

?N2; Socotra.

A little-known species described from material collected on Socotra and in Somalia, and only the syntype is so far known from Somalia.

9. B. eruciformis (Sm.) Griseb. (1853). Fig. 130.

Kule-kule (Som.).

Annual up to 60 cm high, slender, geniculately ascending; leaves linear to narrowly lanceolate, glabrous or pubescent. Inflorescence of 3−14 racemes on an axis 1−8 cm long; racemes 0.5−2.5 cm long, secund, bearing single spikelets imbricate on a triquetrous rhachis. Spikelets elliptic, 1.7−2.7 mm long, pubescent, subacute, without a stipe; lower glume up to 1/5 the length of the spikelet; upper glume adjacent to the lower; upper lemma readily deciduous, smooth and shining, obtuse.

Weed of cultivated fields and wasteland; 300−400 m. S1, 2; Mediterranean region eastwards to India and southwards to South Africa. Beckett 1628; Eagleton 48; Bavazzano 69; Warfa 1014.

10. B. leersioides (Hochst.) Stapf (1919); *Panicum leersioides* Hochst. (1855).

Annual up to 100 cm high, ascending; leaves linear, setaceously acuminate. Inflorescence of 3−14 widely spaced racemes, these spreading horizontally or deflexing at maturity, borne on an axis 3−20 cm long; racemes 1−7 cm long, slender, secund, bearing paired spikelets on a triquetrous rhachis. Spikelets narrowly elliptic, 2−3.5 mm long, glabrous, subacute, without a stipe; lower glume 1/3−1/2 the length of the spikelet, clasping; upper glume separated from the lower by a distinct internode 0.2−0.5 mm long; upper lemma coarsely rugose, subacute.

Open *Acacia* woodland on red sand; 300−1900 m. N1; C1; S1, 3; tropical Africa and Arabia. Glover & Gilliland 1002; Herlocker 439; Moggi & Bavazzano 73; Fiori s.n.

Fig. 130. *Brachiaria eruciformis*. A: habit, × 2/3. B: portion of inflorescence, × 6. — Modified from a drawing by D. Erasmus.

11. B. breviglumis Clayton (1980).

Annual up to 25 cm high, ascending; leaves narrowly lanceolate. Inflorescence of 3—5 racemes appressed to an axis 1—6 cm long; racemes 0.5—1.5 cm long, bearing single spikelets on a triquetrous rhachis. Spikelets broadly elliptic, 2 mm long,

silvery-villous, the hairs concealing the spikelet and extending up to 1 mm beyond its tip, acute; lower glume rotund, very short (0.2 mm); upper lemma granulose, acute.

Acacia-Commiphora shrubland on silty soils; 100—200 m. C2, S1—3; Ethiopia and Kenya. Kuchar 17536A; Thulin & Bashir Mohamed 6976; Kilian & Lobin 6631; Moggi & Bavazzano 1988.

12. B. reptans (L.) C.A. Gardner & C.E. Hubb. (1938); *Urochloa reptans* (L.) Stapf (1920).
    B. prostrata (Lam.) Griseb. (1857).

Annual up to 60 cm high, usually decumbent and rooting from the nodes; leaves narrowly lanceolate to lanceolate. Inflorescence of 5—15 racemes on an axis 1—8 cm long; racemes 1—4 cm long, bearing paired spikelets crowded on a triquetrous rhachis. Spikelets narrowly ovate to broadly elliptic, 1.5—2.2 mm long, glabrous, acute, without a stipe; lower glume up to 1/4 the length of the spikelet, hyaline, clasping, truncate (rarely a little longer and broadly ovate); upper glume adjacent to the lower; upper lemma rugose, subacute, mucronulate.

Weed of cultivated and irrigated land; 0—150 m. C2; S2, 3; tropical Africa and tropical Asia, but an introduced weed throughout the tropics. Kuchar 17593; Hansen 6067; Rose Innes 644.

13. B. leucacrantha (K. Schum.) Stapf (1919).

Annual up to 40 cm high, often prostrate and rooting from the nodes; leaves linear to narrowly lanceolate, velvety-pubescent. Inflorescence of 3—5 racemes on an axis 2—8 cm long; racemes 1—6 cm long, bearing single, loosely contiguous spikelets in 1—2 rows on a triquetrous rhachis. Spikelets narrowly elliptic, 4—5 mm long, pubescent, becoming pilose above with the hairs forming a loose tuft on each side of the midnerve, caudate-acuminate, with a stipe c. 0.5 mm long; lower glume 1/3—1/2 the length of the spikelet; upper lemma papillose, subacute.

Open *Commiphora* bushland on red or orange sand; c. 150—200 m. C2; S2, 3; Uganda, Kenya, Tanzania, Zaire and Mozambique. Kuchar 17639; McLeish 863; Moggi & Bavazzano 1994.

14. B. deflexa (Schumach.) C.E. Hubb. ex Robyns (1932).
    B. regularis (Nees) Stapf (1919).
    B. clavuliseta Chiov. (1926); type: S3, Kismayu, Gorini 19 (FT holo., K iso.).
    Buldorle agar (Som.).

Annual up to 70 cm high, often weak and ascending; leaves broadly linear. Inflorescence of 7—15 racemes on an axis 6—15 cm long; racemes 2—10 cm long, often compound, bearing mostly paired spikelets spreading from the triquetrous rhachis, the inflorescence imitating a panicle; pedicels, or some of them, longer than the spikelets, up to 15 mm long. Spikelets broadly elliptic, 2.5—3.5 mm long, glabrous to pubescent, acute, with a stipe up to 0.5 mm long; lower glume 1/3−1/2 the length of the spikelet; upper glume ± adjacent to the lower; upper lemma rugose, subacute to acute.

Acacia shrubland on red sand, and occasionally on saline soils; 40−400 m. N1; C2; S1−3; tropical and southern Africa, through Arabia to India. Hansen & Heemstra 6177; Thulin & Abdi Dahir 6401; O'Brien 63; Alstrup & Michelsen 63; Kazmi 5214.

15. **B. ovalis** Stapf (1919); type: N Somalia, Thomson s.n. (K syn.), Appleton s.n. (K syn.).

B. glauca Stapf (1919).

B. somalensis C.E. Hubb. (1941); type: N1, near "Buramo", "Warieta" tug, Gillett 4903 (K holo.).

Baldowley, bashi, buldorle agar, caws qaalwaweyn, wiil xaaris (Som.).

Annual up to 50 cm high, often densely pubescent; leaves broadly linear to narrowly lanceolate. Inflorescence of 3−15 spreading or appressed racemes on an axis 4−15 cm long; racemes 3−7 cm long, bearing single or paired spikelets loosely spaced on a triquetrous rhachis; pedicels shorter than the spikelets. Spikelets elliptic, 2.5−4.5 mm long, glabrous or papillose-pubescent, obtuse to subacute, with a stipe 0.3−0.7 mm long; lower glume 1/4−1/2 the length of the spikelet, broadly ovate; upper glume and lower lemma thinly cartilaginous, smooth and glossy; upper lemma granulose to rugulose, subacute to mucronulate.

Acacia bushland on limestone hills, alluvial basins along rivers, and as a weed of cultivation; 150−1600 m. N1−3; C1, 2; S1−3. Sudan, Kenya, Arabia and Pakistan. Farquharson 16; Glover & Gilliland 158; Bally & Melville 15709; Herlocker 404; Kuchar 17536; Eagleton 28; Moggi & Bavazzano 272; Kazmi 5193.

A record of *B. lata* (Schumach.) C.E. Hubb. (syn. *Urochloa insculpta* (Steud.) Stapf) in Chiov., Fl. Somala 2: 445 (1932) and Cuf. Enum.: 1314 (1969) from southern Somalia is based on Gorini 505, which is *B. ovalis*.

16. **B. ramosa** (L.) Stapf (1919).

Panicum petiveri Trin. (1826).

Dalat (Som.).

Annual up to 70 cm high; leaves broadly linear. Inflorescence of 3−15 racemes on an axis 3−10 cm long; racemes 1−8 cm long, simple or the longest with branchlets at the base, bearing mostly paired, loosely contiguous spikelets appressed to the triquetrous rhachis; pedicels shorter than the spikelets, 1−2 mm long. Spikelets elliptic to broadly elliptic, 2.5−3.5 mm long, glabrous or pubescent, acute to cuspidate, with or without a stipe up to 0.5 mm long; lower glume 1/3−1/2 the length of the spikelet; upper glume and lower lemma membranous, or rarely the latter coriaceous; upper lemma rugose, subacute to acute.

Shaded places on silt or clay, often heavily grazed; 150−300 m. C2; S1, 2; tropical Africa southwards to South Africa, through Arabia to tropical Asia. Kuchar 17602; Eagleton 40; Dahir & Rose Innes ODA8.

56. UROCHLOA P. Beauv. (1812)

Annual or perennial; leaves linear to lanceolate; ligule a line of hairs. Inflorescence of racemes along an axis. Spikelets single or paired, abaxial, plano-convex, cuspidate to acuminate; lower glume shorter than the spikelet; upper glume membranous to firmly chartaceous; lower floret male or barren, its lemma similar to the upper glume; upper lemma coriaceous, obtuse with a long mucro housed within the spikelet, its margins inrolled and covering only the edges of the palea; upper palea obtuse. Caryopsis broadly elliptic to subrotund, strongly flattened.

12 species in the Old World tropics, mainly Africa.

1. Lower glume over 2/3 the length of the spikelet .... 3. *U. trichopus*
− Lower glume up to 1/2 the length of the spikelet 2
2. Spikelets warty; palea of lower floret coriaceous about the keels ....1. *U. rudis*
− Spikelets not warty; palea of lower floret hyaline to membranous .... 2. *U. panicoides*

1. **U. rudis** Stapf (1920); type: C1, "Obbia", Drake-Brockman 954 (K holo.).

U. gorinii Chiov. (1926); type: S3, Kismayu, Gorini 35 (FT holo.).

Creeping annual, rooting from the lower nodes, up to 100 cm high; leaves broadly linear to narrowly lanceolate, pubescent to velvety-tomentose. Inflorescence of 2−5 racemes on an axis 2−6 cm long; racemes 2−6 cm long, bearing single spikelets on a narrowly winged rhachis. Spikelets broadly elliptic, plumply plano-convex, 3.5−5 mm long, cuspidate; lower glume cuff-like, up to 1/4 the length of the spikelet, truncate; upper glume tuberculate, the tubercles often with spines or bristles; lower lemma tuberculate; lower palea with hyaline centre (middle 1/3−1/2) flanked by wide coriaceous strips along the keels; upper lemma submuricate, with a mucro 0.5 mm long.

Fixed dunes with grassland or shrubland cover, and limestone grassland; 0−350 m. C1, 2; S2, 3; Kenya, Tanzania and Zanzibar. Thulin & al. 7299; Kuchar 17562; Thulin & Warfa 7115; Gillett & al. 25318A.

2. **U. panicoides** P. Beauv. (1812).

U. helopus (Trin.) Stapf (1920).

U. helopus var. *hochstetterianum* (A. Rich.) Chiov., Fl. Somala 2: 444 (1932).

Agar fowdo, farsho.

Annual up to 100 cm high, often ascending from a

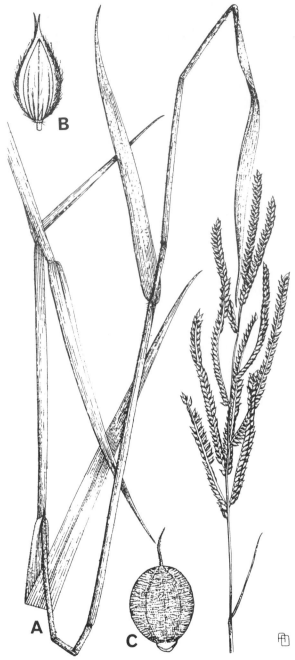

long, acute; lower glume ovate, 1/4−1/2 the length of the spikelet, 3- to 5-nerved, obtuse to subacute; upper glume smooth; lower lemma smooth, sometimes with a setose fringe; lower palea hyaline throughout; upper lemma rugulose, with a mucro 0.3−1 mm long.

Open *Acacia* or mixed shrubland, seasonally flooded depressions, irrigated fields and farms, gardens and wasteland; 50−1500 m. N1; C1, 2; S1−3; East tropical Africa southwards to South Africa and eastwards to India, introduced in Australia. Gillett 3913; Moggi & al. 143; Kuchar 17464; Eagleton 27; Alstrup & Michelsen 61; Kazmi & al. 616.

3. **U. trichopus** (Hochst.) Stapf (1920). Fig. 131.
   Eriochloa trichopus (Hochst.) Benth. (1881).
   Kurdo (Som.).

Coarse annual up to 170 cm high, ascending; leaves broadly linear, hispid. Inflorescence of 3−20 racemes on an axis 4−20 cm long; racemes 1−14 cm long, bearing single spikelets on a narrowly winged rhachis. Spikelets ovate, 2.5−5.5 mm long, acuminate or with a short awn-point; lower glume elliptic-oblong, 2/3−5/6 the length of the spikelet, 3-nerved, often with a tuft of hairs from the middle of the back, obtuse; lower lemma often with a ciliate fringe; upper lemma papillose, becoming rugulose on the flanks, with a mucro 0.5 mm long.

Open deciduous woodland on fixed dunes and silty clays; 0−50 m. S2, 3; tropical Africa, mainly in the east, and Arabia. Rose Innes 877; Gillett & al. 24935.

57. ERIOCHLOA Kunth (1816)

Annual or perennial; leaves linear; ligule a line of hairs. Inflorescence of numerous racemes along a central axis, rarely a panicle contracted about the primary branches. Spikelets single, paired or on short secondary branchlets, adaxial, thinly biconvex, acute to aristate, supported on a bead-like swelling formed from the lowest rhachilla-internode and adherent vestigial lower glume (rarely the latter developed); upper glume as long as the spikelet, often with an awn-point; lower floret male or barren, the lemma resembling the upper glume but a little shorter; upper lemma crustaceous, papillose, obtuse and mucronate, its margins inrolled and clasping only the edges of the palea; upper palea obtuse. Caryopsis oblong-elliptic, dorsally compressed, obtuse.

30 species in the tropics.

1. Lower glume present; upper glume acute; lower floret with a palea; plant perennial...............
   .... 1. *E. meyeriana*
 − Lower glume apparently absent; upper glume attenuate to an awnlet up to 4 mm long; lower floret without a palea; plant annual..............
   .... 2. *E. fatmensis*

Fig. 131. *Urochloa trichopus*. A: habit, × 0.5. B: spikelet, showing lower glume, × 6. C: upper glume, × 10. − Modified from Fl. Trop. E. Afr. (1982). Drawn by A. Davies.

prostrate rooting base; leaves linear to narrowly lanceolate, subamplexicaul, coarse, glabrous or pubescent, the margins tuberculate-ciliate at least near the base. Inflorescence of 2−7 racemes on an axis 1−9 cm long; racemes 1−7 cm long, bearing single or paired spikelets on a narrowly winged rhachis. Spikelets elliptic, (2.5−)3.5−4.5(−5.5) mm

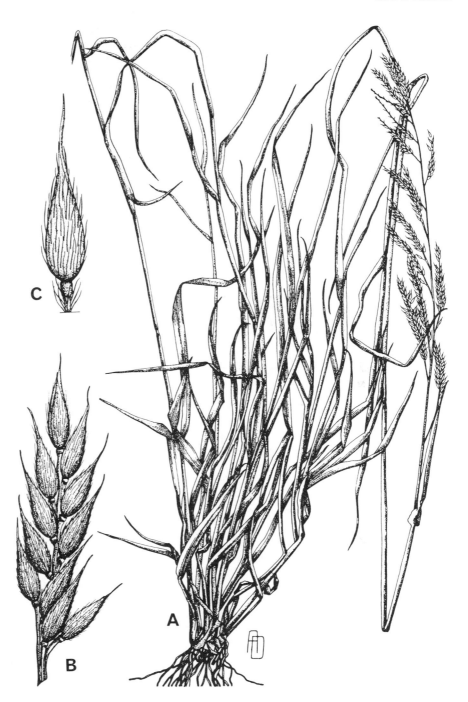

Fig. 132. *Eriochloa fatmensis*. A: habit, × 0.5. B: raceme, × 5. C: spikelet, × 10. – From Fl. Trop. E. Afr. (1982). Drawn by A. Davies.

1. **E. meyeriana** (Nees) Pilg. (1940); *Panicum meyerianum* Nees (1841).

P. schimperianum Hochst. ex A. Rich. (1850).

Caws weyn (Som.).

Perennial up to 200 cm high; stems ascending and rooting from the lower nodes, often scrambling, sometimes almost woody. Inflorescence 8–18 cm long; racemes 2–7 cm long, with puberulous triquetrous rhachis, the spikelets paired or more often densely clustered on short secondary branchlets. Spikelets elliptic, 2.3–3.5 mm long, glabrous or occasionally thinly pubescent; lower glume a truncate or cuspidate cuff up to 0.5 mm long; upper glume acute; lower floret with a palea; upper lemma obscurely mucronate.

Swamps, streamsides and wet wooded grassland; 0–50 m. S2, 3; tropical and South Africa and Arabia. Friis & al. 4624; Hemming & Deshmukh 135.

2. **E. fatmensis** (Hochst. & Steud.) Clayton (1975). Fig. 132.

E. acrotricha (Steud.) Hack. ex Thell. (1907).

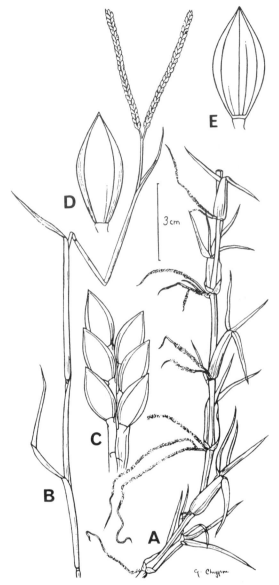

Fig. 133. *Paspalum vaginatum*. A: stolon. B: portion of stem with inflorescence. C: portion of raceme. D: spikelet, showing upper glume. E: spikelet, showing opposite side. — From Fl. Gabon 5 (1962).

E. nubica (Steud.) Hack. ex Stapf & Thell. (1919).
E. ramosa sensu Chiov. non (Retz.) Kuntze.
Agar, buldorle agar, darabo, dhalad (Som.).
Annual up to 120 cm high, erect or geniculately ascending. Inflorescence 3—20 cm long; racemes 1—5 cm long, the rhachis puberulous, triquetrous or narrowly winged, the spikelets paired. Spikelets lanceolate, (2.5—)3—5 mm long, thinly pubescent; lower glume apparently absent; upper glume attenuate to an awnlet 0.5—4 mm long; lower floret without a palea; upper lemma with a mucro 0.3—1 mm long.

Acacia-Commiphora bushland on sandy and alluvial loams, seasonally flooded soils overlying limestone or gypsum, and areas of cultivation; 150—1550 m. N1—3; C1, 2; S1—3; tropical and South Africa and Arabia, a few records also from India and Thailand. Gillett 3913A; Glover & Gilliland 677; Hansen & Heemstra 6305; Wieland 4530; Gillett & al. 22660; Friis & al. 4791; Bavazzano 337; Rose Innes 773.

Very similar to, and perhaps conspecific with, the Asiatic *E. procera* (Retz.) C.E. Hubb. (syn. *E. ramosa* (Retz.) Kuntze). The two species intergrade and are traditionally distinguished by the tip of the spikelet (acute to acuminate in *E. procera*, with an awn up to 4 mm in *E. fatmensis*) and the pedicel (glabrous in *E. procera*, setose in *E. fatmensis*).

58. PASPALUM L. (1759)

Annual or perennial; leaves linear to narrowly lanceolate; ligule a short membrane. Inflorescence of racemes, these conjugate, digitate or on an elongated central axis, rarely solitary; rhachis flat, narrowly or broadly winged, bearing single or paired spikelets in 2—4 rows. Spikelets orbicular to oblong or ovate, mostly plano-convex, abaxial; lower glume absent, rarely represented by a minute scale; upper glume membranous; lower floret barren, without a palea, the lemma resembling the upper glume; upper lemma coriaceous to crustaceous, usually obtuse, its margins inrolled and clasping only the edges of the palea; upper palea obtuse, sometimes acute but not reflexed. Caryopsis plano-convex.

Some 330 species in the tropics, predominantly in the New World.

P. vaginatum Sw. (1788). Fig. 133.
P. distichum sensu Stapf (1907) non L.
Daat, dehi (Som.).
Creeping stoloniferous perennial up to 60 cm high. Inflorescence of 2(—5) conjugate racemes, the spikelets borne singly in 2 rows on a winged rhachis 1—2 mm wide. Spikelets narrowly ovate-elliptic, 3—4.5 mm long, markedly flattened, acute; lower glume very rarely present as a minute scale; upper glume and lower lemma thinly chartaceous, glabrous; upper lemma smooth.

Wet sand, marshes and at the margins of fresh or brackish pools; 600—1100 m. N1—3; throughout the tropics, extending into the subtropics. Gillett 4925; Hemming 1568, 1674.

59. SETARIA P. Beauv. (1812)

Annual or perennial; leaves flat or folded, sometimes pleated; ligule usually a line of hairs. Inflorescence a panicle, open or spike-like, the spikelets subtended by 1 or more bristles which persist on the axis after the spikelets have fallen. Spikelets oblong to ovate, ± plano-convex, awnless; lower glume ovate from a clasping base, shorter than the spikelet; upper glume as long as or shorter than the spikelet; lower floret male or barren, the lemma herbaceous; upper lemma crustaceous, strongly convex on the back, the margins inrolled and clasping only the edges of the palea. Caryopsis oblong-ellipsoid.

Some 100 species in the tropics and subtropics.

1. Panicle spike-like, sometimes ± lobed; leaves neither pleated nor sagittate 2
− Panicle clearly branched; leaves either sagittate or pleated 6
2. Bristles retrorsely barbed, tenaciously clinging to fur or clothing .... 1. *S. verticillata*
− Bristles antrorsely barbed, not clinging 3
3. Nodes pubescent; upper glume mostly 7- to 9-nerved, covering much of the upper lemma, the latter usually punctate 4
− Nodes quite glabrous; upper glume 3- to 5-nerved, much shorter than the upper lemma, this usually rugose 5
4. Perennial ....2. *S. incrassata*
− Annual ....3. *S. acromelaena*
5. Perennial .... 4. *S. sphacelata*
− Annual .... 5. *S. pumila*
6. Leaves, or at least the lower, sagittate at the base, not pleated .... 6. *S. sagittifolia*
− Leaves entire at the base, pleated fanwise..........
.... 7. *S. megaphylla*

1. **S. verticillata** (L.) P. Beauv. (1812). Fig. 134.

S. verticillata subsp. *aparine* (Steud.) T. Durand & Schinz, Consp. Fl. Afric. 5: 775 (1895).

Dheg-dhegle, marabob (Som.).

Annual up to 100 cm high; leaves broadly linear. Panicle contracted, spike-like, linear to untidily lobed, 2−15 cm long, often entangled, the rhachis hispidulous; bristles 3−8 mm long, retrorsely barbed, tenaciously clinging to clothing or fur. Spikelets ellipsoid, 1.5−2.5 mm long; lower glume 1/3−1/2 the length of the spikelet; upper glume as long as the spikelet; lower floret barren; upper lemma finely rugose.

Grassland, shrubland and as a weed of cultivated ground; 60−1800 m. N1−3; C1; S1−3; tropical and warm temperate regions throughout the world. Gillett 3916; Newbould 801; Scortecci s.n.; Herlocker 442; Hemming & Deshmukh 296; Virgo 64; Gorini 506.

2. **S. incrassata** (Hochst.) Hack. (1891).

S. avettae Pirotta (1896).

Tufted perennial arising from a short rhizome; stems up to 200 cm high, pubescent at the nodes; leaves flat or inrolled, attenuate to a filiform tip. Panicle contracted, spike-like, cylindrical or somewhat interrupted, 3−30 cm long, the bristles pallid, often with purple tips, sometimes wholly purple, the

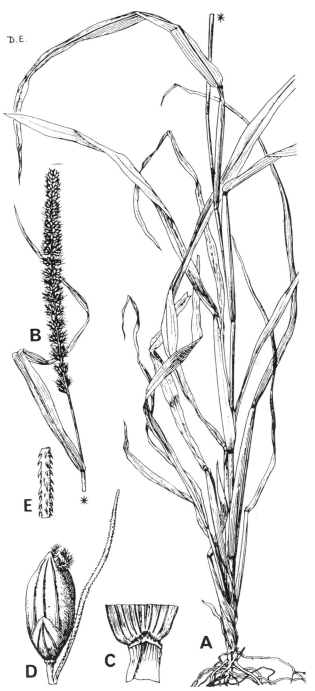

Fig. 134. *Setaria verticillata*. A: habit, × 2/3. B: inflorescence, × 2/3. C: ligule, × 2. D: spikelet and supporting bristle, × 12. E: portion of bristle, × 30. − Modified from Fl. Trop. E. Afr. (1982). Drawn by D. Erasmus.

rhachis tomentellous to sparsely pilose; bristles 2−15 mm long. Spikelets broadly ovate to gibbously suborbicular, somewhat laterally compressed, 2−3 (−4) mm long; lower glume 1/2−2/3 the length of the spikelet; upper glume 2/3 as long to as long as the spikelet, 7- to 9-nerved; lower floret male; upper lemma punctate to obscurely rugose.

Floodplains with clay or alluvial soil. S2, 3; tropical and South Africa. Rose Innes 654; Deshmukh 427.

3. **S. acromelaena** (Hochst.) T. Durand & Schinz (1895).

Annual up to 150 cm high, pubescent at the nodes; leaves broadly linear. Panicle contracted, spike-like, cylindrical, 1–20 cm long, the bristles pallid, often with purple tips, the rhachis pubescent to sparsely hirsute; bristles 3–10 mm long. Spikelets broadly elliptic, strongly gibbous, 2–3 mm long; lower glume 1/2 the length of the spikelet; upper glume 2/3 as long to as long as the spikelet, 7-nerved; lower floret male or barren; upper lemma punctate to finely rugose.

Seasonally inundated grassland and as a weed of sorghum. N1; S1; Sudan, Ethiopia, Uganda, Kenya and Tanzania. Hansen & Heemstra 6143; Terry 3471.

The annual counterpart of *S. incrassata*.

4. **S. sphacelata** Stapf & C.E. Hubb. (1929).
Daraas (Som.).

Tufted, shortly rhizomatous perennial up to 150 cm high, the nodes glabrous; basal sheaths breaking up into fibres; leaves flat or inrolled, acuminate. Panicle contracted, spike-like, cylindrical, 5–35 cm long, the bristles fulvous, the rhachis tomentellous; bristles 1.5–12 mm long. Spikelets elliptic, 1.5–3.5 mm long; lower glume up to 1/2 the length of the spikelet; upper glume 1/3–2/3(–3/4) the length of the spikelet, 3- to 5-nerved; lower floret male; upper lemma rugose.

Along water courses up to c. 1500 m. N1; S3; tropical Africa and Arabia. Drake-Brockman 567; Peveling 228.

A polyploid complex that is apparently uncommon in our area. The complexity of the group is discussed by Hacker, in Austral. J. Bot. 16: 539–544, 551–554 (1968), and Clayton in Kew Bull. 33: 503–506 (1979). The infraspecific taxa recognised by these authors are not all that distinct, but Somali material seems to correspond well with var. *aurea* (A. Braun) Clayton.

5. **S. pumila** (Poir.) Roem. & Schult. (1817).

Annual up to 130 cm high; leaves linear. Panicle contracted, spike-like, cylindrical, 1–10(–20) cm long, the bristles fulvous, the rhachis tomentellous; bristles 3–12 mm long. Spikelets ovate, 1.5–2.5 mm long; lower glume 1/3–1/2 the length of the spikelet; upper glume 1/3–1/2 the length of the spikelet, 3- to 5-nerved; lower floret male or barren; upper lemma rugose to corrugate.

S3; tropical and warm temperate regions of the Old World. Kazmi 5213.

The annual counterpart of *S. sphacelata* from which it can be difficult to distinguish if the basal parts are missing. Collected only once in Somalia.

6. **S. sagittifolia** (Hochst. ex A. Rich.) Walp. (1852).

Weak-stemmed annual up to 80 cm high; leaves broadly linear to narrowly lanceolate, at least the lower falsely petiolate and sagittate, the lobes 2–30 mm long. Panicle narrowly oblong to ovate, 3–15 cm long, the spikelets secund along scaberulous raceme-like primary branches; bristles 2–15 mm long. Spikelets suborbicular, 1.8–2 mm long, laterally compressed; lower glume 1/3 the length of the spikelet; upper glume 1/2 the length of the spikelet; lower floret male or barren; upper lemma conspicuously gibbous, keeled, rugose.

Riverine forest. S3; Sudan and Arabia southwards to South Africa. Kazmi 5192.

The combination of gibbous spikelets and keeled upper lemma was, at one time, considered sufficient to warrant segregation of this and other species into the genus *Cymbosetaria* Schweick. However, they represent only the extreme condition of a continuum, and segregation cannot thus be justified.

7. **S. megaphylla** (Steud.) T. Durand & Schinz (1894).
S. plicatilis (Hochst.) Hack. (1891).

Clump-forming perennial up to 300 cm high, the stems stout and erect or slender and radiating outwards to form a leafy tuft with emergent flowering shoots; leaves broadly linear to narrowly lanceolate, conspicuously pleated, sometimes falsely petiolate. Panicle linear to lanceolate, 10–60 cm long, with short projecting, ascending or appressed densely spiculate branches, the rhachis puberulous; bristles 3–15 mm long, often inconspicuous. Spikelets narrowly ovate to elliptic, 2.2–3.5 mm long; lower glume 1/3–1/2 the length of the spikelet; upper glume 1/2–3/4 the length of the spikelet; lower floret male or barren; upper lemma smooth or obscurely rugose, often shiny, becoming light brown.

Open forest, formerly of *Juniperus*; above 1700 m. N1; tropical Africa and Arabia southwards to South Africa, also in tropical America and India. Bally & Melville 16221; Wood 73/116.

Typically a robust plant, but variants of more slender habit and shorter stature, and with a tendency towards a male lower floret have been segregated as *S. plicatilis*. To what extent these differences are environmentally controlled is impossible to say, especially from herbarium material, but it seems that recognising them as species criteria is perhaps overemphasising them.

60. PASPALIDIUM Stapf (1920)

Annual or perennial, often aquatic; leaves linear; ligule a line of hairs. Inflorescence of several to many short racemes on a common axis and ± appressed to shallow hollows in it; rhachis triquetrous or winged, usually ending in a point, bearing single spikelets in 2 neat rows. Spikelets ovate, dorsally compressed, glabrous, abaxial; lower glume much shorter than the spikelet; upper glume shorter than or almost as long as the spikelet; lower floret male or barren; upper lemma crustaceous, acute, its margins inrolled and clasping only the edges of the palea; upper palea acute, its tip slightly reflexed. Caryopsis elliptic, dorsally compressed.

Some 40 species throughout the tropics.

1. Plant creeping, with spongy stolons; raceme-rhachis narrowly winged, often ciliate............ .... 1. *P. geminatum*
 − Plant tufted from a knotty base or sometimes with hard stolons; raceme-rhachis triquetrous, glabrous or puberulous .... 2. *P. desertorum*

1. **P. geminatum** (Forssk.) Stapf (1920); *Panicum geminatum* Forssk. (1775). Fig. 135.

Sabool (Som.).

Perennial up to 60 cm high, with creeping or floating spongy stolons, the stems prostrate and rooting from the lower nodes. Inflorescence 5−30 cm long; racemes 0.5−4 cm long, contiguous (but the lowermost usually remote), the rhachis narrowly winged, 0.5−1 mm wide, usually shortly ciliate. Spikelets ovate, 1.6−2.6 mm long; lower glume truncate, 1/4−1/3 the length of the spikelet; upper glume 2/3−4/5 the length of the spikelet; upper lemma granulose.

Sandy soil in permanently wet grassland, streamsides and active dunes; 0−1300 m. N1, 3; C1; S2, 3; Old World tropics. Gillett 4926; Lavranos 10290; Hemming 3382; Munro 66; Paoli 310.

2. **P. desertorum** (A. Rich.) Stapf (1920).

Gargaro (Som.).

Tufted perennial from a knotty base, the stems up to 60 cm high, erect or geniculately ascending, sometimes decumbent to form hard stolons. Inflorescence 4−20 cm long; racemes 0.3−3 cm long, contiguous (but the lowermost usually remote), the rhachis triquetrous, 0.2−0.4 mm wide, glabrous to puberulous. Spikelets ovate, 1.7−2.6 mm long; lower glume obtuse or truncate, 1/3−1/2 the length of the spikelet; upper glume 3/4 as long to as long as the spikelet; upper lemma granulose or weakly rugulose.

Seasonally flooded grassland and open bushland, occasionally becoming a weed in cultivated fields; 0−1100 m. N1−3; C1, 2; S1−3; Ethiopia, Sudan, Kenya and Arabia. Gillett 4368; Hemming 2199; Bally & Melville 15827; Wieland 4350; Kuchar 16997; Friis & al. 4839; Alstrup & Michelsen 21; Deshmukh 414.

Fig. 135. *Paspalidium geminatum*. A: habit, × 0.45. B: stolon, × 0.45. C: spikelet, × 9. − Modified from Fl. Trop. E. Afr. (1982). Drawn by A. Davies.

61. HOLCOLEMMA Stapf & C.E. Hubb. (1929)

Annual or perennial; leaves flat; ligule a short membrane with ciliate fringe. Inflorescence a panicle contracted about the short appressed primary branches, the terminal spikelet of each branch subtended by an inconspicuous bristle. Spikelets oblong to lanceolate, ± plano-convex, awnless; glumes both shorter than the spikelet, membranous; lower floret male or barren, the lemma herbaceous, sulcate, the palea hyaline except for the enlarged keels which, at maturity, clasp the sides of the upper

Fig. 136. *Holcolemma inaequale*. A: habit (in 3 parts), ×
0.9. B: spikelet, × 11. − From Thulin & al. 7429. Drawn by
K. Thunberg.

lemma; upper lemma crustaceous, gibbous, rugose or
granular, its margins inrolled and clasping only the
edges of the palea. Caryopsis elliptic-oblong, dorsally
compressed.

Four species in East Africa, southern India, Sri
Lanka and Australia.

H. inaequale Clayton (1978). Fig. 136.

Loosely tufted perennial up to 60 cm high, the
slender stems arising from a knotty rootstock; leaves
linear. Panicle subspiciform, 5−17 cm long, interrupted, with very short appressed branches; bristles
2−5 mm long, but inconspicuous. Spikelets lanceolate, 2.5−3.5 mm long; glumes obtuse to subacute,
the lower 1/4, the upper 1/2 the length of the spikelet;
lower floret male; upper lemma rugose, c. 0.5 mm
shorter than the lower.

Coastal grassland and degraded bushland; near sea
level. C1; Kenya and Tanzania. Herlocker 363;
Thulin, Hedrén & Abdi Dahir 7429.

62. TRICHOLAENA Schult. (1824)
Zizka in Biblioth. Bot. 138: 36−50 (1988).

Perennial, rarely annual; leaves rigid, glaucous, often
inrolled; ligule a line of hairs. Inflorescence a panicle.
Spikelets symmetrically oblong in profile, slightly
laterally compressed, awnless; lower glume small or
suppressed, slightly separated from the upper; upper
glume as long as the spikelet, thinly membranous, not
gibbous, slightly emarginate to acute; lower floret
male, the lemma resembling the upper glume; upper
lemma dorsally compressed, firmly cartilaginous,
smooth and shining, white, readily deciduous, the
margins flat and clasping 1/4−2/3 of the palea, acute.
Caryopsis ovate to elliptic-oblong.

Four species in the Mediterranean region and
Africa to India.

T. vestita (Balf.f.) Stapf & C.E. Hubb., otherwise an
endemic of Socotra, was recorded from Somalia (N3)
in Cuf. Enum.: 1360 (1969). However, the single
basis for this record is Puccioni & Stefanini 600
which is *T. teneriffae*.

T. teneriffae (L.f.) Link (1829). Fig. 137.

Melinis leucantha (A. Rich.) Chiov. (1907); *T.
leucantha* (A. Rich.) Stapf & C.E. Hubb. (1930).

M. somalensis Mez (1921); types: N2, "Las
Khoreh", Hildebrandt 1479 (B K W isosyn) and
others.

T. gillettii C.E. Hubb. (1941); type: N1, "Duwi",
Gillett 4395 (K holo.).

T. setacea C.E. Hubb. (1941); type: N1, "Burmado",
Gillett 4539 (K holo.).

Agar, dahlan fordade, oro jer yer, ramed guriyeh,
tongari (Som.).

Perennial, forming tussocks from a woody rootstock, up to 60 cm high, ascending, the stems wiry

with narrow leaves or herbaceous and leafy; leaves flat or inrolled, glabrous or thinly pubescent. Panicle usually narrowly oblong, 3–15 cm long, fairly dense, fully exserted. Spikelets 2.5–3.5 mm long; lower glume a minute truncate scale (rarely up to 0.5 mm); upper glume ovate, tuberculate-pilose with white hairs extending 0.5–4 mm beyond the tip, but often glabrous in the upper 1/4, acute and usually mucronate.

Sandy, gravelly and alluvial soils and rocky hillsides; 400–1600 m. N1–3; C2; East tropical Africa northwards to the Mediterranean region, thence westwards to Macaronesia and eastwards to India. Hemming 2305; Glover & Gilliland 1066; Thulin & Warfa 5981; Bally 3758.

A variable species with several ill-defined regional facies. One of these, referred to as subsp. *eichingeri* (Mez) Zizka (loc. cit.), is said to occur in Somalia alongside subsp. *teneriffae*; it is rather hard to say how the two differ except that in subsp. *eichingeri* the spikelet hairs are said to be shorter and the stems leafier. Only one specimen cited by Zizka (Collenette 119) has been seen, and the present author doubts the value of attempting to distinguish the taxa.

63. MELINIS P. Beauv. (1812)
Zizka in Biblioth. Bot. 138: 50–134 (1988).
Rhynchelytrum Nees (1836).

Annual or perennial; leaves flat or filiform; ligule a line of hairs. Inflorescence a panicle with capillary branches. Spikelets laterally compressed, glabrous to silky-hairy; lower glume small, sometimes distant from the upper; upper glume as long as the spikelet, firmly membranous to chartaceous or subcoriaceous, sometimes becoming thinner towards the tip, straight or ± gibbous on the back below the middle and sometimes tapering to a beak above, emarginate to bilobed (rarely entire), often awned from the sinus; lower floret male or barren, the lemma resembling the upper glume or narrower and less gibbous, the palea present or absent; upper lemma laterally compressed, often deciduous before the rest of the spikelet, cartilaginous, smooth, its margins flat and ± enfolding the palea. Caryopsis oblong to oblong-ellipsoid.

22 species in the Mediterranean region and from Africa to India.

1. Upper glume gibbous on the back; spikelets 2–12 mm long, villous .... 1. *M. repens*
– Upper glume straight on the back; spikelets 1.5–2(–2.4) mm long, glabrous or thinly hairy .... 2. *M. minutiflora*

1. **M. repens** (Willd.) Zizka (1988); *Rhynchelytrum repens* (Willd.) C.E. Hubb. (1934).
Annual or short-lived perennial up to 90 cm high.; leaves not aromatic. Panicle ovate to oblong, (6–)8–15 cm long, fluffy, silvery-pink or purple. Spikelets

S. Hameed 78

Fig. 137. *Tricholaena teneriffae*. A: plant, × 0.5. B: spikelet, × 8. – Modified from Fl. Pakistan 143 (1982).

ovate, 2–12 mm long, usually villous with hairs extending 1–4 mm beyond the tip; lower glume narrowly oblong, (0.6–)1.5–3(–4.3) mm long, distant from the upper; upper glume thinly 5-nerved, conspicuously gibbous, chartaceous, tapering to a

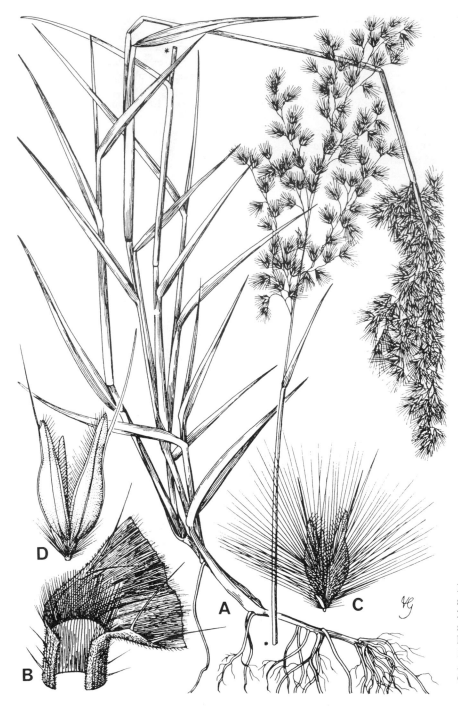

Fig. 138. *Melinis repens* subsp. *repens*. A: habit, × 2/3. B: ligule, × 10. C: spikelet, × 8. D: spikelet with hairs removed to show detail, × 8. — From Fl. Trop. E. Afr. (1982). Drawn by V. Goaman.

glabrous membranous beak 1/4−1/2 its length, emarginate, mucronate or with an inconspicuous awn up to 7(−10) mm long; lower lemma resembling the upper glume but narrower and less gibbous.

1. Spikelets 2−5 mm long; internode between the glumes 0.1−0.7 mm long .... subsp. *repens*
− Spikelets (4−)5−12 mm long; internode between the glumes (0.5−)0.7−1.7(−2) mm long............
.... subsp. *grandiflora*

subsp. **repens.** Fig. 138.

Tricholaena rosea Nees (1837); *R. roseum* (Nees) Stapf & C.E. Hubb. ex Bews (1929).

Rocky hillsides; 1000−1500 m. N1; throughout Africa, but introduced to most other tropical countries. Gillett 4993.

subsp. **grandiflora** (Hochst.) Zizka (1988); *R. grandiflorum* Hochst. (1844); *T. grandiflora* Hochst. ex A. Rich. (1850).

R. villosum (Parl.) Chiov. (1908).

T. ruficoma (Hochst. ex Steud.) Hack. (1888).

Duramo, sadeh-ho (Som.).

Rocky hillsides and sandy soils; 650–1600 m. N1; C2; S1; tropical and South Africa eastwards through Arabia to India. Gillett 4266; Moggi & Bavazzano 768, 1277.

2. **M. minutiflora** P. Beauv. (1812).

Perennial up to 100 cm high, ascending; leaves sometimes smelling strongly of linseed oil. Panicle lanceolate to narrowly ovate, 10–30 cm long, dense, often purplish. Spikelets narrowly oblong, 1.5–2 (–2.4) mm long, glabrous or sometimes thinly hairy; lower glume sometimes almost suppressed, but usually an oblong scale 0.2–0.5 mm long, not distant from the upper; upper glume straight on the back, prominently 7-nerved, the nerves forming raised ribs, obtusely bilobed, with or without a mucro up to 0.5 mm long; lower lemma 5-nerved, the nerves forming raised ribs, acutely bilobed, with an awn up to 15 mm long or sometimes awnless.

Open ground. S2; tropical Africa, introduced throughout the tropics as a fodder plant under the name 'Molasses grass.' Sacco, Sappi & Ariello 128.

64. **DIGITARIA** Hall. f. (1768), nom. cons.

Annual or perennial; leaves mostly linear and flat; ligule a short membranous or scarious rim. Inflorescence composed of racemes, these digitate or borne upon an elongated axis; rhachis flat or triquetrous, bearing the spikelets in suppressed groups of (1–)2–3(–6). Spikelets lanceolate to oblong, elliptic, flattened on the front, convex on the back; lower glume small or appressed; upper glume membranous, as long as the spikelet or shorter and exposing the upper lemma; lower floret barren, the prominently nerved lemma as long as the spikelet (rarely shorter), often hairy, usually with the hairs forming stripes between the 1st and 2nd lateral nerves and along the margin, the palea absent; upper lemma chartaceous to coriaceous, finely longitudinally striate, with its flat hyaline margins enfolding and concealing most of the palea, subacute to acuminate. Caryopsis oblong, plano-convex, acute to subacute.

Some 230 species in tropical and warm temperate regions.

1. Racemes stiffly radiating, plumose below, bare of spikelets in the lower half .... 1. *D. pennata*
   – Racemes not stiffly radiating, not plumose below, bearing spikelets to the base 2
2. Spikelets regularly in groups of 3 3
   – Spikelets in pairs 6
3. Upper glume shorter than the spikelet, exposing the upper lemma, 3-nerved; fruit dark brown; spikelet-hairs clavate .... 2. *D. ternata*
   – Upper glume as long as the spikelet, concealing the upper lemma, 5-nerved; fruit pallid to brown; spikelet-hairs verrucose 4
4. Spikelets (1.5–)2.4–2.9 mm long, covered in copious soft hairs extending up to 1.5 mm beyond the tip .... 3. *D. argyrotricha*
   – Spikelets 1–1.8 mm long, at most pubescent 5
5. Annual or short-lived perennial, erect or prostrate and creeping, but without rhizomes...... .... 4. *D. longiflora*
   – Rhizomatous perennial ....5. *D. brunoana*
6. Plants perennial 7
   – Plants annual 11
7. Spikelets quite glabrous ....6. *D. abyssinica*
   – Spikelets pubescent to villous 8
8. Inflorescence with a central axis much longer than the racemes ....7. *D. rivae*
   – Inflorescence digitate or subdigitate, the central axis shorter than the racemes 9
9. Nerves of the lower lemma scabrid, often also with stiff glassy bristles; rhachis narrowly winged ....8. *D. milanjiana*
   – Nerves of the lower lemma smooth; rhachis triquetrous 10
10. Nodes shortly pilose to villous................... .... 9. *D. macroblephara*
    – Nodes glabrous .... 10. *D. nodosa*
11. Nerves of the lower lemma scaberulous; upper glume 1/3–1/2 the length of the spikelet.......... .... 11. *D. acuminatissima*
    – Nerves of the lower lemma smooth 12
12. Racemes diverging from a central axis, this seldom exceeding the longest raceme, delicate, the spikelets loosely imbricate; spikelets 1.5–2.1 mm long ....12. *D. velutina*
    – Racemes digitate or with a very short central axis in robust specimens, the spikelets closely imbricate; spikelets 2.5–3.3 mm long.......... .... 13. *D. ciliaris*

1. **D. pennata** (Hochst.) T. Cooke (1908); *Panicum pennatum* Hochst. (1855). Fig. 139.

D. pennata var. *pilosa* Chiov., Pl. Nov. Aethiop.: 24 (1928); type: S1, "Baidoa", Puccioni & Stefanini 225 (FT holo.).

Domaar, horogi, jet-jib (Som.).

Tufted perennial up to 100 cm high, the stems wiry, woody, almost suffrutescent, bulbous below and clad in silky-pubescent scales at the base; leaves narrowly lanceolate. Inflorescence of 4–14 long stiff radiating racemes in 1 or 2 whorls, the whole inflorescence breaking off at maturity; racemes 7–25 cm long with slender triquetrous rhachis, the lower half bare of spikelets and plumose, the upper half bearing distant pairs of appressed spikelets. Spikelets lanceolate, 2.5–3 mm long; lower glume a minute hyaline truncate scale; upper glume as long as the spikelet, 3-nerved, pubescent between the nerves; lower lemma coarsely ribbed with 7 evenly spaced nerves,

Fig. 139. *Digitaria pennata*. A: habit, × 0.5. B: spikelet, × 10. — From Fl. Pakistan 143 (1982).

finely pubescent between the nerves (rarely pilose or glabrous); fruit narrowly ellipsoid, chestnut-brown.

Open woodland and shady places on rocky hillsides, ledges, gullies and stony plains; 100−1500 m. N1−3; C1, 2; S1−3; Ethiopia, Kenya and Tanzania, and through Arabia to northern India. Gillett 4904; Bally 11128; Kazmi & al. 5646; Beckett 204; Kuchar 15530; Friis & al. 4817; Paoli 1272; Rose Innes 624.

2. D. ternata (A. Rich.) Stapf (1898).

Annual up to 100 cm high; leaves broadly linear. Inflorescence of 2−11 subdigitate racemes; racemes 3−23 cm long, the spikelets in threes on a ribbon-like winged rhachis with shallowly angular midrib; pedicels nearly always with a crown of stiff hairs 0.2−1 mm long at the tip. Spikelets ovate-elliptic, 1.9−2.7 mm long; lower glume an obscure hyaline rim; upper glume 2/3−3/4 the length of the spikelet, 3-nerved, densely pubescent with clavate hairs; lower lemma 5(−9)-nerved, typically with the 3 central nerves close together and the hairs confined to a

stripe outside of each lateral nerve, but sometimes hairy all over, the hairs clavate, ranging from short and appressed to long (0.2−0.3 mm) and shaggy; fruit elliptic, dorsally flattened, dark brown to black.

?N2; tropical and southern Africa eastwards through Arabia to India, Thailand, China and Indonesia. McKinnon s.n. (label lost; the only specimen known from Somalia).

3. D. argyrotricha (Anderss.) Chiov. (1916).

D. xanthotricha auct. non (Hack.) Stapf.

Annual up to 70 cm high, erect or sprawling and rooting from the lower nodes; leaves broadly linear to narrowly lanceolate. Inflorescence of 2−3(−5) digitate racemes; racemes 6−12(−16) cm long, the spikelets in threes on a ribbon-like winged rhachis with low rounded midrib. Spikelets oblong-elliptic, (1.5−)2.4−2.9 mm long; lower glume a truncate membranous cuff 0.1−0.3 mm long; upper glume and lower lemma similar, 5-nerved, obscured by copious soft white or yellowish fluffy verrucose hairs extending for up to 1.5 mm beyond the tip of the spikelet; fruit ellipsoid, pallid brown or yellowish.

Disturbed sandy soils. S2, 3; Kenya, Tanzania, Mozambique and Ghana. Elmi & Hansen 4105; Kazmi 5083; Moggi & al. 350; Senni 230.

Material from Somalia is slightly different from that of the rest of Africa in that the spikelets are noticeably longer, the leaves slightly narrower and the fruit less strongly coloured. However, the taxonomic significance of these differences is unclear.

4. D. longiflora (Retz.) Pers. (1805).

Annual or short-lived perennial, either erect or prostrate and creeping, up to 60 cm high; leaves broadly linear to narrowly lanceolate. Inflorescence of 2−4 (typically 2) digitate racemes; racemes 1−10 cm long, the spikelets in threes on a ribbon-like winged rhachis with low rounded midrib. Spikelets narrowly ovate-elliptic, 1.2−1.8 mm long; lower glume a minute hyaline rim; upper glume as long as the spikelet, 5-nerved, with short appressed verrucose hairs between the nerves (sometimes glabrous between the 1st and 2nd lateral nerves); lower lemma 7-nerved, pubescent with verrucose hairs or these appressed and barely visible (sometimes glabrous beside the midrib and between the 2nd and 3rd nerves); fruit ellipsoid, pallid, light brown or light grey.

Open dunes and disturbed sandy soils. N1; S2, 3; throughout the Old World tropics; introduced to the New World. Drake-Brockman 150; Thulin, Hedrén & Abdi Dahir 7189; Rose Innes 761.

5. D. brunoana Raimondo (1989); type: S2, Raimondo s.n. (PAL holo., K FT iso). Fig. 140.

Rhizomatous perennial up to 15 cm high; leaves narrowly triangular. Inflorescence of 2(−4) digitate racemes; racemes 0.6−2.7 cm long, the spikelets in threes on a ribbon-like winged rhachis with low rounded midrib. Spikelets ovate-elliptic, 1—1.2 mm long; lower glume ± suppressed; upper glume as long as the spikelet, 5-nerved, with short appressed hairs between the nerves; lower lemma 7-nerved, pubescent between the nerves; fruit ellipsoid, pallid.

Coastal sand; 10—30 m. S2; known only from the type.

6. **D. abyssinica** (Hochst. ex A. Rich.) Stapf (1907).

Panicum scalarum Schweinf. (1894); *D. abyssinica* var. *scalarum* (Schweinf.) Stapf in Bull. Misc. Inform. 1907: 213 (1907).

P. scalarum var. *elatior* Chiov. in Ann. R. Ist. Bot. Roma 6: 166, t. 9 (1896); type: N2, "Habr Awal", Robecchi-Bricchetti 562 (FT holo.).

D. somalensis Chiov. (1917); type: S2, "Merca", Provenzales s.n. (FT holo.).

Domar, garguro, houla, sirdi (Som.).

Creeping perennial with wiry rhizomes, forming mats; stems up to 60 cm high, weak, decumbent below, the basal sheaths usually glabrous, rarely pubescent or villous; leaves broadly linear to narrowly lanceolate. Inflorescence of 2—25 racemes arranged along a short axis 1—9 cm long; racemes 2—11 cm long, the spikelets in pairs on a triquetrous rhachis with or without narrow wings. Spikelets ovate-elliptic to broadly elliptic, plump, 1.5—2.5 mm long; lower glume a ± ovate membranous scale 0.1—0.8 mm long; upper glume 2/3 as long to as long as the spikelet, 3- to 7-nerved, glabrous, the nerves usually prominent; lower lemma 7-nerved, glabrous; fruit ellipsoid, in various shades of grey, light brown or purple.

Sandy loam and black clay, often forming a close sward, an excellent soil-binder but may also be a serious weed of sorghum; up to 1750 m. N1, 2; S2, 3; tropical Africa to South Africa, Arabia, Madagascar and Sri Lanka. McKinnon 280; Hansen & Heemstra 6248; Rose Innes 846; Terry 3396.

7. **D. rivae** (Chiov.) Stapf (1907); *Panicum rivae* Chiov. (1897).

Budiwene, dabasali, domar, garogaro (Som.).

Densely tufted perennial up to 100 cm high, arising from a short knotty rhizome (rarely with slender wiry rhizomes), the basal sheaths silky-tomentose; leaves linear. Inflorescence linear to lanceolate, comprising 6—30 (or more) racemes spreading from or appressed to an elongated axis 4—30 cm long; racemes 1—11 cm long, the spikelets in pairs on a triquetrous rhachis. Spikelets narrowly ovate, 1.7—2.5 mm long; lower glume usually a hyaline ± ovate and obtuse scale 0.1—0.6 mm long; upper glume 2/3—4/5 the length of the spikelet, 3(—5)-nerved, pubescent to pilose; lower lemma 5- to 7-nerved, densely silky-pilose or villous with soft white hairs slightly exceeding the spikelet; fruit ellipsoid, brown.

Grassland on alluvial or sandy soils, often in the

Fig. 140. *Digitaria brunoana*. A: habit, × 1. B: ligule, × 3. C: portion of raceme, × 15. D: upper glume and lower lemma, × 20. − Modified from Giorn. Bot. Ital. 122 (1988).

shade of *Acacia*, a weed of bananas and citrus; 0—1650 m. N1, 2; C1, 2; S1, 2; Djibouti, Ethiopia, Kenya and Tanzania. Hansen & Heemstra 6136; Glover & Gilliland 1076; Wieland 4251; Kuchar 16996; Eagleton 38; Thulin & Hedrén 7165.

8. **D. milanjiana** (Rendle) Stapf (1919).

D. gallaensis Chiov. (1916).

Loosely tufted rhizomatous perennial up to 250 cm high; basal sheaths glabrous or pubescent, rarely bulbously swollen; nodes usually glabrous; leaves linear. Inflorescence of 2—18 digitate or subdigitate racemes (axis up to 6 cm long); racemes 5—25 cm long, slender, stiff, the spikelets in pairs on a triquetrous winged rhachis. Spikelets lanceolate, (1.7—)2.5—3(—3.5) mm long; lower glume distinct, ovate or triangular, 0.2—0.5 mm long; upper glume 1/3—2/3 the length of the spikelet, 3-nerved, ciliate on the margins; lower lemma 7-nerved (the nerves evenly spaced or with the central interspaces up to 1/2 the width of the spikelet), pubescent to glabrescent with shortly ciliate margins, ± scabrid on the nerves and often pectinate with stiff spreading or appressed yellowish or brown glassy bristles; fruit ellipsoid, grey to greyish or pallid brown.

Grassy glades among trees on sandy soils, mostly *Acacia-Commiphora* woodland or shrubland; 0—50 m. S3; tropical and South Africa, introduced elsewhere in the tropics as a fodder grass. Hemming & Deshmukh 141; Moggi & Bavazzano 1530; Rose Innes 661.

9. D. macroblephara (Hack.) Stapf (1919).

Caws cade (Som.).

Perennial up to 100 cm high, forming open tufts from a knotty rootstock, the basal sheaths ± silky-pubescent; nodes shortly pilose to villous; leaves linear. Inflorescence of 2—11 digitate or subdigitate racemes (axis up to 3 cm long); racemes 2—20 cm long, the spikelets in loose pairs on a triquetrous rhachis. Spikelets narrowly elliptic-oblong, 2.2—3.5 mm long; lower glume an ovate scale 0.2—0.3 mm long; upper glume 2/3—3/4 the length of the spikelet, 3-nerved, villous; lower lemma 7-nerved, the nerves smooth, softly villous with copious white hairs, these nearly always slightly exceeding the tip of the spikelet, occasionally interspersed with fine glassy bristles, the central interspaces noticeably glabrous; fruit ellipsoid, grey or light brown.

Low deciduous bushland (mainly *Acacia-Commiphora*) on orange or white sand overlying limestone; 0—350 m. C1, 2; Ethiopia, Sudan, Uganda, Kenya and Tanzania. Thulin & Dahir 6562; Kuchar 17645.

10. D. nodosa Parl. (1842).

D. eriantha auct. non Steud.

Tufted perennial up to 100 cm high, usually bulbous at the base, without rhizomes, the basal sheaths silky-pubescent to tomentose; nodes glabrous; leaves linear. Inflorescence of 4—12 digitate or subdigitate racemes (axis up to 7.5 cm long); racemes 3—15 cm long, the spikelets in approximate pairs on a triquetrous rhachis. Spikelets narrowly elliptic, 2—3 mm long; lower glume an ovate scale 0.1—0.4 mm long; upper glume 2/3 as long to as long as the spikelet, 3-nerved, glabrous to pubescent or villous; lower lemma 7-nerved, glabrous to pilose or woolly-villous; fruit ellipsoid, greyish to light brown.

Grassland on fixed dunes overlying limestone; c. 50 m. N1; C1; S2; Canary Is. and North Africa through

Eritrea, Kenya and Tanzania to Arabia and Pakistan. Appleton s.n.; Herlocker 410; Bigi 7.

A variable species in Africa rather hard to distinguish from *D. macroblephara*. The latter has pilose or villous nodes, but these are often concealed in the sheaths. *D. nodosa* has a tendency towards a tighter raceme than in *D. macroblephara* with the spikelet-hairs appressed rather than forming a spreading fringe and not or scarcely exceeding the tip of the spikelet, but these characters are not very reliable.

11. D. acuminatissima Stapf (1919).

Malaaso (Som.).

Robust annual up to 120 cm high, rooting from the lower nodes; leaves broadly linear. Inflorescence of 2—4 racemes on an axis (0—)1—10 cm long; racemes stiff, 7—25 cm long, the spikelets in pairs on a winged rhachis with triquetrous midrib. Spikelets narrowly lanceolate to narrowly elliptic, 2.5—4 mm long, acuminate; lower glume an ovate scale up to 0.4 mm long; upper glume 1/3—1/2 the length of the spikelet, 3-nerved; lower lemma 7-nerved, the nerves evenly spaced and scaberulous, obscurely and appressedly pubescent; fruit lanceolate, pallid or grey, 0.2—0.5 mm shorter than the lower lemma.

Overgrazed land; c. 50 m. S3; scattered throughout tropical Africa, but nowhere common. Hemming & Deshmukh 125.

Similar to the more temperate *D. sanguinalis* (L.) Scop. but with narrower acuminate spikelets and upper lemma exceeded by the lower.

12. D. velutina (Forssk.) P. Beauv. (1812).

D. horizontalis auct. non Willd.

Annual up to 80 cm high, often decumbent and rambling; leaves broadly linear to lanceolate, thin. Inflorescence of (3—)7—20 racemes diverging from a common axis 1—7 cm long, this seldom exceeding the longest raceme; racemes delicate, 3—13 cm long, the spikelets in pairs and overlapping by less than 1/2 their length on a narrowly winged triquetrous rhachis. Spikelets narrowly ovate-elliptic, 1.5—2.1 mm long, bluntly acute; lower glume obscure or an ovate scale up to 0.2 mm long; upper glume 2/3—4/5 the length of the spikelet, 3-nerved; lower lemma 7-nerved, the nerves evenly spaced and smooth, obscurely appressed-pubescent, occasionally with a ciliate frill; fruit ellipsoid, mostly grey but varying from yellowish to purplish-brown.

Acacia-Commiphora bushland; 200—1350 m. N1; C1; S1—3; East Africa and Arabia southwards to South Africa. Gillett 4980; Herlocker 434; Thulin & Bashir Mohamed 6812; Kazmi 5076; Rose Innes 752.

13. D. ciliaris (Retz.) Koeler (1802); *D. sanguinalis* var. *ciliaris* (Retz.) Domin in Preslia 13/15: 47 (1935).

Caws cade (Som.).

Annual up to 100 cm high, often decumbent at the

base and geniculately ascending; leaves broadly linear. Inflorescence of 2−12 digitate or subdigitate racemes, the axis up to 5 cm long in robust specimens; racemes stiff, 6−22 cm long, the spikelets in pairs and overlapping by 2/3 their length on a winged rhachis with triquetrous midrib. Spikelets narrowly elliptic, 2.5−3.3(−3.7) mm long, sharply acute; lower glume a distinct triangular scale usually 0.2−0.4 mm long; upper glume (1/2−) 2/3−3/4 the length of the spikelet, 3-nerved; lower lemma 7-nerved, the nerves smooth and evenly spaced or with a wide interspace flanking the midrib (sometimes the spikelets of a pair different), appressed-puberulous to silky-pubescent or rarely shortly villous, often with a ciliate frill, sometimes one or both spikelets of a pair beset with glassy bristles; fruit ellipsoid, grey to light brown.

Weedy places and disturbed ground. S2, 3; throughout the tropics. Thulin 6376; Rose Innes 668.

65. TAENIORHACHIS Cope (1993)

Perennial; leaves flat or folded; ligule a short scarious membrane. Inflorescence comprising 2 conjugate deciduous racemes; rhachis triquetrous, broadly winged on the lateral angles, narrowly winged on the central, bearing the spikelets in pairs. Spikelets lanceolate-elliptic, flattened on the front, convex on the back; lower glume small; upper glume membranous, shorter than the spikelet, villous; lower floret barren, the prominently nerved lemma as long as the spikelet, villous, the palea absent; upper lemma chartaceous, finely longitudinally striate, with its flat hyaline margins enfolding and concealing the palea.

One species only.

T. repens Cope (1993); type: S2, 33 km NE of Mogadishu, Thulin, Hedrén & Abdi Dahir 7183 (UPS holo., K iso.). Fig. 141.

Extensively creeping perennial with both rhizomes and stolons, up to 8 cm high; leaves oblong-lanceolate, flat or folded, distichous, rigid, densely pubescent especially below, with white cartilaginous margins. Racemes 1.5−2.5 cm long, the rhachis 2.5−3 mm wide. Spikelets 6−6.5 mm long; lower glume a minute ovate scale 0.5−0.6 mm long; upper glume 2/3 the length of the spikelet, 3-nerved, villous between the nerves; lower lemma 9- to 11-nerved, the nerves prominent and close together, but the outermost more remote, villous in the wider interspaces; fruit ellipsoid, acuminate, pallid.

Dunes of white sand on the coast; 5 m. S2; known only from the type.

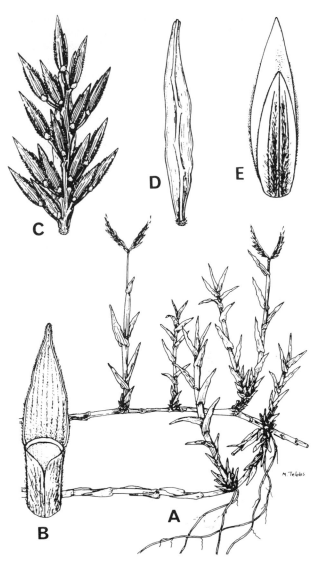

Fig. 141. *Taeniorhachis repens.* A: habit, × 1. B: leaf-blade and ligule, × 6. C: raceme, × 6. D: rhachis, dorsal view, × 6. E: spikelet, dorsal view, × 12. − Modified from Kew Bull. 48 (1993). Drawn by M. Tebbs.

66. PENNISETUM Rich. (1805)
Beckeropsis Fig. & De Not. (1853).

Annual or perennial; leaves linear to lanceolate; ligule a line of hairs, rarely a membrane. Inflorescence spike-like, cylindrical to globose, terminal, rarely axillary but then often gathered into a leafy false panicle, each spikelet or cluster of spikelets subtended by a deciduous involucre of 1−many slender bristles, these free throughout; rhachis rounded or angular, with or without short peduncle-stumps, occasionally the involucre shortly stipitate below the insertion of the bristles. Spikelets narrowly lanceolate to oblong, dorsally compressed, glabrous or almost so; lower glume up to 1/2 the length of the spikelet, sometimes suppressed; upper glume very small to as long as the spikelet; lower floret male or barren, its lemma variable in length, membranous; upper lemma as long as the spikelet or nearly so, membranous to coriaceous, its thin flat margins covering half the palea. Caryopsis oblong and dorsally compressed to subglobose.

Some 80 species throughout the tropics.

1. Bristles solitary below each spikelet; panicles both terminal and axillary .... 1. *P. unisetum*
   - Bristles several in each involucre; panicles terminal on the stems and branches 2
2. Ligule a membrane .... 2. *P. stramineum*
   - Ligule a line of hairs 3
3. Involucral bristles (or at least the inner) ciliate to plumose, 16−65 mm long (if a suffruticose desert grass see 5. *P. divisum*) 4
   - Involucral bristles glabrous (rarely thinly hairy), 4−20 mm long 5
4. Panicle densely ovoid to subspherical; low mat-forming rhizomatous plant .... 3. *P. villosum*
   - Panicle linear; erect, densely tufted plant without rhizomes .... 4. *P. setaceum*
5. Plant woody, suffruticose; spikelets 6.5−8.5 mm long; leaves tightly inrolled, pungent..............
   .... 5. *P. divisum*
   - Plant tufted, sometimes branched above; spikelets 2−6 mm long; leaves flat or inrolled, but not pungent 6
6. Lower glume 1/2 the length of the spikelet; upper glume 2/3−3/4 the length of the spikelet..........
   .... 6. *P. massaicum*
   - Lower glume up to 1/4 the length of the spikelet; upper glume up to 1/2 the length of the spikelet .... 7. *P. macrourum*

1. **P. unisetum** (Nees) Benth. (1881); *Beckeropsis uniseta* (Nees) K. Schum. (1895).

Tufted perennial; stems robust, 0.6−4 m high, 2−15 mm in diameter at the base, rarely smaller and mat-forming; leaves broadly linear to linear-lanceolate, often falsely petiolate; ligule a line of hairs. Panicles numerous, axillary, slender, 2−4 cm long, gathered into a copious false panicle; rhachis with rounded ribs and distinct peduncle-stumps, minutely pubescent; involucre reduced to a single bristle below each spikelet; bristle glabrous, 2.5−9(−40) mm long, often purplish. Spikelets narrowly elliptic, 2−3 mm long; glumes 0.2−0.5 mm long, truncate, emarginate or obtuse, rarely the upper subacute and up to 0.8 mm long; lemmas as long as the spikelet, membranous, scaberulous, the lower barren.

In thickets at c. 1500 m. N2; tropical and South Africa and Arabia. Newbould 879.

2. **P. stramineum** Peter (1930).

Shrubby, tufted perennial; stems up to 120 cm high, hard, wiry, much-branched; leaves linear; ligule a membrane 0.5−1.5 mm long, finely lacerate. Panicle linear, 2−15 cm long, silvery; rhachis with rounded ribs and short peduncle-stumps, glabrous or scaberulous; involucre enclosing 1−2 sessile spikelets, without a distinct stipe; bristles glabrous, the longest 7−12 mm. Spikelets oblong-lanceolate, 4−5.5 mm long, acuminate; lower glume up to 1/4 the length of the spikelet, rarely more; upper glume 1/3−2/3 the length of the spikelet; lower lemma as long as the spikelet, male; upper lemma similar to the lower but a little shorter.

Gypseous plains, exposed rocky ledges and areas of abandoned cultivation; 1300−1500 m. N2; Ethiopia, Kenya, Tanzania and Arabia. Hansen & Heemstra 6240; Hemming & Watson 3273; McKinnon 245.

3. **P. villosum** R. Br. ex Fresen. (1814).

Low, mat-forming rhizomatous perennial; stems loosely ascending, up to 45 cm high; leaves linear, flat; ligule a line of hairs. Panicle densely ovoid to subspherical, 5−10 cm long and wide; rhachis with rounded ribs and short peduncle-stumps, thinly pilose; involucre enclosing 1(−2) subsessile spikelets, with a short villous stipe up to 0.7(−1) mm long; bristles softly plumose below, the longest 35−65 mm. Spikelets lanceolate, 8.5−10(−12) mm long; lower glume vestigial or up to 1 mm long; upper glume 1/3−1/2 the length of the spikelet; lower lemma almost as long as the spikelet; upper lemma as long as the spikelet, coriaceous below.

Damp sandy ground, rocky hillsides and gypseous plains, especially in juniper forest; 1500−2100 m. N2; Ethiopia and Arabia. Glover & Gilliland 595; Hansen & Heemstra 6223; Lavranos 6709.

4. **P. setaceum** (Forssk.) Chiov. (1923).
   Arabjeb, arablab, baldoli (Som.).

Densely tufted perennial up to 130 cm high, the stems usually unbranched; leaves inrolled with noticeably thickened midrib on the upper surface, rigid, harsh, glaucous; ligule a line of hairs. Panicle linear, 6−30 cm long; rhachis with shallowly angular ribs below the stumpless scars, glabrous to pilose; involucre enclosing 1 sessile and 0−1(−2) pedicelled spikelets, borne upon a slender pubescent stipe 1−3 mm long; bristles loosely plumose, the longest 16−40 mm. Spikelets lanceolate, (4.5−)5−7 mm long; lower glume vestigial or up to 2 mm long; upper glume 1/5−2/3 the length of the spikelet; lower lemma as long as the spikelet, male or barren, acuminate; upper lemma similar to the lower.

Rocky hillsides and stony plains; 1200−1800 m. N1 3; North and North-east Africa and South-west Asia. Gillett 4259; Hansen & al. 6495; Thulin & Warfa 6120.

5. **P. divisum** (J.F. Gmel.) Henrard (1938). Fig. 142.
   P. dichotomum Del. (1813), based on *Panicum dichotomum* Forssk. (1775) non L. (1753).
   Dhalan, dungarro (Som.).

Perennial with short woody rhizome; stems woody, suffruticosely branched throughout, up to 1(−2) m high; leaves tightly inrolled, pungent, glaucous; sheaths inflated, the lower readily shedding the short blade. Panicle oblong, 5−12 cm long; rhachis with shallowly angular ribs below the cupular scars, scaberulous; involucre enclosing 1 sessile spikelet, borne upon a short oblong stipe 0.5−1 mm long; bristles glabrous, rarely thinly hairy, the longest 7−20 mm. Spikelets narrowly lanceolate, 6.5−8.5 mm long; lower glume 1/2−3/4 the length of the spikelet; upper glume almost as long as thc spikelet; lemmas similar, the lower male.

Deserts, rocky slopes and dry river beds; 0−1000 m. N1−3; C1; North Africa through South-west Asia to India. Hansen & Heemstra 6144; Hemming 2054; Bally & Melville 15654; Wieland 4408.

6. P. massaicum Stapf (1906).

Irdug (Som.).

Perennial arising from a short woody rhizome; stems up to 90 cm high, much-branched, wiry or woody; leaves flat or folded; ligule a line of hairs. Panicle linear to oblong, 2−10 cm long, dense; rhachis with angular ribs below the scars, scaberulous; involucre enclosing 1 sessile spikelet, the base with a very short oblong stipe; bristles glabrous or rarely sparsely ciliate, the longest 4−10 mm, usually in 2 whorls. Spikelets oblong, 3.5−5.5 mm long; lower glume 1/2 the length of the spikelet; upper glume 2/3−3/4 the length of the spikelet; lemmas similar in texture, acute to subulate, the lower male.

Shallow soil at c. 1300 m. N1; Kenya and Tanzania. Gillett 4133.

7. P. macrourum Trin. (1826).

P. thunbergii auct. non Kunth.

Reed-like perennial from a creeping rhizome; stems up to 1 m high; leaves often inrolled, hard, glaucous; ligule a line of hairs. Panicle linear, 6−20 cm long; rhachis cylindrical with rounded ribs, with or without peduncle-stumps, scaberulous or pubescent; involucre enclosing 1 sessile spikelet, with or without a stipe at the base; bristles glabrous, the longest 5−20 mm. Spikelets narrowly elliptic to narrowly ovate, 2−6 mm long; lower glume up to 1/4 the length of the spikelet; upper glume 1/8−1/4(−1/2) the length of the spikelet, acute to acuminate, sometimes obtuse; lower lemma male or barren, 3/4 as long to as long as the spikelet, acute to cuspidate; upper lemma similar to the lower.

Boggy ground at c. 1500 m. N2; tropical and South Africa, mainly in the east, and Arabia. Newbould 907.

Just one of a cluster of intergrading species from tropical Africa that inhabit watersides and mires, and the determination of the single known Somali specimen as P. macrourum is admittedly somewhat arbitrary; the specimen was misnamed P. thunbergii in Cope (1985).

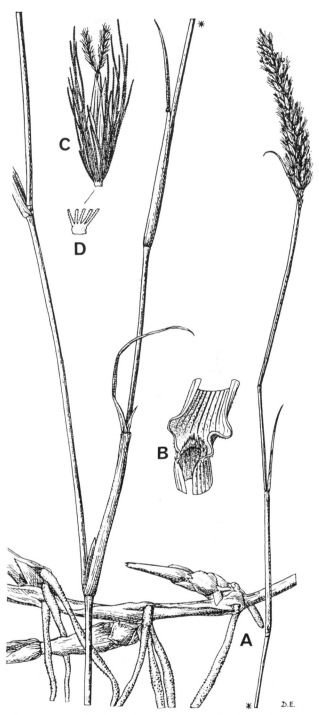

Fig. 142. *Pennisetum divisum*. A: habit, showing rhizome, × 2/3. B: ligule, × 4. C: involucre with spikelet, × 4. D: detail of base of involucre, × 8. − Modified from a drawing by D. Erasmus.

67. CENCHRUS L. (1753)

Annual or perennial; leaves flat or inrolled; ligule a line of hairs. Inflorescence a cylindrical spike-like panicle with angular rhachis, each spikelet or cluster of spikelets enclosed by a deciduous involucre; involucre composed of 1 or more whorls of bristles, those of the innermost whorl flattened and often spiny, connate at the base or for some distance along their length. Spikelets lanceolate to ovate, dorsally compressed, acute to acuminate; lower glume up to 1/2 the length of the spikelet, sometimes suppressed; upper glume a little shorter than the spikelet; lower floret male or barren, its lemma as long as the spikelet, membranous; upper lemma as long as the spikelet, firmly membranous to coriaceous, its thin flat margins covering much of the palea. Caryopsis elliptic to ovoid, dorsally compressed.

22 species throughout the tropics.

1. Bristles of the involucre retrorsely barbellate, tenaciously clinging to clothing or fur............ .... 6. *C. biflorus*
   — Bristles of the involucre antrorsely scaberulous, not clinging to clothing or fur 2
2. Inner bristles flexuous, filiform above 3
   — Inner bristles rigid, flattened, connate for 1/4– 2/3 their length to form a cup 5
3. Inner bristles united only at the base to form a shallow disc 0.5–1.5 mm in diam., occasionally connate for up to 0.5 mm above its rim.......... .... 1. *C. ciliaris*
   — Inner bristles connate for 1–2.5 mm above the rim of the disc 4
4. Annual or short-lived perennial; leaves flat, 2–5 mm wide .... 2. *C. pennisetiformis*
   — Tufted perennial; leaves narrow, inrolled, 1 mm wide .... 3. *C. somalensis*
5. Body of the involucre glabrous; outer bristles commonly suppressed ....4. *C. setigerus*
   — Body of the involucre pubescent; outer bristles numerous ....5. *C. mitis*

1. **C. ciliaris** L. (1771); *Pennisetum ciliare* (L.) Link (1827).

Pennisetum polycladum Chiov. (1896); type: N1, "Habr Awal", Robecchi-Bricchetti 569 (FT holo.).

P. ciliare var. *anachoreticum* Chiov. (1897).

Agar loak, anodug, arapsor, balhorle, garrow, gudomad, gurde agar, harfo, irdug (Som.).

Perennial, often forming mats or tussocks; stems up to 150 cm high, wiry or somewhat woody; leaves flat, 2–13 mm wide. Panicle 2–14 cm long; involucre elongate, 6–16 mm long; inner bristles much exceeding the spikelets, one of them longer and stouter than the rest, united only at the base to form a disc 0.5–1.5 mm in diam. (or sometimes connate for up to 0.5 mm above its rim), sparsely to densely ciliate below, filiform above, flexuous, antrorsely scaberulous; outer bristles filiform. Spikelets 1–4 per burr, 2–5.5 mm long.

Sandy, silty and alluvial soils, rocky hillsides with *Acacia-Commiphora* shrubland, and irrigated farmland; 0–1650 m. N1–3; C1, 2; S1–3; throughout

Africa, extending through South-west Asia to India, widely introduced elsewhere in the Old World. Hemming 191; Bally 10994; Kazmi & al. 5587; Thulin, Hedrén & Abdi Dahir 7272; Kuchar 15681; Tardelli 167; Bavazzano 83; Hemming & Deshmukh 200.

2. **C. pennisetiformis** Hochst. & Steud. (1854). Fig. 143.

Cagaar biyood (Som.).

Annual or short-lived perennial; stems up to 40(–60) cm high; leaves flat, 2–5 mm wide. Panicle 2–6(–8) cm long; involucre elongate, 6–16 mm long; inner bristles much exceeding the spikelets, one of them longer and stouter than the rest, flattened at the base, connate for 1–2.5 mm above the rim of the basal disc to form a cup, almost glabrous to sparsely ciliate below, filiform above, flexuous, antrorsely scaberulous; outer bristles filiform. Spikelets 1–3 per burr, 3–5 mm long.

Dunes, gravel plains and streambeds, limestone hillsides and pavements, and irrigated farmland; 0–1850 m. N1–3; East and North-east tropical Africa eastwards to India, introduced in Australia. Gillett 4785; Hemming & Watson 3171; Thulin & Warfa 5623.

3. **C. somalensis** Clayton (1977); type: N2, "Erigavo", McKinnon 221 (K holo.).

Godo maat (Som.).

Densely tufted perennial up to 45 cm high; leaves inrolled, 1 mm wide. Panicle 2.5–6 cm long; involucre cup-shaped, 6–9 mm long; inner bristles much exceeding the spikelets, one of them longer and stouter than the rest, flattened at the base, connate for 1/2 their length to form a cup, shortly ciliate below, filiform above, flexuous, antrorsely scaberulous; outer bristles filiform. Spikelets 1–2 per burr, 4–5 mm long.

Juniper forest; 1500–2200 m. N2, 3; not known elsewhere. Glover & Gilliland 1094; Scortecci s.n.

4. **C. setigerus** Vahl (1806).

Perennial, forming clumps from a bulbous base, up to 80 cm high; leaves flat, 2–7 mm wide. Panicle 2–12 cm long; involucre cup-shaped, 3–7 mm long; inner spines short, flattened, connate for 1/4–1/2 their length to form a cup, glabrous or obscurely puberulous; outer spines few, short, often suppressed, rarely almost as long as the inner. Spikelets 1–3 per burr, 3–5 mm long.

Sandy and silty soils; 100–950 m. N1, 2; C2; S2; East and North-east tropical Africa eastwards to India; introduced elsewhere in the tropics. Gillett 4325; Glover & Gilliland 702; Kuchar 15827; Kilian & Lobin 6881.

5. **C. mitis** Anderss. (1863).

C. aequiglumis Chiov. (1926); type: S3, Kismayu, Gorini 20 (FT holo., K iso.).

Annual up to 100 cm high; leaves flat, 2−10 mm wide. Panicle 4−18 cm long; involucres globose to ovoid, 6−9 mm long; inner spines flattened, connate for 1/3−2/3 their length to form a cup, pubescent on the face, ciliate on the margins, antrorsely scaberulous, aciculate at the tip; outer bristles filiform, mostly shorter than the inner. Spikelets 2(−3) per burr, 4−6 mm long.

Acacia-Commiphora bushland on reddish soil, and areas of heavy grazing and abandoned cultivation; 0−50 m. S2, 3; Kenya, Tanzania and Mozambique. McLeish 866; Tardelli & Bavazzano 512.

6. C. biflorus Roxb. (1820).

C. barbatus Schumach. (1827).

Annual up to 90 cm high; leaves flat, 2−7 mm wide. Panicle 2−15 cm long; involucre ovoid, 4−11 mm long; inner spines flattened, united at the base to form a shallow disc 2−4 mm in diam., ciliate below, retrorsely barbellate and pungent at the tip; outer spines numerous, acicular, shorter than the inner, often divergent. Spikelets 1−3 per burr, 3.5−6 mm long.

Deciduous bushland on orange sand and areas of overgrazing, disturbance and abandoned cultivation; 0−125 m. C1, 2; S2, 3; tropical Africa, Arabia and India. Elmi & Hansen 4077; Kuchar 17407; Bavazzano 1164; Rose Innes 632.

68. ANTHEPHORA Schreb. (1779)

Annual or perennial; leaves flat; ligule membranous. Inflorescence spike-like, cylindrical, bearing oblong to conical deciduous subsessile clusters of 3−11 spikelets surrounded by an involucre of stiffly coriaceous narrowly elliptic bracts, these free almost to the base and obtuse to awned at the tip. Spikelets dorsally compressed; lower glume absent; upper glume subulate from a broad base with its back facing inwards; lower floret reduced to a hyaline lemma shorter than the spikelet; upper lemma cartilaginous, its thin flat margins covering much of the palea. Caryopsis oblong-ellipsoid, dorsally compressed.

12 species in Africa and Arabia, one species in tropical America.

1. Involucral bracts broadest at the middle, acute; inflorescence obscurely hairy .... 1. *A. nigritana*
 − Involucral bracts broadest below the middle, acute to setaceously acuminate; inflorescence distinctly hairy .... 2. *A. pubescens*

1. A. nigritana Stapf & C.E. Hubb. (1930).

A. lynesii Stapf & C.E. Hubb. (1930).

Tufted perennial up to 75(−150) cm high. Inflorescence 10−25 cm long, the clusters with pubescent peduncle; involucral bracts narrowly elliptic to oblong-elliptic, broadest at the middle, 4−6 mm long, appressed-pubescent, acute; lower lemma glabrous to shortly

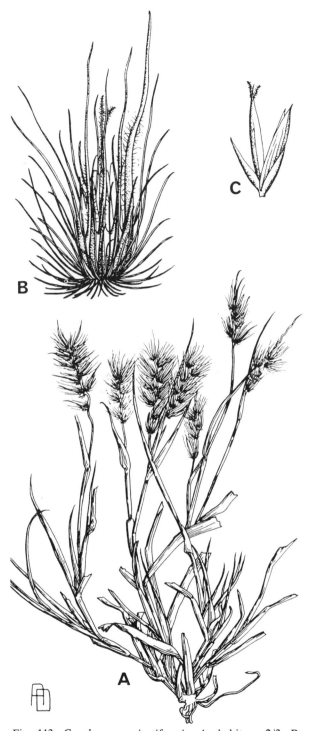

Fig. 143. *Cenchrus pennisetiformis*. A: habit, × 2/3. B: involucre, front view, × 6. C: spikelet, × 6. − Modified from Fl. Trop. E. Afr. (1982). Drawn by A. Davies.

ciliate, the hairs concealed within the involucre.

Acacia woodland on rocky hillsides; c. 1200 m. N1; Niger and Nigeria eastwards to Ethiopia, Kenya and Arabia. Gillett 4994.

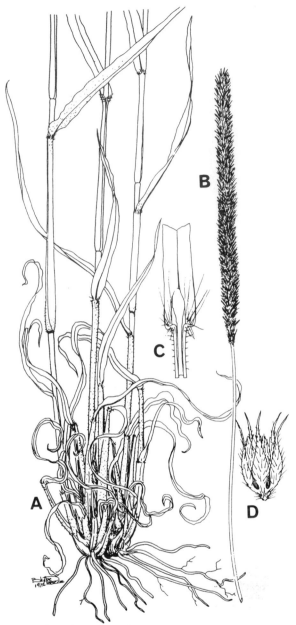

Fig. 144. *Anthephora pubescens*. A: base of plant. B: inflorescence. C: ligule. D: spikelet. — From Mem. Bot. Surv. S. Afr. 58 (1991).

2. **A. pubescens** Nees (1841). Fig. 144.

A. hochstetteri Nees ex Hochst. (1844).

Tufted perennial up to 100 cm high. Inflorescence 5—25 cm long, the clusters with bearded peduncle; involucral bracts lanceolate to narrowly ovate, broadest below the middle, 6—11 mm long, sparsely pilose to densely villous, acute to setaceously acuminate; lower lemma densely ciliate, the hairs c. 1 mm long and projecting from the involucre.

N1; tropical and South Africa and Iran. Drake-Brockman 47, 426.

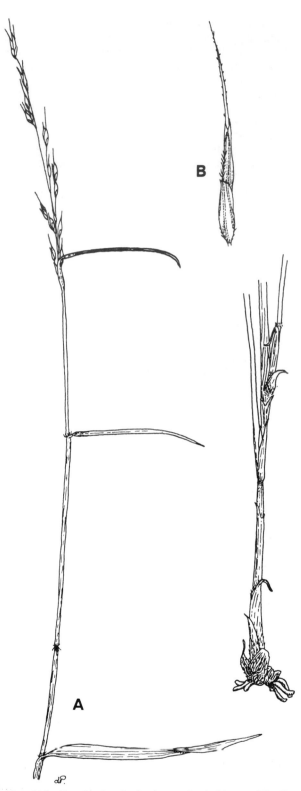

Fig. 145. *Danthoniopsis barbata*. A: habit, × 0.9. B: spikelet, × 5.5. — From Thulin & al. 7972 (Yemen). Drawn by L. Petrusson.

Tribe 15. ARUNDINELLEAE

Ligule usually a line of hairs, sometimes membranous. Inflorescence a panicle, the spikelets all alike, usually immature at emergence and completing their growth on the panicle, often in triads, these sometimes with connate pedicels. Spikelets 2- or rarely 1-flowered without rhachilla-extension, lanceolate, slightly laterally compressed, shedding one or both florets; glumes persistent, the upper about as long as the spikelet, the lower usually shorter, membranous to coriaceous; lower floret male or barren, the lemma resembling the upper glume, often persistent, 3- to 9-nerved; upper floret bisexual, subterete, the lemma thinly coriaceous, often decorated with hair-tufts, 2-toothed or 2-lobed, awned from the sinus; awn geniculate, with flat or terete spiral column, often deciduous. Hilum often linear.

12 genera and about 175 species in the tropics, mainly of the Old World.

1. Lower lemma 5- to 9-nerved; upper lemma with 2 shortly awned lobes and tufts of hair across the back; spikelets usually in triads.................. .... 69. *Danthoniopsis*
— Lower lemma 3-nerved; upper lemma with 2 short acute lobes and without tufts of hair across the back; spikelets solitary or loosely paired .... 70. *Loudetia*

69. DANTHONIOPSIS Stapf (1916)

Perennial, rarely annual. Panicle open or contracted, bearing spikelets in groups of 2 or 3. Spikelets with glabrous glumes; lower lemma 5- to 9-nerved; upper lemma usually with 2−8 transversely arranged tufts of hair, sometimes glabrous, 2-lobed; callus square to oblong, obtuse (narrowly oblong and 2-toothed in Somalia).

About 20 species mainly in central and southern Africa but extending to Guinée and Sierra Leone in the west, and to Arabia and Pakistan in the east.

D. barbata (Nees) C.E. Hubb. (1934); *Tristachya barbata* Nees (1841). Fig. 145.

T. somalensis Franch. (1882); type: N3, "Karin Ossé", Revoil 138 (K iso.).

Dramogale, jabioho, oro jar (Som.).

Tufted perennial with knotty base, up to 1 m high; nodes densely bearded. Leaves distichous, flat, stiff and pungent, pilose and with conspicuous white margins. Panicle contracted, rarely open, the spikelets mostly in triads, usually flushed with purple. Lower glume 5.5−7 mm long; upper glume 8.5−10.5 mm long; lower lemma 8.5−11 mm long; upper lemma, including the 0.6−1.2 mm long 2-toothed callus and 3−5 mm long awned lobes, 8−10.5 mm long, with tufts of hair across the back;

awn rigidly falcate, 15−20 mm long.

Rocky ledges, stony hillsides and lava-plains; 60−1600 m. N1−3; Ethiopia, Sudan and southern Arabia. Gillett 4322; Glover & Gilliland 681; Lavranos & Bavazzano s.n.

70. LOUDETIA Steud. (1854), nom. cons.

Perennial, rarely annual. Panicle open or contracted, bearing single or paired spikelets. Spikelets brown; lower lemma 3-nerved; upper lemma glabrescent to pilose, 2-toothed; callus oblong to linear, truncate, 2-toothed or obliquely pungent.

26 species in tropical and southern Africa, Madagascar and South America.

L. migiurtina (Chiov.) C.E. Hubb. (1934); *Trichopteryx migiurtina* Chiov. (1928); type: N3, between "Hongolo" and "Hariri", Puccioni & Stefanini 738 (FT holo.).

Tufted perennial up to 30 cm high. Panicle lanceolate, loosely contracted. Lower glume 3−5 mm long; upper glume 7.5−8 mm long; lower lemma 8−9 mm long; upper lemma, including the 0.4 mm long oblong callus and 0.7 mm long lobes, 4−4.5 mm long; awn very slender, flexuous, 20−25 mm long.

N3; known only from the type.

Tribe 16. ANDROPOGONEAE

Ligule scarious or membranous, a line of hairs, or absent. Inflorescence composed of fragile (very rarely tough) racemes, these sometimes in a large leafy panicle, but usually solitary, paired or digitate, terminal or axillary and numerous, in the latter case each true inflorescence subtended by a modified leaf-sheath (spatheole) and often aggregated into a leafy false panicle. Racemes bearing spikelets in pairs (rarely singly or in threes, but usually terminating in a triad), nearly always with one sessile and the other pedicelled, these sometimes alike, but usually dissimilar, the sessile being bisexual and the pedicelled male or barren; occasionally with 1 or more of the lowermost pairs (homogamous pairs) alike, infertile and subpersistent. Sessile spikelet 2-flowered, falling entire at maturity with the adjacent internode and pedicel (the pedicelled spikelet usually falling separately);

glumes as long as the spikelet and hardened; lower floret male or barren, the lemma hyaline or membranous and awnless, the palea suppressed if the floret is barren; upper floret bisexual, the lemma membranous or hyaline, awned or awnless. Pedicelled spikelet sometimes resembling the sessile, but usually male or barren; rarely the pedicel absent or fused to the internode. Grain with large embryo and punctiform hilum.

85 genera and about 960 species throughout the tropics, extending into warm temperate regions.

1. Spikelets unisexual with male and female in separate inflorescences ....88. *Zea*
－ Spikelets, or at least one of each pair, bisexual 2
2. Rhachis-internodes and pedicels slender, sometimes thickened upwards but then the upper lemma awned . 3
－ Rhachis-internodes and pedicels stout, thickening upwards; upper lemma awnless 15
3. Pedicelled spikelet similar to the sessile, both fertile ....71. *Saccharum*
－ Pedicelled spikelet differing from the sessile in shape and sex, rarely absent or represented by a barren pedicel 4
4. Inflorescence a panicle with elongated central axis; raceme-internodes never with a translucent median line 5
－ Inflorescence of single or subdigitate racemes, sometimes with an elongated central axis but then the raceme-internodes with a translucent median line 6
5. Lower glume of sessile spikelet dorsally compressed .... 72. *Sorghum*
－ Lower glume of sessile spikelet laterally compressed .... 73. *Chrysopogon*
6. Pedicels and internodes with a translucent median line .... 75. *Bothriochloa*
－ Pedicels and internodes without a translucent median line 7
7. Lower floret of sessile spikelet male and with a palea 8
－ Lower floret of sessile spikelet barren and reduced to a lemma 9
8. Ligule membranous .... 76. *Ischaemum*
－ Ligule a line of hairs .... 77. *Sehima*
9. Upper lemma of sessile spikelet awned from low down on the back ....80. *Arthraxon*
－ Upper lemma awned from the tip, or from the sinus of the bilobed tip 10
10. Lower glume of sessile spikelet 2-keeled; callus inserted in the hollowed tip of the internode 11
－ Lower glume of sessile spikelet convexly rounded without keels; callus applied obliquely to the internode with its tip free 12
11. Racemes not deflexed at maturity, borne upon unequal terete raceme-bases; leaves not aromatic ....78. *Andropogon*
－ Racemes usually deflexed at maturity and borne upon subequal flattened raceme-bases; leaves nearly always aromatic ....79. *Cymbopogon*
12. Upper lemma 2-toothed ....81. *Hyparrhenia*
－ Upper lemma entire 13
13. Raceme with 2 large homogamous spikelet-pairs at the base forming an involucre ....83. *Themeda*
－ Raceme with or without homogamous spikelet-pairs, but if present then not forming an involucre 14
14. Sessile spikelet subterete, the callus pungent ....82. *Heteropogon*
－ Sessile spikelet dorsally compressed, the callus obtuse ....74. *Dichanthium*
15. Pedicels fused to the internodes ....87. *Rottboellia*
－ Pedicels free from the internodes 16
16. Callus of sessile spikelet obtuse to acute, with oblique articulation-scar ....84. *Elionurus*
－ Callus of sessile spikelet truncate, with transverse articulation often reinforced by a central peg 17
17. Raceme silky-villous ....86. *Lasiurus*
－ Raceme glabrous ....85. *Vossia*

71. SACCHARUM L. (1753)
Erianthus Michx. (1803).

Tufted or rhizomatous perennials; leaves linear; ligule scarious or a line of hairs. Inflorescence a panicle, often large and plumose, bearing numerous racemes crowded on its branches; racemes with fragile rhachis, bearing paired similar spikelets, one sessile, the other pedicelled; internodes linear, slender. Spikelets lanceolate, enveloped in long silky hairs from the callus, dorsally compressed; callus very short, truncate; glumes equal, membranous to coriaceous, the lower flat or rounded on the back, 2-keeled; lower floret barren, reduced to a lemma; upper lemma lanceolate, hyaline, awnless or with a straight awn, entire or rarely 2-toothed, sometimes almost suppressed; stamens 2－3. Caryopsis subglobose to narrowly oblong.

35－40 species throughout the tropics and subtropics.

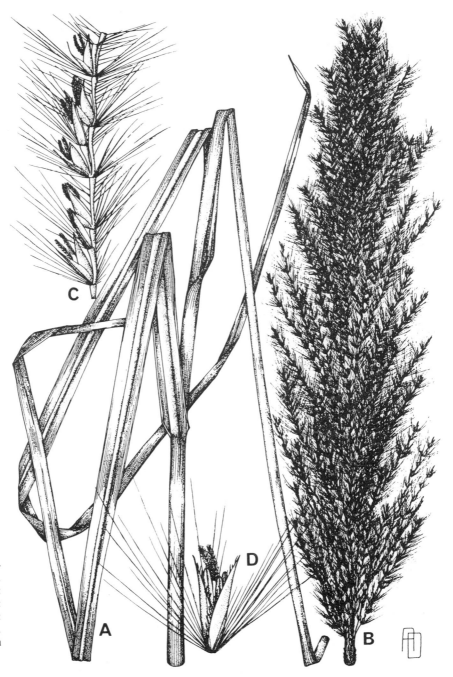

Fig. 146. *Saccharum spontaneum* subsp. *aegyptiacum*. A: leaves, × 0.5. B: inflorescence, × 0.5. C: portion of raceme, × 3. D: spikelet, × 5. — From Fl. Trop. E. Afr. (1982). Drawn by A. Davies.

1. Callus-hairs 2−3 times as long as the spikelet; upper lemma apparently awnless...................
......................................1. *S. spontaneum*
− Callus-hairs scarcely as long as the spikelet; upper lemma conspicuously awned................
......................................2. *S. ravennae*

1. **S. spontaneum** L. (1771).
subsp. **aegyptiacum** (Willd.) Hack. in DC., Monogr. Phan. 6: 115 (1889); *S. aegyptiacum* Willd. (1809). Fig. 146.
Seef-dhaley (Som.).

Rhizomatous perennial up to 5 m high; leaves 5−15(−40) mm wide, the lamina extending to the base; ligule crescent-shaped. Panicle 25−40(−60) cm long, the axis and especially the top of the peduncle hairy; racemes 3−15 cm long, usually much longer than the supporting branches. Spikelets 2.5−5(−7) mm long, the callus bearded with silky white hairs 2−3 times as long as the spikelet; glumes subcoriaceous in the lower third, glabrous on the back; upper lemma very shortly awned, the awn not visible beyond the tips of the glumes.

Riverbanks. N3; S1, 3; tropical and northern Africa

southwards to Malawi, also in Syria. Scortecci s.n.; Tardelli 262; Hemming & Deshmukh 87.

Subsp. *spontaneum* is very similar, differing only in the form of the leaf. The blade is gradually narrowed towards the base until it is scarcely more than a narrow wing on either side of the conspicuous thickened white midrib; the ligule is also triangular rather than crescentic. It is more common in tropical Asia, but does occur in parts of Africa and in Arabia.

2. **S. ravennae** (L.) Murray (1774); *Erianthus ravennae* (L.) P. Beauv. (1812).

E. purpurascens Anderss. (1855); *E. ravennae* var. *purpurascens* (Anderss.) Hack. in DC., Monogr. Phan. 6: 140 (1889).

E. ravennae var. *binervis* Chiov., Fl. Somala 1: 327 (1929); type: N3, "Karin", Puccioni & Stefanini 1026 (FT holo.).

Alalo, chado (Som.).

Tufted perennial up to 4.5 m high; leaves flat, up to 20 mm wide or narrower with thickened midrib. Panicle 25−70 cm long, dense or interrupted, the axis and peduncle glabrous; racemes 1.5−3 cm long, much shorter than the supporting branches. Spikelets slightly heteromorphous, 3−4 mm long, the callus bearded with whitish or greyish hairs scarcely as long as the spikelet; glumes membranous, those of the sessile spikelet glabrous, those of the pedicelled sparsely to moderately hairy on the back; upper lemma with an awn 2.5−10 mm long and well exserted from the glumes.

Along water courses and in swamps; 0−1400 m. N1−3; Mediterranean region and the Sahara eastwards to northern India. Gillett 4583; Hemming 1975; Collenette 180.

72. SORGHUM Moench (1794), nom. cons.

Annuals or perennials, mostly robust, with or without rhizomes; ligule membranous or scarious, rarely a line of hairs. Inflorescence a large terminal panicle with persistent branches bearing short fragile (except in cultivated species) racemes, these with paired dissimilar spikelets, one sessile, the other pedicelled; internodes and pedicels filiform. Sessile spikelet dorsally compressed; callus obtuse; lower glume coriaceous, broadly convex across the back, becoming 2-keeled and narrowly winged near the tip, usually hairy; lower floret reduced to a hyaline lemma; upper lemma hyaline, 2-toothed, with a glabrous awn from the sinus, or awnless. Caryopsis obovoid, dorsally compressed. Pedicelled spikelet male or barren, linear-lanceolate to subulate, usually much narrower than the sessile and awnless.

Some 20 species in the Old World tropics and subtropics.

S. bicolor (L.) Moench (1794) is an important tropical cereal grown in parts of Somalia. Snowden,

The cultivated races of *Sorghum* (1936), recognized 28 species, but it is now generally agreed that these are best regarded as cultivars. The main 'species' noted by Snowden as occurring are:

S. ankilob Stapf
S. durra Stapf
S. nigricans (Ruiz & Pav.) Snowden
S. subglabrescens Asch. & Schweinf.

S. × *drummondii* (Nees ex Steud.) Millsp. & Chase (1903) is a tardily disarticulating element, selected for cultivation as a fodder grass ('Sudan grass'), derived from hybridization between *S. bicolor* and its wild progenitor *S. arundinaceum*. It can be extremely difficult to distinguish from the latter since, being cytogenetically indistinct, they freely hybridize. Some records of *S. arundinaceum* may in fact be partial or complete reversions of *S.* × *drummondii* arising through successive back-crossings with the parent. Indeed, some records of *S.* × *drummondii* itself may be spontaneous hybrids rather than the deliberately sown fodder crop.

1. Nodes glabrous or pubescent 2
− Nodes densely bearded 4
2. Racemes fragile; wild grass ..1. *S. arundinaceum*
− Racemes tough or tardily disarticulating; cultivated or subspontaneous grasses 3
3. Grain large, commonly exposed by the gaping glumes; sessile spikelets persistent; cultivated...
.... *S. bicolor* (see above)
− Grain small, enclosed by the glumes; sessile spikelets persistent or tardily deciduous; cultivated or subspontaneous..............................
.... *S.* × *drummondii* (see above)
4. Sessile spikelet elliptic-oblong, 5−7 mm long; pedicelled spikelet 3−5 mm long 2. *S. versicolor*
− Sessile spikelet lanceolate, 7.5−10 mm long; pedicelled spikelet 6−10 mm long................
.... 3. *S. purpureo-sericeum*

1. **S. arundinaceum** (Desv.) Stapf (1917). Fig. 147.

S. aethiopicum (Hack.) Stapf (1917).
S. lanceolatum Stapf (1917).
S. verticilliflorum (Steud.) Stapf (1917).
S. somaliense Snowden (1955); type: N1, "Burao, Adad", McKinnon 274 (K holo.).

Makadeey (Som.).

Annual or short-lived perennial without rhizomes, up to 4 m high, the nodes glabrous or pubescent. Panicle lanceolate to broadly spreading, 10−60 cm long; primary branches compound, ultimately bearing racemes of 2−7 spikelet-pairs. Sessile spikelet narrowly ovate to elliptic, 4−9 mm long; lower glume glabrescent to white-pubescent, sometimes tomentose or fulvously pubescent; upper lemma awnless or with an awn 5−30 mm long. Pedicelled spikelet linear to lanceolate, male or barren, smaller than the sessile.

Damp alluvial soils, heavy black clays and irrigated farmland; 50–1300 m. N1–3; S1–3; throughout Africa, extending eastwards to Australia. Hemming 2249; Gillett & Beckett 23566; Hansen & Heemstra 6309; Thulin & Bashir Mohamed 6984; Rose Innes & Trump 936; Kazmi 5189.

Chiovenda (1932) recorded *S. virgatum* (Hack.) Stapf from S3, but this seems to have been in error. The species is part of the *S. arundinaceum* complex, but seems to be the only one worth retaining as distinct. It has a narrowly lanceolate sessile spikelet and a narrow, scanty panicle. *S. halepense* (L.) Pers. has also been listed for Somalia, but all records refer to material from Ethiopia (Ogaden).

2. S. versicolor Anderss. (1863).

S. purpureo-sericeum var. *trinervatum* Chiov., Fl. Somala 2: 439 (1932); type: S3, "Osboda", Senni 244 (FT holo.).

Annual or short-lived perennial up to 2.5 m high, without rhizomes, the nodes densely bearded. Panicle narrowly oblong, 5–25 cm long; primary branches whorled, simple (rarely the longer branched), bearing 3–7 spikelet-pairs. Sessile spikelet elliptic-oblong, 5–7 mm long; lower glume coriaceous, glossy, glabrous to loosely hairy, bearded from the callus; upper lemma with an awn 25–40 mm long. Pedicelled spikelet linear to lanceolate, 3–5 mm long.

Coastal bushland. S3; Ethiopia and Kenya southwards to South Africa. Kazmi 5167.

3. S. purpureo-sericeum (Hochst. ex A. Rich.) Asch. & Schweinf. (1867).

Annual up to 1.5 m high, the nodes densely bearded. Panicle narrowly oblong, 5–35 cm long; primary branches whorled, simple (rarely the longer branched), bearing 3–5 spikelet-pairs. Sessile spikelet lanceolate, 7.5–10 mm long; lower glume coriaceous, glossy, glabrous to densely hairy, bearded from the callus; upper lemma with an awn 20–40 mm long. Pedicelled spikelet linear to lanceolate, 6–10 mm long.

In fallow fields on clay soils; up to c. 300 m. S1, 2; tropical Africa and India. Eagleton 43; Peveling G2.

73. CHRYSOPOGON Trin. (1822), nom. cons.

Tufted perennials; leaves linear, often harsh and glaucous; ligule a short membrane or a line of hairs. Inflorescence a terminal panicle with whorls of slender branches bearing terminal racemes; racemes reduced to a triad of 1 sessile and 2 pedicelled spikelets with linear pedicels, often the sessile spikelet pallid with fulvous callus-beard, and the pedicelled spikelet purple. Sessile spikelet laterally compressed; callus elongated, acute to pungent;

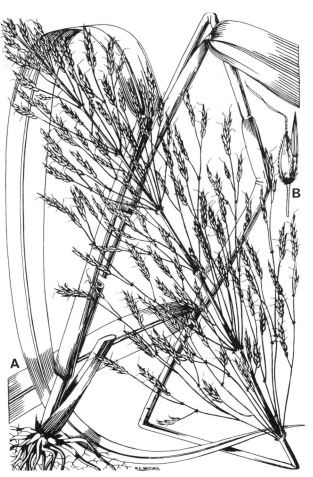

Fig. 147. *Sorghum arundinaceum*. A: habit, × 0.4. B: spikelet-pair, × 2.5. – From Fl. Trop. E. Afr. (1982). Drawn by W. E. Trevithick.

lower glume cartilaginous, rounded on the back, sometimes spinulose on the margins; upper glume often awned; lower floret reduced to a hyaline lemma; upper lemma hyaline, 2-toothed or entire, with a glabrous or pubescent awn. Caryopsis narrowly ellipsoid. Pedicelled spikelets male or barren, narrowly lanceolate, awned or awnless.

26 species in tropical and warm temperate regions of the Old World, mainly in Asia and Australia; one species in the New World.

1. Awn of pedicelled spikelet plumose...............
.... 1. *C. plumulosus*
– Awn of pedicelled spikelet glabrous..2. *C. aucheri*

1. C. plumulosus Hochst. (1847). Fig. 148.

Andropogon quinqueplumis (Hochst. ex A. Rich.) Steud. (1854); *A. aucheri* Boiss. var. *quinqueplumis* (Hochst. ex A. Rich.) Hack. in DC., Monogr. Phan. 6: 561 (1889); *Chrysopogon aucheri* var. *quinqueplumis* (Hochst. ex A. Rich.) Stapf in Bull. Misc. Inform. 1907: 211 (1907).

Fig. 148. *Chrysopogon plumulosus*, habit, × 0.5. B: spikelet-triad, × 7. — Modified from Fl. Trop. E. Afr. (1982). Drawn by A. Davies.

Chrysopogon aucheri var. *pulvinatus* Stapf in Prain, Fl. Trop. Afr. 9: 161 (1917); types: N1, "Golis Range", Drake-Brockman 153 (K syn.), "Upper Sheikh", Drake-Brockman 558 (K syn.), and others.

Daremo (Som.).

Stems wiry, up to 90 cm high; leaves cauline or sometimes forming a compact cushion, glaucous,

glabrous to densely puberulous, with or without tubercle-based hairs; basal sheaths laterally compressed or not. Panicle ovate, 3−7 cm long. Sessile spikelet narrowly oblong, 4−6 mm long; upper glume with a plumose awn 7−15 mm long; upper lemma with a puberulous awn 2−3 cm long. Pedicelled spikelet 4−7 mm long, the lower and often also the upper glume bearing a plumose awn up to 15 mm long, this sometimes glabrous near the tip; pedicels 1/3−1/2 the length of the sessile spikelet.

Sandy and alluvial plains and rocky hillsides, or in lightly wooded grassland, often severely overgrazed; 50−1800 m. N1−3; C1, 2; S2, 3; Djibouti, Ethiopia, Sudan, Kenya, Tanzania and Arabia. McKinnon 10; Glover & Gilliland 186; Thulin & Warfa 5959; Herlocker 370; Rose Innes 840; Thulin & Dahir 6740; Rose Innes 797.

2. **C. aucheri** (Boiss.) Stapf (1907).

Chrysopogon montanus Trin. var. *migiurtinus* Chiov., Pl. Nov. Aethiop.: 16 (1928); *C. fulvus* (Spreng.) Chiov. var. *migiurtinus* (Chiov.) Chiov., Fl. Somala 1: 328 (1929). Type: N3, "Dol-Dol", Puccioni & Stefanini 823 (FT holo.).

Stems slender, up to 60 cm high; leaves shortly and densely pubescent and with tubercle-based cilia on the margins especially below; basal sheaths often silky-villous. Panicle ovate, 5−10 cm long. Sessile spikelet narrowly elliptic to narrowly oblong, 5−8 mm long; upper glume with a shortly pubescent awn 1.5−10 mm long; upper lemma with a shortly pubescent awn 2.5−4 cm long. Pedicelled spikelet (4−)7−10 mm long, the lower and often also the upper glume bearing a glabrous awn 4−7 mm long, this sometimes ciliate at the very base, often quite absent; pedicels 1/3−1/2 the length of the sessile spikelet.

Probably semidesert grassland near the coast. N3; Ethiopia, Arabia, Iran, Afghanistan, Pakistan and India.

Known only from the type of var. *migiurtinus* in Somalia, and at the western and southern limits of its range in neighbouring Ethiopia.

The two species in Somalia are generally distinct enough, but on occasion they tend to merge, with the awn-character becoming rather unreliable. It is doubtful whether they are really distinct at species level, subspecies being perhaps more appropriate given their marked geographical segregation, but problems in Asia as well as in Africa will have to be resolved before a decision can be made. For now, they are kept apart, but *C. aucheri* is apparently very rare in Somalia.

74. DICHANTHIUM Willemet (1796)
Eremopogon Stapf (1917).

Annual or perennial; leaves sometimes aromatic; ligule membranous. Inflorescence terminal or sometimes also axillary, of single or subdigitate racemes, these sometimes pedunculate, with or without homogamous spikelet-pairs at the base; internodes and pedicels linear, solid. Sessile spikelet dorsally compressed; callus very short, obtuse; lower glume chartaceous to cartilaginous, broadly convex to slightly concave on the back, abruptly rounded on the flanks, with or without a circular pit; upper lemma stipitiform, entire or rarely minutely 2-toothed, with a glabrous or puberulous awn. Caryopsis oblong, dorsally compressed. Pedicelled spikelet similar to the sessile, rarely herbaceous.

Some 20 species in the Old World tropics.

1. Lower glume of sessile spikelet pitted; raceme solitary ....1. *D. foveolatum*
   − Lower glume of sessile spikelet not pitted; racemes subdigitate ....2. *D. annulatum*

1. **D. foveolatum** (Del.) Roberty (1960); *Andropogon foveolatus* Del. (1813); *Eremopogon foveolatus* (Del.) Stapf (1917). Fig. 149.

Aus goron, aus koksar, dalan, laah, saren (Som.).

Tufted perennial with silky-hairy basal sheaths and bearded nodes, the stems wiry, up to 80 cm high. Inflorescence of solitary racemes, each subtended by a narrow spatheole, these terminal and loosely aggregated; racemes 1.5−4.5 cm long, with 0−1 smaller homogamous spikelet-pairs at the base. Sessile spikelet narrowly elliptic, 2.5−4 mm long; lower glume cartilaginous, glabrous, shining, with a circular depression in the upper third, acute; upper lemma with an awn 12−18 mm long. Pedicelled spikelet as long as the sessile, with or without a pit.

Open bushland on rocky hillsides, low-lying clay soils and sandy coastal plains; 0−2700 m. N1−3; C1, 2; S1−3; Mali and North Africa eastwards through Arabia to India and Sri Lanka. Gillett 4736; Hansen & al. 6501; Thulin 5555; Thulin & Warfa 5372; Rose Innes 843; Thulin & Bashir Mohamed 6931; Bavazzano 1050; Kazmi 5140.

2. **D. annulatum** (Forssk.) Stapf (1917); *Andropogon annulatus* Forssk. (1775).

Aiyah macareh, aus guran, darer adili, domar, jebin (Som.).

Tufted perennial with conspicuously bearded nodes, up to 100 cm high. Inflorescence composed of (1−)2−15 subdigitate, shortly pedunculate racemes; racemes 3−7 cm long, the spikelets subimbricate, with 0−6 smaller homogamous spikelet-pairs at the base. Sessile spikelet narrowly oblong, 2−6 cm long; lower glume firmly cartilaginous, without a pit, slightly concave, pubescent to villous below, with bulbous-based hairs above, obtuse to subacute; upper lemma with an awn 8−25 mm long. Pedicelled spikelet as long as the sessile, without a pit.

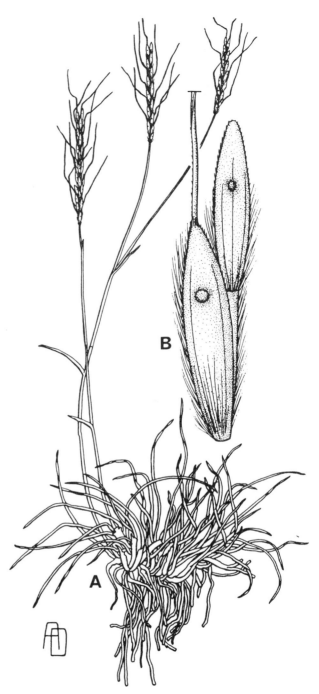

Fig. 149. *Dichanthium foveolatum*. A: habit, × 2/3. B: spikelet-pair, × 10. − Modified from Fl. Trop. E. Afr. (1982). Drawn by A. Davies.

1. Lower glume of sessile spikelet pubescent to pilose below the middle, the bulbous-based hairs mainly confined to the margin above.............. .... var. *annulatum*
   − Lower glume of sessile spikelet pilose to villous, with a distinct subapical fringe of bulbous-based hairs ....var. *papillosum*

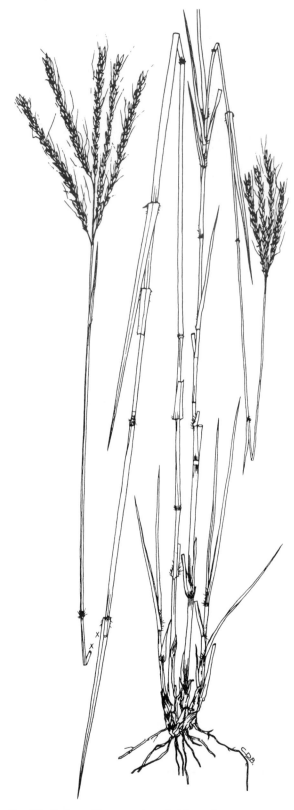

var. **annulatum.**

Diploid or tetraploid (2n = 20, 40).

Hillsides, plains, dunes and riverside alluvium, irrigated, cultivated and disturbed ground; 0−1220 m. N1−3; C1, 2; S1−3; tropical Africa eastwards through Southwest Asia to India and Indonesia; introduced in southern Africa, tropical America and Australia. Godding 9; McKinnon 123; Thulin & Warfa 5988; Wieland 4686; Rose Innes 805; Gillett & Hemming 24748; Dahir & Rose Innes ODA38; Kazmi & al. 714.

var. **papillosum** (Hochst. ex A. Rich.) Pilg. ex de Wet & J.R. Harlan in Boln. Soc. Argent. Bot. 12: 212 (1968); *Andropogon papillosus* Hochst. ex A. Rich. (1850).

Hexaploid (2n = 60).

Rocky streambeds and cultivated soils; 900−1830 m. N1, 2; C2; S2; Ethiopia to southern Africa. McKinnon 87; Collenette 35; Moggi & Bavazzano 529; Hansen 6063.

75. BOTHRIOCHLOA Kuntze (1891)

Perennial; leaves sometimes aromatic; ligule membranous. Inflorescence of digitate or subdigitate racemes, sometimes paniculate with simple or rarely subdivided branches and a long central axis; racemes with more than 8 sessile spikelets, with or without homogamous spikelet-pairs at the base; internodes and pedicels linear, hyaline and translucent between the thickened margins. Sessile spikelet dorsally compressed; callus very short, obtuse; lower glume cartilaginous or firmly membranous, broadly convex to slightly concave on the back, sometimes with 1−3 circular pits, acute; upper lemma entire, with a glabrous awn; caryopsis oblong, slightly dorsally compressed. Pedicelled spikelet well developed, similar to the sessile or smaller.

Some 35 species throughout the tropics.

1. Lower glume of the sessile spikelet pitted..........
   .... 1. *B. insculpta*
   − Lower glume of the sessile spikelet not pitted....
   .... 2. *B. radicans*

1. **B. insculpta** (Hochst. ex A. Rich.) A. Camus (1931). Fig. 150.

B. pertusa auct. non (L.) A. Camus; *Amphilophis pertusa* auct. non (L.) Nash ex Stapf; *Andropogon pertusus* auct. non (L.) Willd.

Aus gudun, aus ma'an, domar, gurguro, magalhed (Som.).

Tufted perennial; stems up to 200 cm high, often becoming decumbent and rambling or developing into stout woody stolons. Inflorescence subdigitate or with a central axis seldom over 3 cm long, bearing 3−20 shortly pedunculate racemes; racemes 2−8 cm long,

Fig. 150. *Bothriochloa insculpta*, habit. − From Mem. Bot. Surv. S. Afr. 58 (1991).

pilose. Sessile spikelet narrowly elliptic, 3—4.5 mm long; lower glume firmly cartilaginous, glabrous oɪ pilose below the middle, glossy, with a deep circular pit; awn 15—25 mm long. Pedicelled spikelet glabrous, with 0—4 pits.

Grassy plains, open woodland and cultivated ground; 300—1650 m. N1, 2; S1—3; throughout Africa and in tropical Arabia. Hansen & Heemstra 6126; Glover & Gilliland 939; Eagleton 33; Rose Innes 871A; Tozzi 308.

Very similar to the Asiatic *B. pertusa* (L.) A. Camus and unreliably separated from it. De Wet & Higgins, in Phyton (Buenos Aires) 20: 205—211 (1963), have shown that the two do not interbreed and that African material resembling the Asiatic species should be regarded as a variant of *B. insculpta*. It is doubtful whether true *B. pertusa* occurs in Africa. The two species would scarcely be worth maintaining as distinct were it not for the apparent genetic as well as geographical isolation.

2. **B. radicans** (Lehm.) A. Camus (1931); *Amphilophis radicans* (Lehm.) Stapf (1917).

Andropogon ischaemum L. var. *somalensis* Stapf in Bull. Misc. Inform. 1907: 210 (1907); type: N1, Appleton s.n. (K holo.).

Aus gurun, gurguro (Som.).

Tufted perennial; stems up to 100 cm high, often fasciculate and branching from the lower nodes, usually ascending. Inflorescence subdigitate with a central axis up to 5 cm long, bearing 5—16 sessile or shortly pedunculate racemes; racemes 3—7 cm long, villous. Sessile spikelet lanceolate, 2.5—4 mm long; lower glume firmly membranous, pilose below the middle, not glossy and not pitted; awn 10—25 mm long. Pedicelled spikelet glabrous, not pitted.

Open *Acacia-Commiphora* bushland on limestone hillsides; 50—350 m. N1—3; S1—3; Ethiopia and Sudan southwards to South Africa and in Saudi Arabia. Glover & Gilliland 14, 985; Bally & Melville 15778; Friis & al. 4808; Bavazzano 1038; Senni 414.

76. ISCHAEMUM L. (1753)

Perennial, sometimes annual; ligule membranous. Inflorescence of paired or occasionally digitate racemes, the former often interlocked back to back, terminal or axillary; internodes and pedicels clavate to inflated, the pedicels often very short. Sessile spikelet dorsally compressed; callus obtuse and inserted into the concave top of the internode; lower glume chartaceous to coriaceous, concave to convex, laterally 2-keeled or the flanks rounded, often rugose, sometimes winged; lower floret male and with a palea; upper lemma bifid, awned from between the teeth or awnless; caryopsis oblong to lanceolate, dorsally compressed. Pedicelled spikelet as large as the sessile or much smaller, often asymmetrical.

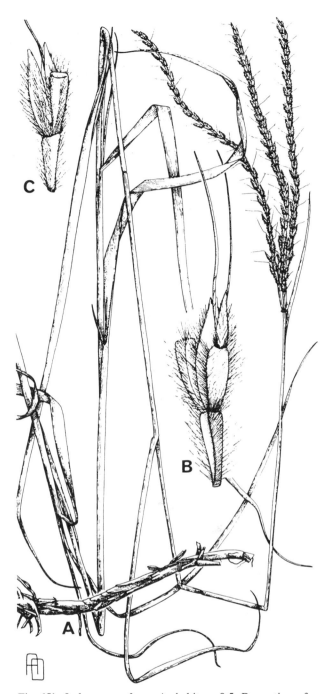

Fig. 151. *Ischaemum afrum*. A: habit, × 0.5. B: portion of raceme, showing pedicelled spikelet, × 4. C: portion of raceme, showing sessile spikelet, × 5. — Modified from Fl. Trop. E. Afr. (1982). Drawn by A. Davies.

Some 65 species throughout the tropics, but mostly in Asia.

I. afrum (J.F. Gmel.) Dandy (1956). Fig. 151.
Aus ma'an, hamashleh wein, rareh biyud (Som.).
Densely tufted perennial with scaly rhizomes; stems

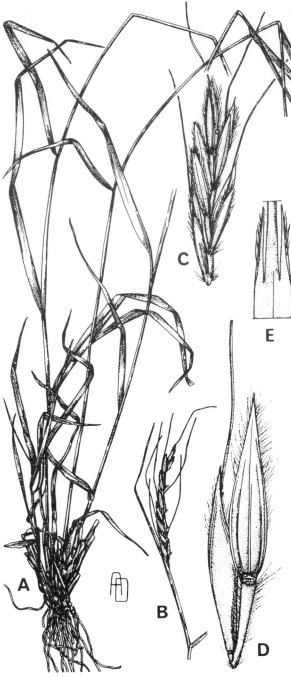

Fig. 152. *Sehima nervosum.* A: base of plant, × 0.5. B: raceme, × 0.5. C: detail of raceme, × 3. D: spikelet-pair, × 5. E: tip of upper lemma, × 10. — Modified from Fl. Trop. E. Afr. (1982). Drawn by A. Davies.

up to 200 cm high. Inflorescence terminal, of (1−)2−5 subdigitate racemes, each 6−20 cm long; internodes and pedicels glabrous to villous, yellowish to purple, subequal, the former clavate, the latter subinflated. Sessile spikelet lanceolate, 5−8 mm long; lower glume chartaceous, 2-keeled along its

length, concave between the keels, glabrous to villous, wingless; upper lemma with an awn 5−20 mm long. Pedicelled spikelet 1−6 mm long, awnless or rarely awned.

Acacia-Commiphora bushland and badly drained grassy plains, and in cultivated ground; 175−1550 m. N1; S1; tropical and South Africa and in India. McKinnon 279; Friis & al. 4870.

77. SEHIMA Forssk. (1775)

Annual or perennial; ligule a line of hairs. Inflorescence a single terminal raceme, exserted from the uppermost sheath; internodes and pedicels stoutly linear to subclavate, ciliate. Sessile spikelet laterally, rarely dorsally, compressed, fitting between the internode and pedicel; callus obtuse and inserted into the concave top of the internode; lower glume coriaceous, concave or with a median groove, laterally 2-keeled or lyrate with the keels becoming dorsal towards the base, scarcely winged, 2-toothed at the tip; lower floret male and with a palea; upper lemma 2-toothed, awned from between the teeth with an awn puberulous to ciliate along the edges of the coils; caryopsis lanceolate-oblong, dorsally compressed, concave on one side. Pedicelled spikelet large, lanceolate, strongly dorsally compressed, distinctly nerved.

Five species in the Old World tropics.

1. Plant perennial; lower glume of sessile spikelet with a membranous tip up to 1/3 the length of the body, shallowly bifid; column of awn minutely ciliolate along the edges of the coils..............
    ....1. *S. nervosum*
−   Plant annual; lower glume of sessile spikelet with a membranous tip as long as the body, deeply bifid; column of awn ciliate along the edges of the coils ....2. *S. ischaemoides*

1. **S. nervosum** (Rottl.) Stapf (1917). Fig. 152.
Ischaemum laxum R. Br. var. *genuinum* Hack. in DC., Monogr. Phan. 6: 245 (1889).
Tufted wiry perennial up to 100 cm high. Raceme 3−12 cm long, gently curved. Sessile spikelet 6−10 mm long, laterally compressed; lower glume narrowly oblong-elliptic, thinly coriaceous with a shallowly bifid membranous tip up to 1/3 the length of the body, 2-keeled, concave between the keels, lyrately nerved with 2−3 closely spaced nerves adjacent to each keel; upper lemma with an awn 2−4 cm long, the column minutely ciliolate along the edges of the coils. Pedicelled spikelet 6−10 mm long, ciliate, the lower glume with a midnerve and 2−3 conspicuous lateral nerves adjacent to each keel.

Rocky hillsides and desert grassland; 1200−1500 m. N1; eastern tropical Africa from Sudan and Ethiopia southwards to Mozambique, and eastwards through

Arabia to China and Australia. Drake-Brockman 473; Gillett 4855; Hemming 2333.

2. S. ischaemoides Forssk. (1775).

Hol (Som.).

Annual up to 60 cm high. Raceme 3–15 cm long, gently curved. Sessile spikelet 9–15 mm long, laterally compressed; lower glume linear, firmly coriaceous with a deeply bifid membranous tip ± as long as the body, 2-keeled above, the keels curving inwards towards the midline below, forming a deep slot between them and obscuring the intercarinal nerves; upper lemma with an awn 4–7 cm long, the column ciliate along the edges of the coils. Pedicelled spikelet 7–15 mm long, ciliate, the lower glume with a midnerve and 2–3 conspicuous lateral nerves adjacent to each keel.

Subdesert grassland. N2; tropical Africa eastwards through Arabia to Pakistan. Glover & Gilliland 1058.

A rare and sporadic grass throughout its range, collected only once from Somalia.

78. ANDROPOGON L. (1753)

Annual or perennial; leaves not aromatic; ligule membranous or reduced to a line of hairs. Inflorescence composed of paired or digitate (rarely solitary) racemes, terminal or axillary, the latter often numerous and crowded into a leafy false panicle; raceme-base terete, rarely deflexed at maturity; racemes without homogamous spikelet-pairs at the base (or these present but scarcely differentiated); internodes and pedicels filiform to obovoid. Sessile spikelet dorsally or laterally compressed; callus obtuse, inserted in the concave top of the internode; lower glume membranous to coriaceous, flat to concave or deeply grooved on the back, 2-keeled, the keels lateral or dorsal, sometimes narrowly winged, with or without intercarinal nerves; lower floret barren, reduced to a hyaline lemma; upper lemma hyaline, bilobed, awned from between the lobes (rarely entire and awnless); caryopsis narrowly lanceolate to oblong, subterete to plano-convex. Pedicelled spikelet male or barren (rarely suppressed), never concave on the back, usually awnless.

Some 100 species throughout the tropics.

1. Leaves tightly inrolled, less than 1 mm wide, with reduced lamina; stems branched throughout forming dense cushions up to 25 cm high .... 1. *A. aridus*
   − Leaves flat or loosely inrolled, 1.5–8 mm wide, with well developed lamina 2
2. Stems profusely branched throughout forming dense bushes up to 2(−3.6) m high and 1.8 m across .... 2. *A. kelleri*
   − Stems branched only at the base (if at all) and plant not at all bushy 3
3. Rhachis-internodes and pedicels cuneate; lower glume of sessile spikelet deeply depressed between the dorsal keels .... 7. *A. pungens*
   − Rhachis-internodes and pedicels filiform to cuneate; lower glume of sessile spikelet flat or slightly convex, but with a shallow median depression between the lateral keels 4
4. Inflorescence a large leafy false panicle.......... .... 5. *A. gayanus*
   − Inflorescence of paired (rarely more) terminal racemes 5
5. Lower glume of sessile spikelet coarsely scabrid, with conspicuous nerves in the depression; densely tufted plant with a cushion of old leaf-sheaths at the base .... 6. *A. leprodes*
   − Lower glume of sessile spikelet smooth, with nerveless depression; loosely or densely tufted plant, sometimes mat-forming, but without a cushion of old leaf-sheaths at the base 6
6. Pedicelled spikelet 1- to 2-awned; sessile spikelet narrowly lanceolate to narrowly elliptic; pedicels linear to slightly clavate .... 3. *A. amethystinus*
   − Pedicelled spikelet awnless; sessile spikelet linear to lanceolate; pedicels filiform to linear .... 4. *A. greenwayi*

1. **A. aridus** Clayton (1977); type: N2, "Hubera", McKinnon 251 (K holo.).

Densely tufted perennial, the stems branched throughout forming cushions up to 25 cm high; leaves 3–5 cm × 1 mm, with much reduced lamina. Racemes solitary or paired, 2–3 cm long, terminal; internodes and pedicels linear, ciliate. Sessile spikelet 6–6.5 mm long including the 1 mm long callus; lower glume narrowly oblong-elliptic, chartaceous, slightly convex on the back, sparsely pilose below the middle, 4-nerved between the keels, 2-toothed at the tip; keels lateral, narrowly winged above; upper glume acute; upper lemma bilobed to the middle, with an awn up to 15 mm long. Pedicelled spikelet c. 5 mm long; lower glume narrowly lanceolate, acute.

Stony hillsides; 1000 m. N2; known only from the type.

2. **A. kelleri** Hack. (1900); *Schizachyrium kelleri* (Hack.) Stapf (1919); type: N1, "Tuyo", Keller 156 (K isosyn.). Fig. 153.

   A. cyrtocladus Stapf (1907); types: N1, Drake-Brockman 43, 44 (K syn.).

   A. bentii auct. non Stapf (1907).

Dulun, durr, toon (Som.).

Suffrutescent rhizomatous perennial forming bushes up to 2(−3.6) m high and 1.8 m across; stems woody below, profusely branched above; leaves 2–10 cm × 2–4 mm, flat. Racemes solitary or rarely paired, 2–3 cm long, terminal on short leafy branches and long-exserted on filiform peduncles; internodes and pedicels filiform, villous. Sessile spikelet 5–6.5 mm long including the 0.5 mm long

Fig. 153. *Andropogon kelleri*. A: fertile branch, × 0.8. B: spikelet-pair, × 5. − From Thulin & Warfa 5828. Drawn by K. Thunberg.

callus; lower glume linear-lanceolate, thinly cartilaginous, shallowly depressed between the thickened lateral keels, nerveless between the keels, acute; upper glume mucronate; upper lemma bifid in the upper 1/4, with an awn 10−25 mm long. Pedicelled spikelet 6−7 mm long; lower glume narrowly lanceolate with a mucro up to 1 mm long (the upper glume similarly mucronate).

Bushland in silty or sandy depressions overlying limestone or gypsum; 180−1520 m. N1−3; C1, 2; S1; Ethiopia. McKinnon 80; Hemming 1358; Beckett 1334; Bally & Melville 15362; Kuchar 17018; Wieland 1143.

Similar to the Socotran endemic *A. bentii* Stapf, but a much larger plant with longer, narrower spikelets and longer racemes.

3. **A. amethystinus** Steud. (1854).

Perennial, either straggling with wiry rhizomes or forming dense tufts, up to 60 cm high; leaves 1−15 cm × 1−4 mm, flat or inrolled. Racemes paired, 2−8 cm long, terminal; internodes and pedicels linear (or the latter slightly clavate), ciliate to villous. Sessile spikelet 5−8.5 mm long including the 1 mm long callus;

lower glume narrowly lanceolate to narrowly elliptic, slightly convex to slightly concave on the back, thinly coriaceous to papery, 2- to 6-nerved between the keels, 2-toothed at the tip; keels lateral, winged or not; upper glume with an awn 2−6 mm long; upper lemma bilobed to halfway, with an awn 10−15 mm long. Pedicelled spikelet 4−8 mm long; lower glume narrowly lanceolate to narrowly ovate, herbaceous, with an awn 1−3 mm long (rarely the upper glume also shortly awned).

Grassland; 200−300 m. N2; tropical and South Africa, Arabia (Yemen) and India (Nilgiri Hills). Hansen & Heemstra 6211.

Known from Somalia from a single collection that is ascribed to the species with some hesitation. The over-mature spikelets have largely been shed and the awns of the pedicelled spikelet are not obvious, nor are the number of racemes and their overall appearance. *A. amethystinus* is sufficiently variable to accommodate this specimen which is far from suitable as the basis of a new taxon.

4. **A. greenwayi** Napper (1963). Fig. 154 A.

Agar, domar (Som.).

Shortly rhizomatous mat-forming perennial up to 60 cm high; leaves 3−10 cm × 1−4 mm, flat, glaucous. Racemes paired (rarely more), 3−7 cm long, terminal; internodes and pedicels filiform to linear, ciliate. Sessile spikelet 6.5−11 mm long including the 1 mm long callus; lower glume linear to narrowly lanceolate, slightly convex on the back and usually with a nerveless median depression, thinly coriaceous with herbaceous or membranous tip, 4- to 10-nerved between the keels, glabrous, sharply 2-toothed at the tip; keels lateral, narrowly winged; upper glume with an awn 0.5−4 mm long; upper lemma bilobed to halfway, with an awn 14−20 mm long. Pedicelled spikelet 5.5−10 mm long; lower glume narrowly lanceolate, acute to acuminate, awnless (the upper glume also awnless).

Forest glades and around the edges of cultivated fields; 1500−2100 m. N1, 2; Ethiopia, Kenya, Tanzania and Arabia (Yemen). McKinnon 31; Hemming 1949.

5. **A. gayanus** Kunth (1833).

Tall tufted perennial up to 2.5 m high; leaves up to 60 cm × 20 mm, often narrowed at the base and falsely petiolate. Inflorescence of paired racemes gathered into a large leafy false panicle; racemes 4−9 cm long; internodes and pedicels cuneate, ciliate on one side only. Sessile spikelet 5−8 mm long including the c. 1 mm shortly oblong callus; lower glume narrowly oblong, ± flat on the back with many intercarinal nerves and a conspicuous median depression, wingless; upper glume muticous or mucronate; upper lemma deeply bilobed, with an awn 10−30 mm long. Pedicelled spikelet 5−8 mm long; lower glume narrowly elliptic with an awn 1−10 mm long (the upper similar).

Deciduous bushland. C1; S1; tropical and South Africa. Paoli 727; Moggi & Bavazzano 89.

6. **A. leprodes** Cope (1995); type: N2, "Erigavo", Hemming & Watson 3286 (K holo.). Fig. 154 B.

Gogane, rammas (Som.).

Densely tufted perennial up to 50 cm high, the unbranched stems arising from a basal cushion derived from the remains of old leaf-sheaths; leaves 2—7 cm × 1—2 mm, flat, green or glaucous. Racemes 2—4, subdigitate, 4—6 cm long, terminal; internodes and pedicels filiform to linear, ciliate. Sessile spikelet 6.5—9 mm long including the 1 mm long callus; lower glume narrowly lanceolate to narrowly elliptic, slightly convex on the back but with a median depression, this with conspicuous nerves within, herbaceous with membranous, sharply 2-toothed tip, thickly 5- to 7-nerved between the keels, densely scabrid; keels lateral, wingless; upper glume with an awn-point 1—1.5 mm long; upper lemma bilobed to halfway, with an awn 15—20 mm long. Pedicelled spikelet 4—7 mm long; lower glume narrowly elliptic, acute to acuminate, awnless (the upper glume also awnless).

Gypsum plains; 1340—1740 m. N2; not known elsewhere. Glover & Gilliland 1026, 1067; Lavranos 6768.

A distinctive species retaining a cushion of old leaf-sheaths around the base of the unbranched stems. The thickly nerved, densely scabrid lower glume is also characteristic, as are the nerves within the median depression of the sessile spikelet.

7. **A. pungens** Cope (1995); type: C1, Herlocker 329 (K holo.).

A. chinensis auct. non (Nees) Merr.

Tufted perennial up to 125 cm high; leaves 6.5—20 cm × 1—5.5 mm, flat, green. Racemes paired, 4.5—7 cm long, gathered into a scanty, false panicle; internodes and pedicels cuneate, ciliate. Sessile spikelet 5—8 mm long, not including the 2—3.5 mm long, very pungent callus; lower glume linear, deeply depressed between the dorsal wingless keels, glabrous, 2-toothed at the tip; upper glume with an awn 3—6 mm long; upper lemma 2-toothed, with a puberulous awn 20—40 mm long. Pedicelled spikelet 5—7 mm long; lower glume lanceolate to narrowly elliptic, glabrous on the back, ciliolate on the margins, with an awn 3—5.5 mm long (the upper glume with an awn 1.5—4 mm long).

Grassy plains, often near the coast. C1; S3; Kenya. Hemming 3335; Rose Innes 765.

79. CYMBOPOGON Spreng. (1815)

Robust perennials, rarely annual; leaves usually aromatic; ligule membranous or scarious. Inflorescence composed of short paired racemes borne on a short common peduncle and ± enclosed by a boat-

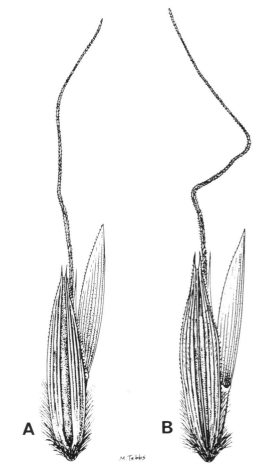

Fig. 154. Spikelet-pairs, × 9, of A: *Andropogon greenwayi*; B: *A. leprodes*. — From Kew Bull. 50 (1995). Drawn by M. Tebbs.

shaped spatheole, these crowded into a leafy false panicle; raceme-bases short, flattened, usually deflexed, the lower bearing a homogamous spikelet-pair at the base; internodes linear, sometimes the pedicel of the homogamous pair swollen and ± fused to the internode. Sessile spikelet dorsally compressed; callus obtuse, inserted in the concave top of the internode; lower glume ± chartaceous, concave, 2-keeled, the keels usually lateral and often winged near the tip, with or without intercarinal nerves; lower floret barren, reduced to a hyaline lemma; upper lemma hyaline or stipitiform, bilobed or rarely entire, with or without an awn from the sinus; caryopsis oblong, subterete to plano-convex. Pedicelled spikelet male or barren, ± as long as the sessile but never concave on the back, awnless.

Some 40 species in the Old World tropics and subtropics.

C. citratus (DC.) Stapf is an important medicinal and culinary herb throughout the tropics. Its awnless spikelets are diagnostic, but it rarely flowers in cultivation; its leaves, however, give off a very strong lemon scent when crushed. It is known only as a

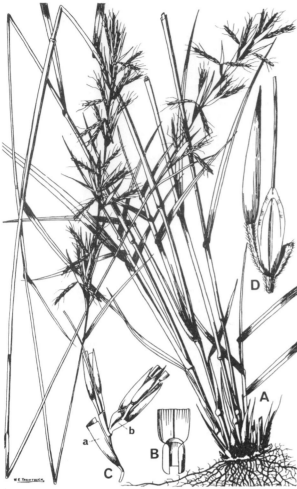

Fig. 155. *Cymbopogon caesius*. A: habit, × 0.4. B: ligule, × 0.8. C: base of raceme-pair, showing base of spatheole (a) and raceme-base (b), × 3. D: spikelet-pair, × 6. − Modified from Fl. Trop. E. Afr. (1982). Drawn by W. E. Trevithick.

cultivated plant, and there is one record of it being grown in Somalia (S2).

1. Lower glume of sessile spikelet flat to shallowly concave, with a fine V-shaped median groove in the lower third, laterally winged on the keels ... 2
− Lower glume of sessile spikelet flat to deeply concave, with the bottom of the depression rounded, wingless on the keels 4
2. Leaves over 8 mm wide, dark green; ligule scarious, rarely over 1 mm long; inflorescence 20−70 cm long; stems robust ..... 3. *C. giganteus*
− Leaves up to 8(−10) mm wide, glaucous; ligule membranous, 1−4 mm long; inflorescence 5−20 (−30) cm long; stems wiry 3
3. Lower glume of sessile spikelet with the groove flanked by brown oil-streaks ....1. *C. nervatus*
− Lower glume of sessile spikelet without oil-streaks ....2. *C. caesius*

4. Sessile spikelet awnless, scarcely winged..........
.... *C. citratus* (see above)
− Sessile spikelet awned 5
5. Lowermost pedicel not swollen...................
....4. *C. pospischilii*
− Lowermost pedicel swollen and barrel-shaped ..6
6. Lower glume of sessile spikelet with the keels sharply inflexed throughout ...5. *C. schoenanthus*
− Lower glume of sessile spikelet with the keels rounded below .... 6. *C. commutatus*

1. **C. nervatus** (Hochst.) Chiov. (1909).
  Annual or short-lived perennial up to 100 cm high; ligule membranous, 1−4 mm long; leaves rounded at the base, up to 10 mm wide, glaucous. False panicle linear to oblong, 15−30 cm long; racemes 10−15 mm long, the lowermost internode and pedicel connate and slightly swollen. Sessile spikelet narrowly elliptic, 4−5 mm long; lower glume thinly membranous, narrowly to broadly winged, shallowly concave on the back and with a shallow V-shaped median groove in the lower 1/2, the two clear intercarinal nerves flanked by brown oil-streaks; awn of upper lemma c. 15 mm long, with distinct column. Pedicelled spikelet narrowly lanceolate, 4.5−5 mm long, with clear closely spaced nerves.
  S3; Ethiopia and Sudan. Tozzi 389.
  The specimen cited, thus far the only record of the species in Somalia, is in a rather poor state of preservation; however, it is possible to see that while the texture of the glume and the presence of oil-streaks indicate *C. nervatus*, the pedicel and internode do not appear to be swollen.

2. **C. caesius** (Nees ex Hook. & Arn.) Stapf (1906). Fig. 155.
  C. connatus (Hochst. ex A. Rich.) Chiov. (1909).
  C. excavatus (Hochst.) Stapf (1919).
  Sandul (Som.).
  Tufted wiry perennial up to 120 cm high; ligule membranous, 1−4 mm long; leaves rounded at the base, up to 8(−10) mm wide, glaucous. False panicle linear to oblong, 5−20(−30) cm long; racemes 10−15 mm long, the lowermost internode and pedicel connate and swollen. Sessile spikelet narrowly elliptic, 3−4.5 mm long; lower glume firmly membranous, narrowly to broadly winged, shallowly concave on the back and with a V-shaped groove in the lower 1/3−1/2, without intercarinal nerves (or these obscure and confined to the tip) or oil-streaks; awn of upper lemma 6−15 mm long, with distinct column. Pedicelled spikelet narrowly lanceolate, 3−5 mm long, with clear closely spaced nerves.
  N1; S3; Sudan and Arabia (Yemen) through eastern Africa to South Africa, and in India and Sri Lanka. Drake-Brockman 463; Tozzi 509.

3. **C. giganteus** Chiov. (1909), based on *Andropogon giganteus* Hochst. (1844) non Ten. (1811).

Loosely tufted robust perennial up to 3 m high; ligule scarious, up to 1(−2) mm long; leaves cordate to subamplexicaul, 8−30 mm wide, dark green. False panicle linear, 20−70 cm long; racemes 10−15 mm long, the lowermost internode and pedicel connate and swollen. Sessile spikelet narrowly elliptic, 3.5−5 mm long; lower glume firmly membranous, narrowly winged, slightly concave and with a V-shaped median groove towards the base, without intercarinal nerves (or these obscure and confined to the tip) or oil-streaks; awn of upper lemma 10−17 mm long, with distinct column. Pedicelled spikelet narrowly lanceolate, 3.5−5 mm long, with clear closely spaced nerves.

Silty soil on riverside levee. S2; tropical Africa. Rose Innes 653.

4. C. pospischilii (K. Schum.) C.E. Hubb. (1949).

C. plurinodis (Stapf) Stapf ex Burtt Davy (1912).

Baila, beli, burburao, caws mulaax (Som.).

Tufted perennial up to 100 cm high; leaves attenuate at the base. False panicle linear to narrowly oblong, 10−30 cm long; racemes 15−35 mm long, the lowermost pedicel oblong but not swollen, free from the adjacent internode. Sessile spikelet narrowly lanceolate, 4.5−7 mm long; lower glume chartaceous, nerveless or with 1−3 short nerves near the tip, concave between the keels, these rounded in the lower 1/2; awn of upper lemma 10−20 mm long, with distinct column. Pedicelled spikelet narrowly lanceolate, 4−7 mm long.

Deciduous bushland and degraded juniper forest on limestone hillsides and plateaus, and subdesert grassland; 40−2130 m. N1−3; C1; eastern Africa from Ethiopia to the Cape, and northern Pakistan to Nepal. McKinnon 7; Glover & Gilliland 1141; Bavazzano s.n.; Thulin & Abdi Dahir 6681.

5. C. schoenanthus (L.) Spreng. (1815).

Aus damer, aus goron, werahr (Som.).

Tufted perennial up to 120 cm high; leaves attenuate at the base. False panicle narrowly oblong, 5−40 cm long; racemes 20−30 mm long, woolly-villous with hairs 2−4 mm long, the lowermost pedicel swollen, its adjacent internode very short. Sessile spikelet narrowly lanceolate, 4−7 mm long; lower glume chartaceous, glabrous, nerveless and concave between the keels, these sharply inflexed throughout; awn of upper lemma 5−9 mm long, almost straight, scarcely differentiated into column and limb. Pedicelled spikelet narrowly lanceolate, 4−7 mm long.

Limestone hillsides and subdesert plains; 425−1475 m. N1−3; C1; Africa north of the Sahara and in Arabia. Gillett 4371; Glover & Gilliland 169; Hemming 1626; Bettini s.n.

Represented in Somalia by subsp. *schoenanthus*. Subsp. *proximus* (Hochst. ex A. Rich.) Maire & Weiller differs in its shorter racemes gathered into dense clusters, and puberulous lower glume of the sessile spikelet; it occurs in tropical Africa south of the Sahara, but the two subspecies overlap in parts of eastern tropical Africa.

6. C. commutatus (Steud.) Stapf (1907).

C. floccosus (Schweinf.) Stapf (1919).

C. divaricatus Stapf (1919); types: N1, Drake-Brockman 410, 411 (K syn.).

Arabjib, aus gudud, carowe, daremo-as, hadaf, lebjir (Som.).

Tufted perennial up to 150 cm high; leaves attenuate at the base. False panicle linear to narrowly oblong, 5−35 cm long; racemes 15−30 mm long, ciliate with hairs up to 2 mm long, the lowermost pedicel swollen, free from the adjacent internode which itself is sometimes swollen. Sessile spikelet narrowly lanceolate, 4−7 mm long; lower glume chartaceous, glabrous or pubescent, nerveless or with 2−3 intercarinal nerves in the upper 1/3, shallowly to deeply concave between the keels, these rounded in the lower 1/2 and often separated by a gibbous swelling at the base; awn of upper lemma 10−20 mm long, geniculate and with a distinct column. Pedicelled spikelet narrowly lanceolate, 4.5−7 mm long.

Grassy plains on limestone and gypsum hills, deciduous bushland and degraded juniper forest; 200−1590 m. N1−3; C1; tropical Africa eastwards through Arabia to Iraq and India. McKinnon 35; Bally & Melville 15987; Thulin & Warfa 5956; Herlocker 422.

80. ARTHRAXON P. Beauv. (1812)

Annual or perennial with slender often trailing stems; leaves broad, usually lanceolate, cordate and amplexicaul at the base; ligule membranous. Inflorescence composed of slender subdigitate racemes, these terminal and axillary but not spathate, their internodes filiform to linear. Sessile spikelet dorsally or laterally compressed; callus truncate; lower glume membranous to coriaceous, usually rounded on the back, with or without lateral keels, often scabrid or muricate; lower floret reduced to a hyaline lemma; upper floret with hyaline subentire lemma bearing a glabrous geniculate awn from low down on the back; caryopsis terete. Pedicelled spikelet variable, from as large as the sessile to quite suppressed.

Some 10 species in the Old World tropics, mainly in India.

A. prionodes (Steud.) Dandy (1956). Fig. 156.

A. serrulatus Hochst. (1856); *A. lanceolatus* (Roxb.) Hochst. var. *serrulatus* (Hochst.) T. Durand & Schinz, Consp. Fl. Afric. 5: 704 (1895).

Loosely tufted wiry perennial arising from a knotty base with silky-villous scales; stems straggling, up to

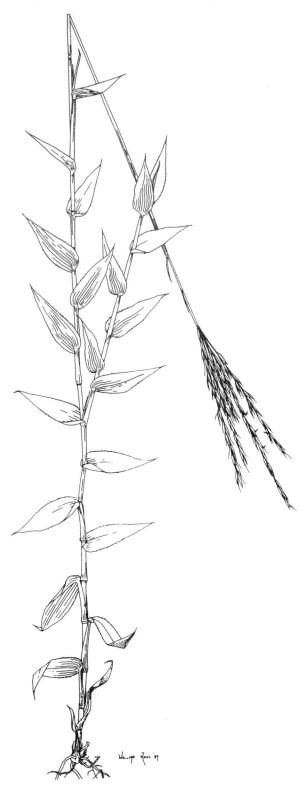

Fig. 156. *Arthraxon prionodes*, habit. — From Mem. Bot. Surv. S. Afr. 58 (1991).

60 cm long. Racemes 2−8, each 4−8 cm long; internodes ciliate to pilose, the hairs increasing in length upwards to 2−3 mm. Sessile spikelet linear to narrowly lanceolate, 5−7 mm long; lower glume chartaceous, strongly convex, pectinate-spinose on the keels and often muricate on the back along the indistinct nerves; anthers 3, 2.5−3.5 mm long; awn of upper lemma 8−15 mm long. Pedicelled spikelet narrowly lanceolate, 4−5.5 mm long, on a pedicel half as long as the internode.

Scrubby hillsides and in damp rock crevices; 1300−1700 m. N1, 2; tropical East Africa eastwards through Arabia to India, China and Thailand. Drake-Brockman 476; Bally & Melville 15930.

A recent revision of the genus by Welzen, in Blumea 27: 255−300 (1981), places this species in *A. lanceolatus*. Although the two species are rather similar, the latter, which is confined to southern India, has the lower glume of the sessile spikelet more or less flat on the back and strongly nerved.

81. HYPARRHENIA E. Fourn. (1886)

Annual or perennial; leaves never aromatic; ligule scarious. Inflorescence composed of paired racemes, each pair supported on a peduncle and subtended by a sheathing spatheole, the latter crowded into a large leafy false panicle; racemes short, slender, each borne upon a short stalk (raceme-base) which is often deflexed at maturity, and with up to 2 of the lowermost spikelet-pairs (homogamous pairs) male or barren, awnless and tardily deciduous; internodes and pedicels linear. Sessile spikelet narrowly lanceolate to lanceolate-oblong, dorsally compressed or terete; callus obtuse to pungent, applied obliquely to the top of the internode with its tip free; lower glume coriaceous, broadly convex across the back and sides, without keels or these developed only in the upper 1/3; upper glume awnless; lower floret reduced to a hyaline lemma; upper lemma stipitiform, 2-toothed, passing between the teeth into a stout awn; caryopsis oblong, subterete. Pedicelled spikelet male or barren, narrowly lanceolate, usually a little longer than the sessile, awnless or aristulate from the lower glume.

55 species, mainly in Africa.

H. baddadae Chiov. is a teratological specimen of unknown identity. It is probably a species of *Elymandra* (and quite possibly *E. grallata* (Stapf) Clayton), but it is excluded from this account pending the gathering of identifiable material.

1. Perennial; raceme-bases unequal, the upper terete, much longer than the lower .....1. *H. hirta*
− Annual; raceme-bases subequal, flattened, the upper scarcely longer than the lower............
...................................2. *H. anthistirioides*

1. **H. hirta** (L.) Stapf (1918). Fig. 157.

Aibarh, aus gurun, bein, deilan, hadaf (Som.).

Tufted, shortly rhizomatous perennial up to 60(−100) cm high. False panicle typically scanty, up to 30 cm long; spatheoles linear-lanceolate, 3−8 cm long, reddish; peduncle about as long as the spatheole, with or without spreading white hairs above. Racemes never deflexed, 2−4 cm long, 8- to 13(−16)-awned per pair; raceme-bases unequal, the upper 2.5−5 mm long, filiform; homogamous spikelet-pairs 1 at the base of the lower or both racemes. Sessile spikelet 4−6.5 mm long, villous; awn 10−35 mm long. Pedicelled spikelet 3−7 mm long, white-villous.

Dry rocky hillsides; 900−2000 m. N1−3; Mediterranean region eastwards through Arabia to Pakistan, and southwards to South Africa, but absent from much of tropical Africa. Gillett 4914; McKinnon 119; Bally & Melville 15828.

2. **H. anthistirioides** (Hochst. ex A. Rich.) Anderss. ex Stapf (1918).

Annual up to 150 cm high, geniculate or ascending and often supported by stilt-roots. False panicle ample, 20−30 cm long; spatheoles lanceolate in profile, 1.8−3.2 cm long, streaked with green, yellow or orange-brown; peduncles 1/4−1/2 the length of the spatheole, bearded above. Racemes 1.1−1.3 cm long, 3- to 4(−5)-awned per pair; raceme-bases subequal, flattened, the upper about 1 mm long, with or without a scarious lobe up to 0.2 mm long; homogamous spikelet-pairs 1 at the base of the lower raceme only, 8−11 mm long. Sessile spikelet 5−6 mm long, glabrous or sparsely pubescent; awn 32−45 mm long. Pedicelled spikelet 5−6 mm long, glabrous except for the ciliolate margins, finely aristulate with an awnlet 3−6 mm long.

Rocky hillsides; c. 1300 m. N1; Ethiopia, Sudan and Tanzania. Hemming 2250.

82. HETEROPOGON Pers. (1807)

Annual or perennial; ligule short, membranous, ciliate on the upper edge. Inflorescence a single raceme, arising terminally and in the axils, sometimes loosely aggregated into a false panicle; racemes with homogamous spikelet-pairs in the lower 1/4−2/3; internodes linear. Sessile spikelet subterete; callus long and pungent; lower glume coriaceous, convex, obtuse; lower floret reduced to a hyaline lemma; upper lemma stipitiform, passing directly into a stout pubescent awn; caryopsis lanceolate, channelled on one side. Pedicelled spikelet male or barren, lanceolate, larger than the sessile, awnless, with a long slender callus functioning as a pedicel (the true pedicel reduced to a tiny stump).

Six species in tropical and warm temperate regions.

Fig. 157. *Hyparrhenia hirta*. A: base of plant. B: ligule. C: inflorescence. D: spikelet-pair. − From Mem. Bot. Surv. S. Afr. 58 (1991).

H. contortus (L.) P. Beauv. ex Roem. & Schult. (1817). Fig. 158.

Tufted perennial with laterally compressed basal sheaths; stems up to 1 m high, erect. Racemes 3−10 cm long, solitary or aggregated into a scanty false panicle, the awns forming a twisted spire; homogamous spikelet-pairs 3−17, resembling the pedicelled spikelets. Sessile spikelet 5.5−10 mm long, including the ferociously pungent, rufously bearded 2−3 mm long callus; lower glume elliptic-

Fig. 158. *Heteropogon contortus*. A: habit, × 0.4. B: ligule, × 2.5. C: pedicelled spikelet, × 3.5. D: sessile spikelet, × 3.5. − Modified from Fl. Trop. E. Afr. (1982). Drawn by W. E. Trevithick.

Fig. 159. *Themeda triandra*. A: habit, × 0.4. B: ligule, × 1.6. C: raceme, showing homogamous (a), sessile (b) and pedicelled (c) spikelets, × 2. − Modified from Fl. Trop. E. Afr. (1982). Drawn by W. E. Trevithick.

oblong, brown, hispidulous; awn 5−8 cm long. Pedicelled spikelet 5−15 mm long with a callus 2−3 mm long.

Sandy, rocky and alluvial soils; 25−1650 m. N1, 2; C1, 2; S3; tropical and warm temperate regions generally. Gillett 4252; Glover & Gilliland 1140; Thulin & Dahir 6565; Moggi & al. s.n.; Kazmi & al. 650.

83. THEMEDA Forssk. (1775)

Annual or perennial; ligule very short, membranous. Inflorescence composed of solitary racemes embraced by sheathing spatheoles, these sometimes single but mostly in fan-shaped bunches on flexuous pedicels and gathered into a leafy false panicle; racemes comprising 2 homogamous spikelet-pairs forming an involucre, and 1−4 sessile spikelets with their pedicelled attendants; internodes linear. Homo-

gamous spikelets all sessile, persistent. Sessile spikelet subterete or dorsally compressed; callus obtuse to pungent; lower glume coriaceous, obtuse; lower floret reduced to a hyaline lemma; upper lemma stipitiform and passing directly into the awn, rarely hyaline and awnless; caryopsis lanceolate, channelled on one side. Pedicelled spikelet male or barren, narrowly lanceolate, awnless, with a long slender callus as long as or longer than the true pedicel (this often reduced to a minute stump.

18 species in the Old World tropics and subtropics, but mainly in Asia.

T. triandra Forssk. (1775). Fig. 159.

T. forskaolii Hack. var. *imberbis* (Retz.) Hack. in DC., Monogr. Phan. 6: 661 (1889).

T. forskaolii Hack. var. *punctata* (Hochst. ex A. Rich.) Hack., op. cit.: 662 (1889).

Daboshobil (Som.).

Tufted perennial up to 2 m high (but usually much less), erect, slender. False panicle up to 30 cm long; spatheole 1.5—3.5 cm long, glabrous to tuberculate-hairy. Raceme with 1 fertile spikelet; involucral spikelets narrowly elliptic, 6—14 mm long, glabrous to tuberculate-hairy. Sessile spikelet 6—11 mm long, including the pungent, rufously bearded 2—4 mm long callus; lower glume brown, smooth except for the appressedly pubescent tip; awn 2.5—7 cm long. Pedicelled spikelet 6—14 mm long, glabrous or tuberculate-hairy; callus 2—3 mm long.

Clearings in juniper forest; 1650—2130 m. N1, 2; tropical and subtropical Old World. Glover & Gilliland 1163; Hansen & Heemstra 6246.

84. ELIONURUS Kunth ex Willd. (1806)

Annual or perennial; ligule a short densely ciliate membrane. Inflorescence a single raceme, these terminal or sometimes axillary and gathered into a leafy false panicle; racemes flexuous, dorsally flattened; internodes columnar to subclavate, not hollowed or rimmed. Sessile spikelet lanceolate to narrowly ovate; callus often large, applied obliquely to the top of the internode; lower glume subcoriaceous to herbaceous, broadly convex, smooth or sometimes toothed on the keels, laterally 2-keeled, the keels ciliately fringed and often bordered with an oil-streak, mostly cuspidate to a bifid tip; lower floret reduced to a hyaline lemma; upper lemma entire and awnless; caryopsis ellipsoid, dorsally compressed. Pedicelled spikelet well developed, muticous or aristulate, the pedicel resembling the internode.

15 species mainly in tropical Africa, America and Australia.

1. Annual; lower glume of sessile spikelet with pectinate margins below, each of the blunt teeth bearing a tuft of long hairs ....1. *E. royleanus*
- Perennial; lower glume of sessile spikelet hairy but without teeth on the margins ... 2. *E. muticus*

1. **E. royleanus** Nees ex A. Rich. (1850).
Annual up to 35 cm high, fastigiately branching. Racemes 2—6 cm long, embraced below by a reddish spathe, gathered into fascicles; internodes bearded towards the top. Sessile spikelet lanceolate; callus obconical, 1 mm long; lower glume with body 5—6 mm long, glabrous to villous on the back, with tufts of long hair arising from blunt teeth along the margins, divided above into 2 flattened tails 4—6 mm long. Pedicelled spikelet narrowly lanceolate, 6—10 mm long, caudate-acuminate to a subulate tip.

Grassy places in open scrub on silty soil. S3; Mauritania to north-western India. Rose Innes 619.

2. **E. muticus** (Spreng.) Kuntze (1898). Fig. 160.
Gat-er-gan (Som.).

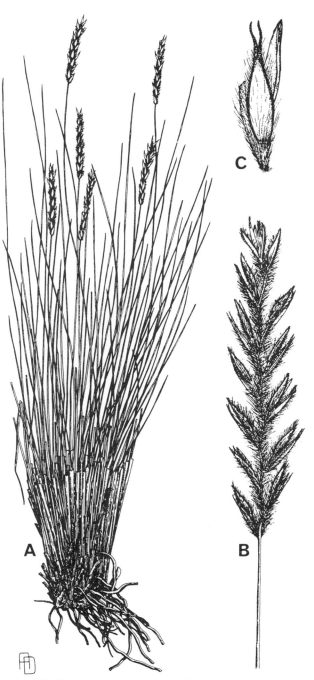

Fig. 160. *Elionurus muticus*. A: habit, x 0.5. B: raceme, x 2. C: spikelet-pair, showing sessile spikelet, x 5. — Modified from Fl. Trop. E. Afr. (1982). Drawn by A. Davies.

Densely tufted perennial up to 25(—100) cm high. Racemes 4—14 cm long, terminal or sometimes 1—2 axillary, silvery to grey or purple; internodes villous. Sessile spikelet lanceolate; callus broadly cuneate, 1—1.5 mm long; lower glume with body 4—8 mm long, villous on the back, entire or divided at the tip into 2 teeth up to 7 mm long. Pedicelled spikelet

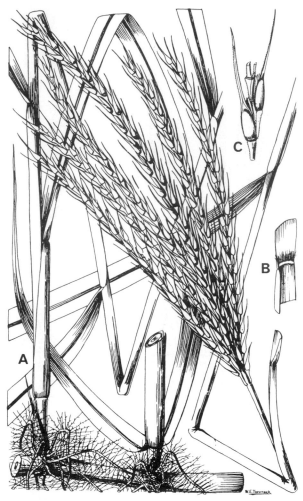

Fig. 161. *Vossia cuspidata*. A: habit, × 0.4. B: ligule, × 0.6. C: spikelet-pair, × 0.8. — Modified from Fl. Trop. E. Afr. (1982). Drawn by W. E. Trevithick.

lanceolate, 4−7 mm long, pubescent to villous, acuminate or with a short awn-point.

Heavily grazed hillsides. N1; tropical Africa southwards to South Africa and eastwards to Arabia, tropical and subtropical America. Glover & Gilliland 1149A.

85. VOSSIA Wall. & Griff. (1836), nom. cons.

Aquatic perennial; ligule a short pilose membrane. Inflorescence composed of digitate racemes, borne terminally; racemes ± flattened, tardily disarticulating; internodes and pedicels thickened, clavate, hollowed at the tip. Sessile spikelet flat or slightly convex across the back; callus truncate with irregular central convexity but without a pronounced central peg; lower glume coriaceous, 2-keeled, smooth except for the scabrid keels, narrowly winged above and drawn out into a long linear flattened tail; lower

floret male with hyaline lemma and palea; upper lemma entire and awnless. Pedicelled spikelet resembling the sessile, its pedicel free.

One species only.

V. cuspidata (Roxb.) Griff. (1851). Fig. 161.

Stems submerged or floating, up to 7 m long and 1 cm in diam., spongy, with fibrous roots from the nodes, standing 1−2 m out of the water; leaves 30−100 cm × 6−18 mm. Racemes 1−12, each 10−30 cm long. Sessile spikelet narrowly ovate; lower glume 2−4 cm long, the body 6−8 mm long and yellowish, the tail green. Pedicelled spikelet a little smaller than the sessile.

In or near water, often floating. S2, S3; tropical Africa and India. Moggi & Bavazzano 2751; Rose Innes 721.

86. LASIURUS Boiss. (1859)

Perennial; ligule a line of hairs. Inflorescence a single raceme, these terminal and axillary; racemes fragile, dorsally compressed, with 2 sessile and 1 pedicelled spikelet at each node (rarely 1 sessile and 2 pedicelled) or 1 sessile and 1 pedicelled; internodes stoutly clavate. Sessile spikelet flat across the back; callus truncate with a central peg; lower glume subcoriaceous, 2-keeled, densely ciliate on the back and keels, with a short 2-toothed or entire tail; lower floret male with hyaline lemma and palea; upper lemma entire and awnless; caryopsis oblong, terete. Pedicelled spikelet similar to the sessile or smaller, without a tail; pedicel free, resembling the internode.

One species only.

L. scindicus Henrard (1941). Fig. 162.

Saccharum hirsutum Forssk. (1775) non *Rottboellia hirsuta* Vahl, nom. illeg., based on *Triticum aegilopoides* Forssk., non *Lasiurus hirsutus* (Vahl) Boiss., nec *Elionurus hirsutus* (Vahl) Munro ex Benth.

Darif, dungara, sefar (Som.).

Stems often woody below, up to 90 cm high, simple or suffruticose, erect from a thick woody rhizome covered with firm, imbricate, often silky cataphylls. Racemes up to 10 cm long, silky-villous from the internodes, pedicels and glumes. Sessile spikelet 6−13 mm long; lower glume lanceolate, often caudate, 2-toothed at the tip with divergent teeth, often spreading horizontally at maturity or when dry. Pedicelled spikelet usually 5−7 mm long.

Sandy and gravelly desert soils, often amongst shrubs; 0−1000 m. N1−3; from Mali eastwards through Arabia and South-west Asia to North-east India. Hemming 2079; Bally 11217; Beckett 576.

Fig. 162. *Lasiurus scindicus*. A: habit, × 0.5. B: ligule, ×
3.2. C: portion of inflorescence, × 2.4. D: spikelet-pair, ×
3.2. − Modified from a drawing by J. C. Webb.

87. ROTTBOELLIA L.f. (1779), nom cons., non Scop. (1777)

Annual; ligule a very short membrane. Inflorescence
a single raceme, these axillary and often gathered into
a leafy false panicle; racemes cylindrical, fragile;
internodes flattened or semi-cylindrical, partly or
wholly fused to the adjacent pedicel. Sessile spikelet
sunk in the internode; callus truncate, with prominent
central peg; lower glume coriaceous, broadly convex,
2-keeled, smooth, narrowly winged at the tip; lower
floret male with hyaline lemma and palea; upper
lemma entire and awnless; caryopsis ovate in
face-view, crescentic in side-view. Pedicelled spikelet

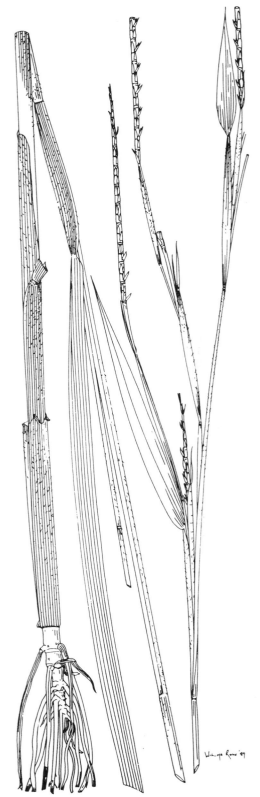

Fig. 163. *Rottboellia cochinchinensis*, habit. − From Mem.
Bot. Surv. S. Afr. 58 (1991).

269

herbaceous or scarious, a little smaller than the sessile; pedicel oblong, flattened, scarcely distinguishable from the internode.

Four species in the Old World tropics, introduced in the New World.

R. cochinchinensis (Lour.) Clayton (1981). Fig. 163.
R. exaltata L.f. (1781) non (L.) L.f. (1779).
Stems up to 3 m high, supported below by stilt-roots, the basal sheaths painfully hispid; leaves up to 45 × 2 cm. Racemes 3—15 cm long, glabrous, terminating in a tail of reduced spikelets, gathered into a leafy false panicle. Sessile spikelet oblong-elliptic, pallid; lower glume 3.5—5 mm long. Pedicelled spikelet narrowly ovate, 3—5 mm long, herbaceous, green; pedicel shorter than the internode.

Woodland clearings, cultivated ground and urban wasteland; 0—300 m. N2; S1—3; Old World tropics. Glover & Gilliland 994; Wieland 1518; Hansen 6062; Kazmi 5187.

88. ZEA L. (1753)

Robust broad-leaved monoecious annual (rarely perennial). Female inflorescence axillary, a single raceme wrapped in several spathes; internodes fused into a polystichous woody cob bearing paired sessile spikelets at each node; spikelets shallowly inserted in the surface of the cob, with short chaffy glumes exposing the grain; lower floret barren; style single, very long, silky, pendulous from the tip of the inflorescence. Male inflorescence terminal, of digitate or paniculate racemes; internodes tough, narrow, bearing paired spikelets, one of them on a slender, free pedicel; both florets male.

Four species in Central America.

Z. mays L. (1753) is the familiar maize, Indian corn or sweet-corn (known locally as ara-bighi) introduced from tropical America and cultivated either for its edible grain or for fodder at least in N1 and N2. A single specimen is known, collected in the area of Ijara (N1) where it is said to be cultivated, and sent as a 'curio'. McKinnon 66.

168. ARECACEAE (PALMAE)

by M. Thulin

Cuf. Enum.: 1495—1500 (1971); Fl. Trop. E. Afr. (1986); Uhl & Dransfield, Genera Palmarum (1987).

Trees, shrubs or climbers, sometimes armed, bisexual, polygamous, monoecious or dioecious; stems usually prominently ringed with leaf-scars, usually unbranched or dichotomously branched. Leaves usually with sheathing base; petiole sometimes armed with spines (modified leaflets or epidermal emergences); blade pinnate, bipinnate, palmate or costapalmate (with the petiole extending into the leaf-blade as a well-defined central axis); leaflets induplicate or reduplicate, composed of one or more folds. Inflorescences axillary, single or occasionally grouped, often complex with bracteate branches of different orders, the ultimate branches bearing bracts subtending flowers, singly or in pairs, triads or small groups. Flowers usually small, bisexual, unisexual or sterile male, usually 3-merous; tepals sometimes similar, sometimes differentiated in calyx and corolla, free or fused. Stamens 3—many, free or united; anthers basi- or dorsifixed, straight or twisted; staminodes often present in female flowers. Carpels usually 1—3, free, or forming usually 3-celled ovary; ovule solitary in each carpel or cell; stigmas erect or recurved; pistillode often present in male flowers. Fruits usually 1-seeded, more rarely 2—10-seeded, usually with distinct epicarp (outer layer), mesocarp (middle layer) and endocarp (inner layer).

Some 200 genera and 2700 species, mainly in the moister tropics and subtropics, few in arid regions.

Borassus aethiopum Mart. was recorded from Somalia in Cuf. Enum.: 1499 (1971) on the basis of a vague statement by Warburg in Engl. Pflanzenw. Ost-Afr. C: 130 (1895) about its occurrence in "Somalitiefland". However, an occurrence in Somalia has never been substantiated and the species is therefore excluded here.

1. Leaves pinnate 2
- Leaves palmate or costapalmate 3
2. Petiole armed with spines ....2. *Phoenix*
- Petiole unarmed ....4. *Cocos*
3. Trunk simple, tall; petiole base unsplit; petiole triangular in cross-section ....1. *Livistona*
- Trunk dichotomously branched, short to tall or underground; petiole base with a conspicuous central triangular cleft; petiole semi-circular in cross-section ....3. *Hyphaene*

1. LIVISTONA R. Br. (1810)
Dransfield & Uhl in Kew Bull. 38: 199—200 (1983).
Wissmannia Burret (1943).

Palms slender to robust, solitary, bisexual or rarely dioecious; trunk unbranched, erect. Leaves palmate or costapalmate; petiole unarmed or armed with spines; blade divided to varying depths into single-fold induplicate segments, filaments sometimes present at the sinuses. Inflorescences branched to 5 orders, pedunculate. Flowers usually cream-coloured; tepals united, in 2 whorls of 3. Stamens 6, inserted on the inner tepals, the filaments united to form a fleshy ring; anthers medifixed. Ovary of 3 free carpels that are united above to form a single slender style with apical stigma; ovule basal. Fruit usually developing from 1 carpel; stigmatic remains apical; epicarp smooth; mesocarp thin or thick, usually easily separated from the stony endocarp.

Some 28 species, one in the Horn of Africa and southern Arabia, the others in south-eastern Asia to Australia.

L. carinensis (Chiov.) J. Dransf. & N.W. Uhl (1983); *Hyphaene carinensis* Chiov. (1929); *Wissmannia carinensis* (Chiov.) Burret (1943); type: N3, "Carin", at "Uncud", Puccioni & Stefanini 1027 (FT holo.). Fig. 164.

Monod in Bull. Inst. Franç. Afr. Noire, ser. A, 17: 338−358 (1955); Bazara'a, Guarino, Miller & Obadi in Edinb. J. Bot. 47: 375−379 (1990).

Daban, madah (Som.).

Palm 20−40 m tall. Petiole long, triangular in cross-section, armed with 7−19 mm long sharp spreading spines; leaflets numerous, c. 80 × 4−4.5 cm. Inflorescences up to 2 m long, tomentose, pedunculate. Flowers bisexual, c. 2 mm long, tomentose with branched hairs; inner tepals much longer than the outer. Fruits ± globose, c. 6 mm in diam., brownish, glabrous.

Along wadis or near springs; c. 800−1075 m. N3; Djibouti, Yemen (Hadramaut). Barbier 972; Lavranos & Carter 24835; Bavazzano s.n.

This is a threatened species in Somalia of which about 50 trees remained in the area around "Karin" and "Galgala" in 1986, but only 11 in January 1995. They have been much used for house building, drainage pipes etc. and regeneration is prevented as the leaves of young plants are grazed or used for the production of mats and baskets.

Some recent authors have, by mistake, cited Chiovenda as the collector of the type of *L. carinensis*.

2. PHOENIX L. (1753)

Palms solitary or clustered, dioecious. Leaves pinnate; petiole armed with narrow spines (modified leaflets); blade divided into numerous single-fold induplicate leaflets. Inflorescences branched to 1 order; flowers borne singly in a spiral along the axis, each subtended by a small bract. Male flowers with united tepals in 2 whorls. Stamens usually 6, inserted on the inner tepals; pistillode absent or rudimentary. Female flowers with 3 outer tepals united into a cup, 3 inner tepals free, imbricate. Staminodes 6, minute.

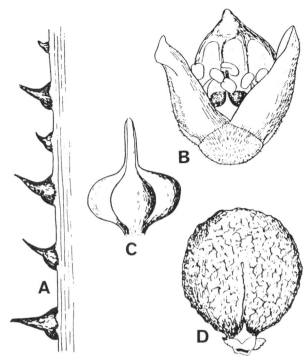

Fig. 164. *Livistona carinensis*. A: margin of petiole, × 0.75. B: flower, × 20. C: pistil, × 60. D: fruit, × 5. − From Bull. Inst. Franç. Afr. Noire, ser. A, 17 (1955).

Carpels 3, free, with short recurved fleshy stigmas, usually only one carpel developing to fruit. Fruits with epicarp smooth, mesocarp fleshy, endocarp thin, membranous. Seed 1, deeply grooved longitudinally.

Some 15 species in the Old World tropics and subtropics.

1. Plant without aerial stem ....2. *P. caespitosa*
− Plant with a distinct ± erect trunk 2
2. Plant usually clustered and forming thickets; trunk slender, up to c. 15 cm in diam.; ripe fruit up to 2 cm long ....1. *P. reclinata*
− Plant usually occurring as solitary individuals; trunk often more robust; ripe fruit 4−7 cm long .... 3. *P. dactylifera*

1. **P. reclinata** Jacq. (1801). Fig. 165.

P. reclinata var. *somalensis* Becc. in Chiov., Result. Scient. Miss. Stefanini-Paoli: 176, 230 (1916); types: S1, near "Bardera", Paoli 824, S3, "Uagadi" near "Bulo Nassib", Paoli 430 & "Zingibar", Paoli 385 (all FT syn.).

Usually clustered palm, often forming dense thickets with trunks up to 10 m tall and c. 15 cm in diam. Leaves up to 2.5 m long, bright green, curved; leaflets numerous, up to 25 × 2 cm. Inflorescences 20−50 cm long, pedunculate. Male flowers with inner tepals acute. Female flowers globose; inner tepals rounded. Fruit 13−20 × 9−13 mm, pale yellow to orange or dull red. Seed 10−12 × 6−8 mm.

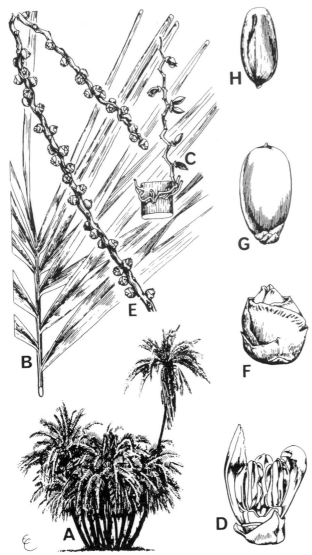

Fig. 165. *Phoenix reclinata*. A: habit. B: leaf-tip, × 2/3. C: fragment of male inflorescence, × 2/3. D: male flower with 1 tepal removed, × 4. E: part of female inflorescence, × 2/3. F: female flower, × 4. G: fruit, × 2. H: seed, × 2. — Modified from Fl. Trop. E. Afr. (1986). Drawn by E. Catherine.

Along watercourses; 0−450 m. C2; S1, 3; ?Djibouti, Ethiopia, East Africa, widespread in tropical Africa.

P. abyssinica Drude in Ethiopia is probably conspecific.

2. **P. caespitosa** Chiov. (1929); types: N3, between "Ereri Jelleho" and "Martisor Dinsai", Puccioni & Stefanini 659 (FT syn.) & "Scorasar" valley, Puccioni & Stefanini 672 (FT syn.).

Mairu, mayro (Som.).

Clustered palm, forming thickets, without aerial stem. Leaves to 3 m long, curved; leaflets numerous,

up to 50 x 1.5 cm. Inflorescence up to 35 cm long, pedunculate. Male flowers with inner tepals obtuse. Female flowers globose; inner tepals rounded. Fruit 10−14 × 8−10 mm, deep orange when ripe.

In wadis; up to 900 m. N2, 3; ?Djibouti. Bally & Melville 15690; Collenette 58; Gillett 23055.

Fruits edible. *P. caespitosa* is close to *P. arabica* Burret in southern Arabia and further studies may show the two to be conspecific. However, *P. caespitosa* is the older name.

3. **P. dactylifera** L. (1753).

Date palm (Eng.), timir (Som.).

Usually solitary palm, never thicket-forming, with trunk up to 20 m tall (in Somalia much shorter) and up to 40−50 cm in diam. Leaves up to 3 m long, glaucous, with stiff rhachis and scarcely curved; leaflets numerous, up to 30 × 2 cm. Inflorescences up to 60 cm or more long, pedunculate. Male flowers with inner tepals obtuse. Female flowers globose; inner tepals rounded. Fruit 40−70 × 20−30 mm, yellow to orange-brown or almost black. Seed c. 24 × 6−8 mm.

Cultivated at least in N1−3 for its edible fruits and sometimes naturalised; origin unknown but possibly from Arabia, widely cultivated in the drier tropics and subtropics. Sacco s.n.

3. **HYPHAENE** Gaertn. (1788)
Beccari, Palme Borass.: 18−49 (1924).

Palms small to robust, solitary or clustered, dioecious; trunk unbranched, or usually dichotomously branched (forking), sometimes prostrate and underground. Leaves costapalmate; petiole base with a conspicuous central triangular cleft; petiole armed with spines; blade divided to c. 1/3 of the length into single-fold induplicate segments, filaments often conspicuous at the sinuses. Inflorescences basically similar, but the male ones often more slender and highly branched than the female. Male flowers in groups of 3, embedded in hairs, one flower exposed at a time; outer tepals 3, united into a tube below; inner tepals 3, united into a tube with imbricate lobes, reflexed at anthesis. Stamens 6, inserted at the base of the inner tepals; anthers basifixed; pistillode minute. Female flowers solitary in the axil of each bract, much larger than the male, outer and inner tepals similar, 3, free, imbricate, rounded; staminodal ring with 6 teeth, each tipped with minute empty anthers. Ovary globose with 3 apical triangular ± sessile stigmas, 3-celled, usually only 1 ovule developing. Fruit borne on the enlarged pedicel; stigmatic remains basal; epicarp sometimes pitted, mesocarp fibrous, often edible, endocarp hard, stony.

Some 40 species though probably much fewer, widespread in tropical and subtropical Africa, Arabia and southern Asia.

Dransfield in Fl. Trop. E. Afr. (1986) drastically reduced the number of species of *Hyphaene* (doum palms) in East Africa and some further names are reduced to synonyms below. However, the number of good collections from Somalia is very limited and the present account has to be regarded as provisional.

All species in Somalia are much used by man, for thatching and for making baskets and mats, and for the wood, and many populations appear to be decreasing.

1. Stems prostrate, creeping, flattened 1. *H. reptans*
   − Stems erect or ascending, not flattened 2
2. Fruits irregularly knobbly .... 2. *H. thebaica*
   − Fruits variable in shape, but not knobbly 3
3. Fruits at least 6 cm long .... 1. *H. compressa*
   − Fruits less than 6 cm long .... 3. *H. coriacea*

1. H. compressa H. Wendl. (1878). Plate 4 E.

H. mangoides Becc. (1908); type: Somalia, without precise locality or collector (FI holo.).

H. benadirensis Becc. (1908); types: S3, "Margherita", Macaluso s.n. (?PAL syn., not found), near "Giumbo", Pantano s.n. (FI syn.).

Baar (Som.).

Tree, up to 20 m tall, but in Somalia usually much smaller, solitary or clustered, dichotomously branched up to 4−5 times. Leaves long persistent if not burnt off, glaucous green; petiole armed with upcurved black spines to 20 × 6 mm. Mature fruit (6−)7−10(−12) × (4−)5−8(−9) cm, oblong to obovoid, pitted, usually compressed on two faces.

Along watercourses, coastal dunes; up to c. 500 m. C1, 2; S2, 3; coastal parts of East Africa, southwards to Mozambique. Thulin, Hedrén & Dahir 7271; Paoli 529; Senni 647.

2. H. thebaica (L.) Mart. (1838).

Timir (tree); garow (fruit) (Som.).

Tree, up to 10 m or more tall, solitary, dichotomously branched; similar to *H. compressa*, but fruits irregularly knobbly.

Gypseous, gullied slope, also planted; 5 m. N2; Djibouti, Eritrea, Sudan, Egypt. Gillett & Watson 23873; Thulin, Dahir & Hassan 9197.

Gillett & Watson 23873, from 10°43'N, 46°38'E, is probably from a native tree, while Thulin, Dahir & Hassan 9197 is from an originally planted stand near Laasqoray. The Somali plants have smaller fruits (c. 4.5−5.5 cm long) than normal in the species.

The relationship between *H. thebaica* and *H. dankaliensis* Becc. (1906) in Eritrea needs further study.

3. H. coriacea Gaertn. (1788). Fig. 166.

H. sphaerulifera Becc. (1908); type: S2, "Brava", Pantano s.n. (FI holo.).

H. sphaerulifera var. *gosciaensis* Becc. in Agric. Col. 2: 170 (1908); *H. pyrifera* var. *gosciaensis*

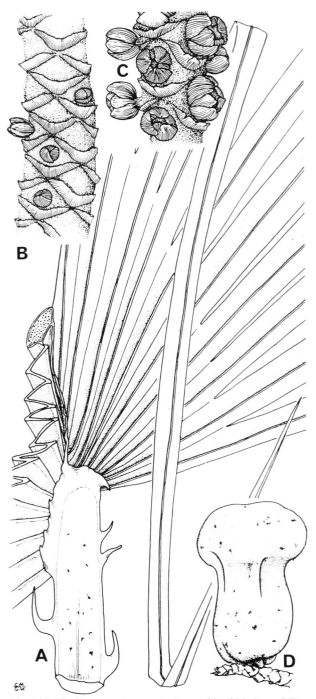

Fig. 166. *Hyphaene coriacea*. A: base of leaf-blade, × 2/3. B: part of male inflorescence, × 2. C: part of female inflorescence, × 2. D: fruit, × 2/3. − Modified from Fl. Trop. E. Afr. (1986). Drawn by C. Grey-Wilson.

(Becc.) Becc., Palme Borass.: 38 (1924); type: S2, "Goscia", Gioli s.n. (FI holo.).

H. pyrifera Becc. (1908); type: S3, "Ghescud", Pantano s.n. (FI holo.).

H. pleuropoda Becc. (1908); type: S2, "Brava", Pantano s.n. (FI holo.).

H. oblonga Becc. (1908); type: S3, "Giumbo", Pantano s.n. (FI holo.).

H. pyrifera var. *arenicola* Becc., Palme Borass.: 37 (1924); type: S3, between "Giumbo" and "El Sai", Paoli 288 (FI holo., FT iso.).

H. pyrifera var. *margaritensis* Becc., Palme Borass.: 37 (1924); type : S3, "Margherita", Gioli s.n. (FI holo.).

Baar (Som.).

Palm usually clustering at the base, tending to form shrubby thickets, rarely more than 5 m tall; stems decumbent, suckering, dichotomously branched once or twice, rarely more. Leaves long persistent, glaucous; petiole armed with upcurved black spines to 10 mm long. Mature fruit 3−6 × 2.5−4 cm, variable in shape, not deeply pitted.

Coastal sand dunes; 0−100 m. S2, 3; coastal parts of Kenya and Tanzania and southwards to South Africa and Madagascar.

4. H. reptans Becc. (1908). Plate 4 F.

H. migiurtina Chiov. (1929); type: N3, "Scorasar" valley, Puccioni & Stefanini 671 (FT holo.).

Adu, au daror (Som.).

Stem prostrate, flattened, dichotomously branched, on the surface of the ground or underground. Leaves to c. 1 m long, long persistent, glaucous; petiole armed with upcurved black spines. Fruit subglobose to ovoid or pear-shaped, 4−6 × 3−5 cm.

In wadis; 100−900 m. N2, 3; S Arabia. Collenette 223; Gillett 23018; Bally 10901.

Beckett & White 1790 from S1 (3°18'N, 44°08'E) is a sterile plant with "underground rootstock" collected in a red sandy loam over limestone at 500 m altitude (vern. name: dumbal). This may possibly be *H. reptans*, but could also be a form of *H. compressa* or *H. coriacea* without aerial stem.

4. COCOS L. (1753)

Palms solitary, unarmed, monoecious. Leaves pinnate; blade divided into numerous single-fold reduplicate leaflets. Inflorescences branched to 1 order; flowers in groups of 3 at the base of the axis and solitary or paired in upper part. Male flowers with 2 whorls of 3 tepals. Stamens 6; pistillode small. Female flowers much larger than the male, subglobose; staminodal ring inconspicuous. Ovary large, ± globose, with trifid sessile stigma at apex. Fruit massive, with thick fibrous mesocarp and very hard endocarp with 3 basal pores, usually 1-seeded; endosperm hard with central cavity partly filled with liquid.

One species only, the coconut.

C. nucifera L. (1753).

Harries in Bot. Rev. 44: 265−319 (1978).

Qumbe (Som.).

Palm of varying size, trunk usually somewhat swollen at the base. Leaves up to 4−5 m long; petiole up to 2 m long; leaflets up to 100 × 2 cm, bright green above, somewhat paler beneath. Inflorescences with male flowers in upper part. Fruit obovoid, obscurely trigonous, up to 25 × 20 cm, filled by the seed; embryo next to the functional pore.

Cultivated near the coast at least in S2 and S3; of unknown origin, possibly western Pacific, widely cultivated in the lowland tropics. Sacco, Sappa & Ariello 200.

169. PANDANACEAE

by M. Thulin

Fl. Trop. E. Afr. (1993).

Trees, shrubs or woody climbers, dioecious, often dichotomously branched and often with stilt-roots. Leaves simple, rigid, evergreen, long and narrow, glabrous, usually with spiny margins, alternate and usually aggregated at the ends of the branches. Flowers massed together and mostly difficult to distinguish. Perianth absent or rudimentary. Stamens numerous, apparently in spikes or panicles. Ovary superior with 1−several cells; ovules 1−several. Fruit compound, made up of woody or fleshy units.

Family of three genera and some 675 species in the Old World tropics.

PANDANUS Parkinson (1773)

Plants usually dichotomously branched and with stilt-roots; trunk with prominent ring-like leaf-scars. Leaves basically arranged in 3 rows which are twisted so that the leaves appear conspicuously spirally arranged; margins and undersides of midrib spiny. Stamens in elongate spikes. Female flowers aggregated into large spherical or oblong heads. Carpels 1-ovulate. Fruit a cone-like head of drupes or clusters of partly fused drupes.

Perhaps 600 species in the Old World tropics, "screw-pines", poorly represented in Africa.

P. kirkii Rendle (1894). Fig. 167.

Tree; trunk unarmed or with a few blunt knobs; stilt-roots from lower part of trunk, with small sharp spines in longitudinal rows. Leaves c. 0.9−1.5 m long

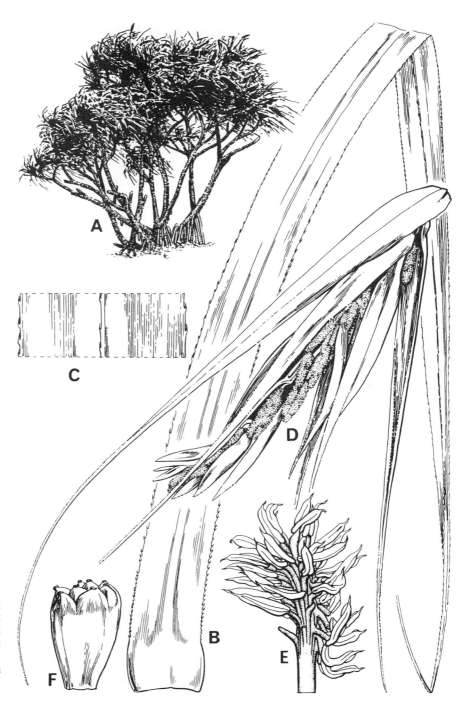

Fig. 167. *Pandanus kirkii*. A: habit. B: leaf, × 1/3. C: detail of leaf undersurface, × 1. D: male inflorescence, × 1/6. E: group of stamens, × 4. F: cluster of drupes, × 0.5. — Modified from Fl. Trop. E. Afr. (1993). Drawn by E. Catherine.

or more, clasping at the base, attenuate into a long flagella at the apex; spines at margins and on underside of midrib c. 1−3 mm long. Male inflorescences with 9−12 spadices. Fruiting heads almost round, 8−18 × 5−16 cm, made up of up to 38 clusters of partly fused drupes, each cluster composed of 5−11(−14) carpels with distinct apices separated by grooves, yellow-brown to orange when ripe.

Seashore. S3; E Kenya, E Tanzania. Moggi & Bavazzano s.n.

First record for Somalia of a screw-pine, based on a seedling and a cluster of drupes (the dispersal unit), from "Sar Uanle", 15−20 km SW of Kismaayo at 0°30'S, 42°26'E.

APPENDIX. ADDITIONS TO FLORA OF SOMALIA, VOL. 1

During the fieldwork in north-eastern Somalia in January 1995 (see Preface) a number of species (and genera) were found that are additional to the accounts in vol. 1. As far as identified and dealt with these are listed and briefly described below, along with a forgotten species of *Microcharis* that recently turned up in a herbarium bundle. The same sequence for families and genera as in vol. 1 is used.

26. BRASSICACEAE

ERUCA Mill. (1754)

Annuals or perennials. Leaves pinnatifid. Sepals erect, unequal. Petals yellow or whitish, clawed. Fruit a siliqua with a long beak; valves 1-veined; seeds in 2 rows in each cell.

Some five species in the Mediterranean Region and NE Africa.

E. vesicaria (L.) Cav. (1802).
subsp. **sativa** (Mill.) Thell. in Hegi, Ill. Fl. Mitteleur. 4(1): 201 (1918).
Annual, usually hispid, 20−100 cm tall. Leaves with a large terminal lobe and 1−5 narrow lateral lobes on each side. Sepals soon falling. Petals 15−24 mm long, whitish yellow with violet veins. Siliqua 12−35(−40) × 3−6 mm, erect, with a sword-like beak.
Weed in garden near sea level. N3; Mediterranean Region, cultivated as a salad plant and naturalized elsewhere. Thulin, Dahir & Hassan 9241.
First record for Somalia (Boosaaso at 11°17'N, 49°11'E).

ARABIDOPSIS (DC.) Heynh. (1842)

Annuals or perennials; branched hairs often present. Leaves entire to pinnatifid. Sepals not or only slightly saccate at the base. Petals white, yellow or violet. Stamens usually 6. Fruit a siliqua; valves 1-veined.

Some 10 species, mainly in Europe. *Arabidopsis* is one of several poorly circumscribed segregates of *Arabis* L.

A. thaliana (L.) Heynh. (1842); *Arabis thaliana* L. (1753).
Annual up to 20 cm tall or more; stems hairy below. Leaves hairy with mostly branched hairs, entire or somewhat toothed, spathulate, in a rosette, up to c. 4 × 1.5 cm, upper leaves smaller. Racemes without bracts, elongating in fruit; pedicels spreading, 4−9 mm long. Petals longer than the sepals, c. 2−3 mm long, white, spathulate. Siliqua linear, straight or almost so, c. 6−20 × 0.7 mm; style c. 0.3 mm long. Seeds c. 0.7 × 0.4 mm.
In grassy patches in evergreen bushland; 1650 m. N2; Eritrea, Ethiopia, eastern tropical Africa, N Africa, Macaronesia, Europe and much of Asia, also introduced elsewhere. Thulin, Dahir & Hassan 8972.
First record for Somalia (Karin Xaggarood at 10°58'N, 48°52'E).

CARDAMINE L. (1753)

Annuals or perennials, glabrous or with simple hairs. Leaves simple to pinnate. Petals white, cream, pink or purple. Stamens usually 4−6. Fruit a siliqua; valves coiling spirally from the base at dehiscence, without prominent nerves.

Some 130 species in temperate and tropical montane areas all over the world.

C. hirsuta L. (1753).
Annual to 20 cm high or more; stems glabrous or sparsely hairy: leaves glabrous or hairy above, pinnate with a terminal leaflet and 2−6 pairs of smaller lateral leaflets, the lower in a rosette with reniform terminal leaflet and rounded lateral leaflets. Racemes without bracts, elongating in fruit; pedicels ascending, 3−8 mm long in fruit. Petals longer than the sepals, c. 2−3 mm long, white, spathulate. Siliqua erect, linear, straight, 12−27 × c. 1 mm; style c. 0.5−1 mm long. Seeds c. 1 × 0.8 mm.
In grassy patches in evergreen bushland; 1550 m. N2; Eritrea, Ethiopia, East Africa, almost cosmopolitan. Thulin, Dahir & Hassan 8978.
First record for Somalia (Karin Xaggarood at 10°58'N, 48°52'E).

30. CRASSULACEAE

Umbilicus horizontalis (Guss.) DC. (1828).
Rootstock subglobose, covered by fibrous roots. Basal leaves peltate; petiole up to 10 cm long; blade up to 5 cm in diam., with crenate margin. Raceme erect, up to at least 30 cm high, usually unbranched; flowers horizontal; pedicels up to 2.5 mm long. Calyx-lobes lanceolate, c. 1.5 mm long. Corolla greenish-white, tubular, c. 6−7 mm long, with lanceolate-acuminate lobes c. 1.5 mm long.
Evergreen bushland, in holes of limestone rocks;

1350 m. N2; Socotra, Mediterranean Region, SW Asia. Thulin, Dahir & Hassan 9131.

First record for Somalia (near Ragad, 11°00'N, 48°29'E).

Further material of the species treated as *U.* sp. = Bally & Melville 15772 in Flora of Somalia vol. 1 has

been collected as well (Thulin, Dahir & Hassan 8940 from 1500 m altitude in N2). This has convinced me that the plant is best regarded as a form of *U. tropaeolifolius* Boiss. (previously known from Turkey, Iran and Iraq), and that *U. paniculiformis* Wickens from Sudan is also conspecific.

32. CARYOPHYLLACEAE

Polycarpaea basaltica Thulin (1996); type: N2, 6 km W of Ceelayo, 11°16'N, 48°50'E, Thulin, Dahir & Hassan 9019 (UPS holo., FT K iso.). Fig. 168.

Perennial herb, forming 10–20 cm high cushions, the base covered by persistent leaf-bases; stems glabrous, brittle. Basal leaves in dense rosettes, spathulate, pale green, blade up to 40 × 17 mm, acute to shortly acuminate at the apex, tapering below into an ill-defined up to 20 mm long petiole; cauline leaves whorled, similar to the basal ones but with shorter petioles; stipules triangular to ovate, acuminate, c. 1.5–2 mm long, scarious with a darker base, entire or toothed. Inflorescences ± capitate, consisting of 2–4 dense spike-like c. 10–15 mm long silvery cymes; peduncles 15–55 mm long; bracts scarious. Sepals lanceolate, 5–6.5 mm long, medially brownish and with broad scarious margins. Petals linear-lanceolate, 2.5–3 mm long, slightly lacerate. Stamens 5, c. 2/3 as long as the sepals; anthers c. 0.8 mm long. Ovary c. 2 mm long; style 2.5–3 mm long including linear c. 0.5–0.6 mm long stigma-lobes. Capsule 3–4 mm long. Seeds many, ovoid, c. 0.4 mm long, pale brown.

Steep slope near the sea, among basalt rocks; 10–20 m. N2; not known elsewhere.

This belongs to a group of species including *P. guardafuiensis* in Somalia and a number of other species in Socotra, Abd al-Kuri and Oman, but is very distinctive by its large, broad leaves, silvery inflorescences, and large flowers.

POLYCARPON Loefl. ex L. (1759)

Similar to *Polycarpaea*, but with keeled sepals.
Genus of some 16 species, cosmopolitan.

P. tetraphyllum (L.) L. (1759).

Annual or short-lived perennial. Leaves mostly in whorls of 4, ovate-spathulate, c. 7–12 × 3–7 mm. Inflorescence spreading, much-branched. Sepals up to c. 2 mm long. Petals emarginate. Stamens 3–5. Seeds c. 0.5 mm long, minutely warty.

In grassy patches in evergreen bushland; 1550 m. N2; Eritrea, Ethiopia, widespread in both the Old

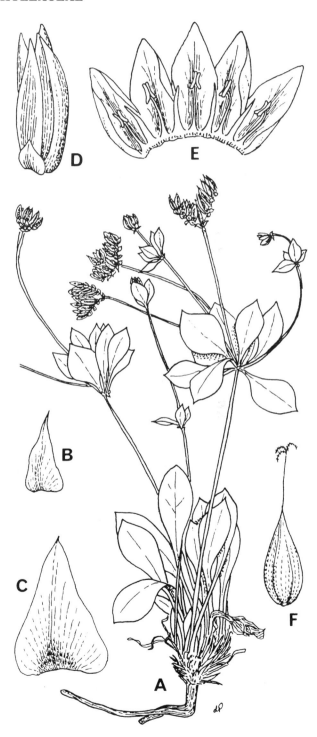

Fig. 168. *Polycarpaea basaltica*. A: habit, × 1. B: stipule, × 12. C: bract, × 12. D: flower, × 6. E: flower, opened up, × 6. F: pistil, × 6. – From Thulin, Dahir & Hassan 9019. Drawn by L. Petrusson.

and the New World, perhaps of Mediterranean origin. Thulin, Dahir & Hassan 8976.

First record for Somalia (Karin Xaggarood at 10°58'N, 48°52'E).

Minuartia hybrida (Vill.) Schischkin (1936).

Slender annual, ± densely glandular-pubescent, or glabrous. Leaves 4−16 mm long, linear-subulate. Inflorescence lax, with slender 5−20 mm long pedicels. Sepals 3−4 mm long, linear-lanceolate. Petals shorter than sepals, white. Capsule as long as or longer than the sepals.

In grassy patches in evergreen bushland; 1650 m. N2; Mediterranean Region and SW Asia. Thulin, Dahir & Hassan 8971.

First record for Somalia and tropical Africa (Karin Xaggarood at 10°58'N, 48°52'E). The Somali material is poor and consists of entirely glabrous plants without fruits. It probably represents subsp. *hybrida*.

CERASTIUM L. (1753)

Annual or perennial herbs. Flowers in cymose inflorescences, or sometimes solitary. Sepals free. Petals white, emarginate, sometimes absent. Stamens 5−10. Styles (3−)5(−6). Capsule cylindric, longer than sepals, with twice as many teeth as styles.

Genus of some 60 species, almost cosmopolitan.

C. glomeratum Thuill. (1799).

Annual up to at least 30 cm tall, hairy with eglandular and glandular hairs. Leaves up to 30 mm long, obovate to ovate or elliptic, obtuse. Flowers in ± compact inflorescences; bracts herbaceous. Sepals 4−5 mm long, lanceolate, hairy. Petals ± equalling to shorter than sepals, or absent. Stamens 5−10. Styles 5. Capsule 6−10 mm long. Seeds c. 0.5 mm long,

finely warty.

Evergreen bushland, usually in grassy areas; 1400−1650 m. N2; Europe, North Africa, Asia, introduced in America. Thulin, Dahir & Hassan 8952, 9073.

First record for Somalia and tropical Africa (Karin Xaggarood at 10°58'N, 48°52'E, and near Mirci at 11°01'N, 48°27'E). The Somali plants lack petals and have 5 stamens only.

Silene burchellii Otth. ex DC. (1824).

Perennial herb, covered with short eglandular hairs in all aerial parts; stems erect, up to at least 50 cm tall. Leaves crowded towards the base of the plant, linear to narrowly oblanceolate, 12−50 × 1−5 mm in Somali material, acute or subacute at the apex. Cymes raceme-like, unbranched or forked at the base, 1−7-flowered. Calyx tubular-clavate, 20−25 mm long with 3−5 mm long teeth in Somali material. Petals reddish-brown, purple, pink or white. Styles 3. Seeds rounded-reniform, c. 1.6 mm in diam. in Somali material, dorsal surface deeply grooved with marginal wings, surface with fine radiating rugulose lines, dark brown.

Rocky limestone slopes; 1600−1900 m. N2, 3; Ethiopia and southwards to South Africa, and in tropical Arabia. Thulin & Warfa 6058; Thulin, Dahir & Hassan 9172.

In the account of *Silene* in Flora of Somalia vol. 1 only the glandular-hairy and perennial *S. flammulifolia* and the ephemeral *S. apetala* were included. However, the collection Thulin, Dahir & Hassan 9172 convinced me that there is also a third taxon present. This is perennial with short eglandular hairs in all parts, and seems best regarded as a form of the widespread *S. burchellii*. Furthermore, the illustration of *S. flammulifolia* in vol. 1 (Fig. 56) is actually based on material of *S. burchellii*.

38. CHENOPODIACEAE

Chenopodium schraderianum Schult. (1820).

Annual up to 1 m or more tall, erect, usually unbranched, glandular and strongly aromatic. Leaves mostly 1−5(−8) × 0.5−3(−5) cm, ± obtuse at the apex, pinnately divided into 3−5 obtuse, usually entire lobes at each side; glands on lower side of leaves sessile, not accompanied by hairs. Flowers in axillary cymes, 0.5−1 mm in diam.; perianth-

segments 5, each with a toothed keel on the back, glandular. Stamens 1−2. Seeds 0.7−0.8 mm in diam., black with blunt margin; testa very minutely pitted.

In open grassy patches in evergreen bushland; 1400 m. N2; Ethiopia, Sudan and southwards to the Cape, rarely introduced elsewhere. Thulin, Dahir & Hassan 9043.

First verified record for Somalia (near Moon at 11°01'N, 48°25'E).

42. POLYGONACEAE

EMEX Campderá (1819), nom. cons.

Annual herbs, monoecious. Ochreae without bristles at the margin. Female flowers at base of inflore-

scence. Tepals 6, free in male flowers, united in female flowers, the outer 3 spinescent and hardened in fruit. Stamens 4−6. Styles 3. Nut trigonous, enclosed in the persistent perianth.

Two species, the second, *E. australis* Steinh., mainly in South Africa.

E. spinosa (L.) Campderá (1819).

Plant glabrous with erect or ascending stems 15−50 cm long. Leaves c. 4−8 x 2−5 cm, ovate, subacute at the apex, truncate to subcordate at the base, petiolate. Male flowers in terminal and axillary pedunculate clusters. Female flowers axillary, sessile, in fruit with outer tepals ending in spreading to reflexed spines and inner tepals erect.

Open ground in evergreen bushland; 1150 m. N2; Eritrea, Socotra, Mediterranean Region, SW Asia, also introduced elsewhere. Thulin, Dahir & Hassan 8915.

First record for Somalia (Madarshon at 11°02'N, 48°58'E).

57. CUCURBITACEAE

Zehneria somalensis Thulin (1996); type: N2, Karin Xaggarood, 10°58'N, 48°52'E, Thulin, Dahir & Hassan 8975 (UPS holo., FT K iso.). Fig. 169.

Plant herbaceous, with stems up to several m long, angular with green edges, trailing or hanging, glabrous. Leaf-blade broadly ovate-triangular in outline, 3−8 x 2.5−6.5 cm, scabrid with minute scattered stiff hairs above, glabrous or almost so beneath, palmately deeply (3−)5-lobed, cordate at the base, lobes elliptic to lanceolate or ovate, acute, the central the largest; petiole 1.2−4.5 cm long, glabrous or almost so. Monoecious. Male flowers 3−6 in axillary clusters; pedicels 10−14 mm long; hypanthium campanulate, 2.5−3.2 mm long; sepals triangular c. 0.4 mm long; petals white, 1.6−2.8 mm long. Female flowers solitary, coaxillary with the male flowers; pedicels 25−40 mm long; ovary fusiform, c. 6.5 mm long, glabrous; perianth as in male flowers. Fruits elliptic, c. 13−17 x 6−9 mm, apiculate, red when ripe, c. 4−12-seeded. Seeds c. 4.4 x 3.2 x 1.5 mm.

Evergreen bushland, hanging over vertical rock-face in shade; 1600 m. N2; not known elsewhere.

Differs from *Z. scabra* by being monoecious, and by its distinctly palmately lobed leaf-blades.

69. EUPHORBIACEAE

Euphorbia peplus L. (1753).

Glabrous slender annual, branched from near the base. Leaves 5−20 x 3−12 mm, petiolate, ovate to suborbicular, entire; ray-leaves like the cauline, but with shorter petioles; raylet-leaves smaller, slightly obliquely ovate. Rays 3, dichotomous. Glands with 2 filiform horns. Capsule 2 x 2 mm, shallowly sulcate, smooth. Seeds 1.1−1.4 mm long, sulcate ventrally and pitted dorsally, pale grey, darker in the depressions.

Shady places among rocks in evergreen bushland; 1150 m. N2; almost cosmopolitan weed, perhaps of Mediterranean origin, nearest in Eritrea. Thulin, Dahir & Hassan 8921.

First record for Somalia (Madarshon at 11°02'N, 48°58'E).

72. FABACEAE

Acacia cernua Thulin & Hassan (1996); type: N2, escarpment S of Laasqoray, 11°03'N, 48°16'E, Thulin, Dahir & Hassan 9188 (UPS holo., FT K iso.). Fig. 170.

Cabab (Som.).

Slender tree, 2−5 m tall, with characteristically hanging terminal branches and a glaucous foliage; bark smooth, ash grey to white, not flaking; young branchlets purplish-brown, glabrous or pubescent with ± appressed hairs and soon glabrescent. Stipular spines up to 15 mm long, slender, straight. Leaves: pinnae (1−)2−4 pairs; leaflets 4−11 pairs, 2−4.5 (−7.5) x 0.8−2(−3) mm, glabrous or sparsely ciliate when young. Flowers pale yellow, in heads 5−5.5 mm in diam.; peduncles 6−15 mm long (up to 25 mm in fruit), with involucel in lower half. Calyx c. 1.2 mm long. Corolla c. 2 mm long, glabrous. Pods straight or curved slightly upwards in upper part, linear-oblong, dehiscent, 3.5−7 x 0.8−1.3 cm, purplish-brown, with oblique veins which converge to become longitudinal in the middle of the valves, sparsely to densely puberulous with appressed hairs. Seeds elliptic to narrowly ovate or almost rhombic in outline, c. 7−8 x 3.5−4 mm; areole central, c. 4.4−4.8 x 1.4−1.6 mm.

Rocky slopes and on rocks along wadis; 400−625 m. N2; not known elsewhere. Fagg & Styles 62; Thulin, Dahir & Hassan 9024.

Close to *A. etbaica*, but easily distinguished by its

N2; Djibouti, Eritrea, Ethiopia, Sudan. Thulin, Dahir & Hassan 9116.

First record for Somalia (near Markat at 10°59'N, 48°30'E).

The Somali plants have decumbent stems and appear to be short-lived perennials.

Lotus torulosus (Chiov.) Fiori (1912).

Procumbent annual to 40 cm long, pilose. Leaf rhachis 2—15 mm long; leaflets ± cuneate-obovate, up to 20 × 12 mm, the basal pair much wider at the base than the others. Umbels 2—5-flowered; peduncle up to 4 cm long. Calyx-tube c. 3 mm long; lobes 3—4.5 mm long. Corolla pink, 7—8 mm long. Style 4—5 mm long. Pod up to 30—50 × 2 mm, ± torulose, glabrous.

Grassy patches in evergreen bushland or in disturbed deciduous bushland; 650—1450 m. N2; Eritrea, N Sudan. Thulin, Dahir & Hassan 8934, 9222.

First record for Somalia (Karin Xaggarood at 10°58'N, 48°52'E, and S of Xidid at 11°01'N, 48°37'E).

The Somali plants have pods that are scarcely torulose at all, but still they appear to be conspecific with *L. torulosus*.

Medicago orbicularis (L.) Bartal. (1776).

Glabrous or sparsely hairy, procumbent annual. Leaflets obovate-cuneate; stipules laciniate. Racemes 1—5-flowered. Corolla yellow, 2—5 mm long. Pod 10—17(—20) mm in diam., in a spiral of 4—6 turns, lens-shaped, glabrous or almost so, not spiny; transverse veins with a few, usually weak anastomosing branches.

Open grassy areas in evergreen bushland; 1150—1350 m. N2; Eritrea, Ethiopia, Mediterranean Region, extending to SW Asia and India. Thulin, Dahir & Hassan 8919, 9080.

First record for Somalia (Madarshon at 11°02'N, 48°58'E, and near Ragad at 11°00'N, 48°29'E).

Trigonella falcata Balf. f. (1882).

Annual, aromatic herb, with ascending glabrous stems to c. 20 cm long. Leaves pinnately 3-foliolate, sparsely hairy; leaflets obovate-cuneate, up to 14 × 9 mm, toothed; stipules semisagittate, toothed at the base. Racemes axillary, up to c. 15 mm long, shorter than the leaves, 2—7-flowered. Flowers soon deflexed; calyx c. 2 mm long, sparsely hairy, with lobes shorter than the tube. Corolla yellow, c. 4—5.5 mm long. Pod up to c. 20 × 1 mm, linear, strongly curved, sparsely hairy when young, later subglabrous, with raised reticulate venation, c. 12-seeded, dehiscent.

Open grassy area in evergreen bushland; 1350 m. N2; Socotra. Thulin, Dahir & Hassan 9093.

First record for Somalia (near Ragad at 11°00'N, 48°29'). *T. falcata* was previously known only from Socotra.

INDEX TO VERNACULAR NAMES

INDEX TO SCIENTIFIC NAMES

Italicized names are either synonyms or have been casually referred to in the text.